Electromagnetic Interference Shielding Technologies

Electromagnetic Interference Shielding Technologies: Theory and Applications offers a comprehensive evaluation of recent advancements in materials and technologies utilised for electromagnetic interference (EMI) shielding.

This book provides a substantial theoretical foundation. In addition to discussing the synthesis and characterisation of different EMI shielding materials, it offers a comprehensive examination of the current progress, constraints and resolutions in EMI shielding strategies across various sectors including space, defence, electronics and medical. Chapters are split by application area as opposed to material type for increased readability.

This accessible book will be of interest to scientists and researchers working on satellite development, electronic warfare systems, future automobile systems, medical device production and electronic devices. Engineers and postgraduate students will benefit from the incorporation of theory in this book.

Series in Materials Science and Engineering

The series publishes cutting-edge monographs and foundational textbooks for interdisciplinary materials science and engineering. It is aimed at undergraduate and graduate level students as well as practising scientists and engineers. Its purpose is to address the connections between properties, structure, synthesis, processing, characterisation and performance of materials.

Advances in 3D Bioprinting
Roger J. Narayan, Ed.

Hydrogels: Fundamentals to Advanced Energy Applications
Ram K. Gupta and Anuj Kumar, Eds.

Emerging Energy Materials: Applications and Challenges
Govind Nair, H. Nagabhushana, Nirupama Dhoble and Sanjay J. Dhoble, Eds.

Handbook of 2D Materials for a Sustainable Future Vol 1
Mangalaraja Ramalinga Viswanathan, Arulraj Arunachalam, Pandiyarajan Thangaraj and Ernesto Chicardi Augusto, Eds.

Handbook of 2D Materials for a Sustainable Future Vol 2
Mangalaraja Ramalinga Viswanathan, Arulraj Arunachalam, Arun Thirumurugan and P. Sakthivel, Eds.

Unravelling the Power of 2D Materials: Revolutionizing Flexible Devices
Ram K. Gupta, Ed.

Flow Batteries: Fundamentals, Technological Advancement and Challenges
Ram K. Gupta, Ed.

Electromagnetic Interference Shielding Technologies: Theory and Applications
Reenu Elizabeth John, Steffy Sara Varghese, Anoop Chandran and N.V. Unnikrishnan

Series Preface

The series publishes cutting-edge monographs and foundational textbooks for interdisciplinary materials science and engineering.

Its purpose is to address the connections between properties, structure, synthesis, processing, characterisation, and performance of materials. The subject matter of individual volumes spans fundamental theory, computational modelling, and experimental methods used for design, modelling and practical applications. The series encompasses thin films, surfaces and interfaces, and the full spectrum of material types, including biomaterials, energy materials, metals, semiconductors, optoelectronic materials, ceramics, magnetic materials, superconductors, nanomaterials, composites and polymers.

It is aimed at undergraduate and graduate level students, as well as practising scientists and engineers.

Proposals for new volumes in the series may be directed to Carolina Antunes, Commissioning Editor at CRC Press, Taylor & Francis Group (Carolina.Antunes@tandf.co.uk).

Electromagnetic Interference Shielding Technologies
Theory and Applications

Edited by
Reenu Elizabeth John
Steffy Sara Varghese
Anoop Chandran
N.V. Unnikrishnan

CRC Press
Taylor & Francis Group
Boca Raton London New York

CRC Press is an imprint of the
Taylor & Francis Group, an **informa** business

First edition published 2026
by CRC Press
2385 NW Executive Center Drive, Suite 320, Boca Raton FL 33431

and by CRC Press
4 Park Square, Milton Park, Abingdon, Oxon, OX14 4RN

CRC Press is an imprint of Taylor & Francis Group, LLC

ISBN: 978-1-032-84356-8 (hbk)
ISBN: 978-1-032-84358-2 (pbk)
ISBN: 978-1-003-51236-3 (ebk)

DOI: 10.1201/9781003512363

Typeset in Minion
by SPi Technologies India Pvt Ltd (Straive)

In loving memory of our daughter
***Miriam Georgy Puliyelil**, forever 6.*
A melody too beautiful for this world
A little light that still illuminates our lives in big ways.

Contents

Preface

IN TODAY'S INTERCONNECTED WORLD, THE IMPORTANCE OF ELECTROMAGNETIC INTERFERENCE (EMI) shielding has gained unprecedented significance. With the rapid advancement and widespread integration of compact, high-frequency and multifunctional electronic systems, ranging from everyday gadgets like smartphones and smart appliances to critical technologies in aerospace, defence and healthcare, the potential hazards posed by EMI have grown manifold. These interferences no longer remain mere nuisances; they have serious implications for system performance, data integrity, operational safety and even national security.

This book is a consolidated effort to bring under one umbrella the essential theoretical foundations, recent advancements in materials and contemporary approaches to EMI shielding, with a focus on both academic and practical perspectives. It is intended to cater to a broad audience comprising students, researchers and industry professionals. The content spans from basic principles of electromagnetic theory and shielding mechanisms to the most recent developments in materials science, including metal-based systems, polymer composites, nanostructured materials and emerging 2D materials like MXenes. Special attention is given to multifunctional solutions that can simultaneously address thermal, mechanical and environmental constraints.

Each chapter is thoughtfully structured to reflect the growing and dynamic needs of various sectors such as aerospace, automotive, defence, biomedical electronics and flexible and wearable devices. In addition, the book discusses standard measurement methodologies, evolving regulatory frameworks and the relevance of sustainable practices. The book also attempts to explore the forward-looking trends in EMI shielding, particularly in the backdrop of 5G communication, artificial intelligence and the Internet of Things (IoT).

It is our earnest hope that this volume will not only serve as a valuable academic and technical reference but also stimulate further research and innovation in the domain of electromagnetic compatibility. By blending theoretical exposition with practical insights, we seek to contribute to the development of safer, robust and high-performing electronic systems for the future.

About the Editors

Reenu Elizabeth John is currently an Assistant Professor and Head of Department of Physics at Saintgits College of Engineering, Kerala. She was awarded a doctorate degree from Mahatma Gandhi University, Kerala, in 2019, for her research on 1D transition metal oxides. She has published in reputable international journals and has also co-authored book chapters and edited one book. Her areas of interest include one-dimensional nanomaterials, ferroelectric nanomaterials and conducting polymer nanocomposites. She has received project funding from the Centre for Engineering Research and Development (CERD) under the Research Seed Money (RSM) scheme for her work on EMI shielding.

Steffy Sara Varghese is currently an Assistant Professor in the School of Engineering, Architecture and Interior Design at Amity University, Dubai. She earned her Ph.D. in 2019 from the Indian Institute of Geomagnetism (IIG), Mumbai, for her research on nonlinear wave structures in Earth's magnetosphere. She later pursued postdoctoral research at the Space and Planetary Science Center, Khalifa University, UAE. Her research interests span solar-terrestrial interactions, plasma wave theory and materials science in space. She has published extensively in reputed international journals and has also served as a reviewer for international journals. In recognition of her contributions, she received the Young Scientist Award from the International Union of Radio Sciences (URSI-GASS) in 2023. She is a member of the Institute of Electrical and Electronics Engineers (IEEE).

Anoop Chandran is an Assistant Professor and Research Advisor in the Physics Department at St. Cyril's College in Adoor. He earned a Ph.D. for his research on 1D semiconductor nanomaterials in 2015. Later that year, the Dr. D. S. Kothari Postdoctoral Fellowship was granted to him for his proposal on semiconductor-CNT/graphene hybrids. He is also a member of the American Chemical Society. He has published extensively in a number of reputable international journals, co-authored several book chapters and edited one book. He has also served as a reviewer for international journals. 1D nanomaterials, dielectrics, photovoltaics, polymer nanocomposites, hybrid conducting composites for EMI shielding and membrane-based water purification are some of his current research interests.

N.V. Unnikrishnan is currently a Visiting Professor at Mahatma Gandhi University, Kottayam. He earned his doctoral degree from Rohtak University, Delhi for his work on mechanisms in optical materials under pulsed laser excitations. In 2003 he became a Professor at Mahatma Gandhi University and served as an Emeritus Professor in the same university from 2014 to 2019. Prof. Unnikrishnan, during his 32-year teaching career, has supervised 32 PhDs and six postdoctoral fellows. He has completed ten major projects funded by agencies like UGC, CSIR, DAE-DRNS, TWAS among others, and published over 290 research papers. He is a former member of the American Chemical Society.

Contributors

C. K. Aanandan
Department of Electronics
Cochin University of Science and Technology
Cochin
India

Yedla Venkata Durga Anudeep
Department of Chemical Engineering
Indian Institute of Technology Delhi
Hauz Khas, Delhi
India

Farhad Aslani
Materials and Structures Innovation Group
School of Engineering
The University of Western Australia
Perth, WA
Australia

Ayse Aytac
Polymer Science and Technology Programme
Engineering Faculty
Department of Chemical Engineering
Kocaeli University
Kocaeli
Türkiye

Bárbara Bermúdez-Reyes
Departamento de Ingeniería Aeroespacial
Facultad de Ingeniería
Unidad de Tecnología Avanzada, UNAM
Ciudad de México
México

Anoop Chandran
Department of Physics
St. Cyrils College
Adoor, Kerala
India

Jamal Chaouki
Chemical Engineering Department
Polytechnique Montréal
Montréal
Canada

Mohsin Ali Fakhar
School of Chemical and Materials Engineering
National University of Science and Technology
Islamabad
Pakistan

Muhammad Irfan
School of Chemical and Materials Engineering (SCME)
National University of Sciences and Technology (NUST)
Islamabad
Pakistan

Reenu Elizabeth John
Department of Physics
Saintgits College of Engineering
Kottayam, Kerala
India

Rejo Mathew Joseph
Department of Physics
St. Cyril's College
Adoor, Kerala
India

Kamran Ahmed Khan
Department of Aerospace Engineering
Advanced Research and Innovation Center
Khalifa University of Science and Technology Abu Dhabi
UAE

Deepti Das Krishna
Department of Electronics
Cochin University of Science and Technology
Cochin
India

Raghvendra Kumar Mishra
Department of Chemical Engineering
Indian Institute of Technology Delhi
Hauz Khas, Delhi
India

Lakshmi Omana
Department of Physics
St. Cyril's College
Adoor, Kerala
India

Jonathan Tersur Orasugh
Department of Chemical Sciences
University of Johannesburg
Doorfontein, Johannesburg
South Africa

and

Centre for Nanostructures and Advanced Materials
DSI-CSIR Nanotechnology Innovation Centre
Council for Scientific and Industrial Research
Pretoria
South Africa

Immanuvel Paul
College of Engineering and Physics
Khalifa University
Abu Dhabi
UAE

Shannen Chacko Rajan
School of Engineering
Newcastle University
Newcastle upon Tyne
UK

Anilkumar Ramakrishnan
Department of Electronics
Cochin University of Science and Technology
Cochin
India

and

Viswajyothi College of Engineering and Technology
Muvattupuzha, Ernakulam
India

Suprakash Sinha Ray
Department of Chemical Sciences
University of Johannesburg
Doorfontein, Johannesburg
South Africa

and

Centre for Nanostructures and Advanced Materials
DSI-CSIR Nanotechnology Innovation Centre
Council for Scientific and Industrial Research
Pretoria
South Africa

Ramy Sadek
Chemical Engineering Department
Polytechnique Montréal
Montréal
Canada

Soleyman Sahebi
Materials and Structures Innovation Group
School of Engineering
The University of Western Australia
Perth, WA
Australia

Margarita Tecpoyotl-Torres
Centro de Investigación en Ingeniería y Ciencias Aplicadas
CIICAp
Universidad Autónoma del Estado de Morelos
Chamilpa, Cuernavaca
México

Lesego Tabea Temane
Department of Chemical Sciences
University of Johannesburg
Doorfontein, Johannesburg
South Africa

and

Centre for Nanostructures and Advanced Materials
DSI-CSIR Nanotechnology Innovation Centre
Council for Scientific and Industrial Research
Pretoria
South Africa

Paulbert Thomas
The Cochin College
Ernakulam
India

Bedriye Ucpinar
Engineering Faculty
Department of Chemical Engineering
Ege University
Izmir
Türkiye

N.V. Unnikrishnan
School of Pure and Applied Physics
MG University
Kottayam, Kerala
India

Rafael Vargas-Bernal
Departamento de Ingeniería Automotriz
Universidad Politécnica de Guanajuato
Cortazar, Guanajuato
México

Steffy Sara Varghese
School of Engineering
Architecture and Interior Design
Amity University
Dubai
UAE

CHAPTER 1

Introduction to Electromagnetic Interference Shielding and Its Significance in Present and Future

Reenu Elizabeth John, Anoop Chandran, Lakshmi Omana, and Steffy Sara Varghese

1.1 INTRODUCTION

The demand for advanced electronics is rapidly escalating over the years and their pervasive presence in our daily lives continues to grow at an unprecedented rate. This surge shows no sign of slowing down in the near future. From smartphones and wearable fitness trackers to smart home systems and autonomous vehicles, electronic devices are becoming more compact and multifunctional. However, the advent and proliferation of high-frequency electronic systems have given rise to a new form of electronic pollution called electromagnetic interference (EMI). EMI refers to the disturbance generated by external or internal electromagnetic fields that predominantly affect the performance of electronic circuits and communication systems [1]. It can cause communication disruptions, data loss and even lead to device failure. Moreover, EMI poses serious risks in sensitive environments like hospitals, aerospace and defence applications. This unintended by-product of electronic advancements not only disrupts the optimal functionality of electronic and communication devices but can also pose serious and complex detrimental effects to human health and wellbeing [2]. The impact of these adverse effects has spurred researchers to develop materials with good EMI shielding properties.

The mechanism behind EMI shielding is diverse and requires a thorough theoretical and material-specific knowledge. Primarily, reflection leads to the repulsion of electromagnetic waves. Dielectric losses and magnetic losses play a crucial role in determining the electromagnetic wave absorption capacity of the material [3]. Additionally, thermal stability of the material is essential as it influences the long-term durability of the material.

DOI: 10.1201/9781003512363-1

Hence the design and selection of materials for EMI applications require a careful balance of electrical, magnetic, mechanical and thermal properties of the material.

Traditionally used shielding materials like metals and metallic composites are restricted to a large extent in modern technology due to their high density, corrosiveness, poor mechanical stability and high cost of processing [4]. Therefore, for the purpose of application in new generation smart electronic equipment, carbon- and polymer-based composites have emerged as a promising alternative due to their favourable features like high conductivity, low density, enhanced flexibility and resistance to corrosion. Even for these composites, several factors such as filler concentration, distribution and orientation of the filler and the choice of matrix material play a significant role in determining EMI shielding properties [5].

In addition to shielding effectiveness, modern electronic equipment demands EMI shielding materials with multifunctional properties such as enhanced thermal stability, mechanical flexibility and durability, self-healing and self-cleaning property, tunable electrical and mechanical sensitivities and even shape memory behaviour [6]. Such innovations are essential to meet the increasing complexity of application in flexible electronics, aerospace structures, automotive systems and bio-integrated devices. This chapter aims to provide a succinct overview of the theory, materials and future perspective of EMI shielding. The chapter discusses the underlying principles of the electromagnetic interaction with matter. It also outlines the design principles that influence the shielding properties of materials followed by highlighting the recent materials used as EMI shielding applications. Finally, the chapter touches upon the future directions of EMI shielding, thereby highlighting the critical role of EMI shielding materials in modern electronics.

1.2 THEORETICAL BACKGROUND

EMI attenuation occurs through three different mechanisms: multiple reflections, absorption and reflection. The presence of mobile charge carriers, such as electrons or holes, that interact with electromagnetic radiation is necessary for the principal mechanism, reflection. Absorption is the secondary EMI shielding method. The shield must have electric and/or magnetic dipoles that interact with the electromagnetic fields of the radiation to absorb a sizeable amount of it. Multiple reflections are the final shielding method and they require a high shield surface area or interface region. In this section, a summary of each mechanism is provided.

When an electromagnetic (EM) wave impinges on the front interface between a shielding material and free space, reflection occurs due to the impedance mismatch between the two media. The portion of the wave that is not reflected propagates into the material, where its intensity decreases exponentially due to absorption. Upon reaching the rear interface, a part of the wave transmits through the material, while the remaining portion is reflected back, resulting in multiple reflections between the two interfaces. The schematic diagram illustrating the interaction between the EM wave and the shielding material is shown in Figure 1.1.

For simplification, consider a normally incident plane wave striking a homogeneous and isotropic material sheet, following the Schelkunoff theory based on the transmission line

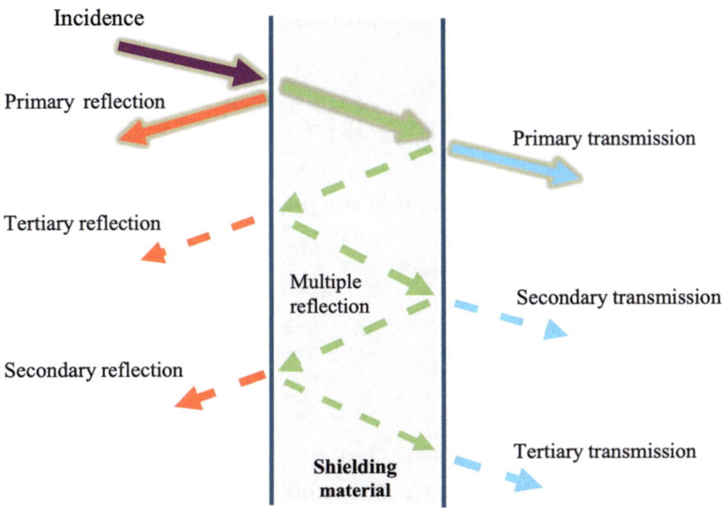

FIGURE 1.1 Diagram showing the propagation of an EM wave through a material sheet.

model of shielding materials and standard shielding effectiveness (SE) measurement methods. According to Schelkunoff theory, shielding effectiveness, typically expressed in decibels on a logarithmic scale, is the sum of reflection loss (SE_R), absorption loss (SE_A) and multiple reflection loss (SE_M),

$$SE(\text{dB}) = 10\log_{10}\left(\frac{P_i}{P_t}\right) = 20\log_{10}\left(\frac{E_i}{E_t}\right) = 20\log_{10}\left(\frac{H_i}{H_t}\right) \\ = SE_R + SE_A + SE_M \tag{1.1}$$

where P_i (E_i or H_i) and P_t (E_t or H_t) represent the power (intensity of the electric or magnetic field) of the incident and transmitted waves, respectively [7–10].

In order to obtain the shielding efficiency (SE) of medium, it is essential to discuss about the characteristic impedance and propagation constant in the medium.

The characteristic impedance (η) of medium is,

$$\eta = \sqrt{\frac{j\omega\mu}{\sigma + j\omega\varepsilon}} \tag{1.2}$$

where μ, ε and σ represents the magnetic permeability, electric permittivity, electrical conductivity of the medium, respectively. ω is the angular frequency of the wave and j is the imaginary unit. The characteristic impedance of free space (Z_0) is equivalent to the square root of the ratio of the permeability of free space (μ_0) to the permittivity of free space (ε_0).

That is, $Z_0 = \sqrt{\frac{\mu_0}{\varepsilon_0}} \approx 377\ \Omega$.

The propagation constant (commonly denoted as γ) is a complex number that represents the changes a wave experiences as it travels through a medium. It describes the changes in both the amplitude and phase of the wave as it propagates through the medium.

The propagation constant is generally represented by its real part (α) and its imaginary part (β):

$$\gamma = \alpha + j\beta\sqrt{j\omega\mu(\sigma + j\omega\varepsilon)}, \tag{1.3}$$

where α is the attenuation constant and β is the phase constant.

When $\sigma >> \omega\varepsilon$ (for good conductors), $\eta = \sqrt{\frac{j\omega\mu}{\sigma}}$ and $\gamma = \sqrt{j\omega\mu\sigma}$, whereas for semiconductors and dielectric materials ($\sigma << \omega\varepsilon$), $\eta = \sqrt{\frac{\mu}{\varepsilon}}$ and $\gamma = j\omega\sqrt{\mu\varepsilon}$.

1.2.1 Reflection Loss

Reflection loss at an interface between two media is indeed related to the difference in their characteristic impedances. When a wave encounters an interface between two different media, part of the wave is reflected back into the first medium and part of it is transmitted into the second medium [8, 11]. The reflection coefficient R and the transmission coefficient T at the interface are given by the impedances Z_1 and Z_2 of the two media:

$$R = \frac{Z_2 - Z_1}{Z_1 + Z_2} \tag{1.4}$$

and

$$T = \frac{2Z_2}{Z_1 + Z_2} \tag{1.5}$$

Let E_i be the electric field strength of the incident wave, then the reflected electric field strength (E_r) and transmitted electric field strength (E_t) are,

$$E_r = RE_i \frac{Z_2 - Z_1}{Z_1 + z_2} E_i \tag{1.6}$$

$$E_t = TE_i = \frac{2Z_2}{Z_1 + z_2} E_i \tag{1.7}$$

respectively, and when the media are infinite in thickness $T = 1 + R$.

When an electromagnetic wave traverses a shielding material, it interacts with two distinct boundaries between the material and free space. The reflection and transmission coefficients at the front and rear surfaces are, $R_1 = \frac{Z_2 - Z_1}{Z_1 + Z_2}$, $T_{21} = 1 + R_1$, $R_2 = -R_1$, and $T_{12} = 1 + R_2$, respectively [10].

Let E_1 be the electric field strength of the transmitted wave through the first interference, then the electric field strength of the transmitted wave through the second interference is,

$$E_t = T_{12}E_1 = T_{12}T_{21}E_i \tag{1.8}$$

Based on the definition of shielding effectiveness, the reflection loss can be expressed as:

$$SE_R = 20\log_{10}\left(\frac{E_i}{E_t}\right) = 20\log_{10}\left(\frac{1}{T_{21}T_{12}}\right) \tag{1.9}$$

which can be rewritten in terms of Z_1 and Z_2 as,

$$SE_R = 20\log_{10}\left(\frac{(Z_1 + Z_2)^2}{4Z_1Z_2}\right) \tag{1.10}$$

For conductive monolithic materials $Z_2 \ll Z_1$, then $SE_R = 20\log_{10}\left(\frac{Z_1}{4Z_2}\right)$

The reflection loss, expressed in decibels, is determined by the inverse of the product of the transmission coefficients at the two interfaces. This indicates that the reflection loss is solely a function of the impedance mismatch between the interacting media.

1.2.2 Absorption Loss

As an electromagnetic wave propagates through a material, its amplitude attenuates exponentially. This attenuation is attributed to the material's dielectric, magnetic and conductive losses [8]. Consequently, the electric field intensity E at a depth t within the material can be expressed as follows:

$$E = E_i e^{-\gamma t} \tag{1.11}$$

The depth at which the amplitude of an electromagnetic wave diminishes to $\frac{1}{e}$ of its original magnitude is known as the skin depth (δ_s). The absorption loss for a material with thickness t is expressed as,

$$SE_A = 20\log_{10}\left|\frac{E_i}{E_t}\right| = 20\log_{10}\left|e^{\gamma t}\right| \tag{1.12}$$

Absorption loss is strongly influenced by the electromagnetic constitutive parameters; permeability (μ), conductivity (σ) and permittivity (ε) as well as the material thickness. Physically, it represents the attenuation of the transmitted wave due to intrinsic losses within the shielding material, particularly under conditions where the material's impedance is matched with that of the surrounding medium, thereby eliminating reflections at both interfaces [10].

1.2.3 Multiple Reflection Loss

When the shielding material is thin, the wave reflected at the second interface undergoes successive internal reflections, first at the initial interface, then again at the second interface, resulting in a series of multiple reflections within the material [9]. Consequently, overall shielding effectiveness is influenced not only by reflection and absorption mechanisms but also by the cumulative effect of these multiple internal reflections. The multiple reflection loss is given as

$$SE_M = 20\log_{10}\left|1 - R_1^2 e^{-2\gamma t}\right| \tag{1.13}$$

The contribution of multiple reflections to the overall effectiveness of the shielding can be considered negligible (i.e., $SE \approx 0$) when the thickness of the shielding material exceeds the depth of the skin or when the absorption loss (SE_A) is greater than 10 dB. In such cases, the amplitude of the electromagnetic wave reaching the second interface is significantly attenuated and may be disregarded [10].

1.2.4 Determination of EMI Shielding Effectiveness using Scattering Parameters

Electromagnetic wave interaction with a material sample can be experimentally analysed using transmission and reflection measurement techniques, commonly facilitated by a vector network analyser (VNA) to obtain scattering parameters. In a two-port VNA, the reflected and transmitted waves are represented by complex scattering parameters. S_{11} and S_{22} describe reflections at ports 1 and 2, while S_{21} and S_{12} represent transmission between the ports. These parameters are related to the material's reflectivity ($\Re$) and transmissivity (τ).

$$\Re = \left|\frac{E_r}{E_i}\right|^2 = \left|S_{11}\right|^2 = \left|S_{22}\right|^2$$

$$\tau = \left|\frac{E_t}{E_i}\right|^2 = \left|S_{21}\right|^2 = \left|S_{12}\right|^2$$

The reflection coefficient R_1 (with $\left|R_1\right| \leq 1$) at the front interface of the material can be expressed as:

$$R_1 = \chi \pm \sqrt{\chi^2 - 1} \tag{1.14}$$

where the intermediate variable χ is given by:

$$\chi = \frac{S_{11}^2 - S_{21}^2 + 1}{2S_{11}} \tag{1.15}$$

The propagation factor $c^{-\gamma t}$, representing attenuation within the material, can be determined using:

$$c^{-\gamma t} = \frac{S_{11} + S_{21} - R_1}{1 - \left(S_{11} + S_{21}\right) R_1} \tag{1.16}$$

These equations (Eqs. 1.14 and 1.16) can be used to calculate the reflection loss, absorption loss and multiple reflection loss [10].

1.3 MATERIAL CHARACTERISTICS INFLUENCING EMI SHIELDING PERFORMANCE

As discussed in the aforementioned section, EMI shielding fundamentally relies on core mechanisms like reflection, absorption and multiple reflections. These mechanisms are governed by principles of electromagnetic theory. However, in real-world scenarios, shielding effectiveness of a material is closely linked to the physical and chemical properties of the material used. Moreover, when designing new shielding materials, it has to be kept in mind that shielding performance not only depends on electrical conductivity but also on the complex interplay of factors including magnetic permeability, dielectric properties, structural morphology and environmental stability. This section explores the critical material parameters that play a pivotal role in enhancing shielding properties necessary for the design of materials with multifunctional properties.

1.3.1 Electrical Conductivity and Shielding by Reflection

Electrical conductivity is one of the fundamental properties that influences a material's ability to shield EMI, particularly through the mechanism of reflection [12]. When an incident electromagnetic wave impinges on a conductive surface, free electrons within the material rapidly respond to the alternating electric field and generate currents. These induced currents effectually generate secondary electromagnetic waves in the opposite direction, effectively bouncing the original wave back. Further, this reflective behaviour is most effective at high frequencies and is prominent in metals like copper and aluminium due to their high free electron densities. However, high conductivity alone does not guarantee comprehensive shielding across the entire frequency spectrum, as materials must also be tailored to manage wave penetration and absorption. To address this issue, material researchers are putting their efforts towards developing engineered materials with good electrical performance and structural adaptability. In recent years efforts have been dedicated towards the development of advanced composites and conductive polymers to mimic metallic conductivity while offering advantages such as flexibility, lightweight and processability, thereby broadening the design space for reflective EMI shielding materials [13]. While reflection is the dominant shielding mechanism in materials with high conductivity, moderate conductivity can promote partial penetration of waves and facilitate energy dissipation via ohmic losses, dielectric loss and magnetic loss mechanisms, making absorption a significant contributor in certain composites. The Simon's formula is used to estimate the EMI shielding effectiveness (SE) of conductive materials. It considers reflection and absorption losses, but not multiple reflections within the shield [14].

$$SE_T = 50 + 10\log\left(\frac{\sigma}{f}\right) + 1.7\sqrt{\sigma f} \tag{1.17}$$

where σ represents the electrical conductivity $\left(\frac{\text{s}}{\text{cm}}\right)$ of the material, f represents the frequency of the EM field (MHz), and t represents the thickness of the material (cm).

The electrical conductivity of EMI shielding materials is reliant on the base material composition. For instance, metals exhibit high conductivity, whereas polymers and ceramics are insulating. To overcome the limitations of insulating matrices, conductive fillers like graphene, carbon nanotubes and metallic nanoparticles are incorporated, forming conductive networks once the percolation threshold is reached. Filler morphology and aspect ratio play key roles in achieving conductivity at lower loadings, while uniform dispersion and proper alignment ensure consistent conductive pathways and minimise performance degradation. However, it is observed that in practical design, exceedingly high conductivity can lead to secondary electromagnetic pollution due to excessive reflection, whereas a moderate conductivity level supports better impedance matching with free space (377 Ω), promoting absorption and minimising secondary reflection. For real-world applications, maintaining stable electrical performance under varying environmental conditions and scaling production processes to ensure uniform conductivity are essential. In summary, electrical conductivity is a foundational property that shapes the EMI shielding behaviour of materials, and its precise tuning through material selection, filler engineering and processing techniques enables the development of shields tailored to both reflection- and absorption-dominant environments.

1.3.2 Dielectric Properties and Absorption Mechanisms

Dielectric loss plays a pivotal role in EMI shielding by facilitating the absorption of electromagnetic wave energy within shielding materials. This energy dissipation occurs primarily through two mechanisms: electric dipole polarisation and charge carrier motion [15]. Under an alternating electromagnetic field, electric dipoles in the material attempt to align with the oscillating field; however, the lag in their realignment leads to polarisation loss, converting part of the electromagnetic energy into heat. Simultaneously, charge carriers such as electrons and ions migrate or hop between localised sites, resulting in conduction loss and further energy dissipation. These mechanisms are frequency-dependent, with different polarisation and conduction processes responding across various frequency ranges. Additionally, multiple internal reflections within the shielding material enhance the path length of electromagnetic waves, increasing their interaction with dipoles and charge carriers and thus improving overall absorption. The extent of dielectric loss is characterised by the material's complex permittivity which contains the real part ε' and the imaginary part ε''

$$\varepsilon' = \varepsilon_\infty + \frac{\varepsilon_s - \varepsilon_\infty}{1+\left(2\pi f^2\right)\tau^2} \tag{1.18}$$

$$\varepsilon'' = \frac{2\pi f\left(\varepsilon_s - \varepsilon_\infty\right)}{1+\left(2\pi f^2\right)\tau^2} + \frac{\sigma}{2\pi f \varepsilon_0} \tag{1.19}$$

where ε_0, ε_s and ε_∞ are dielectric constants in the vacuum, static and infinity frequency. Also, τ is the relaxation time (s). Here relaxation time refers to a specific time for a polarised species that moves from a polarised to a normal state.

The real part is related to the charge storage while the imaginary part (ε'')correlates with energy dissipation capability during polarisation. Equation (1.19) indicates that high conductivity enhances dielectric losses. Materials with high ε'' values are thus highly effective at converting EM energy into heat, making them ideal for absorption-dominant EMI shields. This absorption-focused mechanism is crucial for reducing both transmitted and reflected electromagnetic waves, thereby minimising secondary electromagnetic pollution and enhancing the overall shielding performance.

1.3.3 Magnetic Permeability and Low-Frequency Shielding

Similar to dielectric features, magnetic properties of a material also play a pivotal role in the design of EMI shielding. The magnetic permeability μ is a complex quantity given by $\mu' = \mu' - j\mu''$, where the real part μ'' represents magnetic energy storage and the imaginary part, μ'' implies magnetic energy dissipation or magnetic losses. Both these parameters are beneficial in attenuating incident electromagnetic waves. The magnetic loss mechanisms include natural resonance, eddy current loss, hysteresis loss, domain wall resonance and residual loss. However, an inherent drawback of magnetic materials is that they are inherently limited in their effectiveness for shielding high-frequency electromagnetic waves, particularly in the gigahertz (GHz) range. The effectiveness of magnetic materials in high-frequency EMI shielding is fundamentally constrained by Snoek's limit, which describes an inverse relationship between magnetic permeability and resonance frequency [16]. When field frequency nears the material's natural resonance the permeability begins to drop sharply. This drop is directly influenced by the material's saturation magnetisation. As a result, the material's ability to absorb or reflect electromagnetic waves is significantly reduced beyond the resonance point.

In highly conductive magnetic materials such as iron and its alloys (Fe, FeNi, FeCo), alternating magnetic fields induce strong eddy currents, which contribute to energy dissipation at lower frequencies. However, at higher frequencies, these eddy currents impair skin effects, confining the magnetic response to the material's surface and further reducing permeability in the GHz range. Although ferrites are advantageous due to their semiconducting nature, minimising eddy current losses their inherently lower saturation magnetisation limits their magnetic performance at microwave frequencies [17].

Even though using nanoscale magnetic particles can effectively decrease eddy current losses by minimising particle dimensions, this approach does not circumvent the intrinsic limitations set by Snoek's limit. Therefore, as frequency increases, magnetic permeability still decreases, weakening the magnetic loss component that is essential for effective EMI shielding. This makes magnetic materials more suitable only for low-frequency shielding applications. To overcome these intrinsic physical constraints and improve their performance at microwave frequencies, creative nanostructuring, composite integration and customised material engineering techniques are required. Ultimately, the integration of magnetic properties with dielectric characteristics permits a synergistic enhancement in EMI shielding effectiveness thereby reducing secondary electromagnetic reflection. Overall, developing high-performance EMI shielding materials for modern electrical and communication technologies requires careful manipulation of magnetic characteristics, loss processes and structural aspects.

1.3.4 Thickness and Multiple Internal Reflections

Material thickness plays a pivotal role in determining the performance of EMI shielding materials by influencing key mechanisms such as reflection, absorption and transmission of electromagnetic waves. Thicker materials generally exhibit higher surface conductivity or dielectric response, which enhances reflection but may lead to reflection-dominant shielding and secondary electromagnetic pollution. Conversely, thinner materials which are strategically engineered, are capable of enhancing absorption by promoting multiple internal reflections and internal energy dissipation. Furthermore, it can be seen that advanced designs such as gradient or multilayered structures allow for the fine-tuning of absorption mechanisms even with reduced thickness. Minimising transmission while still maintaining effective shielding can be effectively realised using porous or layered designs that create additional loss pathways within thinner materials.

To overcome the challenges related to material thickness, several optimisation strategies like layer-by-layer assembly enable precise control of individual layer thicknesses and compositions, thereby facilitating frequency-selective shielding and impedance matching. Multi-component or sandwich structures with varied layer thicknesses can target specific frequency bands and enhance broad-spectrum shielding by adjusting resonance peaks. Although highly reflective and effective at shielding, thin conductive films require the addition of porous or absorbing components to reduce reflection dominance and enhance absorption. This balance between reflection and absorption becomes critical in designing thin and effective shields.

1.3.5 Thermal, Mechanical and Environmental Stability

Thermal stability and environmental resistance are critical material properties required for EMI shielding. These essential features guarantee consistent performance of the material under stringent conditions such as high temperatures, humidity, chemical exposure and mechanical stress. High thermal stability allows materials like basalt fibre/aramid nanofibre nanocomposite papers and aerogels to retain structural integrity and electrical properties at high temperatures so as to maintain a good shielding property. Furthermore, heat produced during electromagnetic absorption can be dissipated using materials with high thermal conductivity, avoiding thermal degradation and preserving functionality. Long-term endurance is ensured by environmental resistance, which includes chemical stability in acidic and alkaline situations. This is particularly important in military or aerospace contexts. Resistance to moisture and humidity, attained by hydrophobic or superhydrophobic coatings, maintains impedance matching and loss mechanisms by preventing corrosion and changes in dielectric characteristics. Materials with robust interfaces, such as CNT composites, demonstrate stability through repeated thermal cycling and mechanical stress, preventing issues like cracks or delamination. Moreover, corrosion and oxidation-resistant matrices or coatings help protect conductive fillers like graphene or MXenes, while self-healing capabilities further enhance longevity by repairing microcracks caused by environmental stress. Practical examples, such as paper nanocomposite of basalt and aramid nanofibres with carbon nanotube and reduced graphene oxide/liquid metal films, showcase the importance of these properties in maintaining high shielding efficiency under extreme conditions [18, 19]. Overall, thermal stability and environmental resistance are critical for the reliable, long-term performance of EMI shielding materials in demanding applications.

1.3.6 Surface Roughness and Morphology

Electromagnetic waves primarily come in contact with the surface of a shielding material. Hence surface roughness and also their morphology are critical factors influencing the EMI shielding performance of materials, as they directly affect the balance between reflection and absorption of electromagnetic waves. Smooth surfaces typically promote reflections, cause unwanted secondary electromagnetic pollution. In contrast, rough or textured surfaces enhance surface scattering and internal reflections, increasing electromagnetic wave absorption by enhancing the wave path length and facilitating energy dissipation [20]. Another advantage of rough surfaces is that they reduce impedance mismatch, allowing greater wave penetration for absorption-dominated shielding [21]. Complex surface morphologies which include hierarchical and porous structures act as traps for electromagnetic waves, thereby further improving absorption by enabling multiple internal reflections and boosting dielectric losses through dipole and interfacial polarisation effects [22]. Furthermore, advanced fabrication techniques like additive manufacturing and surface engineering allow for precise control over morphological features, enabling the creation of materials with tailor-made surface roughness and porosity. Such morphological precisions are especially important for achieving effective impedance matching, which minimises reflection and improves absorption efficiency [23]. The integration of MXene and carbon-based fillers into polymers, coupled with strategic morphological engineering, offers a robust pathway for the development of surface engineered EMI shielding materials [24, 25]. Ultimately, strategic design and engineering of surface roughness and morphology play a vital role in optimising EMI shielding materials by promoting absorption-dominant mechanisms, reducing reflections and ensuring functional adaptability across diverse environments.

1.4 PRESENT MATERIALS FOR EMI SHIELDING

EMI and its effective shielding is receiving more scholarly attention, particularly in the context of ensuring a clean and sustainable lifestyle. The rapid development of technology has made our lives much simpler; however, the pursuit of such living standards introduces considerable technical complexity. For example, the use of electronic devices, especially today's multifunctional sensors, has increased exponentially for making our lives smarter and simpler. At the same time, technical challenges and adverse effects caused by these sophisticated complex electronics, such as EMI, must be addressed properly. Thus, the development of advanced materials by tailoring their intrinsic properties emerged as a promising field. Current research efforts are focused on the development of "smart materials" with enhanced properties capable of addressing multiple functions. The recent advances in shielding materials, with a particular emphasis on the classification according to the various functional properties that influence the shielding mechanisms, are discussed here.

1.4.1 Electrically Conductive Materials

Electrical conductivity plays a significant role in the EMI shielding mechanism through reflection or absorption mechanisms [26, 27]. Enhancing electrical characteristics and EMI shielding effectiveness in thermoplastic elastomeric polymer blends by utilising the selective distribution of conductive black. The attenuation of electromagnetic radiations through reflection is coupled with the electrical conductivity of the shielding materials. Metals, used as conventional shields favouring reflection [28], were replaced by polymer composites with carbon-based or metallic nanoparticles for increased durability, lightweight and cost effectiveness [11]. Since the reflection-dominant shielding mechanism was the reason behind the secondary electromagnetic pollution [29, 30]. In response, there has been a paradigm shift towards the use of alternative materials such as carbon materials [31], MXenes [32] and intrinsically conducting polymers [33]. These materials are particularly advantageous as they predominantly absorb, rather than reflect, electromagnetic radiation, thereby mitigating the issue of secondary pollution. However, the challenges in composite fabrication involve problems such as poor dispersion of fillers and diminished mechanical strength.

The improved electric conductivity in polymer-based shields is due to the formation of conducting networks [34], which arise once the filler concentration surpasses the percolation threshold. Shape, aspect ratio, morphology, compatibility and distribution of the fillers are the controlling factors for achieving percolation threshold [35]. Thus, efficient material engineering is essential in optimising the better performance of shielding materials. For instance, Chen et al. [36] reported the use of reduced graphene oxide-carbon fibre (rGO-CF) composites. They could achieve a shielding effectiveness of 37.8 dB in the X-band frequency range, with an electrical conductivity of 7.13 S/m. Later, Verma et al. [37] tried to increase the conductivity to the same range using different polymer matrices but couldn't yield a corresponding increase in shielding effectiveness. This shows that while electric conductivity is a significant factor in EMI shielding performance, the overall effectiveness depends upon a complex interplay of multiple material properties and structural features. Notably, Singh et al. [38] observed that even materials with relatively modest electrical conductivity can achieve high shielding effectiveness values, such as 42 dB, highlighting the multifaceted nature of EMI shielding mechanisms.

A highly flexible and expandable polymer with silver nanowires (AgNW) has been developed by Wei et al. [39] with an EMI shielding effectiveness of 40 dB for a 1 mm thick layer. Shielding performance is further enhanced by increasing the weight percentage of AgNW reaching up to 111.5 dB. Furthermore, a study on Ag@PDMS/silicone (APS) was conducted by Duan et al. [40] revealed that by increasing the percentage of conductive fillers can enhance the interaction of free charge carriers and thereby increase the shielding. Evidently, they could achieve a shielding of 50–70 dB due to its improved compressibility and conductive features. Thus, it is made clear that various factors influence shielding effectiveness directly or indirectly, including conductivity, structural design, filler ratio, etc.

1.4.2 Magnetically Permeable Materials

Materials having high magnetic permeability are capable of interacting with the magnetic component of the incident electromagnetic radiation and can contribute towards shielding mechanism [41]. Composites comprising ferrites [42] and magnetic nanoparticles [43] can initiate magnetic loss mechanism facilitated by the increased magnetic permeability. This can favour absorption and reduce the unwanted reflection of the incident electromagnetic radiations.

The activation of magnetic loss mechanism can be achieved by the incorporation of magnetic fillers like Fe_3O_4, $NiFe_2O_4$ [44]. This type of inclusion not only alters the impedance matching characteristics of the composite but also enhances its energy dissipation capabilities. For example, Ba-Pb hexaferrite particles have been employed to increase the magnetic permeability of nanocomposites, thereby improving their EMI shielding capabilities [45]. In a similar way, nanostructured ferrites such as $CoFe_2O_4$, when decorated on microspheres [46], have been synthesised to function as lightweight and highly efficient microwave absorbers. Furthermore, the integration of $NiFe_2O_4$ (nickel ferrite) nanocomposites into flexible sheets offers both mechanical flexibility and superior shielding performance. Likewise, the incorporation of $NiCoFe_2O_4$ (nickel-cobalt ferrite) nanoparticles into carbon black/poly(vinyl alcohol) (PVA) composites has been explored for microwave absorption and EMI shielding applications [47].

Additionally, magnetic graphene, or graphene sheets decorated with magnetic nanoparticles such as Fe_3O_4, has gathered considerable attention [48]. Sharif et al. [49] prepared PMMA/graphene nanocomposites without and with magnetic nanofillers. The shielding effectiveness of 1.1 vol% PMMA/G was 20.7 dB. But the addition of 0.5 vol% of magnetic filler increased the total shielding effectiveness of PMMA/G to 29.3 dB. Qin et al. [50] prepared Fe_3O_4 foams by optimising the complex permittivity to achieve high impedance matching and thereby a better shielding effectiveness even at a low density of 0.05078 g/cm^3. These materials facilitate combined magnetic and dielectric loss mechanisms, thereby improving the absorption of incident electromagnetic waves and enhancing the overall shielding performance of the composites.

1.4.3 Dielectric and Lossy Materials

In EMI shielding mechanism, dielectric loss property of the shield can attenuate incident radiation through energy dissipation. This is because of the polarisation dynamics such as dipole, electronic and interfacial polarisation of the dielectric material [51]. It can convert electromagnetic energy into thermal energy via resistive losses, thereby facilitating absorption-dominated shielding mechanisms. Structural engineering strategies such as controlled doping, the design of layered heterostructures and the synthesis of core-shell architectures enable precise modulation of permittivity and loss tangents are used for optimising dielectric performance [52]. For instance, nanostructured materials including porous foams, aerogels and composite systems incorporating high-permittivity fillers enhance dielectric loss by promoting interfacial polarisation and charge accumulation at heterojunctions. Such tailored nanostructures can amplify energy dissipation pathways, thereby improving the overall shielding effectiveness (SE) of the material.

Oxides like SnO_2, TiO_2, ZrO_2, ZnO and Al_2O_3 are widely used for their high dielectric constants and energy dissipation capabilities and their incorporation into composites enhances dielectric losses, aiding in broadband EMI shielding. Graphene oxide sheets decorated with $BaTiO_3$ and Fe_3O_4 nanoparticles demonstrate synergistic dielectric and magnetic losses, leading to improved shielding performance over a broad frequency range [53] employing a conductivity gradient within the multilayered structures can enhance wave absorption and minimise impedance mismatch, and thereby elevate EMI shielding. The use of vapour-phase polymerisation to control PEDOT density on the nanofibre surface resulted in the creation of distinct conductivity layers. This method significantly enhanced the absorption of EM waves compared to uniform-conductivity designs. Peng et al. [54] fabricated electrodeless plated Ni-coated melamine foam (Ni@MF) modified with aligned multiwalled carbon nanotubes (MWCNTs), temperature-sensitive microspheres and polydimethylsiloxane (PDMS). This porous structure exhibits a SE of 45.3 dB at 3 wt% loading of MWCNTs due to enhanced reflection, induced electric/magnetic dipoles and ohmic losses.

1.4.4 Hybrid and Multicomponent Materials

Hybrid materials for EMI shielding are composite structures that combine multiple types of nanomaterials such as conductive, magnetic and dielectric components. These materials are designed to enhance electromagnetic wave attenuation through multiple mechanisms, including electric conductivity, dielectric polarisation, magnetic losses and multiple-interface effects. Common approaches involve combining magnetic and dielectric nanoparticles, such as decorating magnetic nanoparticles on dielectric substrates or embedding them within polymer matrices. These hybrid structures facilitate improved complex permittivity and permeability, resulting in higher SE across wider frequency ranges. Examples include conductive hybrid structures composed of conductive materials like carbon nanotubes or graphene with magnetic nanoparticles like Fe_3O_4, and multilayered or core-shell configurations that prevent nanoparticle agglomeration and optimise electromagnetic interactions [28].

Additionally, the incorporation of magnetic nanoparticles (e.g., Fe_3O_4, CoNi nanostructures) with conductive nanomaterials (such as graphene or CNTs) leverages the complementary effects of magnetic and dielectric losses. This will result in improved attenuation of electromagnetic waves. Core-shell architectures [55], demonstrated by Fe_3O_4 coated multiwalled carbon nanotubes (Fe_3O_4@MWNT), effectively reduce nanoparticle agglomeration while combining magnetic and dielectric loss mechanisms. Furthermore, multilayered or gradient structures composed of alternating magnetic and dielectric materials facilitate impedance matching and broaden the effective shielding bandwidth. Composites with graphene-decorated magnetic nanoparticles can maximise complex permittivity and permeability through enhanced interfacial polarisation and heterostructure interactions. Ultra-flexible films were prepared by Liu et al. [21] using MXene sediment (MS) and WPU at a thickness of 0.51 mm, showcasing an EMI SE of 45.3 dB for 70% loading of MS. The multiple interfaces between the MS and WPU result in the development of interfacial polarisation of incident EMW. The dipolar polarisation caused by localised defects and MS surface termination also triggered the loss of incident EMW.

1.5 FUTURE OF EMI SHIELDING TECHNOLOGY

The extensive proliferation of intelligent and interconnected systems, ranging from industrial automation, next-generation telecommunications and bio-integrated devices, has made EMI a growing concern. To foster collaborative efforts aimed at meeting the demands of next-generation electromagnetic technologies, the field has drawn scholars from diverse domains such as physics, chemistry, materials science, information technology and artificial intelligence to work towards next-generation shielding materials. With the escalating demand for compact, sensitive and ubiquitous electronic devices, it is obligatory that EMI shielding solutions ensure device stability and performance. Devices functioning in close proximities are prone to electromagnetic disturbances from both internal and external sources which poses a serious threat to the performance and reliability of devices, particularly in high-stakes fields such as defence, aerospace and communication technologies. Hence it is mandatory that emerging trends in EMI shielding emphasise multifunctionality, adaptability and sustainability, transforming it from a passive protective measure to an intelligent system component.

1.5.1 Hyperconnected Environments

The rise of 6G and terahertz systems is set to revolutionise wireless connectivity, enabling ultra-fast, low-latency communication essential for a future hyperconnected world. The major outcome of the proliferation of Internet of things (IoT) and Industrial Internet of things (IIoT) devices in digital urban infrastructures is the dramatic skyrocketing of electromagnetic congestion. The billions of connected sensors, actuators and communication modules pose significant risk of device perturbation due to EMI. The basic strategy to mitigate electromagnetic radiations is to reduce or minimise reflection of electromagnetic waves back to the environments. Consequently, tech-enabled city networks not only need compact and lightweight shielding materials but also absorption-dominant materials, an essentiality in meeting the modern electromagnetic environment. According to Isari et al. [29] internal scattering and multilayered configurations are promising directions for enhancing EM wave dissipation in such dense, multi-device ecosystems. Optimising material characteristics such as electrical conductivity, thickness, dielectric constant and pore size contributes to the development of shielding materials with minimised surface reflections. Intelligent surface impedance engineering can serve as a new paradigm for developing absorption-dominant materials. Recently, Meng et al. exploited surface engineering technique to fabricate absorption-dominant multifunctional graphene aerogel with a SE of 90 dB in X-band, Ku-band and K-band of frequencies [56]. Moreover, the resulting graphene films demonstrate excellent thermal conductivity (up to 476.4 Wm^{-1}/K), efficient heat dissipation, electro-thermal conversion, heat storage and flame retardancy. However, to fully bring out the potential of similar innovative materials more methodical research is required. A thorough and systematic theoretical investigation linking structural features with electromagnetic wave interactions is essential for the design of innovative, absorption-dominant materials.

1.5.2 Advanced Computing Systems

The advent of advanced computing systems, such as quantum computers, neuromorphic processors and AI-accelerated chips, has transformed nearly every aspect of modern electronics. These next-generation platforms operate at high frequencies and exhibit extreme sensitivity to external disturbances. As the demand for further miniaturisation and functionally dense devices continues to grow, these systems become increasingly susceptible to even the slightest electromagnetic noise and environmental complexities. Unlike traditional electronic systems, advanced computing platforms require tunable, lightweight and multifunctional EMI shielding materials capable of responding dynamically to varying electromagnetic environments. These materials must offer high SE while maintaining compatibility with compact form factors and thermal management demands. Structural innovations such as gradient conductivity layers [57], multilayered dielectrics [58] and nanoporous configurations [59, 60] are emerging as promising solutions, providing enhanced absorption-dominant shielding without compromising device performance. As the field progresses towards exascale computing and integrated AI architectures, ensuring robust EMI control will remain central to enabling precision, scalability and energy efficiency.

1.5.3 Autonomous Systems and Electric Vehicles (EVs)

Next-generation mobility technologies such as electric vehicles, automated cars and unmanned systems rely extensively on advanced electronic systems. Among these, LIDAR (Light Detection and Ranging) is used for precise 3D mapping and object detection through pulsed laser light, while GPS (Global Positioning System) provides real-time positioning and navigation via satellite signals. Additionally, these vehicles integrate numerous wireless communication modules, including vehicle-to-vehicle (V2V) and vehicle-to-infrastructure (V2I) systems, to ensure safe and efficient operation.

However, the integration of these components into increasingly miniaturised and densely packed devices makes them highly susceptible to EMI. To overcome this challenge, researchers are developing lightweight, thermally stable and mechanically robust EMI shielding materials that align with the structural and performance requirements of modern automotive systems. In this context, carbon fibre-reinforced composites (CFRCs) have been explored as promising candidates for electric vehicle battery enclosures. A multilayered configuration incorporating copper-coated polyester fabric and melamine foam demonstrated impressive performance, with a specific tensile strength of 346.8 MPa/(g/cm^3) and electromagnetic SE reaching up to 88.27 dB within the 30 MHz to 3 GHz range [61]. Furthermore, the integration of graphite particles into the matrix improved thermal properties, enabling the structure to attain a surface temperature of 68.3 °C under a 5 V DC supply within 180 seconds successfully fulfilling structural, shielding and thermal management criteria.

1.5.4 Flexible, Wearable and Implantable Electronics

Electronic miniaturisation is the cornerstone of modern technology. The integration of flexible electronics into fields such as wearables, healthcare, human–machine interfaces and environmental sensing has created an urgent need for EMI shielding materials that are not only effective but also lightweight, flexible and suitable for human-centred applications. The widespread use of devices such as smartphones, smartwatches, fitness trackers, biomedical sensors and flexible displays in everyday life demands shielding materials that can maintain high performance under conditions of repeated bending, stretching and exposure to varying environmental factors. To address the growing demand for flexible and wearable electronics, polymer-based nanocomposites have emerged as a promising platform for EMI shielding applications. By incorporating conductive and magnetic nanofillers into the polymer matrix, these composites achieve enhanced electrical conductivity, electromagnetic SE and mechanical flexibility. Materials such as MXenes, CNTs and graphene-based structures have shown exceptional performance when embedded within polymer matrices, offering a balance of flexibility and shielding capability. However, a significant challenge lies in the trade-off between filler content and material performance. While increasing the concentration of nanofillers typically improves SE, it often compromises mechanical strength, flexibility and structural integrity, thereby limiting their suitability for practical applications. To overcome this limitation, segregated and porous nanocomposite architectures have been explored [62–64]. This approach maintains the desired mechanical properties and stretchability, making the materials more viable for real-world wearable and flexible electronic systems.

1.5.5 Energy and Power Electronics

With the growing global demand for sustainable energy solutions, the emergence of renewable energy technologies necessitates the development of multifunctional systems capable of energy harvesting, conversion and storage. The use of high-power electronic systems such as solar inverters, smart grids, wind turbines and energy storage systems has intensified the need for effective EMI. These systems operate under high voltages and currents, leading to both radiated and conducted electromagnetic noise. Unmanaged EMI can disrupt system communication, degrade power quality and potentially cause hardware failures. The shielding materials used in this context should serve two primary purposes: suppressing electromagnetic noise across a broad frequency spectrum and facilitating efficient thermal management to prevent overheating during continuous or peak-load operations. To achieve this, low-impedance, magnetically active composites are particularly promising, as they can effectively manage both thermal and electromagnetic flux simultaneously. In this context, Shao et al. [65] and his coworkers developed a novel Janus-structured multifunctional membrane (FCFe/M) that combines EMI shielding with personalised thermal management. Using a vacuum filtration technique, they fabricated a membrane consisting of a polytetrafluoroethylene (PTFE)-carbon nanotube (CNT) Fe_3O_4 layer, coated with a highly conductive MXene layer. With a thickness of 84.9 μm, the membrane achieved an SE of 44.56 dB in the X-band and demonstrated excellent thermal performance, reaching up to 145.7°C under 3 V or light exposure. With its strong electrical conductivity, flame retardancy and mechanical stability, this membrane shows great potential for applications in wearable electronics, thermal regulation and energy storage.

1.5.6 3D Printed and On-chip Shielding Solutions

The next wave of technological advancements in the field of EMI shielding lies in additive manufacturing, particularly 3D printing. This new technology permits the fabrication of complex, lightweight and customisable shielding structures that can be directly integrated into electronic devices. Moreover, they effectively address the challenges faced during miniaturisation of devices and simultaneously enhancing the growing demand for efficient EMI protection. The adoption of 3D printing in EMI shielding enables the progress of absorption-dominant materials through the creation of intricate architectures, such as gradient structures or strategic layer-by-layer printing of materials onto conductive or non-conductive substrates. Additionally, 3D printing has also significantly advanced the development of on-chip and conformal EMI shielding solutions. For instance, Shi et al. [66] and his coworkers developed carbon-based EMI shielding modules composed of graphene and CNT nanoparticles using 3D printing. These modules had a very low density as low as 0.076 g/cm^3, on the other hand achieving high SE up to 61.4 dB. This direct integration of this material in devices not only enhances electromagnetic compatibility but also offers space-saving advantages, all without compromising the performance of the underlying electronic systems. In conclusion, bridging the gap between experimental materials research and commercial product development will be essential for widespread adoption of shielding materials. Hence, a continued interdisciplinary research and material innovation will be key to unlock the full potential of EMI shielding in the era of advanced and sustainable electronics. Although this section provides only a brief overview of upcoming developments in EMI shielding technology, a more comprehensive and in-depth investigation into the latest advancements and potential paths forward is presented in Chapter 12. It delves deeply into the evolving landscape of EMI shielding materials, offering valuable insights into cutting-edge developments and emerging applications.

1.6 CONCLUSIONS

The intense demand of high-end electronic devices resulting in enhanced intricacy of modern communication systems has brought consequences of EMIs to the forefront of today's technological landscape. More than being a technical inconvenience, EMI has now become a potential threat to device performance, safety and reliability of electronic equipment, used in sensitive and harsh environments. As electronic devices have become more compact, interconnected and operate at higher frequencies, the demand for effective EMI shielding remains a priority.

This chapter provides a comprehensive overview of the fundamental mechanisms underlying EMI shielding. A brief preview on reflection and absorption losses, and internal scattering has been included. Also, a snapshot of the wide spectrum of materials and their developments for EMI shielding has also been included. While traditional metallic materials and highly conductive polymers continue to play a vital role in reflection-dominant shielding, recent research has shifted focus towards absorption-centric and multifunctional materials. The emergence of advanced composites, carbon-based nanomaterials such as graphene and carbon nanotubes, MXenes and environmentally friendly bio-based materials rightly prompts to field's dynamic response to evolving technological and sustainability demands.

Despite all advancements, a lot of significant challenges persist. Some of the major challenges in this field include the need for a large-scale production of advanced shielding materials, balancing the cost effectiveness and sustainability without compromising on the performance of the material and ensuring long-term stability in adverse environments. Additionally, achieving a consistent SE across a wide range of frequencies and integrating novel materials in existing technology is also an issue of serious magnitude. As we chart the path ahead, the future of EMI shielding relies on the effective inclusion of novel material systems with advanced structural design strategies that ensures not only high SE but also environmental sustainability. Hence the development of multifunctional materials with good shielding properties and thermal management, mechanical strength and flexibility is of paramount importance.

In summary, EMI shielding stands as a pivotal intersection of technological innovation and practical necessity. As our world becomes increasingly hyperconnected and technologically dense, the importance of robust, sustainable and multifunctional EMI shielding solutions will only intensify. Continued interdisciplinary research and collaboration among materials scientists, electrical engineers, chemists and industry stakeholders will be crucial to overcoming current challenges and achieving the next generation of EMI shielding materials and systems.

REFERENCES

1. Xiao-Yun Wang, Si-Yuan Liao, Yan-Jun Wan, Peng-Li Zhu, You-Gen Hu, Tao Zhao, Rong Sun, and Ching-Ping Wong. Electromagnetic interference shielding materials: recent progress, structure design, and future perspective. *Journal of Materials Chemistry C*, 10(1):44–72, 2022.
2. Dimitris J Panagopoulos and George P Chrousos. Shielding methods and products against man-made electromagnetic fields: protection versus risk. *Science of the Total Environment*, 667:255–262, 2019.
3. Wan-Cheng Yu, Ting Wang, Ya-Hui Liu, Zhi-Guo Wang, Ling Xu, Jian-Hua Tang, Kun Dai, Hong-Ji Duan, Jia-Zhuang Xu, and Zhong-Ming Li. Superior and highly absorbed electromagnetic interference shielding performance achieved by designing the reflection-absorption-integrated shielding compartment with conductive wall and lossy core. *Chemical Engineering Journal*, 393:124644, 2020.
4. Jiahao Sun, Zhenliang Gao, Qiang Chen, Yang Li, Chuntai Liu, Changyu Shen, and Xianhu Liu. Interlocking strategy in metal/polymer composites for enhanced anti-corrosion and sustained EMI shielding performance. *Progress in Organic Coatings*, 207:109419, 2025.
5. Jiahui Hong, Ning Luo, Zhengmao Zhang, Liang Zhang, Guanghui Zhang, Lijun Ye, Suprakas Sinha Ray, and Yongjin Li. Regulated orientation and exfoliation of flaky fillers by close packing structures in polymer composites for excellent thermal conduction and EMI shielding. *Composites Part B: Engineering*, 275:111357, 2024.
6. Junchen Luo, Ling Wang, Xuewu Huang, Bei Li, Zheng Guo, Xin Song, Liwei Lin, Long-Cheng Tang, Huaiguo Xue, and Jiefeng Gao. Mechanically durable, highly conductive, and anticorrosive

composite fabrics with excellent self-cleaning performance for high-efficiency electromagnetic interference shielding. *ACS Applied Materials & Interfaces*, 11(11):10883–10894, 2019.

7. Sergei Alexander Schelkunoff. Electromagnetic waves. *(No Title)*, 1943.
8. Henry W Ott. *Noise Reduction Techniques in Electronic Systems*, volume 442. Wiley, New York, 1988.
9. RB Schulz, VC Plantz, and DR Brush. Shielding theory and practice. *IEEE Transactions on Electromagnetic Compatibility*, 30(3):187–201, 2002.
10. Mengyue Peng and Faxiang Qin. Clarification of basic concepts for electromagnetic interference shielding effectiveness. *Journal of Applied Physics*, 130(22), 2021.
11. Lei Wang, Zhonglei Ma, Yali Zhang, Lixin Chen, Dapeng Cao, and Junwei Gu. Polymer-based EMI shielding composites with 3D conductive networks: a mini-review. *SusMat*, 1 (3):413–431, 2021.
12. Deborah Duen Ling Chung. Electromagnetic interference shielding effectiveness of carbon materials. *Carbon*, 39(2):279–285, 2001.
13. K Namsheer and Chandra Sekhar Rout. Conducting polymers: a comprehensive review on recent advances in synthesis, properties and applications. *RSC Advances*, 11(10):5659–5697, 2021.
14. Yue Sun, Yunting Su, Ziyuan Chai, Lei Jiang, and Liping Heng. Flexible solid-liquid bi-continuous electrically and thermally conductive nanocomposite for electromagnetic interference shielding and heat dissipation. *Nature Communications*, 15(1):7290, 2024.
15. Fanbin Meng, Huagao Wang, Fei Huang, Yifan Guo, Zeyong Wang, David Hui, and Zuowan Zhou. Graphene-based microwave absorbing composites: a review and prospective. *Composites Part B: Engineering*, 137:260–277, 2018.
16. Vineeta Shukla. Review of electromagnetic interference shielding materials fabricated by iron ingredients. *Nanoscale Advances*, 1(5):1640–1671, 2019.
17. Adrian Radoń, Lukasz Hawelek, Dariusz Lukowiec, Jerzy Kubacki, and Patryk Wlodarczyk. Dielectric and electromagnetic interference shielding properties of high entropy (Zn, Fe, Ni, Mg, Cd) Fe_2O_4 ferrite. *Scientific Reports*, 9(1):20078, 2019.
18. Shunxi Song, Linghao Li, Dexian Ji, Junfan Zhao, Qi Wu, and Qianyu Wang. Flexible basalt fiber/aramid nanofiber/carbon nanotube electromagnetic shielding paper with outstanding environmental stability and joule heating performance. *ACS Applied Materials & Interfaces*, 15(29):35495–35506, 2023.
19. Yue Sun, Xiao Han, Pu Guo, Ziyuan Chai, Jingyi Yue, Yunting Su, Shengda Tan, Xu Sun, Lei Jiang, and Liping Heng. Slippery graphene-bridging liquid metal layered heterostructure nanocomposite for stable high-performance electromagnetic interference shielding. *ACS Nano*, 17(13):12616–12628, 2023.
20. V Gareyan and Zh Gevorkian. Impact of surface roughness on light absorption. *Physical Review A*, 109(1):013515, 2024.
21. Yue Liu, Na Wu, Sinan Zheng, Yunfei Yang, Bin Li, Wei Liu, Jiurong Liu, and Zhihui Zeng. From mxene trash to ultraflexible composites for multifunctional electromagnetic interference shielding. *ACS Applied Materials & Interfaces*, 14(44):50120–50128, 2022.
22. Yi Liu, Yuanjun Liu, and Xiaoming Zhao. Mxene composite electromagnetic shielding materials: the latest research status. *ACS Applied Materials & Interfaces*, 16(31):41596–41615, 2024.
23. Qian-Ming He, Jun-Ru Tao, Dian Yang, Yi Yang, and Ming Wang. Surface wrinkles enhancing electromagnetic interference shielding of copper coated polydimethylsiloxane: a simulation and experimental study. *Chemical Engineering Journal*, 454:140162, 2023.
24. Aamir Iqbal, Tufail Hassan, Shabbir Madad Naqvi, Yury Gogotsi, and Chong Min Koo. Mxenes for multispectral electromagnetic shielding. *Nature Reviews Electrical Engineering*, 1(3):180–198, 2024.
25. Christos Pavlou, Maria Giovanna Pastore Carbone, Anastasios C Manikas, George Trakakis, Can Koral, Gianpaolo Papari, Antonello Andreone, and Costas Galiotis. Effective EMI shielding

behaviour of thin graphene/pmma nanolaminates in the THz range. *Nature Communications*, 12(1):4655, 2021.

26. Waleed Ghafoor, Mashhood Ahmad, Muhammad Raffi, and Saif Ullah Awan. Role of electrical conductivity in the shielding effectiveness of composite polyvinyl alcohol/multiwall carbon nanotube nanofibers for electromagnetic interference applications. *ACS Omega*, 9(3):3070–3077, 2024.
27. Sreeja Nath Choudhury, Jasomati Nayak, Palash Das, Aparajita Pal, Ankur Katheria, Pallab Banerji, and Narayan Ch Das. Enhancing electrical characteristics and electromagnetic interference shielding effectiveness in thermoplastic elastomeric polymer blends by utilizing the selective distribution of conductive black. *Functional Composite Materials*, 5(1):4, 2024.
28. Krithika Bhaskaran, Rajesh Kumar Bheema, and Krishna C Etika. The influence of Fe_3O_4@GNP hybrids on enhancing the EMI shielding effectiveness of epoxy composites in the x-band. *Synthetic Metals*, 265:116374, 2020.
29. Ali Akbar Isari, Ahmadreza Ghaffarkhah, Seyyed Alireza Hashemi, Stefan Wuttke, and Mohammad Arjmand. Structural design for EMI shielding: from underlying mechanisms to common pitfalls. *Advanced Materials*, 36(24):2310683, 2024.
30. MK Aswathi, AV Rane, AR Ajitha, S Thomas, and M Jaroszewski. EMI shielding fundamentals. In Maciej Jaroszewski, Sabu Thomas, and Ajay V. Rane, eds., *Advanced Materials for Electromagnetic Shielding: Fundamentals, Properties, and Applications*, pages 1–9. John Wiley & Sons, Inc., 2018.
31. Yingying Yu, Yaxi Zhang, Yurong Zhou, Jiajia Xia, Minghui Chen, Huli Fu, Yufang Cao, Tao Wang, Cao Wu, Zhenmin Luo, et al. Enhanced electromagnetic interference shielding properties of CNT/carbon composites by designing a hierarchical porous structure. *Nanomaterials*, 14(13):1099, 2024.
32. Li Lang, Yixuan Zou, Yaru Wang, Jingyi Men, Jiawei Cheng, Xinwei Yan, Jianyu Zhu, Yaqing Liu, Shengrui Zhang, Zhizhou Li, et al. A review of recent advances in MXenes/polymer-based electromagnetic interference shielding materials. *Polymer Composites*, 45(13):11541–11559, 2024.
33. Edina Rusen, Alexandra Mocanu, Oana Brincoveanu, Adina Boldeiu, Cosmin Romanitan, Martino Aldrigo, Sergiu Iordănescu, Aurel Diacon, Gabriela Toader, and Raluca Gavrila. Mxenes and polymeric colloids nanocomposites for EMI shielding. *Journal of Materials Chemistry C*, 12(30):11586–11593, 2024.
34. Li-Chuan Jia, Ding-Xiang Yan, Cheng-Hua Cui, Xin Jiang, Xu Ji, and Zhong-Ming Li. Electrically conductive and electromagnetic interference shielding of polyethylene composites with devisable carbon nanotube networks. *Journal of Materials Chemistry C*, 3(36):9369–9378, 2015.
35. Zi Ping Wu, Ming Mao Li, Ying Yan Hu, Ye Sheng Li, Zhi Xiang Wang, Yan Hong Yin, Yi Sheng Chen, and Xiao Zhou. Electromagnetic interference shielding of carbon nanotube macrofilms. *Scripta Materialia*, 64(9):809–812, 2011.
36. Juan Chen, Jiaming Wu, Heyi Ge, Dan Zhao, Cong Liu, and Xiufeng Hong. Reduced graphene oxide deposited carbon fiber reinforced polymer composites for electromagnetic interference shielding. *Composites Part A: Applied Science and Manufacturing*, 82:141–150, 2016.
37. Meenakshi Verma, Sampat Singh Chauhan, SK Dhawan, and Veena Choudhary. Graphene nanoplatelets/carbon nanotubes/polyurethane composites as efficient shield against electromagnetic polluting radiations. *Composites Part B: Engineering*, 120: 118–127, 2017.
38. Sandeep Kumar Singh, Mohammad Jaleel Akhtar, and Kamal K Kar. Hierarchical carbon nanotube-coated carbon fiber: ultra lightweight, thin, and highly efficient microwave absorber. *ACS Applied Materials & Interfaces*, 10(29):24816–24828, 2018.
39. Jianhong Wei, Zhiqiang Lin, Zuomin Lei, Yadong Xu, Zeyu Zhao, Tao Zhao, Yougen Hu, Hua Miao, and Rong Sun. Lightweight and highly compressible expandable polymer microsphere/silver nanowire composites for wideband electromagnetic interference shielding. *ACS Applied Materials & Interfaces*, 14(4):5940–5950, 2022.

40. Yingjie Duan, Luhui Zhang, Dingkun Tian, Siyuan Liao, Yong Wang, Yadong Xu, Rong Sun, and Yougen Hu. Modulus matching strategy of ultra-soft electrically conductive silicone composites for high performance electromagnetic interference shielding. *Chemical Engineering Journal*, 472:144934, 2023.
41. Ismayadi Ismail and Raba'ah Syahidah Azis. A review of magnetic nanocomposites for EMI shielding: synthesis, properties, and mechanisms. *Journal of Materials Science*, 59 (13):5293–5329, 2024.
42. Praveen Manjappa, Hari Krishna Rajan, Mamatha Gowdaru Mahesh, Karthikeya Gulur Sadananda, Manjunatha Channegowda, Girish Kumar Shivashankar, and Nagabhushana Bhangi Mutt. Effective attenuation of electromagnetic waves by synergetic effect of α-Fe_2O_3 and MWCNT/graphene in LDPE-based composites for EMI applications. *Materials*, 15(24):9006, 2022.
43. Min Wang, Yanlin Zhang, Chengjun Dong, Gang Chen, and Hongtao Guan. Preparation and electromagnetic shielding effectiveness of cobalt ferrite nanoparticles/carbon nanotubes composites. *Nanomaterials and Nanotechnology*, 9:1847980419837821, 2019.
44. Sourav Biswas, Sujit S Panja, and Suryasarathi Bose. Physical insight into the mechanism of electromagnetic shielding in polymer nanocomposites containing multiwalled carbon nanotubes and inverse-spinel ferrites. *The Journal of Physical Chemistry C*, 122(34):19425–19437, 2018.
45. T Kaur, S Kumar, SB Narang, and AK Srivastava. Radiation losses in microwave ku region by conducting pyrrole/barium titanate and barium hexaferrite based nanocomposites. *Journal of Magnetism and Magnetic Materials*, 420:336–342, 2016.
46. Bin Zhang, Jun Wang, Junpeng Wang, Huajun Duan, Siqi Huo, and Yushan Tang. Coprecipitation synthesis of hollow poly (acrylonitrile) microspheres@ $CoFe_2O_4$ with graphene as lightweight microwave absorber. *Journal of Materials Science: Materials in Electronics*, 28:3337–3348, 2017.
47. Gopal Datt, Chetan Kotabage, and AC Abhyankar. Ferromagnetic resonance of $NiCoFe_2O_4$ nanoparticles and microwave absorption properties of flexible $NiCoFe_2O_4$–carbon black/poly (vinyl alcohol) composites. *Physical Chemistry Chemical Physics*, 19(31):20699–20712, 2017.
48. Shital Patangrao Pawar, Mounika Gandi, and Suryasarathi Bose. High performance electromagnetic wave absorbers derived from PC/SAN blends containing multiwall carbon nanotubes and Fe_3O_4 decorated onto graphene oxide sheets. *RSC Advances*, 6(44):37633–37645, 2016.
49. Farbod Sharif, Mohammad Arjmand, Aref Abbasi Moud, Uttandaraman Sundararaj, and Edward PL Roberts. Segregated hybrid poly (methyl methacrylate)/graphene/magnetite nanocomposites for electromagnetic interference shielding. *ACS Applied Materials & Interfaces*, 9(16):14171–14179, 2017.
50. Ming Qin, Limin Zhang, Xiaoru Zhao, and Hongjing Wu. Lightweight Ni foam-based ultra-broadband electromagnetic wave absorber. *Advanced Functional Materials*, 31 (30):2103436, 2021.
51. Yang Zhang, Tong Pan, and Zhangjing Yang. Flexible polyethylene terephthalate/polyaniline composite paper with bending durability and effective electromagnetic shielding performance. *Chemical Engineering Journal*, 389:124433, 2020.
52. Yanhu Zhan, Jian Wang, Kaiye Zhang, Yuchao Li, Yanyan Meng, Ning Yan, Wenkang Wei, Fubin Peng, and Hesheng Xia. Fabrication of a flexible electromagnetic interference shielding Fe_3O_4@ reduced graphene oxide/natural rubber composite with segregated network. *Chemical Engineering Journal*, 344:184–193, 2018.
53. Sourav Biswas, Yudhajit Bhattacharjee, Sujit Sankar Panja, and Suryasarathi Bose. Graphene oxide co-doped with dielectric and magnetic phases as an electromagnetic wave suppressor. *Materials Chemistry Frontiers*, 1(6):1229–1244, 2017.
54. Qiang Peng, Meng Ma, Qindan Chu, Hao Lin, Wenting Tao, Wenqin Shao, Si Chen, Yanqin Shi, Huiwen He, and Xu Wang. Absorption-dominated electromagnetic interference shielding

composite foam based on porous and bi-conductive network structures. *Journal of Materials Chemistry A*, 11(20):10857–10866, 2023.

55. Yudhajit Bhattacharjee and Suryasarathi Bose. Core–shell nanomaterials for microwave absorption and electromagnetic interference shielding: a review. *ACS Applied Nano Materials*, 4(2):949–972, 2021.
56. Xiaqing Meng, Jun Wang, Qian Chen, Zicheng Xuan, Zelin Wu, Yan Zhang, Xifan Chen, Junying Wang, and Junzhong Wang. Boosting flexible thermal management and absorption-dominant electromagnetic shielding through surface engineering of electrochemically exfoliated graphene. *Chemical Engineering Journal*, 511:162262, 2025.
57. Sol Lee, Daeyoung Kim, Nam Khanh Nguyen, Wonkyo Kim, Minje Kim, and Junghyo Nah. 3D-printed gradient conductivity and porosity structure for enhanced absorption-dominant electromagnetic interference shielding. *Carbon*, 231:119759, 2025.
58. Cheng Dong, Ai Ping Zhang, Hai Lan Lin, Bi You Peng, Ke Yang Ni, Ke Cheng Yang, Jun Bian, Dai Qiang Chen, et al. Polymer-based composites for electromagnetic interference shielding: principles, fabrication, and applications. *Composites Part A: Applied Science and Manufacturing*, 108927, 2025.
59. Hai Fu, Yu'an Bai, Shuqian Duan, Hongfu Zhou, and Wei Gong. Structure design of multi-layered ABS/CNTScomposite foams for EMI shielding application with low reflection and high absorption characteristics. *Applied Surface Science*, 624:157168, 2023.
60. Li Ma, Mahdi Hamidinejad, Linfeng Wei, Biao Zhao, and Chul B Park. Absorption-dominant EMI shielding polymer composite foams: microstructure and geometry optimization. *Materials Today Physics*, 30:100940, 2023.
61. Shi Hu, Dan Wang, Josef Ve˘cerńık, Dana Krěmenáková, and Ji˘r´ı Militkỳ. Electro-magnetic interference (EMI) shielding and thermal management of sandwich-structured carbon fiber-reinforced composite (CFRC) for electric vehicle battery casings. *Polymers*, 16(16):2291, 2024.
62. Wanyu Tang, Suilin Liu, Xiaohan Wang, Bo Wang, Fangfang Zou, Guangxian Li, and Xia Liao. Improved electromagnetic interference shielding performance of silicone rubber/CNTs composites by porous structure design via green supercritical CO_2 foaming. *Composites Communications*, 46:101808, 2024.
63. Xin Yan, Hexin Zhang, Weihua Gu, and KeunByoung Yoon. Construction of self-healing polycaprolactone composite foams with segregated structure for superior electromagnetic interference shielding. *Journal of Materials Science and Technology*, 227:96–107, 2025.
64. Heng Pan, Shuyi Qing, Pei Lyu, Jie Ren, Qi Ran, Jun Hou, Zilong Wang, Liangjun Xia, Xiaofang Zhang, Xin Liu, et al. Multifunctional porous carbon fibers-based porous stacking for electromagnetic interference shielding. *Carbon*, 233:119907, 2025.
65. Runze Shao, Guilong Wang, Jialong Chai, Jun Lin, Guoqun Zhao, Zhihui Zeng, and Guizhen Wang. Multifunctional janus-structured polytetrafluoroethylene-carbon nanotube-Fe_3O_4/MXene membranes for enhanced EMI shielding and thermal management. *Nano-Micro Letters*, 17(1):1–22, 2025.
66. Shaohong Shi, Yuheng Jiang, Hao Ren, Siwen Deng, Jianping Sun, Fangchao Cheng, Jingjing Jing, and Yinghong Chen. 3D-printed carbon-based conformal electromagnetic interference shielding module for integrated electronics. *Nano-Micro Letters*, 16(1):85, 2024.

CHAPTER 2

Electromagnetic Interference Shielding

Theory and Measurement

Anilkumar Ramakrishnan, Paulbert Thomas,
Deepti Das Krishna, and C. K. Aanandan

2.1 INTRODUCTION

In the modern world, people live and work amidst a growing sea of electromagnetic radiation (EMR). The widespread adoption of electronic devices, communication tools and compact digital systems across various industries has transformed the way we interact with technology. As these innovations evolve, electronic devices have become smaller, faster and more densely integrated. However, this technological progress brings with it an unintended consequence – electromagnetic interference (EMI), an increasingly prevalent and undesirable form of environmental pollution. High-frequency electromagnetic waves, such as those emitted by mobile phones, can interfere with the operation of nearby electronic systems, including computers and audiovisual equipment. Mutual interference among televisions, radios, personal computers and cell phones can degrade device performance, often manifesting as audio clicks or visual distortions. One of the most common sources of EMI is electrostatic discharge. Even non-technical users can easily observe its effects, such as the familiar flickering of a television screen or the popping noise in audio systems during interference episodes (Hu and Kim 2008).

EMI presents significant risks not only to the performance of electronic devices but also to human health. If unmitigated, EMI can result in equipment malfunction and has been associated with adverse health outcomes, including migraines, asthma, cardiovascular complications and even an elevated risk of cancer due to prolonged exposure to EMR (Zhang 2014). EMI shielding refers to the technique of preventing EMR from penetrating materials by either reflecting or absorbing the radiation, typically through the use of conductive or magnetic substances that serve as protective barriers (Al-Ghamdi et al. 2016). EMI is commonly classified based on its propagation mechanism and frequency characteristics. There are two principal types: conducted EMI, where interference travels

DOI: 10.1201/9781003512363-2

through electrical connections such as power or signal lines, and radiated EMI, which is emitted into the air by electronic devices, cables or associated wiring. Additionally, EMI may be characterised as either narrowband or broadband, depending on its spectral width. Broadband EMI typically spans the 1–8 GHz range (Saini and Choudhary 2013), encompassing multiple frequency bands, each supporting specific applications. For instance, the L-band is used for wireless communication and low-earth orbit satellite systems; the S-band supports mobile phones and television broadcasting; the C-band enables long-distance telecommunications and Wi-Fi connectivity; the X-band is essential for weather monitoring and air traffic control (8–12 GHz); and the Ku-band facilitates satellite communication via very small aperture terminals (12–18 GHz). As EMI shielding requirements vary across these frequency bands, the choice of shielding materials must be tailored to the operational bandwidth (Joshi and Datar 2015).

EMI shielding plays a critical role across various industrial domains, with its importance escalating in tandem with society's increasing reliance on electronics and the ubiquity of radio frequency (RF) radiation (Rathi et al. 2022). In the automotive sector, electronic subassemblies are the backbone of modern vehicles, which now integrate a vast array of microprocessor-controlled components. Malfunctions induced by EMI in critical systems such as anti-lock braking, airbag deployment, cruise control and engine management can severely compromise vehicle performance and passenger safety.

Metals are traditionally regarded as the most effective materials for EMI shielding, owing to their high electrical conductivity and ability to reflect or absorb EMR. Common approaches include using self-supporting full-metal enclosures or applying conductive metal coatings such as nickel on plastic substrates. An alternative strategy involves incorporating conductive fillers such as stainless steel powder or fibres into polymer matrices. These composite structures are advantageous in terms of formability and material integration.

However, several drawbacks hinder the broader applicability of conventional metal-based shielding. High-density metals increase component weight, posing limitations in aerospace, automotive and portable electronics applications where weight reduction is critical. Additionally, metals are prone to corrosion, which diminishes shielding effectiveness (SE) over time. The production of metal coatings involves multiple costly processing steps, including substrate preparation and coating application. Achieving uniform coatings on components with complex geometries remains a significant challenge, as uneven thickness and poor adhesion can lead to performance degradation under mechanical or environmental stress (Eswaraiah et al. 2011).

Moreover, conventional conductive fillers such as metal flakes, stainless steel fibres or nickel-coated carbon fibres require high loading levels to achieve acceptable conductivity, which significantly increases the cost of the composite material. These limitations constrain their widespread adoption in commercial applications and underscore the urgent need for more efficient, lightweight and cost-effective EMI shielding solutions.

2.2 SHIELDING PHENOMENA

An EMI shield functions as a physical barrier that controls the transmission of electromagnetic waves through a medium. Typically implemented as an enclosure surrounding a complete electronic device or specific sensitive components, the shield prevents performance degradation caused by external EMR. In power electronics, this protective enclosure is commonly referred to as a shield. EMI shielding is essential for maintaining the integrity of electronic circuits and minimising the emission of EMR from the device itself. The process involves the use of specially engineered materials that either reflect or absorb electromagnetic energy, particularly in the high-frequency range, such as the radio waves emitted by mobile phones (Gupta et al. 2014). Effective shielding thus serves a dual purpose: protecting internal components from external noise and containing emissions to prevent interference with surrounding devices.

The shielding mechanism operates through three primary phenomena: *reflection, absorption and multiple reflections*. Reflection occurs at the surface of the shielding material when incident electromagnetic waves encounter a discontinuity in impedance. Absorption takes place within the bulk of the material, where the electromagnetic energy is dissipated as heat. Multiple reflections involve the internal bouncing of waves within the material or between internal surfaces, further attenuating the signal before it can transmit through the shield.

These three phenomena – reflection shielding effectiveness (SER), absorption shielding effectiveness (SEA) and multiple-reflection effectiveness (SEM) – combine to determine the total SE of a material. This can be expressed mathematically as:

$$SE_T = SE_A + SE_R + SE_M \tag{2.1}$$

As shown in Figure 2.1, an incident electromagnetic wave is partially reflected at the front surface of the material, partially absorbed within the medium, and a portion of the remaining energy undergoes additional reflections before either being attenuated or transmitted through the rear surface (Al-Saleh and Sundararaj 2009). This multi-stage attenuation significantly enhances the overall shielding capacity, particularly in materials with high conductivity and magnetic permeability.

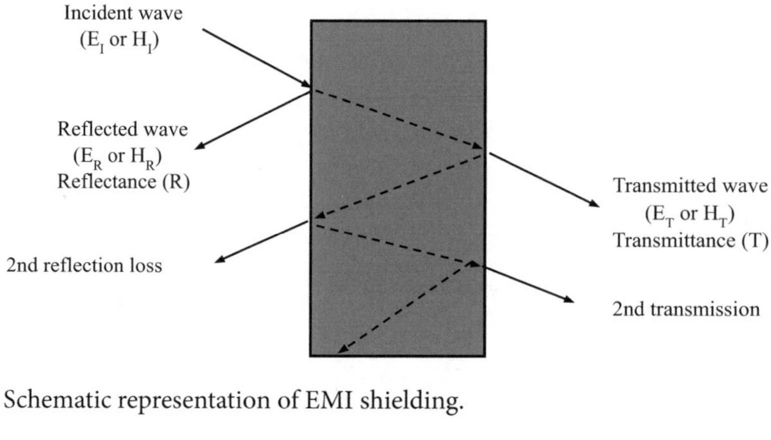

FIGURE 2.1 Schematic representation of EMI shielding.

Measured in decibels, the shielding effectiveness (SE) of any material is defined as the ratio of the output (transmitted) power (P_T) to the input (incident) power (P_1). Lesser energy passes through a shielding material at higher decibel levels of EMI shielding effectiveness (SE) (Eswaraiah et al. 2011).

$$SE_T\left(\text{dB}\right) = 20\ \log_{10}\left(\frac{P_T}{P_1}\right) \tag{2.2}$$

$$SE_T\left(\text{dB}\right) = 20\ \log_{10}\left(\frac{E_T}{E_1}\right) = 20\ \log_{10}\left(\frac{H_T}{H_1}\right) \tag{2.3}$$

where P_T (E_T or H_T) is transmitted electric- or magnetic-field intensity, and P_1 (E_1 or H_1) is incident electric- or magnetic-field intensity (Gelves et al. 2011).

Equations (2.4) and (2.5) establish a correlation between the incident and transmitted waves, as well as the reflectance (R) and transmittance (T) coefficients.

$$T = \left|\frac{E_T}{E_I}\right|^2 \tag{2.4}$$

$$R = \left|\frac{E_R}{E_I}\right|^2 \tag{2.5}$$

This leads to absorbance (A), which can be defined as,

$$A = \left(1 - R - T\right) \tag{2.6}$$

and reflection as

$$\hat{R} = \left(1 - R\right) \tag{2.7}$$

Effective absorbance is defined as

$$A_{\text{eff}} = \frac{\left(1 - R - T\right)}{\left(1 - R\right)} \tag{2.8}$$

Hence, the *SE* due to reflection and absorption loss can be deduced as

$$SE_R\left(\text{dB}\right) = 10\log_{10}\left(1 - R\right) \tag{2.9}$$

$$SE_A\left(\text{dB}\right) = 10\log_{10}\left(\frac{T}{\left(1 - R\right)}\right) \tag{2.10}$$

If SE_A is greater than 10 dB, SE_M can be neglected (Ohlan et al. 2008; Park et al. 2009; Wu et al. 2011).

2.2.1 Theory of Shielding Effectiveness

As an electromagnetic (EM) wave propagates through a shielding material, it undergoes three main processes: reflection, absorption and multiple reflections. These mechanisms work in combination to attenuate the energy of the incident wave. Every EM wave consists of two orthogonal field components: the electric field (E) and the magnetic field (H). For plane waves, these fields are mutually perpendicular and also perpendicular to the direction of wave propagation. This orthogonal relationship defines the wave's orientation and energy transmission characteristics.

The space through which EMR travels – from the emission source to the point of interaction – is referred to as the radiating region. This region is typically divided into three zones based on the distance r from the source relative to the wavelength (Ott 2011) ($\lambda / 2\pi$), as illustrated in Figure 2.2:

- Near-field region: $r < \dfrac{\lambda}{2\pi}$
- Far-field region: $r >> \dfrac{\lambda}{2\pi}$
- Transition region: $r \approx \dfrac{\lambda}{2\pi}$

Understanding these spatial domains is essential for accurately evaluating SE, as the behaviour of EM fields varies dramatically across them.

A key concept in analysing EM shielding is wave impedance (Celozzi et al. 2007), defined as the ratio of the amplitudes of the electric and magnetic fields:

$$Z = \frac{E}{H} \tag{2.11}$$

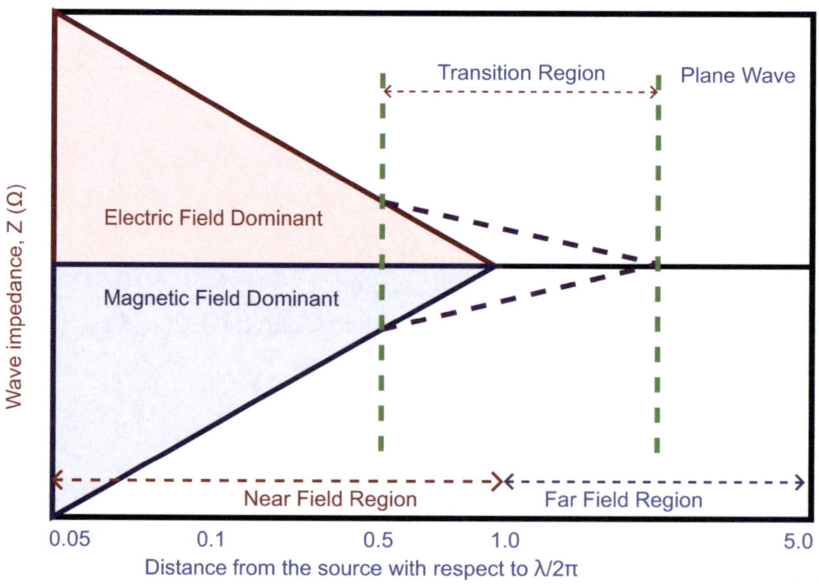

FIGURE 2.2 Dependence of wave impedance on distance from source normalised to $\lambda/2\pi$.

This quantity, also referred to as intrinsic impedance, differs between regions:

- In the far field, $Z = 377\ \Omega$ (the impedance of free space), and this value remains constant.
- In the near field, the impedance is not constant and varies with the source type:
- Electric-field dominant for low-current, high-voltage sources (capacitive sources)
- Magnetic-field dominant for high-current, low-voltage sources (inductive sources)

When an EM plane wave in free space encounters a shielding material with a different intrinsic impedance, part of the wave is reflected due to the impedance mismatch, and part is transmitted into the material. As the transmitted wave enters the shield, its amplitude decays exponentially due to absorption losses. Upon reaching the opposite surface of the material, a portion of the wave exits, while another portion is internally reflected, contributing to the multiple-reflection loss.

The total shielding effectiveness (SE) of a material is influenced by:

- Material thickness d
- Skin depth δ
- Impedance mismatch between the incident medium and the shielding material (i.e., Z_0 vs. Z)

The SE_R, SE_A, SE_M and SE_T in Equation (2.1) can be deduced as:

$$SE_R = 20\log\frac{\eta}{4\eta_0} \tag{2.12}$$

$$SE_A = 20\log e\frac{d}{\delta} \tag{2.13}$$

$$SE_M = 20\log\left(1-e^{2\frac{d}{\delta}}\right) \tag{2.14}$$

and hence,

$$SE_T = 20\ \log\ \frac{\eta}{4\eta_0} + 20\log\ e^{\frac{d}{\delta}} + 20\ \log\left(1-e^{2\frac{d}{\delta}}\right) \tag{2.15}$$

The skin depth (δ) is the depth at which the field decreases to $1/e$ from its primary value and is a function of frequency (f), permeability (μ) and conductivity (σ):

$$\delta = \frac{1}{\sqrt{(\pi f \mu\sigma)}} \tag{2.16}$$

As the thickness of the shield increases and the skin depth decreases, it absorbs more energy. Increasing shield conductivity, magnetic permeability and electromagnetic wave frequency results in a decrease in skin depth.

Impedance of a material is the relation between the electric and magnetic fields.

$$\eta = \left|\frac{E}{H}\right| = \left(\frac{j2\pi f \mu}{\alpha + j2\pi f \varepsilon}\right)^{\frac{1}{2}} \tag{2.17}$$

where j is an imaginary value and ε stands for complex permittivity.

For the conductive medium, $\sigma = \infty$ and conductivity of material >> $2\pi f \varepsilon$, Equation 2.16 can be written as

$$\text{for free space,} \quad \eta = \left(\frac{\mu}{\varepsilon}\right)^{\frac{1}{2}} \tag{2.18}$$

$$\text{for a conducting medium,} \quad \eta_0 = \left(\frac{j2\pi f \mu}{\sigma}\right)^{\frac{1}{2}} \tag{2.19}$$

When a shielding material is thinner than the skin depth, multiple reflections can degrade its overall SE. In contrast, if the material is thicker than the skin depth, multiple reflections become negligible, as the conductive medium effectively absorbs the waves reflected from its inner surface. Thus, ensuring the shield is thicker than the skin depth is sufficient to eliminate overlapping effects. Consequently, there are only two possible mechanisms for electromagnetic (EM) shielding: absorption and reflection.

$$SE = \left[39.5 + 10\log\left(\frac{\sigma}{2\pi f \mu}\right)^{\frac{1}{2}}\right] + \left[8.5d\left(\pi f \mu \sigma\right)^{\frac{1}{2}}\right] \tag{2.20}$$

Equation (2.20) demonstrates that as shield conductivity increases, shielding through absorption and reflection also improves. In contrast to absorption-based shielding (the second term in brackets in Equation 2.20), which increases as product $(f\mu)$ rises or, in other words, increases with frequency, reflection-based shielding (the first term in brackets in Equation 2.20) diminishes as the product $(f\mu)$ increases or decreases with an increase in frequency.

2.3 SE MEASUREMENT METHODS

Measuring SE is facilitated by specialised instruments, most notably the network analyser, which assesses both reflection and absorption losses of electromagnetic waves. Two primary types of network analysers are employed: scalar network analysers (SNAs) and vector network analysers (VNAs). SNAs measure only the amplitude of signals, making them suitable for simpler measurements. In contrast, VNAs can evaluate both the magnitude and phase of transmitted and reflected signals. This capability enables the analysis of complex signal behaviour, which is essential for accurately determining material properties such as complex permittivity and permeability. Consequently, VNAs are the most commonly used instruments, despite being more expensive than SNAs.

Numerous standardised testing methods exist to evaluate the SE of materials, tailored to specific applications and frequency ranges. ASTM D4935 outlines the standard procedure for assessing the SE of planar materials through a coaxial transmission line. IEEE 299 (formerly MIL-STD-285) assesses the SE of full enclosures using waveguide techniques. Meanwhile, IEEE 1128 uses TEM cell methods to measure SE at lower frequencies (DC to approximately 1 GHz). These SE standards specify sample size, testing setup, frequency range and measurement techniques, ensuring results are both consistent and comparable.

The SE can be measured using a two-port VNA and can be quantitatively characterised by utilising complex scattering parameters, commonly referred to as $S-$ parameters, thereby allowing for the quantification of the incident and transmitted waves. A VNA, a sample holder, coaxial cables and an adaptor constitute the complete setup for SE measurement. For a given test setup, the VNA evaluates S-parameters (S_{11} or S_{22} and S_{21} or S_{12}), indicating the reflection and transmission of electromagnetic energy through the sample. In SE testing, the signal strength is assessed with and without the material to determine its effectiveness. After calibrating the two ports of the VNA (Short, Open, Load, Thru) to eliminate errors from cables, connectors and setup, the S_{21} (reference) – the baseline transmission without the material – is recorded. Subsequently, S_{11} and S_{21} (with the sample) are measured to calculate the attenuation caused by the sample, representing the material's SE.

2.3.1 Coaxial Transmission Line Method

This is one of the most widely accepted methods for characterising EMI shielding performance in the frequency range of 30 MHz to 1.5 GHz (Park 2022). As shown in Figure 2.3, the fixture (as per ASTM D4935) consists of two symmetrical flanged coaxial lines. A planar material under test (MUT) is inserted into the coaxial holder. It measures the far-field shielding properties using the transverse electromagnetic (TEM) mode in the circular coaxial transmission line (TL). Testing is conducted using a small, doughnut-shaped sample, which is placed in the coaxial TL fixture and connected to a VNA.

2.3.2 Waveguide Method

The waveguide method is a well-established technique for assessing the SE of materials and enclosures at microwave frequencies, typically between 1 and 18 GHz, depending on the waveguide size used. A metallic waveguide is utilised to direct electromagnetic waves along a specific path, as shown in Figure 2.4. The material sample is precisely cut to fit the aperture of the waveguide, forming a barrier between the transmitting and receiving ends. Transmitting and receiving probes connect to a VNA, which assesses the S_{21} parameter to measure the transmitted signal strength through the material.

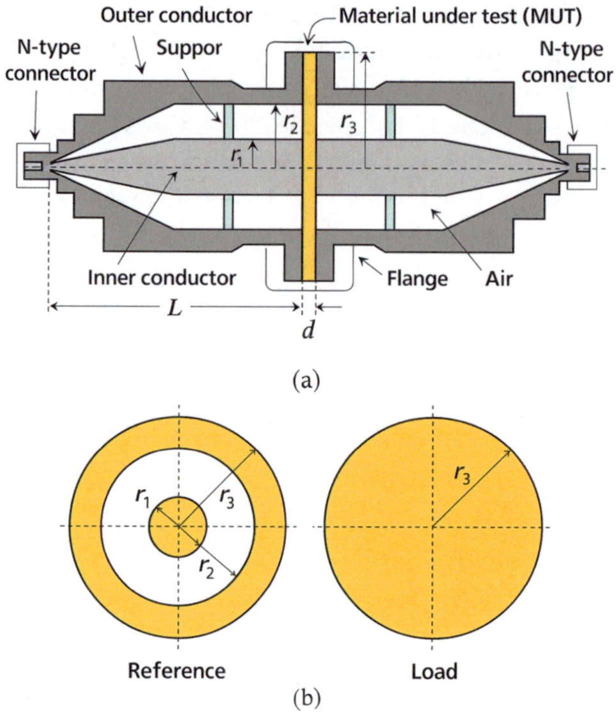

FIGURE 2.3 (a) Configuration of the coaxial line fixture. (b) Reference (left) and load (right) specimens (Park 2022)

2.3.3 Space Method or Open Field

Figure 2.5 illustrates the open-field or free-space method, which is suitable for large, irregular or flexible materials and involves a non-contact, non-destructive form of testing (i.e., without sample cutting or fitting into a waveguide). Two horn antennas (transmitter and receiver) face each other in a line-of-sight configuration. Let d represent the spacing between the sample and the horn antenna. When d is too large the received signal weakens, resulting in excessive noise generated by the receiving horn. If d is too small, the sample will be in the near field of the horn antennas, leading to non-uniform illumination of the sample. Typically, the separation satisfies $d > 2D^2/\lambda$, where D is the aperture of the horn antenna. This method involves transmitting an electromagnetic wave through air and measuring the attenuation of the signal by a test material between the source and the receiver. The test sample is positioned in the path of the wave between the antennas, and the *SE* is calculated by comparing the transmitted signal with and without the test material using a VNA.

2.3.4 Shielded Box Method

The Shielded Box method effectively evaluates the shielding efficiency of materials, gaskets or enclosures by placing the test sample in a metallic or RF-shielded box. The testing setup includes a metal container with an electrically sealed seam, a receiving antenna

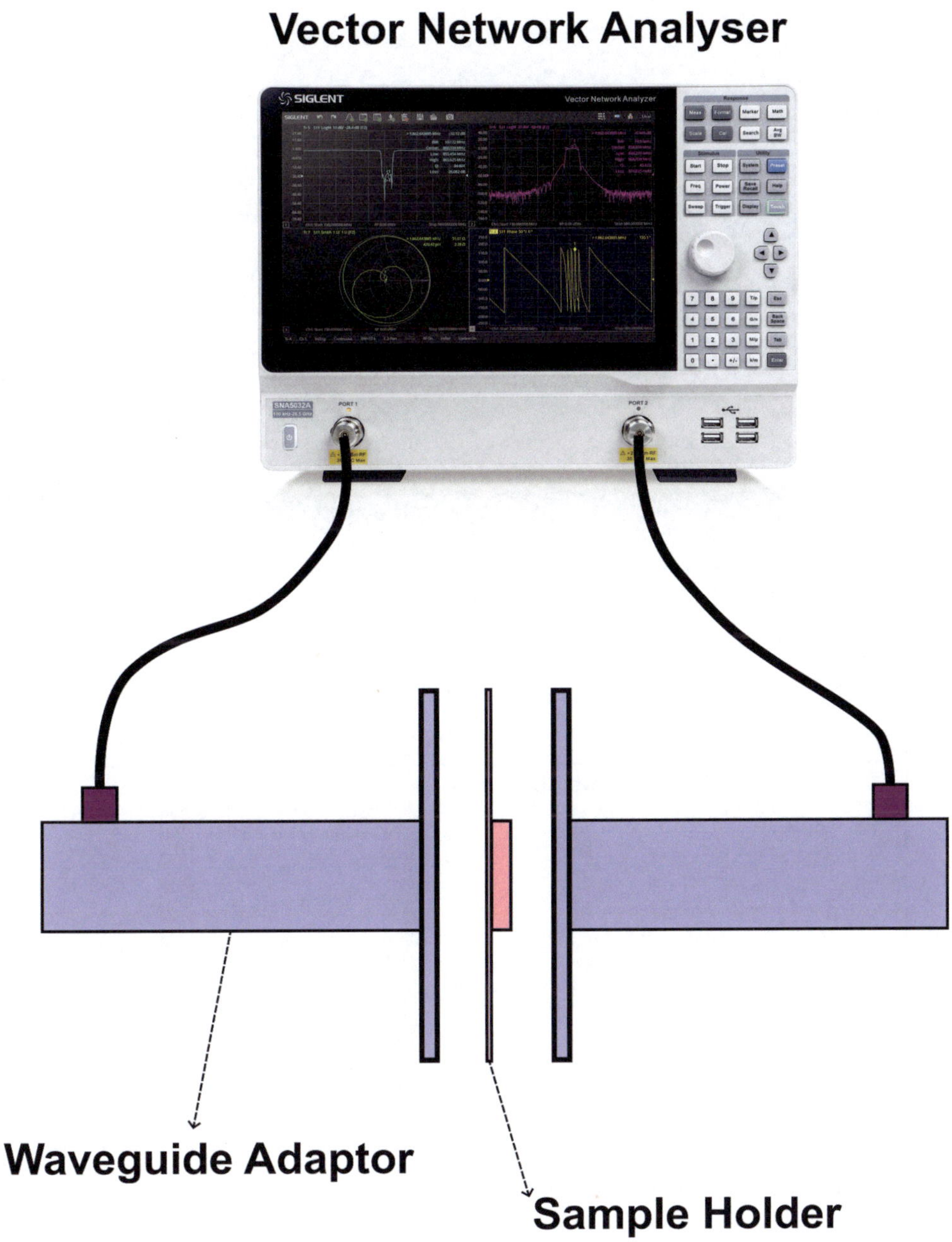

FIGURE 2.4 Waveguide measurement setup for X-band reading.

and a sample port on one wall. Signal strength is measured by positioning a transmitting antenna outside the box and collecting data through the open port, both with and without a test specimen covering it. An example of this method is shown in Figure 2.6. A notable challenge with this method is ensuring proper electrical contact between the test specimens and the metal box. This technique covers only a frequency range of approximately 500 MHz, and the results can vary significantly across different laboratories.

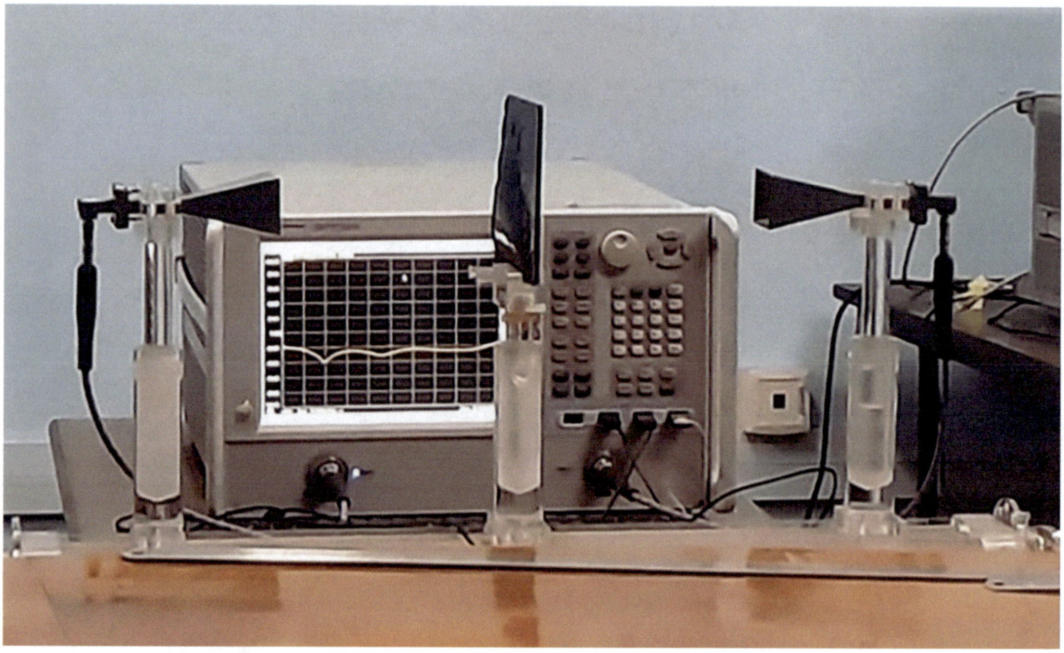

FIGURE 2.5 Photograph of free space measurement system (Anilkumar et al. 2024).

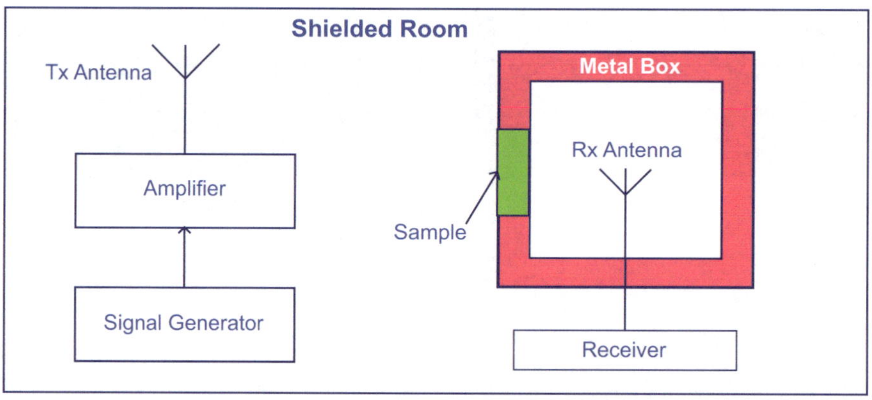

FIGURE 2.6 Shielded box SE measurement (Vicente and Dip 2011).

2.3.5 Electromagnetic Transverse Cells (TEM)

The transverse electromagnetic cell (TEM cell) type of testing environment produces a uniform electromagnetic field. It is a controlled test environment designed to produce a uniform electromagnetic field for evaluating EMI and electromagnetic compatibility (EMC) of electronic devices. The TEM cell was originally designed by Myron L. Crawford in 1974 to produce a consistent electromagnetic field within a controlled environment (Al Takach et al. 2018). TEM cells made from sheet metal typically undergo a skin-loss phenomenon. They generally feature tapering sections at both ends and comprise closed, rectangular,

FIGURE 2.7 Aluminium-based low-cost TEM cell (Hariyawan et al. 2020).

coaxial lines as shown in Figure 2.7. These distribute the single TEM technique across the measurement bandwidth. The three primary components of TEM cells are connectors, transition elements and rectangular coaxial lines. The choice of which TEM cell to use depends on the specific application, the frequency of interest and the size of the item being evaluated. This apparatus can be utilised for measuring emission radiation, testing E-field immunity radiation and calibrating E-field probes. Testing small RF devices for EMI typically occurs in a lab setting with TEM cells (Arezoomand et al. 2016).

When developing TEM cells for specific applications, several approaches have been considered. 3D-printed, aluminium-tape-wrapped TEM cells can operate at frequencies of up to 1.6 GHz (Al Takach et al. 2018). The distribution of electromagnetic fields within the cells, particularly in the transverse section, does not significantly affect the high-order mode (Paffi et al. 2012). By employing multi-step and stepwise linear tapering methods, the conductors in open TEM cells are designed to generate a standard electromagnetic field that spans a frequency range from 1 MHz to 1 GHz. A small strip-line TEM cell can be used to detect electromagnetic compatibility (EMC) on integrated circuit (IC) components, which is not feasible with conventional TEM cells.

Figure 2.7 illustrates an affordable 960 MHz aluminium TEM cell. The design results of this TEM cell can be applied to test radiation emissions from small electronic devices, such as mobile phones, and measure the emissions from cell phones. The EMI response range includes frequencies from 170 to 270 MHz, 400 to 510 MHz, 630 to 740 MHz and 870 to 960 MHz. Figure 2.8 illustrates two variants of the standard transverse electromagnetic (TEM) cell: the open TEM cell and the closed TEM cell. In the closed TEM cell, the device under test (DUT) is inserted through an aperture, ensuring it is fully shielded from external EMI. In contrast, the open TEM cell leaves the DUT exposed to EMI from all directions.

Testing for low-frequency radiated emissions and radiated immunity can be conducted using special TEM cells that operate from 1 kHz to several MHz. Notably, the measurement results obtained from a closed TEM cell closely resemble those from an Open Area Test Site (OATS). Only closed TEM cells are considered suitable for conducting EMC testing to achieve final compliance (Satav and Agarwal 2008). An automated transverse

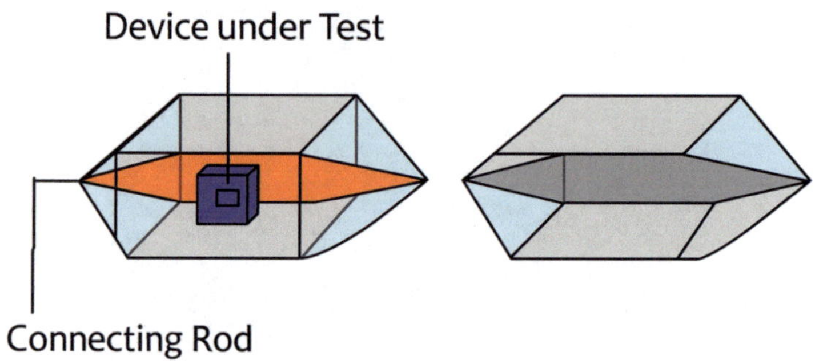

FIGURE 2.8 A closed and open TEM cell (Satav and Agarwal 2008).

electromagnetic (TEM) cell is a space- and labour-saving addition to any home laboratory. Unlike an OATS, an antenna is not necessary to generate electromagnetic fields within a TEM cell. Consequently, TEM cells provide a cost-effective, repeatable and precise laboratory apparatus for on-site EMC testing.

2.4 EM ASPECTS AND CONVERSION TECHNIQUES

The electromagnetic shielding behaviour of a material is fundamentally governed by its complex permittivity and complex permeability, both of which influence how effectively the material can reflect, absorb and attenuate EMR.

2.4.1 Complex Permittivity

When an external electromagnetic (EM) field is applied, it induces two types of electric currents within a material:

- Conduction current: associated with free electron movement
- Displacement current: associated with bound charge polarisation

The complex dielectric permittivity $(\varepsilon = \varepsilon' - j\varepsilon'')$ (Gupta et al. 2014) comprises:

- ε': the real part, representing the material's ability to store electric energy (through polarisation)
- ε'': the imaginary part, representing dielectric losses due to conduction and dipolar relaxation (Bhattacharya et al. 2012)

A higher ε'' enhances microwave absorption, making it crucial for EMI shielding applications. The dielectric loss tangent provides a measure of the energy dissipation capability of a material and is defined as:

$$\tan d\delta_e = \frac{\varepsilon''}{\varepsilon'} \tag{2.21}$$

2.4.2 Complex Permeability

Similarly, complex magnetic permeability is given by $\mu = \mu' - j\mu''$:

- μ': real part, indicates magnetic energy storage capability
- μ'': imaginary part, corresponds to magnetic losses – particularly due to domain wall motion, spin rotation and magnetocrystalline anisotropy in magnetic fillers such as ferrites (Saini and Choudhary 2013).

As the frequency increases, the magnetisation vector, $\bar{M}$, lags behind the applied field, $\bar{H}$, leading to increased magnetic loss μ''. This latency contributes significantly to EMI wave absorption, especially in materials with magnetic loading.

2.4.3 Interaction of EM Waves with Shielding Materials: *S*-Parameter Approach

The interaction of EM waves with a shielding material can be quantitatively analysed using scattering parameters (*S*-parameters) measured by a VNA. These parameters describe how an incident wave is reflected, transmitted or absorbed when encountering a material. Figure 2.9 illustrates how EM waves interact with a shielding material.

Let:

- S_{11} and S_{22}: reflection coefficients at the input and output ports
- S_{21} and S_{12}: transmission coefficients in forward and reverse directions

Then, the reflection coefficient (R) and transmission coefficient (T) are given by:

$$R = \left|\frac{E_R}{E_T}\right|^2 = |S_{11}|^2 = |S_{22}|^2 \tag{2.22}$$

$$T = \left|\frac{E_T}{E_R}\right|^2 = |S_{12}|^2 = |S_{21}|^2 \tag{2.23}$$

The absorbance (A), or fraction of incident power absorbed within the material, is:

$$A = (1 - R - T) = \left(1 - |S_{11}|^2 - |S_{21}|^2\right) \tag{2.24}$$

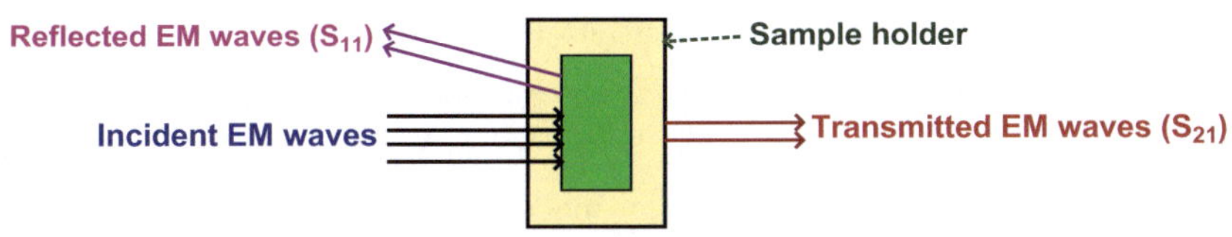

FIGURE 2.9 EM wave interaction with shielding material and the corresponding *S*-parameters.

To determine the effective absorbance when accounting for multiple internal reflections, Equation (2.8) can be used. This allows calculation of the SE due to reflection (SE_R), absorption (SE_A) and consequently the total shielding effectiveness (SE_T) using Equations (2.9), (2.10) and (2.1). This method offers a comprehensive approach to quantifying EMI shielding, linking material properties to measurable parameters using a VNA setup (Hong et al. 2003; Faisal and Khasim 2013; Thomassin et al. 2013; Singh et al. 2014).

2.4.4 Nicholson-Ross-Weir (NRW) Technique

Microwave-based techniques for electromagnetic material characterisation are broadly categorised into resonant and non-resonant methods. Resonant methods determine the dielectric properties of materials at a discrete frequency or within a narrow frequency band and are highly sensitive but limited in bandwidth (Nicolson and Ross 1970).

In contrast, non-resonant methods – such as transmission/reflection and reflection-only techniques – allow for broadband characterisation across a wide frequency spectrum. These methods estimate key electromagnetic parameters, including complex permittivity and complex permeability, by analysing the partial reflection and transmission of incident EM waves at the interface between materials with differing electromagnetic properties.

A TL is employed to direct microwave signals through the MUT and collect the resulting reflected and transmitted signals. This line may be implemented using coaxial lines, planar structures or metallic/dielectric waveguides, depending on frequency range and application. When a sample is inserted into the TL, it alters the impedance profile, causing impedance discontinuities and reflections. These impedance changes enable the extraction of relative permittivity, relative permeability and other electromagnetic characteristics based on the wave's interaction with the material. The S-parameters described in Table 2.1 show various conversion procedures.

Among non-resonant methods, the Nicholson-Ross-Weir (NRW) (Weir 2005) technique is one of the most widely adopted due to its robustness and simplicity. It allows accurate estimation of complex permittivity and permeability from scattering parameters (*S*-parameters) measured using a VNA. In this technique, the S_{11} (reflection coefficient) and S_{21} (transmission coefficient) are used as inputs to calculate the electromagnetic constants. The NRW method assumes the material is homogeneous, isotropic and of finite thickness. It applies closed-form analytical expressions to solve for the complex intrinsic impedance and propagation constant of the MUT (Vicente and Dip 2011).

TABLE 2.1 Commonly Used Non-Resonant Characterisation Techniques and Their Features

Conversion Technique	Input *S*-Parameters	Output Attributes
Nicolson-Ross-Weir (NRW)	S_{11}, S_{21}, S_{12} and S_{22} or and S_{11} and S_{21} (or S_{22} and S_{12})	ε_r and μ_r
NIST iterative	S_{11}, S_{21}, S_{12} and S_{22} S_{11} and S_{21} (or S_{22} and S_{12})	ε_r and $\mu_r = 1$
New non-iterative	S_{11}, S_{21}, S_{12} and S_{22} or and S_{11} and S_{21} (or S_{22} and S_{12})	ε_r and $\mu_r = 1$
Short circuit line (SCL)	S_{11} or S_{22}	ε_r

The fundamental principle behind the NRW technique is that the MUT alters the phase and amplitude of incident waves due to differences in wave impedance between the material and the surrounding environment. This variation is manifested in the S-parameters, which represent the ratio of incident to reflected or transmitted electric fields at the material's boundaries. From these coefficients, the NRW method derives the intrinsic impedance Z and complex propagation constant $\gamma = \alpha + j\beta$, from which permittivity and permeability are subsequently calculated.

The advantages of the method include:

- Broad frequency applicability
- No requirement for resonance conditions
- Accurate for both lossy and low-loss materials
- Compatible with standard coaxial or waveguide setups
- Closed-form expressions reduce dependence on iterative algorithms

The NRW technique remains a standardised and efficient method for characterising electromagnetic materials, particularly useful for testing microwave absorbers, EMI shielding materials and dielectric composites.

Mathematical Formulation

Consider a rectangular material neatly fitting the cross section of a rectangular waveguide as shown in Figure 2.10, having permeability $\mu = \mu_0 \mu_r$, permittivity $\varepsilon = \varepsilon_0 \varepsilon_r$ and thickness L.

Step 1: Wave Propagation in the Medium

The technique starts with the wave propagation in a waveguide filled with the MUT. In such a case, the propagation constant (γ) is given by,

$$\gamma^2 = K_c^2 - K^2 \tag{2.25}$$

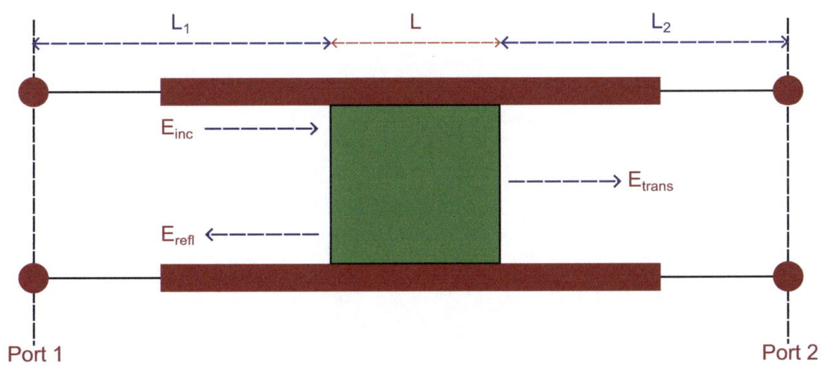

FIGURE 2.10 A rectangular waveguide with a slab of material under test inserted.

where

$K = \omega\sqrt{\mu\varepsilon}$: wavenumber in the material

$K_c = \dfrac{2\pi}{\lambda_c}$: cut-off wavenumber of the waveguide

Thus γ becomes,

$$\gamma = \sqrt{\frac{\omega^2 \mu_r \varepsilon_r}{c^2} - \left(\frac{2\pi}{\lambda_c}\right)^2} \tag{2.26}$$

This shows how γ depends on the material's relative permittivity (ε_r) and permeability (μ_r).

Step 2: Impedance, Reflection Coefficient and Permeability

To determine how much wave is reflected or transmitted at the boundary between air and MUT, we use:

Wave impedance, $\eta = \dfrac{j\omega\mu}{\gamma}$

Reflection coefficient at the air-material interface, $\Gamma = \dfrac{\eta_{\text{mut}} - \eta_0}{\eta_{\text{mut}} + \eta_0}$

Substituting the impedance, we get, $$\Gamma = \frac{\mu_r\gamma_0 - \gamma}{\mu_r\gamma_0 + \gamma} \tag{2.27}$$

This expression ties Γ directly to γ, μ_τ and γ_0 (the free-space propagation constant).

Solving for permeability, $$\mu_r = \left(\frac{1+\Gamma}{1-\Gamma}\right) \cdot \frac{\gamma}{\gamma_0} \tag{2.28}$$

Step 3: Permittivity Calculation

From Equation (2.26) we get:

$$\varepsilon_r = \frac{c^2}{\omega^2 \mu_r}\left[\left(\frac{2\pi}{\lambda_c}\right)^2 - \gamma^2\right]$$

Using $\lambda_0 = \dfrac{2\pi c}{\omega} \Rightarrow \dfrac{c^2}{\omega^2} = \dfrac{\lambda_0^2}{4\pi^2}$ and after further simplification,

$$\varepsilon_r = \frac{\lambda_0^2}{\mu_r}\left[\frac{1}{\lambda_c^2} - \left(\frac{\gamma}{2\pi}\right)^2\right]$$

Define: $\dfrac{1}{\Lambda} = \dfrac{\gamma}{j2\pi} \Rightarrow \left(\dfrac{\gamma}{2\pi}\right)^2 = \dfrac{1}{\Lambda^2}$ and substitute into the above equation.

Finally, we get, $$\varepsilon_r = \frac{\lambda_0^2}{\mu_r}\left[\frac{1}{\lambda_c^2} + \frac{1}{\Lambda^2}\right] \tag{2.29}$$

where

λ_0: Free-space wavelength at frequency f

λ_c: Cut-off wavelength of the waveguide

Λ: Related to the measured propagation constant γ; phase unwrapping may be required to determine this accurately.

Step 4: Ray Tracing and S-Parameter Equations

The wave enters the material and is partially reflected at the first interface due to impedance mismatch. The transmitted part propagates through the material, attenuated and phase-shifted by Z, the propagation factor or transmission factor. The propagation factor in the material slab is defined as

$$Z = e^{-\gamma L} \Rightarrow Z = e^{-(\alpha + j\beta)L} \tag{2.30}$$

At the second boundary, it again reflects and repeats this process. The total reflection and transmission are infinite geometric sums of all these paths.

Let, $R_1 = e^{-\gamma L_1}$, $R_2 = e^{-\gamma L_2}$: propagation correction factors from calibration planes to sample boundaries

De-embedded S-parameters are defined at the MUT interfaces as

$$\text{De-embedded Reflection coefficient,} \quad S_{11}^{D} = \frac{-\Gamma\left(1 - Z^2\right)}{1 - \Gamma^2 Z^2} \tag{2.31}$$

where,

Γ: base reflection at the front face

$1 - Z^2$: contribution from forward and backward travelling waves in the slab

$1 - \Gamma^2 Z^2$: geometric sum of repeated internal reflections

$$\text{De-embedded Transmission coefficient,} \quad S_{21}^{D} = \frac{Z\left(1 - \Gamma^2\right)}{1 - \Gamma^2 Z^2} \tag{2.32}$$

where

Z: propagation through the material

$(1 - \Gamma^2)$: net transmission considering first pass and infinite series of back-and-forth reflections.

In real experiments, the VNA measures at calibration planes located at distances L_1 and L_2 from the slab's boundaries. So, correction factors are reintroduced:

$$S_{11} = R_1^2 .\ S_{11}^D \tag{2.33}$$

$$S_{21} = R_1 R_2 R_{21}^D \tag{2.34}$$

where $R_i = e^{-\gamma L_i}$, accounting for propagation loss between sample and ports.

This results in actual measurable S-parameters from the VNA:

$$S_{11} = R_1^2 .\ \frac{-\Gamma\left(1-Z^2\right)}{1-\Gamma^2 Z^2} \tag{2.35}$$

$$S_{21} = R_1 R_2 .\ \frac{-Z\left(1-\Gamma^2\right)}{1-\Gamma^2 Z^2} \tag{2.36}$$

Step 5: Solving the Core NRW System

Define auxiliary variables using de-embedded Equations (2.32) and (2.33):

$$V_1 = S_{21}^D + S_{11}^D \text{ and } V_2 = S_{21}^D - S_{11}^D$$

$$\text{Then compute} \quad \kappa = \frac{1-V_1V_2}{V_1 - V_2} \tag{2.37}$$

From which a quadratic equation is solved as:

$$\Gamma = \kappa \pm \sqrt{\kappa^2 - 1} \tag{2.38}$$

and choose only the root that satisfies $|\Gamma| < 1$. The magnitude of the reflection coefficient $|\Gamma|$ must be less than 1 for passive, lossy materials – which is always the case for real-world EMI shielding materials.

Step 6: Phase Ambiguity and Its Resolution

$$\text{Since,} \quad Z = e^{-\gamma L} \Rightarrow \gamma = -\frac{1}{L}\ln Z$$

The argument of $\ln Z$ is ambiguous due to its 2π periodicity.

In most practical scenarios, phase unwrapping is the standard method for resolving ambiguity over a broadband frequency range. However, when high accuracy is needed, particularly for dispersive or magnetic materials, the two-sample disambiguation technique provides more reliable results.

2.5 VECTOR NETWORK ANALYZER (VNA)

In order to quickly and accurately measure the phase and amplitude of a microwave signal across a large frequency range, a network analyser is an indispensable tool. A network analyser can display the complex transmission coefficients and reflection coefficients of a microwave signal. When it comes to microwave engineering, network analysers are invaluable tools for evaluating intricate systems, materials and circuits.

Depending on the model and make of a VNA, high-frequency stimulus-response studies can be carried out from 300 kHz to 110 GHz or 320 GHz. The three main components of a VNA are the signal source, the receiver and the display, as in Figure 2.11. With a single-frequency signal, the source bombards the testing material with one frequency at a time. A receiver tuned to that frequency can pick up both the material's reflected and transmitted signals. The measured response generates data on the magnitude and phase of the reflection/transmission coefficients. This measurement is repeated for different frequencies to obtain the reflection and transmission response as a function of frequency.

For EM characterisation of a material using a VNA, a measuring fixture is required to set up the sample in a two-port configuration. Figure 2.12(a) presents a photograph of the VNA with the waveguide fixture, while Figure 2.12(b) shows the X-band rectangular waveguide adaptor, sample holder, pelletised sample and pelletising die. Two waveguide adaptors with the sample holder connect the sample holder to the VNA test port cables. A list of dimensions for rectangular waveguide sample holders in different microwave frequency bands is provided in Table 2.2.

2.5.1 Source of Errors in a Network Analyser

A network analyser circuit has three types of error sources: systematic, drift and random. As the name suggests, random errors unpredictably vary over time. Several repeated measurements and averaging the results help to minimise such errors. However, these randomly occurring errors cannot be eliminated with a standard calibration setup, as they stem entirely from instrumental sources. Even after calibration, drift errors arise due to changes in system performance over time, primarily caused by temperature variations. A temperature-controlled environment can help mitigate these errors. Systematic errors, on

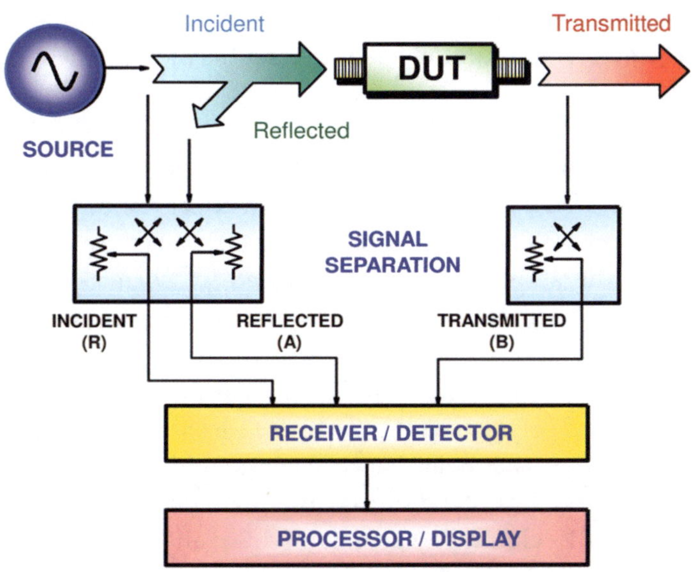

FIGURE 2.11 Graphical representation of a vector network analyser (adapted from www.keysight.com).

(a) (b)

FIGURE 2.12 (a) Photograph of two-port VNA measurement setup for waveguide method; (b) X-band rectangular waveguide adaptor, sample holder, pelletised sample and pelletising die.

TABLE 2.2 List of Dimensions of Rectangular Waveguide Sample Holders for Microwave Frequency Bands

	Sample Holder Dimensions (mm)	
Sample Holder	**Length**	**Width**
S-band	72.14	34.04
X-band	22.86	10.16
Ku-band	15.80	7.90

the other hand, result from flaws in the setup, such as signal leakage, signal reflections and inaccuracies in frequency response. A calibration technique helps identify these errors, allowing them to be mathematically corrected during the measurement process.

2.5.2 Calibration of a Network Analyser

Systematic errors are accounted for in the calibration process using known values of reference standards. Measured data facilitates the mathematical removal of these errors. Two types of error correction are employed in the calibration: response correction and vector correction. Response calibration, though easy to implement, corrects only a limited portion of systematic errors. Vector error calibration is performed in systems where both magnitude and phase measurements are available and systematic errors are the primary concern. This method requires more calibration standards. Thru-Reflect-Line (TRL) calibration is used to achieve error-free measurements. Compared to Short-Open-Load-Thru (SOLT) calibration, TRL calibration is simpler, more accurate and more consistent. Fabricating and characterising the three TRL standards is easier than doing so for the four SOLT standards. The accuracy of TRL calibration depends on the quality and repeatability of the TRL standards. Physical discontinuities, such as bends or air gaps, can significantly degrade its performance. De-embedding *S*-parameter data typically requires TRL calibration.

2.5.3 Network Analyser Measurement

For the measurements carried out at the Department of Electronics at Cochin University of Science and Technology, India, a Keysight VNA (N5227A, 67 GHz) is used. A copper sample holder is attached to the waveguide flanges and the waveguide adaptor of the network analyser. The 2-mm-thick sample can be placed into the waveguide adaptor, and the scattering parameters can be collected through the VNA. A complete two-port calibration is conducted in conjunction with the sample holder to prevent any potential loss or power redistribution.

Figure 2.13 shows the schematic of the two-port measurement of *S*-parameters for the MUT on a VNA. Here, S_{11} and S_{21} correspond to the reflected and transmitted energies, respectively, during the forward transmission of the signal. The complex electromagnetic characteristics, such as the sample's permittivity and permeability, are computed from the measured *S*-parameters using the NRW method explained previously. Furthermore, the fraction of reflected energy (S_{11}) can be directly measured using the single-port shorted line approach, where no transmitted signal is present ($S_{21} = 0$).

$$\begin{pmatrix} b_1 \\ b_2 \end{pmatrix} = \begin{pmatrix} S_{11} & S_{12} \\ S_{21} & S_{22} \end{pmatrix} \begin{pmatrix} a_1 \\ a_2 \end{pmatrix} \tag{2.39}$$

The authors have conducted complex permeability and permittivity measurements on a two-port waveguide using a VNA. Figure 2.12(a) illustrates this process. Using a custom-made die, as shown in Figure 2.12(b), the samples were cut into rectangular X- and Ku-band shapes to facilitate the attachment of the waveguides. To calculate reflection loss, the relation in Eq 2.40 (Naito and Suetake 1971) was used to estimate the absorber's wave impedance (Z_{in}) from ε and μ.

$$Z = Z_0 \sqrt{\frac{\mu}{\varepsilon}} \text{ than } h \left(\frac{-2\pi jfd\sqrt{\mu\varepsilon}}{c} \right) \tag{2.40}$$

where c represents the incident microwave speed and f denotes the incident microwave frequency. Z_0 is the open-space wave impedance. Using the formula from TL theory

FIGURE 2.13 Scattering parameters representation for two-port measurements on VNA.

TABLE 2.3 Relation Between Reflection Loss *RL* (dB) and % Microwave Absorption

RL (dB)	% MW Absorption
−10	90
−20	99
−30	99.9
−40	99.99

(Naito and Suetake 1971), one can calculate the reflection loss (RL) of a metal-backed sample in decibels (dB).

$$RL(\text{dB}) = -20\log_{10}\left(\frac{z - z_0}{z + z_0}\right) \tag{2.41}$$

RL (in dB) can be directly calculated from the S_{11} parameter with the following expression:

$$RL = -20\log\left(S_{11}\right) \tag{2.42}$$

As illustrated in Figure 2.13, one end of the TL was connected to port 1 of the VNA, while the other end was short-circuited to directly measure the reflection loss (RL). To ensure accurate measurements on the VNA, all connectors were two-port calibrated before the test by measuring the reflection loss of the load, open and short terminators. To measure the transmitted and reflected responses, two phase-matched wires were connected together. Another way to express reflection loss, represented as *RL* (dB), is in terms of the percentage of microwave absorption. The *RL* (dB) values, converted to percentage microwave absorption, are shown in Table 2.3.

2.6 CASE STUDIES

As illustrated in Figure 2.5, Anilkumar et al. (2024) developed a nanocomposite material composed of PANI/CNT/Fe_2O_3 and natural rubber, designed for EMI shielding. Its performance in the X-band was evaluated using the waveguide method, achieving a SE of 45 dB. Additionally, the same composite was investigated in the Ku-band using the free-space method, which demonstrated a total SE of 55 dB. Figure 2.14 presents the SE results of the sample.

Further, Punnakkal et al. (2025) developed a cost-effective method for synthesising polypyrrole/silver (Ppy/Ag), polypyrrole/graphene (Ppy/Gr) and polypyrrole/silver/graphene (Ppy/Ag/Gr) composites in a single step, enabling the incorporation of varying amounts of graphene. The total shielding effectiveness (SE_T) of the ternary composites in the Ku-band reaches an impressive 30.86 dB at 12.7 GHz for the composite containing 5 wt.% graphene, as measured using the waveguide method. Figure 2.15 displays the experimental results for the SE of the sample.

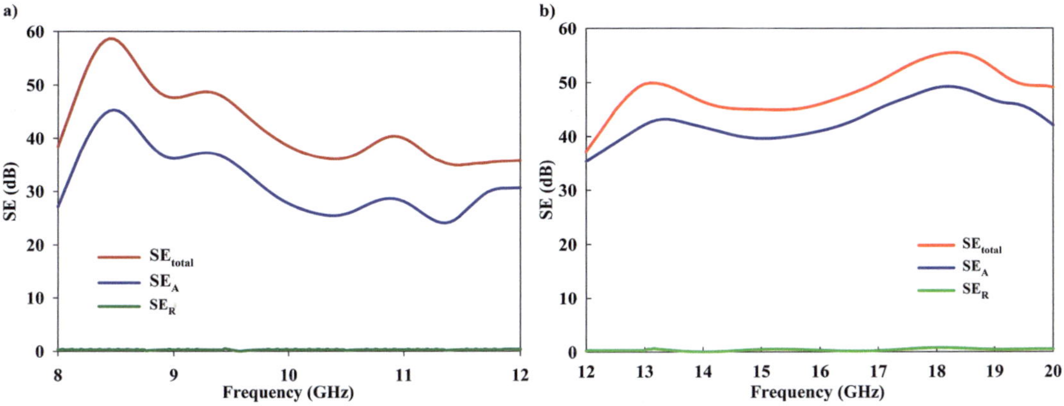

FIGURE 2.14 Shielding effectiveness PANI/CNT/Fe_2O_3/NR in (a) X- and (b) Ku-band.

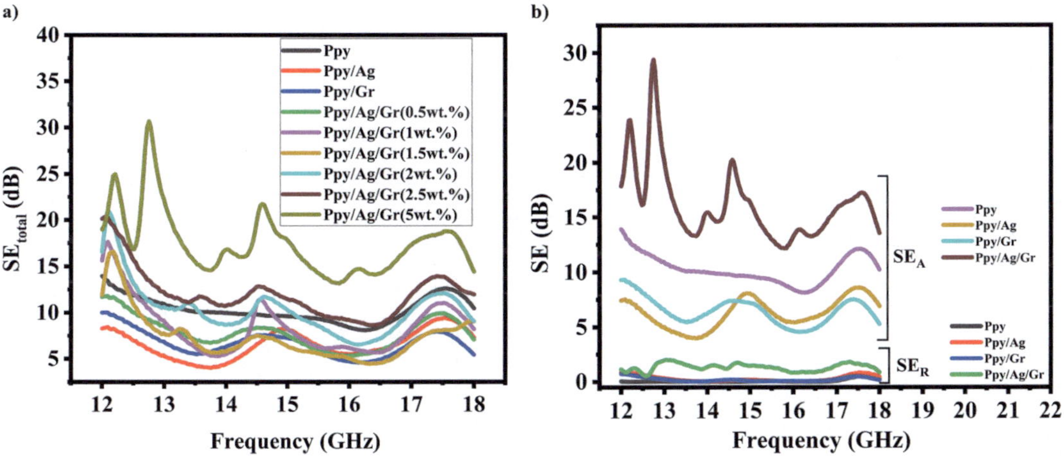

FIGURE 2.15 (a) Total shielding effectiveness (SE_T) of all the synthesised samples; (b) Shielding effectiveness due to reflection (SE_R) and absorption (SE_A) of Ppy, Ppy/Ag, Ppy/Gr and Ppy/Ag/Gr (5 wt.%).

In another work, Anilkumar et al. (2023) proposed a flat-design microwave-absorbing material composite using rice husk (RH), an agricultural waste, to be used for enhancing microwave absorption by incorporating nanoparticles such as graphene oxide (GO). A 91–94% absorption was measured in the frequency range of 2–18 GHz using the free-space method, and it is plotted in Figure 2.16.

Sunil et al. (2025) developed a self-powered vibration sensor that demonstrates EMI shielding efficiency. This sensor utilises lightweight polystyrene/dedoped polyaniline (PS/DPANI) nanofibres, which are produced through electrospinning. It features a copper sheet as the electrode and FEP as the tribonegative layer. The PS composite, with a filler concentration of 3 wt.%, achieved a maximum shielding efficiency of 90 dB when tested using the waveguide method, as shown in Figure 2.17. These findings suggest the potential for commercial applications of the composite. The SE of the sample is illustrated in Figure 2.17.

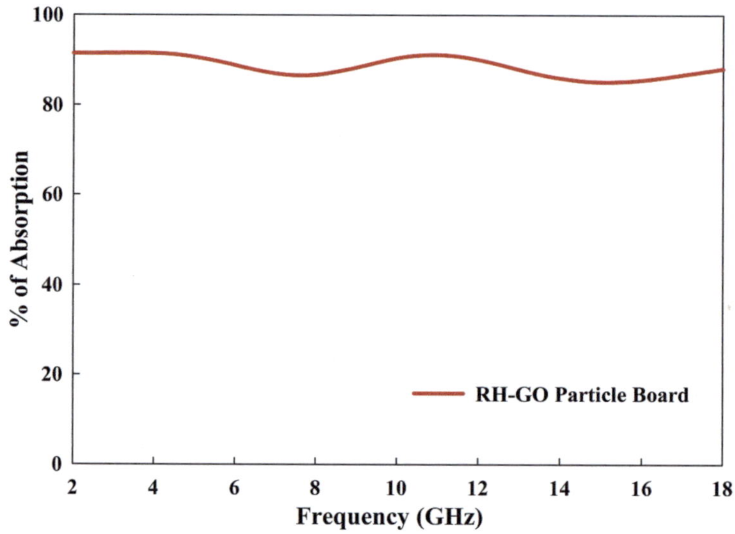

FIGURE 2.16 Microwave absorption plot at 2–18 GHz frequency.

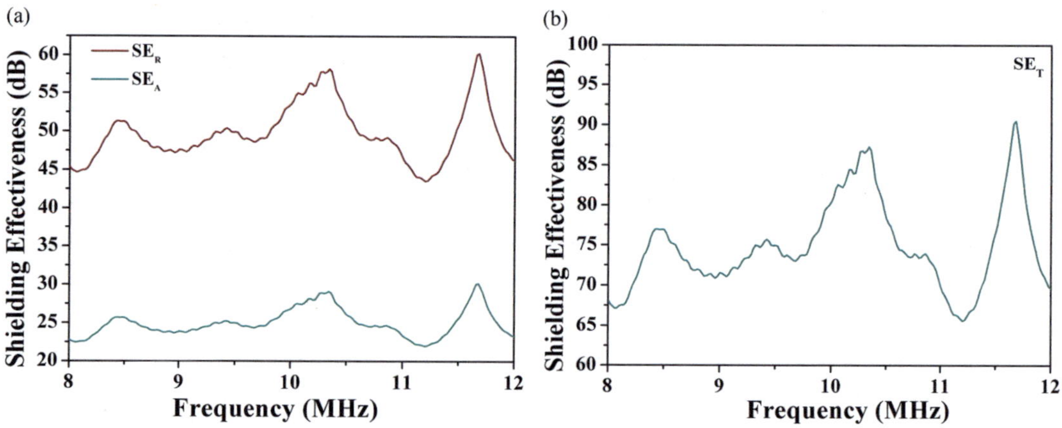

FIGURE 2.17 EM performance of PS/DPANI composites (a) SE_A and SE_R (b) SE_T.

2.7 CONCLUSION

A crucial component of research in electromagnetic shielding involves developing and characterising the performance of EMI shielding materials to protect both electronic devices and biological entities from harmful EM radiation. The identification of novel materials with improved shielding performance is another significant aspect. A detailed theory of mitigating EMI is presented, along with theoretical concepts such as reflection, absorption and multiple reflections. The theory and methodologies for measuring the shielding efficiency (SE) of EMI shields are explained. Various measurement techniques, along with the equipment such as a VNA, are discussed. The waveguide method is the most commonly used method due to its high accuracy and minimal measurement errors. Another approach is the free-space method, which requires free-space calibration and time gating in the PNA for accurate results. Different methods for evaluating the electromagnetic shielding performance of material composites are discussed, along with relevant case studies.

ACKNOWLEDGEMENT

The authors would like to acknowledge the Department of Polymer Science and Rubber Technology at Cochin University of Science and Technology, India, for their support in preparing sample EMI shielding materials.

REFERENCES

Al Takach, A., Ndagijimana, F. & Al-Husseini, M., 2018. 3D-printed low-cost and lightweight TEM cell. International Conference on High Performance Computing & Simulation.

Al-Ghamdi, A. A. et al., 2016. Correlation between electrical conductivity and microwave shielding effectiveness of natural rubber-based composites, containing different hybrid fillers obtained by impregnation technology. *Materials Sciences and Applications*, 7(9), pp. 496–509.

Al-Saleh, M. H. & Sundararaj, U., 2009. Electromagnetic interference shielding mechanisms of CNT/polymer composites. *Carbon*, 47(7), pp. 1738–1746.

Anilkumar, R., Abhilash, A. P., Indhu, K. K. and Aanandan, C. K., 2023. Microwave absorption property of rice husk and graphene oxide composite particle board. *Nano World Journal* 9(S5), pp. S36–S39.

Anilkumar, R., Abhilash, A. P., Indhu, K. K., Joseph, S., John, H., Das Krishna, D. & Aanandan, C. K., 2024. Microwave absorption of natural rubber composites based on PANI/CNT/Fe_2O_3 for flexible shielding applications. *Journal of Electronic Materials*, 53(11), pp. 7122–7131.

Arezoomand, M., Meybodi, M. K. & Noori, N., 2016. Design of a TEM cell using both multi-step and piecewise linear tapering. 8th International Symposium on Telecommunications.

Bhattacharya, P., Das, C. K. & Kalra, S. S., 2012. Graphene and MWCNT: potential candidate for microwave absorbing materials. *Journal of Materials Science Research*, 1(2), p. 126.

Celozzi, S., Araneo, R. & Lovat, G., 2007. *Electromagnetic Shielding*. Wiley.

Eswaraiah, V., Sankaranarayanan, V. & Ramaprabhu, S., 2011. Functionalized graphene–PVDF foam composites for EMI shielding. *Macromolecular Materials and Engineering*, 296(10), pp. 894–898.

Faisal, M. & Khasim, S., 2013. Broadband electromagnetic shielding and dielectric properties of polyaniline-stannous oxide composites. *Journal of Materials Science: Materials in Electronics*, 24, pp. 2202–2210.

Gelves, G. A., Al-Saleh, M. H. & Sundararaj, U., 2011. Highly electrically conductive and high performance EMI shielding nanowire/polymer nanocomposites by miscible mixing and precipitation. *Journal of Materials Chemistry*, 21(3), pp. 829–836.

Gupta, T. K. et al., 2014. MnO_2 decorated graphene nanoribbons with superior permittivity and excellent microwave shielding properties. *Journal of Materials Chemistry A*, 2(12), pp. 4256–4263.

Hariyawan, M. Y., Darwis, R. S. & Habibi, B. A., 2020. Low-cost Transverse Electromagnetic (TEM) cell design for radiated emission measurement. *IOP Conference Series: Materials Science and Engineering*, 1073, p. 012040.

Hong, Y. K. et al., 2003. Method and apparatus to measure electromagnetic interference shielding efficiency and its shielding characteristics in broadband frequency ranges. *Review of Scientific Instruments*, 74(2), pp. 1098–1102.

Hu, Q. and Kim, M. S., 2008. Electromagnetic interference shielding properties of CO_2 activated carbon black filled polymer coating materials. *Carbon Letters*, 9(4), pp. 298–302.

Joshi, A. & Datar, S., 2015. Carbon nanostructure composite for electromagnetic interference shielding. *Pramana*, 84, pp. 1099–1116.

Naito, Y. and Suetake, K., 1971. Application of ferrite to electromagnetic wave absorber and its characteristics. *IEEE Transactions on Microwave Theory and Techniques*, 19(1), pp. 65–72.

Nicolson, A. M. & Ross, G. F., 1970. Measurement of the intrinsic properties of materials by time-domain techniques. *IEEE Transactions on Instrumentation and Measurement*, pp. 377–382.

Ohlan, A., Singh, K., Chandra, A. & Dhawan, S. K., 2008. Microwave absorption properties of conducting polymer composite with barium ferrite nanoparticles in 12.4–18 GHz. *Applied Physics Letters*, 93(5), p. 053114.

Ott, H. W., 2011. *Electromagnetic Compatibility Engineering*. John Wiley & Sons.

Paffi, A. et al., 2012. A TEM cell system for in vivo exposure at 2.45 GHz. 6th European Conference on Antennas and Propagation.

Park H. H., 2022. Electromagnetic shielding analysis of planar materials using ASTM D4935 standard fixture. *IEEE Transactions on Electromagnetic Compatibility*, 64(5), pp. 1767–1778.

Park, S. H., Theilmann, P. T., Asbeck, P. M. & Bandaru, P. R., 2009. Enhanced electromagnetic interference shielding through the use of functionalized carbon-nanotube-reactive polymer composites. *IEEE Transactions on Nanotechnology*, 9(4), pp. 464–469.

Punnakkal, V. S., Ramakrishnan, A., Pradeep, A., Hossain, A., John, M. P. and Anila, E. I., 2025. Polypyrrole/silver/graphene ternary composites for high-performance Ku-band electromagnetic interference shielding. *Synthetic Metals*, 312, p. 117873.

Rathi, V., Prasad, B., Mishra, V. & Kanojia, R., 2022. Carbon based nanocomposite to enhance electromagnetic compatibility of medical equipment. *Materials Today: Proceedings*, 60, pp. 2071–2075.

Saini, P. & Choudhary, V., 2013. Enhanced electromagnetic interference shielding effectiveness of polyaniline functionalized carbon nanotubes filled polystyrene composites. *Journal of Nanoparticle Research*, 15, pp. 1–7.

Satav, S. M. & Agarwal, V., 2008. Do-it-yourself fabrication of an open TEM cell for EMC precompliance. *IEEE EMC Society Newsletter*, 218, pp. 66–71.

Singh, A. P. et al., 2014. Encapsulation of γ-Fe_2O_3-decorated reduced graphene oxide in polyaniline core–shell tubes as an exceptional tracker for electromagnetic environmental pollution. *Journal of Materials Chemistry A*, 2(10), pp. 3581–3593.

Sunil, M., Ramakrishnan, A., Saji, K. J. and John, H., 2025. Polystyrene/dedoped polyaniline composite nanofiber membrane based triboelectric nanogenerator for high performance EMI shielding and self-powered vibration sensing applications. *Materials Today Communications*, 42, p. 111227.

Thomassin, J. M. et al., 2013. Polymer/carbon based composites as electromagnetic interference (EMI) shielding materials. *Materials Science and Engineering: R: Reports*, 74(7), pp. 211–232.

Vicente, A. N. & Dip, M. J., 2011. The step by step development of NRW method. BMO/IEEE MTT-S International Microwave and Optoelectronics Conference (IMOC 2011), pp. 738–742).

Weir, W. B., 2005. Automatic measurement of complex dielectric constant and permeability at microwave frequencies. *Proceedings of the IEEE*, 62(1), pp. 33–36.

Wu, Z. P. et al., 2011. Electromagnetic interference shielding of carbon nanotube macrofilms. *Scripta Materialia*, 64(9), pp. 809–812.

Zhang, L., 2014. Research on preparation and property of alternating multilayer polymer electromagnetic shielding materials. *Applied Mechanics and Materials*, 443, pp. 634–638.

CHAPTER 3

Electromagnetic Interference Shielding for Space Applications

Rafael Vargas-Bernal, Bárbara Bermúdez-Reyes, and Margarita Tecpoyotl-Torres

3.1 INTRODUCTION

Because electronic devices and systems have greatly reduced their size, increased their speed of operation and become more powerful and sensitive, operation with other devices and systems has become complicated due to the possibility of interference between them.[1] Electromagnetic interference (EMI), also known as radio frequency interference or radio interference, is a disturbance caused by a source of internal or external electromagnetic radiation that interrupts, degrades or limits the performance of an electronic component, circuit or system.[2] Sources of EMI can be human-made such as rapidly varying electrical currents or natural such as solar effects and the northern lights. Interference between electromagnetic signals causes the resulting signal between two or more signals to reinforce or cancel each other, which must be avoided as non-coherent results are produced. Manufacturers of these devices and systems must properly choose the components and materials that shield against EMI to prevent catastrophic consequences of its presence.[3] Among the consequences of EMI are the interruption and degradation of the performance of devices and systems. The sources of interference come from nature but have mainly been produced by man. Although performance degradation may be considered a nuisance, the implications in high-performance applications can have irreversible and dire effects on the operation of the device and systems can generate considerable damage. In space applications, EMI produces disruption of operation or partial or permanent electronic damage to infrastructure and vehicles, electromagnetic pollution, and altering the health and safety of luggage and passengers on board.[1–5]

Space applications such as space shuttles, space stations, telescopes, rockets, satellites and other space vehicles and facilities are subject to challenging environments that make them capable of withstanding outgassing, vacuum, harsh conditions and radiation.[1, 2, 4] Essential special considerations such as the need for engineered materials to provide an

DOI: 10.1201/9781003512363-3

adequate thermal interface and high value-added EMI shielding are commonplace. In real applications, materials used in space experience low outgassing and abrupt temperature changes and are lightweight.

The main purpose of this chapter is to review the advances in commercially offered materials and those currently in the laboratory phase used to develop EMI shielding for space applications. This chapter has been divided as follows: in Section 3.2 the basic concepts of EMI shielding for aerospace applications are described. Next, conventional industrial EMI shielding solutions for space applications are discussed in Section 3.3. In Section 3.4, different material proposals for the development of laboratory-grade EMI shielding developed for aerospace applications are presented. Subsequently, future research directions related to EMI shielding for aerospace applications are discussed in Section 3.5. Finally, the conclusions of this research are given in Section 3.6.

3.2 BASIC CONCEPTS OF EMI SHIELDING FOR SPACE APPLICATIONS

When electromagnetic waves are transmitted through space, they travel freely unless they interact with radiation or matter which stops or fades them. Therefore, an electromagnetic wave sent into space will propagate indefinitely until it finds some medium that reflects, absorbs or transmits it in another direction. Electromagnetic waves do not require a medium to propagate, since they can travel through the vacuum of space.

EMI or radio frequency interference can be defined as unwanted interference or noise present in an electrical path, device, circuit or system caused by an external electromagnetic source that disturbs its performance. EMI leads to poor operation, malfunction or complete stoppage of electronic devices and systems. Interference can be caused by human-made or naturally occurring electromagnetic sources. Until now, engineers have reduced the impact of interference through modern error correction, electrical shielding and high-quality electronics. A classic example of EMI occurs when a cell phone is brought close to powered speakers or audio equipment and acoustic noise or a series of beeps are produced, which disturb the performance of the elements involved.

Electricity and magnetism are closely related to EMI as described later. Electric currents produce magnetic fields, and moving magnetic fields give rise to electric currents. These processes are the principle of operation of electric motors and generators. For their part, electrical conductors serve as radio antennas. Radio frequency sources and highly powered electrical sources cause interference even to distant devices. Advances such as size reduction, high sensitivity, higher operating speed and compact packaging have made them more susceptible to EMI.

Among the natural sources that can produce EMI are: first, lightning that can produce magnetic pulses and strong electrostatic discharges; second, solar flares and storms that emit highly charged particles that prevent terrestrial and satellite communication; and third, cosmic radiation that causes bit changes or bit flips in digital electronic circuits. Human-made EMI comes from different sources, such as high-power electrical and radio frequency systems, malfunction or inappropriate design of electronic devices or systems, signal jammers and electromagnetic pulses or intentionally induced faults.

Because radio waves allow safe and easy communication with a spacecraft, antennas must be designed with two core elements for developing the transmission and reception of electromagnetic signals. These radio waves require a broad frequency spectrum that can be used in radio navigation, remote sensing and telecommunications. In this way, the performance of these antennas must be optimised so that both wanted and unwanted signals are always evaluated, and possible EMI must be avoided. Therefore, devices and systems must achieve electromagnetic compatibility so as not to decrease their performance. EMI as time progresses becomes increasingly important to take advantage of next-generation broadband telecommunications satellites, which exploit multiple spot beams to operate at higher frequencies.

In both infrastructure and space vehicles, electromagnetic signals must be transmitted and received, however, care must be taken that the electrical currents in electrical and electronic devices and systems do not generate magnetic and electrical fields that can cause EMI thereby degrading their performance. The electromagnetic space environment has two sources: the first source comes from processes such as transmission, propagation, interaction and reception of electromagnetic signals, while the second source is caused by the effects coming from the orbital environment. To solve the electromagnetic pollution problem of electronic systems, metals such as silver, iron and copper were initially used as materials to achieve EMI shielding.[2] Unfortunately, these materials presented corrosion problems, processing difficulties and had high weight, limiting their exhaustive use. Recently, carbon-based nanomaterials such as graphene and its derivatives and carbon nanotubes were considered an alternative to avoid corrosion and excessive weight. However, these are difficult to disperse in the matrices that contain them and achieve low shielding effectiveness (SE). One of the most recent options is the application of MXenes, made up of transition metal nitrides and carbides which offer chemical activity, hydrophilicity, high electrical conductivity and high SE. These materials have disadvantages such as low mechanical properties and susceptibility to oxidation. A commercial EMI shielding material must achieve at least a SE of 20 dB, whose value is achievable for electrical conductivities greater than 1 S/m.[1]

Antennas are the most sensitive devices to EMI since they must convert electric currents into a magnetic field during transmission and vice versa during reception. Antenna performance must consider adjacent electric and magnetic fields from electronic devices or systems outside the radio frequency range, reflective surfaces and interaction with other radio frequency sources. Antennas are used in the transmission or reception of communication systems, as precise position sensors in navigation systems or as remote sensors in environmental probes, and in all these applications they must interact with the natural environment in a complex way. They also allow telecommand and tracking to establish the link between satellites and ground stations to develop telemetry.

Engineers must know how materials and technology will work in space for the success of any mission. Before sending a mission, engineers must design, test and debug the durability and performance of materials and technology subjected to harsh space environments. These environments include debris impact, high-energy radiation, atomic oxygen, ultra-high vacuum and exposure to extreme thermal heat and cold cycles. Various technologies and materials are found inside and outside space vehicles and infrastructure, including paints, coatings, 3D printed materials, sensors and electronic devices and systems. The exterior walls of space transportation and infrastructure are exposed to a wide range of hygrothermal climatic conditions including heating, drying, cooling, humidity, thawing and freezing. The space environment naturally contains particulates, gases, plasmas and high-energy particle radiation, which when subjected to electromagnetic signals can produce reflections, multiple reflections, absorptions and transmissions of electromagnetic waves.

Radiation has three sources of origin: galactic cosmic rays, explosive events from the sun and Earth's radiation hoods. It damages humans, sensors and electronic components. In addition, plasma causes the accumulation of electrostatic charges on the surface of the spacecraft which are dissipated during excesses as electrostatic discharges in the spacecraft's components. Atmospheres can act as a means of friction for the flow of matter, and in this way they can be used to slow down spacecraft during landings. Both natural and human-made space debris such as micrometeoroids and small space debris when colliding with spacecraft or infrastructure can damage or even break or penetrate them. Therefore, engineers who design and implement spacecraft must study the effects of these harsh environments using models and computational tools with real space data.

In addition to the previous effects, other electromagnetic factors impair the performance or integrity of space infrastructure and vehicles, as described later. High electrical voltages on a space element under a plasma environment can accelerate the accumulation of electrostatic charges that, when discharged, generate an electrical lightning bolt that damages space equipment. Systems travelling in low Earth orbit (LEO) such as satellites can be attacked by energetic particle radiation and collide with micrometeorites and space debris, which generate disruption or damage to electrical systems. The presence of atomic oxygen in the LEO orbit erodes some space materials. The interaction of electromagnetic waves with the natural spatial environment must be fully exploited to optimise the performance of radio links and use the type of information recoverable by using remote sensing in determining, for example, the state of the weather.

3.2.1 Types of Electromagnetic Interference

The element susceptible to EMI is known as the victim and is the receiver of the disturbance. There is always some element that acts as a source or transmitter and there is a medium or path that allows it to reach the receiver. EMI from the source to the receiver can be transmitted through different methods and types of paths (see Figure 3.1) as described later. EMI can be coupled through materials either magnetically or inductively through the impedances of inductors or transformers, capacitively through capacitors or conductively through the electrical resistance of an electrical conductor. It can even be radiatively coupled when the electromagnetic wave is radiated through a vacuum or air via an antenna.

The purpose of an electromagnetic shield is to attenuate or eliminate the transmission of electromagnetic waves from or to a device that generates or receives them.[4] SE against EMI measures the attenuation produced by the material to prevent this transmission in dB, which is better when a greater attenuation is achieved. This effect is not met in antennas

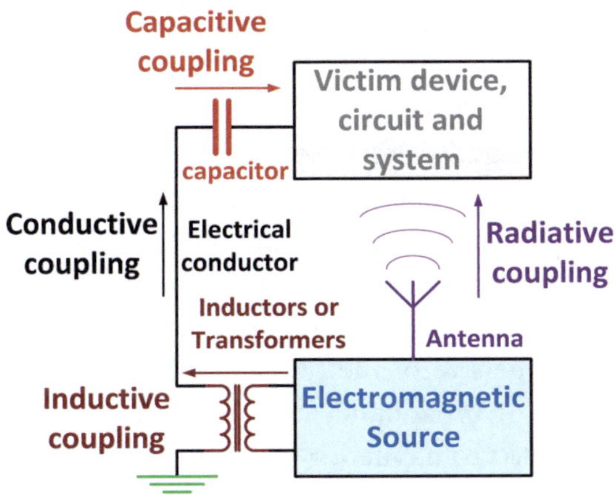

FIGURE 3.1 Main types of electromagnetic interferences.

which must be able to transmit or receive electromagnetic signals to achieve communication, however other devices, circuits or systems must include these shielding capabilities. According to the MIL-STD-188-125-1 standard issued by the United States Department of Defense, the materials used as shielding against EMI in the range of 1 kHz to 1.5 GHz must guarantee SE with a minimum attenuation value of 80 dB.[3]

When an electromagnetic wave is incident on a material, three phenomena can occur: reflection, absorption and multiple reflections, as illustrated in Figure 3.2. Reflection will reflect the incident signal away from the material, absorption will trap the wave inside the material and multiple reflections will keep the wave bouncing around inside the material. The undesired effect is that the wave will leave the material and pass to another material, modifying its behaviour. Therefore, shielding materials are designed to eliminate or attenuate this transmission and are regularly used as coatings placed on the material to be protected.

As described in the following sections, the mechanisms that materials present to behave as a shield against EMI depend not only on their chemical composition but also on their structure, properties and percentage of the different components used in their manufacture.

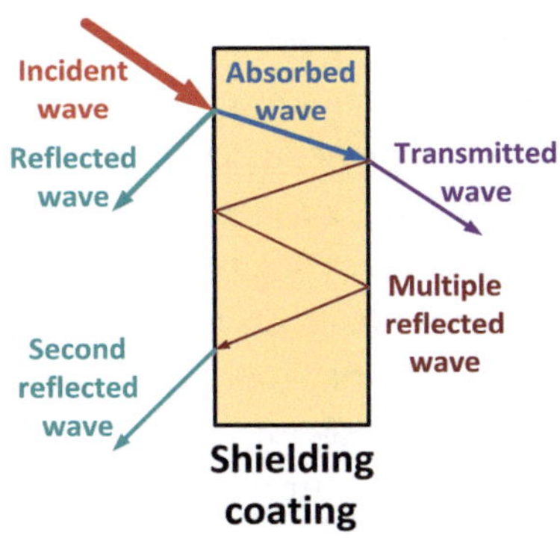

FIGURE 3.2 Main mechanisms of EMI shielding.

The behaviour of volumetric materials and that based on nanomaterials concerning EMI shielding properties differ.

3.2.1.1 Radiated Electromagnetic Interference

This type of interference occurs when an electronic device or system (source) produces a radio frequency that is received by one or more receivers, which are very far apart and that modifies the performance of these devices. Examples of this type of interference are a computer restarting due to broken microwave equipment or Wi-Fi dropping on old cordless phones. Depending on the size of the frequency band, this type of interference can be classified as narrowband or broadband. Narrowband interference is generated when the radio transmitter emits electromagnetic waves that perturb a specific radio frequency. Broadband interference occurs due to equipment malfunction and disrupts a large portion of the electromagnetic spectrum.

3.2.1.2 Conducted Electromagnetic Interference

Conducted EMI transports electromagnetic waves through physical electrical paths connecting the source and receiver. This is very common in power transmission lines or power to motors that demand large amounts of electrical current. Examples of this type of interference include turning on clothes dryers or treadmills, which can reset a computer or other lower-demand electrical device connected to the same circuit.

3.2.1.3 Coupled Electromagnetic Interference

Coupled interference occurs when the disturbance is induced by the proximity between transmitter and receiver even though there is no electrical connection between them. This is transmitted through capacitive or inductive means. Inductive or magnetic-type coupling occurs because the magnetic field in one conductor induces an unwanted electric current in a nearby conductor. A classic example occurs when a power cable close to an audio cable induces a magnetic field that is heard as a hum. Capacitive coupling occurs when two conductors are closely paralleled and store an unwanted electrical charge between them, which occurs on printed circuit boards or in groups of closely packed wires paired over long runs.

3.2.1.4 Intentional Electromagnetic Interference

Intentional EMI occurs when an external source emits electromagnetic signals with the intention of harming the victim or recipient, as with electronic warfare. Electronic warfare implies that a user attacks the adversary's energy spectrum to reduce, exploit or prevent the use of the user's transmission spectrum to use it for their benefit. Radio jamming is the deliberate blocking of the radio signal that causes interference with wireless communications. The disturbance seeks to reduce the signal-to-noise ratio of the signal used in the communication link and therefore the flow of information is affected. To achieve intentional jamming, the transmitter must be tuned to the same frequency as the opponent's receiving equipment and use the same type of modulation with enough power to override any signal received to jam their system.

Intentional signal jamming is achieved when the following signals are used: sweep through, gulls, stored sounds, spark, pulse, rotary, tone, random key modulated continuous wave, warbler, step tones, random pulses and random noise. These signals are classified as obvious and subtle.

3.2.1.5 Unintentional Electromagnetic Interference

Unintentional EMI occurs when an external source accidentally emits electromagnetic signals outside its assigned spectrum or is used to reveal sensitive information about the attacked user and gives rise to an undesirable effect on a third party. Examples of these types of interference are electrostatic discharges, induction motors and switching systems, medical equipment etc. Unintentional radio jamming occurs when a system transmits on a busy frequency without checking first to see if it is in use, or without having heard stations using this frequency. Another possibility arises from accidental radiation of a signal emitted on an aircraft emergency frequency.

3.2.2 Frequencies for Space Radio Communication

According to Recommendation ITU-R V.431-8 of August 2015 of the International Telecommunications Union (ITU), the frequencies used for space radio communications are designated by letters: *L*-Band (1–2 GHz), *S*-Band (2–4 GHz), *C*-Band (4–8 GHz), *X*-Band (8–12 GHz), *Ku*-Band (12–18 GHz), *K*-Band (18–27 GHz), *Ka*-Band (27–40 GHz) and *V*-Band (> 40 GHz). Electronic devices and systems used in infrastructure or space vehicles are susceptible to emitting or receiving electromagnetic waves in these frequency ranges, so EMI shielding must be designed to avoid interference from these elements. Telecommunications in deep space are regularly carried out in the *X*-band. The possible radio frequencies that can be used in space communications have been summarised in Figure 3.3. Different coloured arrows have been used to illustrate how variables such as attenuation, antenna size, bandwidth and performance are modified according to the radio frequency used.

In the *X*-band, frequencies between 7.25 and 7.75 GHz are used for space-to-Earth communications, while frequencies between 7.9 and 8.4 GHz are used for Earth-to-space communications. Applications range from Fixed Satellite Service (FSS) to Mobile Satellite Service (MSS).

3.3 INDUSTRIAL SOLUTIONS AGAINST EMI SHIELDING FOR SPACE APPLICATIONS

For protection against EMI to be effective, companies such as Parker Chomerics (www.parker.com/us/en/divisions/chomerics-division/solutions/emi-shielding.html) offer a variety of solutions, including laminated materials, gaskets, coatings and paints. Specific finished products for EMI shielding that can be found on the market include metal foil tapes, shielding laminates, grounding, cable shields, shielded honeycomb air vents, shielded windows and gaskets. Among the raw materials to achieve shielding from EMI are conductive plastics, inks, greases, sealants, coatings and conductive adhesives on the market. These materials must meet diverse requirements to ensure optimal reliability and performance while offering durability, robustness and cost-effectiveness.

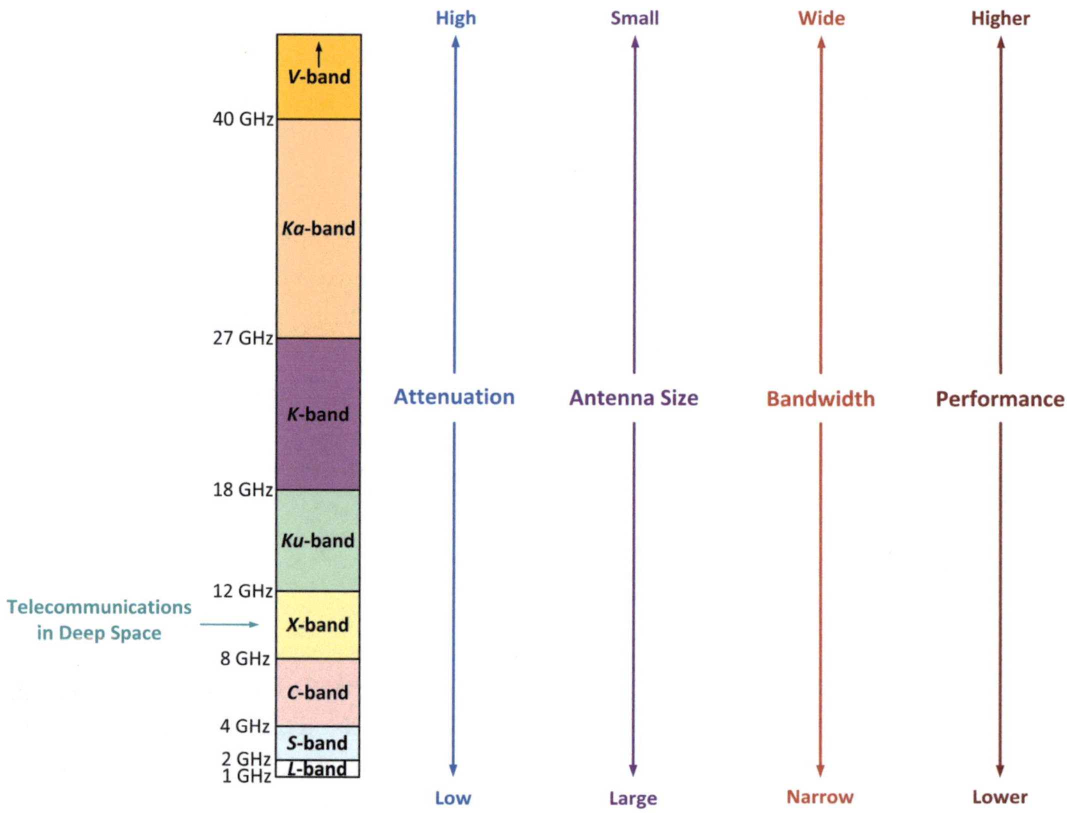

FIGURE 3.3 Radio frequencies used for space communications.

3.3.1 EMI Shielding Gaskets

EMI shielding gaskets are intended to prevent electromagnetic waves from leaving or entering an electronic enclosure. Depending on the application, there are different options and materials on the market including traditional metal-based gaskets, fabric-over-foam, locally shaped materials and electrically conductive elastomers. Among the categories of gaskets, the following can be identified: woven wire meshes, expanded metal films, oriented wire gaskets, shielding mesh gaskets, gaskets based on electrically conductive foams, electrically conductive shaped-in-place material gaskets and electrically conductive elastomer gaskets. Woven aluminium meshes are pressure-sealed using silicone, neoprene or fluorosilicone elastomer. Metal-based gaskets enable lightning protection and shielding against electromagnetic pulses (EMP) and EMI using expanded metal, silicone-oriented wire or woven wires. Shielded mesh gaskets against EMI are used as seals or choke flanges with slightly round or rectangular shapes on small housings, shafts or waveguides. Electrically conductive foam-based and fabric-on-foam gaskets can be used in indoor EMI shielding and grounding applications with low clamping force urethane foam or fabric and a mesh, wound or woven electrically conductive material typically being used. The minimum operating temperature ranges from −40 to 5 °C and the maximum operating temperature ranges from 40 to 150 °C. The interference shielding level ranges from 70 to 109 dB. Form-in-place gaskets can be robotically dispensed for applications with small cross-sections and complex patterns, and at the lowest cost, offering electrically conductive or non-conductive options. Electrically conductive elastomer-based gaskets allow EMI shielding, environmental sealing and cost-effective corrosion resistance. These materials are offered as set parts or moulded shapes, they can be extruded into strips or offered as ropes or splices of selectable lengths. Filling materials in these gaskets can include silver-plated nickel, silver-plated glass, silver-plated copper, silver-plated aluminium, silver, nickel-plated graphite, nickel-plated aluminium/graphite, nickel-plated aluminium, nickel-graphite or carbon. Polymer materials such as silicone foam or rubber, fluorosilicone rubber, fluorocarbon rubber or ethylene-propylene rubber can be used.

3.3.2 Laminated Shielding Materials

Laminated materials for electrical grounding and isolation and EMI shielding are based on metals, dielectrics and adhesives bonded together to form a sandwich. Various metals or conductive materials are available: nylon rip-stop, nylon taffeta, nickel-plated silver, nickel-plated copper, tinned copper, copper or aluminium. Among the dielectric materials available are Kapton®, polypropylene (Formex™), PVC and mylar. Adhesive options such as silicone, general-purpose acrylic, flame-retardant conductive acrylic and electrically conductive acrylic are available. The lengths of these laminates range from 50 to 100 feet, a width from 12 to 24 inches and a thickness from 0.002 to 0.01 inches. The electrical resistance ranges from 0.005 Ω/sq to 0.1 Ω/sq.

3.3.3 Electrically Conductive Coatings and Paints

Electrically conductive coatings and paints can be applied to plastic and composite housings to prevent EMI in electronic components and aircraft fuselage components. Different linker configurations and electrically conductive fillers can be chosen to achieve high-performance, reliable and durable EMI shielding. These raw materials have a minimum operating temperature of −60 to 5 °C and a maximum operating temperature between 40 and 100 °C. These materials achieve a surface resistivity of 0.01–10 Ω-cm. The film thickness ranges from 1 to 76 μm. The filler material can be nickel, silver or a silver-copper alloy. The polymers that can be used are thermoset polyurethane, polyester, epoxy resin or acrylic resin. The interference shielding level ranges from 50 to 109 dB.

3.3.4 Heteropolymers Reinforced with Electrically Conductive Fillers

Heteropolymers including electrically conductive fillers are also available for EMI shielding due to their unique physical, mechanical and electrical properties. Two categories are available: conductive plastic pellet materials and injected moulded plastic parts. Injected moulded plastic parts can replace metals in electromagnetic shielding through low-weight, low-cost formulations. Heteropolymers are based on polycarbonate (PC) and acrylonitrile butadiene styrene (ABS) or based on polybutylene terephthalate (PBT) and polycarbonate (PC). Applications for these composite materials focus on metal-to-plastic, high-temperature housings and faceplates for advanced telecommunications and computing architecture (ATCA). The maximum operating temperature ranges from 40 to 200 °C and the surface resistance ranges from 0.25 to 10 Ω-cm.

3.3.5 Safety and Environmental Issues

The use and handling of these materials used in EMI shielding must comply with the Restrictions on Hazardous Substances (RoHS) of the United States and the Safety, Health, and Environmental Regulations of the European Union and other environmental organisations around the world for electrical and electronic equipment. Among the chemical substances that should be avoided are contents of cadmium (Cd), lead (Pb), mercury (Hg), hexavalent chromium (Cr VI), polybrominated biphenyls (PBB), polybrominated diphenyl ethers (PBDE), bis(2-ethylhexyl) phthalate (DEHP), benzyl butyl phthalate (BBP), dibutyl phthalate (DBP) and di-*N*-isobutyl phthalate (DnBP). Some of these products may contain chemicals that may cause birth defects or other reproductive harm. For more information, please consult your provider.

Using good-quality electronic devices and systems supplied by reputable companies is the best strategy to avoid generating EMI. The Federal Communications Commission (FCC) in the United States is the institution that establishes the standards that must be met to test the effectiveness of shielding electronic devices from EMI. Other countries have different institutions that perform this verification or conform to FCC rules. Unfortunately, there are counterfeit, cheap or poorly manufactured devices on the market that have not been properly tested or do not contain electromagnetic shielding, which makes them susceptible to producing EMI and therefore affects the performance of other devices when they are used in commercial, industrial or military equipment. Filters and digital error correction systems can reduce or eliminate EMI from nearby sources.

3.3.6 Design Considerations

High-speed networks must consider EMI by separating data lines from power lines using twisted pairs or shielded cables to provide signal integrity. Fibre optic cables are not susceptible to interference and can replace copper cables. To ensure electromagnetic shielding in a wireless network design, high-power transmission lines, nearby radio sources and the power and number of transmission sites must be considered.

Medical (hospitals, in-patient devices etc.), energy (gas stations, electrical substations etc.) and transportation (aircraft, trains etc.) facilities may be susceptible to experiencing EMI from cell phones or laptop computers which affect and modify and paralyse performance unless they are turned off when they are in their surroundings. The design of high-speed electronic systems on printed circuit boards must consider the location and routing of the components in the design. Some devices need conductive tapes or metal shielding cans to block EMI from components sensitive to electric and magnetic fields. In extreme cases, a Faraday cage is used as a shield against EMI for devices or systems. For example, radio telescopes are deployed in isolated areas far from cities to reduce EMI from multiple electromagnetic sources.

Electronic engineers should review the design of their systems against EMI from the design phase because it is much faster and less expensive than trying to solve them when they appear.

During the design phase, solutions often range from relocating components to another location or orientation on the printed circuit board, the use of shorter electrical paths that reduce electrical impedance and a better overall design that considers all the factors to which the system, circuit or device is sensitive to possible sources of EMI. In the case of a system purchased or designed by another that is not feasible to modify, adaptations such as optical fibres, isolation transformers, ferrite beads, decoupling capacitors and electronic filters are some options. Reprogramming of electronic systems can be used to reduce work speed when feasible. In other cases, the use of shielding systems including the use of Faraday cages, changes of sensitive components, activation at zero voltages in the case of systems operating on alternating current (AC) and relocation of the system work site are other options.

3.3.7 Susceptibilities of Different Radio Frequencies

When radio transmissions were based on amplitude modulation (AM), the systems did not distinguish between unwanted in-band signals and desired signals, that is, they were selective even by using omnidirectional antennas. Improvements in selectivity were introduced by applying digital radio systems where error correction techniques were used in Wi-Fi technology. Frequency hopping and spread spectrum techniques were much more resistant to EMI since they used a highly directional receiver based on an antenna diversity receiver or satellite dish to pick out one signal in space and exclude the others. Recently, one of the alternatives that allow very efficient use of the spectrum is digital spread spectrum signalling (DSS) known as ultra-wideband (UWB), which, although it uses long sections of the spectrum, requires low amplitudes to transmit digital data with high bandwidth.

Shielding against EMI is of great interest in research and new materials are continually being incorporated into the range of available possibilities. Novel composite materials for absorbing microwaves and radio frequency signals include ferrites and two-dimensional materials incorporated into polymers.

3.4 SOLUTIONS AT THE LABORATORY LEVEL AGAINST EMI SHIELDING FOR SPACE APPLICATIONS

Nanomaterials can be used in diverse applications thanks to their unique physicochemical properties. Unfortunately, they are rarely used individually, so they are regularly used as composite materials, heterostructures or hybrid materials, where multifunctional contributions can be achieved by taking advantage of the synergy between the included components. For example, two-dimensional nanomaterials such as MXenes have good electrical conductivities but poor mechanical or chemical properties. To improve their properties, they must be combined with other materials such as inclusion in a polymer matrix, and in this way it is possible to optimise their properties. In this way, it is possible to achieve multifunctionality. Examples of materials applied as shields to EMI in the aerospace area are described later.

3.4.1 Composites Based on Two-Dimensional Materials

A composite material based on glass fibre and liquid thermoplastic resin Elium® and co-cured with graphene nanoplatelet (GNP) films has been synthesised as material to achieve

EMI shielding for space applications.[1] The composite material was assembled by placing resin films with graphene as outer layers and a resin film with glass fibre inside using co-infusion and co-curing. This combination of materials offers multifunctional properties since the material is subjected to impact, so not only the effectiveness of the shielding (> 32 dB) but also the impact energy absorption (105%), the stiffness (45%) and strength (48%) must be increased in comparison with the values achieved by glass fibre-reinforced polymer composites (GFRPC). These polymeric matrices allow for ease of processing, high impact resistance, repair, bonding freedom, formability and recyclability. Fibre-reinforced composite materials offer lightness, structural stability, corrosion resistance and high specific strength which are feasible to the aerospace industry. Graphene contributes to low weight and high electrical conductivity. To achieve commercial SE levels, the concentration of graphene must be high at levels that exceed the percolation threshold, which increases the viscosity and aggregation of the composite material and the cost, which must be resolved to achieve its commercial application. The solution to these problems was reported by embedding the graphene nanosheets through a hot mechanical press above the glass transition temperature of the resin to achieve dispersion of the nanofiller and using a lower concentration of this in the composite material. The composite material obtained was tested in the frequency range of the *X*-band which goes from 8.5 to 12.5 GHz considering the electromagnetic contributions by both absorption and reflection. The purpose of the composite material is to design lightweight and low-cost enclosures to protect space systems against EMI, radiation and impact.

Composite films based on polyimide and the MXene $Ti_3C_2T_x$ were proposed to achieve EMI shielding in harsh environments.[2] This composite material could withstand harsh environments such as abrupt temperature changes (−196 to 250 °C), flame retardancy and ultrasonic and hydrothermal environments while maintaining its EMI shielding capacity and adequate mechanical properties. Polyimide (PI) is an antioxidant. It increases tensile strength (114.9 MPa) and fracture deformation (3% increase). MXene provides electrical conductivity to achieve a SE of 37 dB in the *X*-band range for a composite thickness of 15 μm. The manufacture of this composite material using a thermal imidisation process in a vacuum environment allows the interpenetration between PI and MXene to form layered materials with high absorption of internal electromagnetic waves and multiple external reflections making it attractive for wearable applications and aerospace.

Transition metal dichalcogenides (TMDs) comprise a broad family of two-dimensional materials with the general chemical formula MX_2 with one transition metal M^{4+} and two chalcogens X^{2-}. Molybdenum disulfide (MoS_2) is the most investigated dichalcogenide for shielding against EMI. This material as a shield against EMI improves the impedance matching with free space due to its low electrical conductivity and dielectric constant. The operating frequency range of this material within polymer matrices goes from 9.6 to 13.76 GHz with a SE of 38.42 dB. When reinforcements based on reduced graphene oxide (rGO) and molybdenum disulphide (MoS_2) heterostructures are used as nanofillers in polymer matrix composites, the operating frequency range varies from 10.72 to 16.44 GHz with SE of 60 dB. The EMI shielding mechanisms for two-dimensional materials are outlined in Figure 3.4.

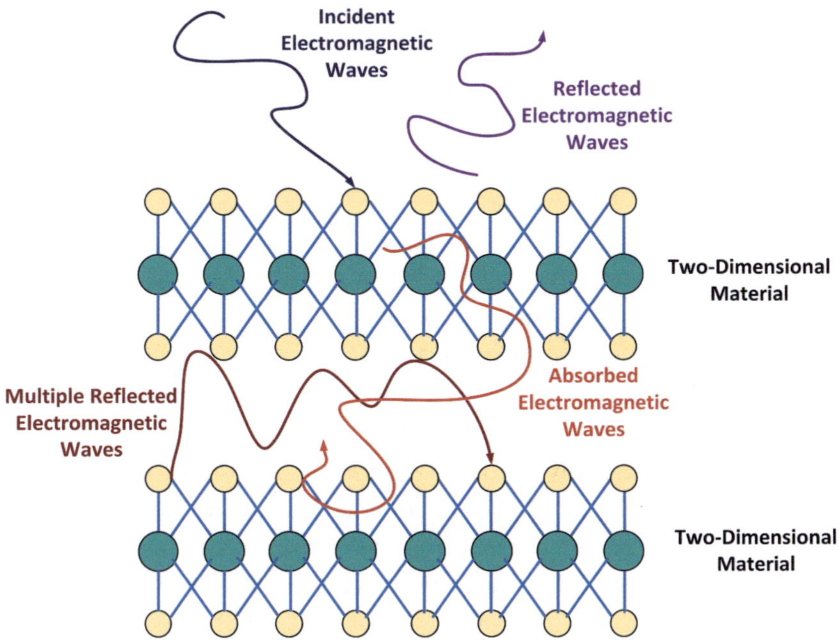

FIGURE 3.4 Mechanism of electromagnetic shielding in two-dimensional materials.

3.4.2 Sustainable Polymer Matrices

Recently, researchers have explored the possibility of using sustainable biopolymers as a shielding coating against EMI.[6] Materials such as cellulose can be exploited due to their flexibility, high aspect ratio, low density, good thermal properties, excellent mechanical strength, biocompatibility and biodegradability in EMI shielding applications in the aerospace industry. In particular, the biocompatibility and sustainability of nanocellulose as a matrix material containing conductive nanofillers give rise to nanocomposites that minimise the insulating gaps between the nanofillers, increase the carrier transfer efficiency and thus the electromagnetic shielding performance is increased.

3.4.3 Porous Matrices

Porous materials such as two-dimensional materials and carbon nanotubes allow the implementation of lightweight materials even if they are thick and favour the attenuation of electromagnetic waves by absorption by exploiting the impedance mismatch that occurs between the available interfaces and the air-filled pores.[7] Furthermore, thin film-based shielding coatings containing multiple nanofiller components and layers of the same shielding material exploit the reflections through the material to achieve EMI shielding. Two-dimensional materials such as MXenes, TMDs, double-layer hydroxides (DLH), hexagonal boron nitride (*h*-BN), graphitic carbon nitride (g-C_3N_4), graphene and phosphorene (black phosphorus, BP) are achieving outstanding EMI SE. These nanomaterials offer large surface areas, low density and high electrical conductivity, which give them superior EMI shielding properties.

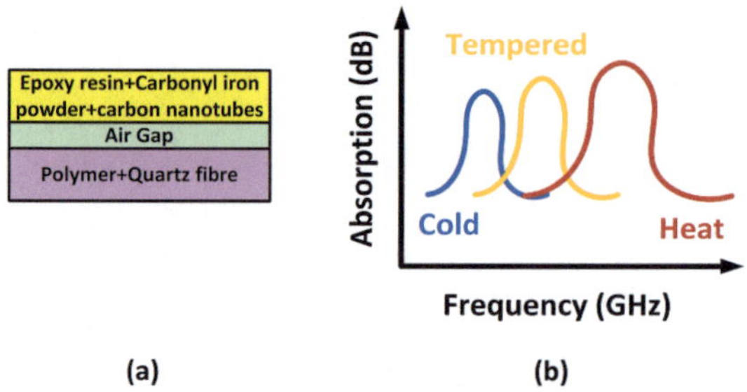

FIGURE 3.5 Operating principle of a bimorph electromagnetic wave absorbing actuator: (a) device cross-section and (b) absorption-versus-frequency curve of the device at different temperatures.

3.4.4 Tunable Approaches

Traditional techniques commonly used as external coatings at a fixed distance acting as electromagnetic wave absorbing materials with a narrow bandwidth are now being redesigned. Dynamically tunable approaches are now implemented through multifunctional nanocomposites that absorb electromagnetic waves in a much broader bandwidth and with unique absorption characteristics using bimorph actuators including magnetic materials and electrical materials.[8] An example of such materials are nanocomposites based on an epoxy resin layer reinforced with carbonyl iron powder (magnetic material) and carbon nanotubes (electrical material), a quartz fibre-reinforced polymer layer and between them a resonant air gap, as shown in Figure 3.5(a). The layer with nanomaterials is responsible for attenuating electromagnetic waves, supporting the mechanical stress of the actuator and conducting heat to produce dynamic resonant tuning of the actuator, thus providing multifunctionality. When heat is applied, the peak absorption magnitude and the oscillation frequency are shifted due to the dielectric loss layer relative to a ground plane using the thermal bimorph actuator arms. Therefore, the electromagnetic absorption capabilities are greater when the temperature at which the bimorph actuator material system operates is increased, as illustrated in Figure 3.5(b).

Scientific research and technological development fields such as wearable electronics and aerospace require solutions based on multifunctional, lightweight and miniature materials.[9] Nanocomposites based on multifunctional thermoplastic polymers such as ABS and graphene nanosheets can be designed to offer EMI shielding materials for wide frequency ranges from microwaves to terahertz, as well as high electrical conductivity and suitable thermal management. For graphene concentrations of 5% by weight, an electromagnetic SE of 80 dB was achieved for material samples with a thickness of 1 mm, a thermal conductivity of 1.8 W/(m·K) and an electrical conductivity of 188 S/m. The main mechanism of suppression of electromagnetic waves is electromagnetic absorption.

3.4.5 Gel- and Foams-Based Approaches

For low-temperature space environments, the possibility of using three-dimensional cellular films assembled by microgels by filtration-embedding two-dimensional materials such as graphene and MXenes, subsequently applying a freeze-casting process, has been proposed.[10] These lightweight films offer superior energy storage capacity and effective shielding against EMI. The advantage of MXene nanosheets over graphene nanosheets is that those based on MXenes are easier to handle to develop lightweight composite materials to achieve electromagnetic shielding. Furthermore, MXene nanosheets possess high electrical conductivity and hydrophilicity that can be exploited to implement lightweight foamed materials with excellent mechanical strength using ice crystals as porous templates to produce hydrogel films for electromagnetic shielding. The aggregation of MXene nanosheets is induced using monovalent metal ions such as K^+ thanks to the surface electrochemical active sites on MXene. These materials exploit the mechanism of multiple reflections to attenuate the effect of electromagnetic waves and thereby achieve electromagnetic shielding.

Among the recent technological developments for developing materials used for EMI shielding is the solid-state foaming of carbon materials by assembling short vermicelli rods and bonding them together using polymers during carbonisation in the temperature range of 200–250 °C.[11] Polymeric materials that have been used to achieve solid-state foaming include polyesters, poly (phenylene sulfide)/poly (ether sulfone) blends, polysulfone, ABS, poly (lactic acid), polypropylene, epoxy resin, poly (methyl methacrylate) and polycarbonate. Carbon foams can provide EMI shielding through absorption and multiple reflection mechanisms with SE in the range of 47.5–65.6 dB in the *X*-band (8.2–12.4 GHz) for deep-space telecommunications. In addition, these foams offer low thermal conductivity. Porous materials used in electromagnetic shielding are preferred over their dense counterparts because electrical impedance matching increases the ease of penetration of electromagnetic waves into the material due to the multiple reflections produced which absorb the electromagnetic waves. Other advantages of these materials include their corrosion resistance, low cost and ease of fabrication.

3.4.6 Bioinspired Approaches

The possibility of using renewable bio-based materials in the design of EMI shielding materials has also been promoted.[12] For example, the application of hybrid nanocomposites with heterostructure based on polypyrrole (PPy), chrome-tanned collagen fibres (CF) and multiwalled carbon nanotubes (MWCNTs) has been proposed since they exhibit high electrical conductivity and achieve a SE of 30 dB in the *X*-band for materials with a thickness of 0.48 mm thanks to the synergy between these materials. The mechanism to shield against EMI in this material is centred on the multiple reflections that lead to the absorption of electromagnetic waves within the composite material obtained.

An interesting alternative is a proposal to use bioinspired double-lattice structural materials to develop EMI shielding materials with SE of 100 dB that exploit the use of cellulose nanofibres (CNF) and carbon nanotubes (CNTs) to achieve specific strength of 146 MPa·cm^3/g, high electrical conductivity (600–2820 S/m) and lightweight.[13] In addition, this material offers easy processability, excellent chemical resistance and corrosion resistance, which makes it interesting because it provides unique multifunctional properties compared with traditional products found on the market.

3.4.7 Heterostructures-Based Approaches

Heterostructures of two-dimensional materials such as graphene and molybdenum disulfide (MoS_2) nanoplatelets can be used as EMI shielding materials.[14] The first material provides electrical conductivity while the second material is used as a dispersant for graphene. The material achieved a SE of 24 dB using the multiple reflection mechanism to absorb electromagnetic waves that produce interference. A sandwich structure is obtained to exploit the absorption mechanism produced by multiple cavities formed between the layered materials.

Emerging EMI shielding applications require materials with mechanical integrity, functional properties, reliability and high flexibility.[15] The prolonged exposure to electromagnetic radiation in the GHz range is considered carcinogenic and it affects the central nervous system of humans, leading to physical/cognitive abnormality, learning/memory impairment and sleep disorders. These frequency values involve wavelengths in the nanometre range, so nanomaterials with suitable properties must be used to develop shielding against EMI. 2D materials present foldability, twistability and bendability due to their inherent flexibility. Among the emerging two-dimensional materials that will offer EMI shielding including the properties mentioned earlier are phosphorene and borophene. The two-dimensional material that has been most successful in this decade in the implementation of shielding against EMI is MXene.[16] This nanomaterial has been used in multilayer composite materials, foams and aerogels, taking advantage of the absorption mechanism by leveraging multiple reflections between the layers. Initially, this material was unstable and had a high cost. Fortunately, these disadvantages were gradually resolved and now its use tends to be directed towards mass application.

3.4.8 Multifunctional Approaches

Some applications require that materials used in wireless communications can simultaneously transmit visible light and avoid the transmission of electromagnetic waves in the GHz range.[17] Using first-principles calculations, researchers have identified that from a group of 7000 two-dimensional monolayer materials, there is a group of 26 materials with excellent visible light transmission properties with values greater than 98% and an EMI SE of approximately 70%. The highest-ranked material in the study mentioned earlier was silver diselenide ($AgSe_2$), which allows 98.53% visible light transmission and 94% attenuation of EMI, these values are better than those achieved by graphene. This study also found that thicker monolayers do not necessarily produce better EMI shielding and often reduce visible light transmission. Materials such as MXenes with hydroxyl (OH^-) functional groups on their surface have better SE against EMI without reducing their visible light transmission capabilities.

The Sixth Generation (6G) wireless communication systems will produce electromagnetic pollution in the terahertz band, so emerging materials based on two-dimensional materials are proposed to achieve EMI shielding.[18] One approach is to bend MXene nanosheets using alkanes such as octane (C_8H_8) and dodecane ($C_{12}H_{26}$) grafted onto the surface terminals. Alkanes exceed the bond length of traditional surface functionalities in MXenes such as oxide (O_2^-), hydroxyl (OH^-) and fluorine (F^-) groups, leading to intra- and inter-sheet deformations. These deformations lead to shifts in the Raman peaks and wider line widths, which can be exploited to increase the multiple reflections of electromagnetic waves and thus ensure the absorption and consequent blocking of EMIs. The EMI SE of dodecane-modified MXenes was 35 dB and for octane-modified MXenes was 26 dB, higher than that of natural MXene at 22 dB.

The advantageous properties of MXenes such as exceptional thermal stability, ease of processability, surface chemical tunability, increased water dispersibility and most importantly superior electrical conductivity, are driving them as one of the emerging materials with the greatest potential use within the two-dimensional materials family.[19] To increase the shielding efficiency against EMI that MXenes naturally possess and exploit in numerous applications, they must be combined with other materials such as metal-organic frameworks (MOFs), nanoferrites, carbon-derived materials, fibres and polymers. The minimum EMI SE ranges from 10 to 30 dB, but those materials that offer a SE greater than 30 dB are considered relevant for industrial applications. The electrical conductivity required for a material to act as a shield against EMI is 1 S/cm. In the case of MXene, its conductivity ranges from 5 to 20,000 S/cm. Each layer of MXene located on a thin film of material used as shielding attenuates the energy of electromagnetic waves, producing their absorption, thus preventing their effects from reaching the material they are protecting, even in coatings with a thickness of 2.3 nm.

3.4.9 Multilayered Approaches

One of the strategies to minimise reflection is to match the impedance between the substrate and the air where the electromagnetic waves strike.[20] Three-dimensional microcellular scaffolds based on aerogels and foams in a polymeric matrix are used to increase the absorption of electromagnetic waves produced by interference. These scaffolds must include combinations of electrical, magnetic and dielectric materials to achieve the loss mechanisms that provide electromagnetic shielding properties. Electrical materials with very high electrical conductivity can decrease the SE against EMI, so it is advisable to include dielectric and magnetic materials so that materials with intermediate electrical conductivities can be used in the shielding design. Through aerogels and foams, it is possible to distribute nanofillers in a porous self-assembly structure that increases the electromagnetic absorption of interference. The SE for these materials ranges from 30 to 60 dB in the *X*-band depending on the components and their percentages within the material using carbon nanofillers such as nanotubes and graphene. Magnetite (Fe_3O_4 or $FeO{\cdot}Fe_2O_3$), Ni oxides and Co oxides are used as magnetic materials. Polymers such as epoxy resin, poly benzoxazine (PBz) resin, phenolic resin, polyaniline (PANI), polydimethylsiloxane (PDMS), ABS rubber and poly (3,4-ethylenedioxythiophene)-poly (styrene sulfonate) (PEDOT:PSS) have been used.

Three-dimensional multilayered structures can exploit the addition of multiwalled carbon nanotubes (MWCNTs) on polycarbonate (PC) and magnetite-decorated reduced graphene oxide (rGO-Fe_3O_4) on polyvinylidene difluoride (PVDF) as outer layers.[21] A three-dimensional printed polyurethane-based film, foam or mesh containing carbon nanotubes was sandwiched. This material as a multilayer structure has been proposed as a shielding to EMI. This material achieved a SE of 39 dB in the *K*-band and if silver (Ag) is sputtered onto the polyurethane a SE of 50 dB in the *X*-band is achieved. The mechanism used to achieve EMI shielding is based on absorption.

The touch screens transmit electromagnetic waves that harm other devices and human health.[22] These screens must include EMI shielding features to prevent the transmission of electromagnetic waves to devices not involved with their operation. A similar case occurs with smart windows with electrochromic properties to protect humans from dangerous electromagnetic waves, heat from infrared radiation and ultraviolet light radiation. Electromagnetic shielding properties can be incorporated into textiles to generate alternative and efficient energy in inaccessible locations in space.

3.5 FUTURE RESEARCH DIRECTIONS

In this century, researchers will continue to develop scientific research to ensure that shielding against EMI achieves optimal performance and can generate stable and safe environments in different fields of application with a special emphasis on space applications. To achieve this goal, companies, governments, universities and research centres must invest in convenience and safety to achieve sustainable development and improve the quality of life of living beings in all environments where EMI may occur.

Due to the strategic advantages of nanomaterials and two-dimensional materials, these materials will ultimately contribute to achieving effective absorption and reflection of electromagnetic waves in the full range of the frequency spectrum used in space applications. Different research groups will continue to explore and bring to mass application various materials that exploit nanomaterials, in particular two-dimensional materials, for use as EMI shielding materials in composite materials, heterostructures and hybrid materials by exploiting the unique quantum effects in these materials.

Studies that exploit computational screening using deep machine learning models for the study of shielding properties against EMI should be developed as they would represent a relevant contribution to optimise the design and experimentation time required so that the multifunctionality of the materials can be exploited with much better performance in different applications with an important emphasis on the aerospace area.

Advances in materials that enable high levels of shielding against EMI are outstanding. Still, emerging needs drive the need for materials with a wide bandwidth, multifunctional, flexible, durable, lightweight, non-corrosive, facile processes and low cost. Furthermore, current and future systems demand huge amounts of compact, multifunctional devices that integrate new functions into the electromagnetic shielding property. These multifunctional properties can be naturally occurring or custom-designed by engineers. Additional features include heat dissipation, temperature sensing, humidity sensing, strain sensing, gas sensing, transparency, thermal conductivity, flame retardancy, hydrophobicity etc. Properties such as self-healing, self-cleaning, triboelectric and shape memory must be incorporated into new material designs for electromagnetic shielding. The huge challenge is not that materials have multiple properties but that they work synergistically without overshadowing each other and achieve the required values depending on the various possible applications. As a result, scientific research related to EMI shielding is expected to develop new scientific and technological solutions and yet to resolve fundamental issues not theoretically understood.

3.6 CONCLUSIONS

The extensive use of electronic devices that can produce electromagnetic pollution that affects other electronic systems and the health of living beings will continue to require researchers to design and implement emerging alternative materials capable of attenuating or eliminating EMI. Among the materials that are revolutionising the possibilities of achieving this goal of EMI shielding that promise high functionality and reliability are carbon nanotubes and two-dimensional materials such as MXenes, graphene, DLH, phosphorene, TMDs and hexagonal boron nitride (*h*-BN) by exploiting their unique nanofiller capabilities of high electrical conductivity and surface area. The possibilities of avoiding malfunctions, data loss and poor performance due to EMI are now feasible. The use of conventional materials and materials under technological development or scientific research based on nanomaterials to shield EMI will continue to be of interest. Therefore, this engineering design problem will remain a topic of research interest in this century.

ACKNOWLEDGEMENT

The author is grateful for the support of the Universidad Politécnica de Guanajuato for developing this scientific research. The first author thanks the researchers who shared their scientific publications to carry out this research work.

REFERENCES

1. Khan, T. et al., 2023. Co-cured GNP films with liquid thermoplastic/glass fiber composites for superior EMI shielding and impact properties for space applications. Compos. Commun., 44, pp. 101767.
2. Tan, B. et al., 2023. Ultrathin and flexible polyimide/$Ti_3C_2T_x$ MXenes composite films for electromagnetic interference shielding with harsh environment tolerance. *J. Mater. Chem. C*, 12(1), pp. 276–285.
3. United States of America Department of Defense, 1988. High-altitude electromagnetic pulse (HEMP) protection for ground-based C^4I facilities performing critical time-urgent missions part 1 fixed facilities [MIL-STD-188-125-1].
4. Kim, S. and Byun, J., 2024. Research trends in electromagnetic shielding using MXene-based composite materials. *J. Powder Mater.* 31(1), pp. 57–76.
5. Kumari, S. et al., 2023. Emerging two-dimensional materials for electromagnetic interference shielding application. *Int. J. Mol. Sci.* 24, pp. 12267.
6. Velayudhan, P. et al., 2024. Exploring the potential of sustainable biopolymers as a shield against electromagnetic radiations. *ACS Appl. Bio Mater.* 7(6), pp. 3568–3586.
7. Oliveira, F.M. and Gusmão, R., 2020. Recent advances in the electromagnetic interference shielding of 2D materials beyond graphene. *ACS Appl. Electron. Mater.* 2(10), pp. 3048–3071.
8. Charles, A.D.M. et al., 2020. Actuated dielectric-lossy screen for dynamically suppressing electromagnetic interference. *ACS Appl. Electron. Mater.* 2(12), pp. 3923–3935.
9. Zeranska, K. et al., 2022. Graphene-based thermoplastic composites as extremely broadband and frequency-dependent EMI absorbers for multifunctional applications. *ACS Appl. Electron. Mater.* 4(9), pp. 4463–4470.
10. Fan, Z., 2020. Lightweight three-dimensional cellular MXene film for superior energy storage and electromagnetic interference shielding. *ACS Appl Energy Mater* 3(9), pp. 8171–8178.

11. Saraswathy, R. et al., 2023. Carbon foams by solid state foaming of short vermicelli bonded by phenol-formaldehyde for thermal insulations and electromagnetic interference shielding. *ACS Appl. Eng. Mater.* 1(3), pp. 913–923.
12. Zhang, J. et al., 2020. Conductive skeleton-heterostructure composites based on chrome shavings for enhanced electromagnetic interference shielding. *ACS Appl. Mater. Interfaces*, 12(47), pp. 53076–53087.
13. Guan, Q.-F. et al., 2021. Sustainable double-network structural materials for electromagnetic shielding. *Nano Lett.* 21(6), pp. 2532–2537.
14. Sajid, H.M. et al., 2022. Design of multilayered 2D nanomaterial composite structures for EMI shielding analysis. *ACS Omega*, 7 (40), pp. 35586–35594.
15. Cheng, J., 2022. Recent advances in design strategies and multifunctionality of flexible electromagnetic interference shielding materials. *Nano-Micro Lett.* 14, pp. 80.
16. Isari A.A. et al., 2024. Structural design for EMI shielding: from underlying mechanisms to common pitfalls. *Adv. Mater.* 36(24), pp. 2310683.
17. Yost, D.C. et al., 2022. 2D monolayers for superior transparent electromagnetic interference shielding. *ACS Nano*, 16(6), pp. 9498–9509.
18. Zhao, Z. et al., 2024. Curved nanoflakes of alkane-grafted $Ti_3C_2T_x$ MXene thin films for enhanced terahertz electromagnetic interference shielding. *ACS Appl. Nano Mater.* 7(8), pp. 9609–9615.
19. Verma, R. et al., 2023. A review on MXene and its composites for electromagnetic interference (EMI) shielding applications. *Carbon* 208, pp. 170–190.
20. Chhetri, S. and Kuila, T., 2024. Polymer composites with 3D graphene architectures as high-performance EMI shielding materials: a review. *RSC Appl. Polym.* 2(4), pp. 507–533.
21. Sushmita, K. et al., 2022. Multi-layered composites using polyurethane-based foams and 3D-printed structures to curb electromagnetic pollution. *Mater. Adv.* 3(11), pp. 4578–4599.
22. Nguyen, Q.-D. and Choi, C.-G. (2024). Recent advances in multifunctional electromagnetic interference shielding materials. *Heliyon*, 10, pp. e31118.

CHAPTER 4

Electromagnetic Interference Shielding for Electronic Applications

Muhammad Irfan

4.1 INTRODUCTION

The connected time-varying magnetic and electric fields travelling with the speed of light, called electromagnetic (EM) waves, are everywhere around us. The entire spectrum of electromagnetic (EM) waves is utilised in various applications, from telecommunication using radio waves at the low-frequency end to cooking food with microwaves, Figure 4.1. In the mid-range of the spectrum, visible light allows us to see the colourful world around us, while higher-frequency X-rays are employed to detect fractures in bones. Advancements in mobile telecommunications, point-to-point communication, broadcasting, satellite technology and fibre optics over the past century have profoundly transformed human life. These innovations have revolutionised the way we connect, share information and access knowledge, enabling a more interconnected and rapidly evolving world.

The socio-economic benefits of the electromagnetic (EM) spectrum have driven electronic devices to become deeply embedded in today's human life. More recent developments of fifth-generation telecommunication technologies (5G networks) have brought the concepts of internet-of-things (IoTs) to realisation [1]. This widespread use of EM waves, however, has introduced a new kind of environmental pollution called electromagnetic interference (EMI). EMI pollution not only threatens socio-economic development but also human health [2]. Consequently, addressing EMI is of critical importance and remains a key focus of ongoing research.

4.2 ORIGINS OF EMI

Principally, any oscillating charge body is a source of EM waves [3]. As a result, oscillating charges in an antenna continuously emit electromagnetic (EM) waves, like an electric motor operating at high, varying voltages, which also generates EM emissions. These emitted waves can propagate through conduction, induction or radiation [3]. Conductive mode involves EM wave propagation through the conductor or power line while radiation

DOI: 10.1201/9781003512363-4

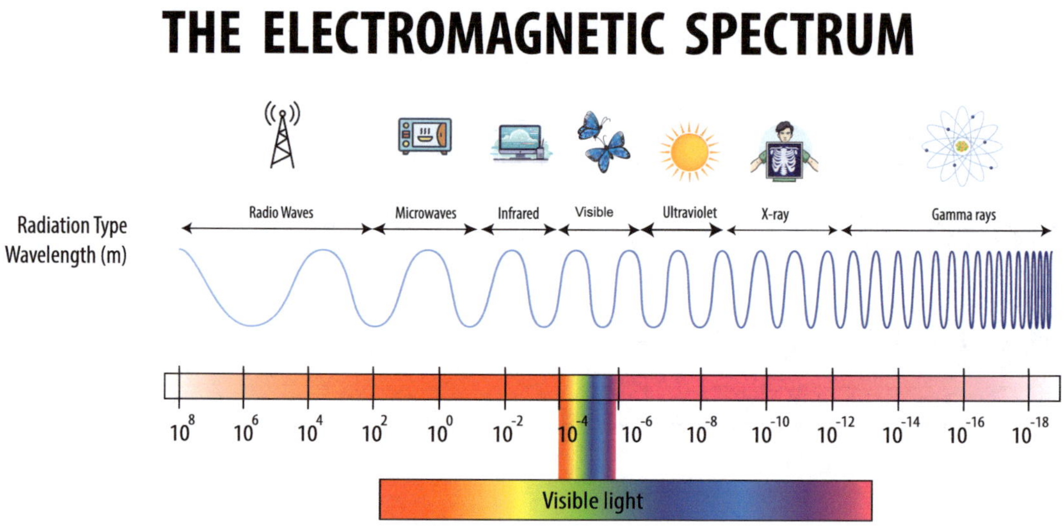

FIGURE 4.1 The electromagnetic spectrum.

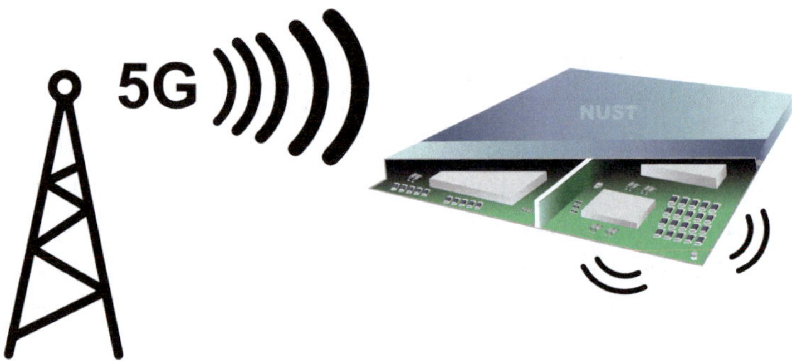

FIGURE 4.2 An electronic device under its internal and external EMI environment.

involves propagation through air. The induction mode requires no physical contact like the radiative one and the EM wave is propagated when a conductor is exposed to a varying magnetic field. A similar mechanism occurs in capacitive coupling. Therefore, any electronic device can experience EM waves generated within itself or from external devices or radiating sources [4, 5] (Figure 4.2).

4.2.1 External Sources

4.2.1.1 Natural Sources

External sources of EM waves can be natural or human-caused. Terrestrial natural sources (located in the atmosphere) emit radio waves of frequency less than 30 MHz [6] and these include thunderstorms. These radio signals are intermittent in nature and vary with the location and season on Earth. These sources not only can interfere through EM waves but can also damage electronic devices as well as humans, animals and structures due to lightning strikes.

Galactic or extraterrestrial natural sources of EM waves that extend beyond the Earth's atmosphere include the sun, cosmic background and planetary noise. The frequency of the EM waves emitted from these sources ranges from 20 MHz to 10 GHz [6].

4.2.1.2 Human-Caused Sources

The human-caused sources of EM radiation can be extraterrestrial like satellites or other space vehicles, and terrestrial like communication electronics, power generation or distribution, tools and machines, ignition systems and industrial and consumer equipment.

Communication sources: the communication electronic devices that emit EM waves can be classified as follows.

Broad band communication sources: these sources include electronic emitters used for radio or television communications. Amplitude modulation band 535–1605 kHz, frequency modulation band 88–108 MHz and television band is used in very high frequency (VHF: 54–216 MHz) and ultra-high frequency (UHF: 470–806 MHz).

Non-relay communication sources: these sources like land mobile transmitters emit EM waves in the frequency range of 14 kHz to 960 MHz.

Relay communication sources: satellite relay transmitters operate in the 399.9 MHz to 275 GHz range while microwave relays operate in the 2.1–12.2 GHz frequency range.

Navigation (non-radar) sources:

Source	**Frequency Range**
Land radio navigation	1605–1800 kHz
Tactical air navigation (TACAN)	960–1215 MHz
instrument landing systems (ILS) Localiser	08–112 MHz
Glide path	328.6–335.4 MHz
Altimeters	4.2–4.4 GHz
Direction finders	405–415 kHz
Maritime radio navigation	285 kHz to 5.65 GHz

Radars: these are the largest producers of EMI in terms of very high frequencies (225 MHz to several GHz), short pulses of high-energy and greater bandwidth and harmonic radiation. These are used in weather stations, harbours, airports, highways, search and rescue operations and defence purposes.

Restricted sources of EM waves: restricted sources emit EM waves that are used for applications other than communications. These include arc welding machines, wireless microphones, induction heaters, eddy current testers etc.

Incidental sources of EM waves: these are sources that incidentally emit radio frequency waves. These sources include power cables, lamps, automotive ignitors, electric motors and switching devices etc.

4.2.2 Internal Sources

Increasing density of electronic components is continuously rising due to the miniaturisation of these devices, and hence large power input in these devices results in large production of EM waves. Figure 4.3 shows various internal sources of EM radiation including integrated circuits (ICs), clock oscillators, switch mode power converters and input/output device cables [7]. The emission potential of any source is determined by its energy spectrum and the ability of its flux to couple with other components of the device. Digital circuits are characterised through their broad spectral range while a clock oscillator demonstrates a comb-like spectrum. Data transmitting signals have continuous energy distribution characteristics. ICs, switching devices and interconnects are the major sources of noise in an electronic system [7]. In this multisource and complex EM wave environment, effective mitigation of these noise sources is required to avoid malfunctioning of signals and the device itself.

High-performance digital circuits require very fast time-varying currents, and these fast-varying electromagnetic fields result in radiated noise which propagates through the power plane of the PCB, Figure 4.4. During the simultaneous operation of multiple outputs of a PCB, simultaneous switching noise (SSN) is induced in the power plane [8]. The following expression is used to quantify the magnitude of switching noise voltage (V_n):

$$V_n = NL_{\mathrm{eq}} \frac{di}{dt} \tag{4.1}$$

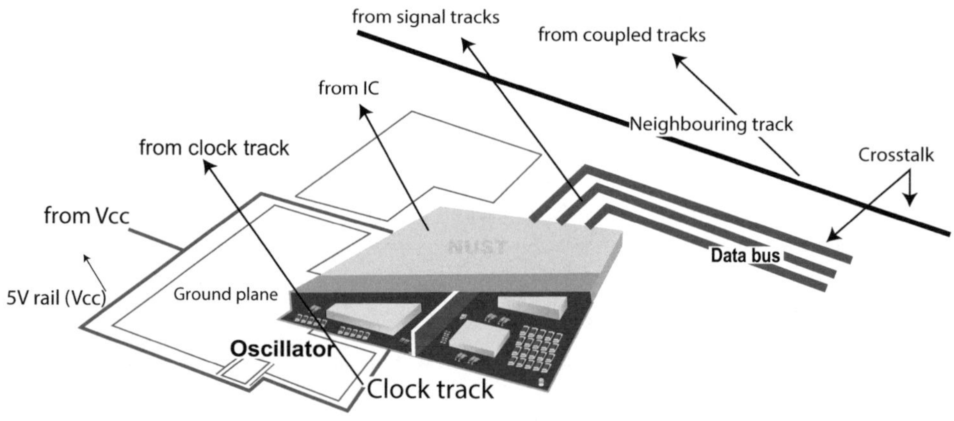

FIGURE 4.3 Internal sources of EM of an electronic device.

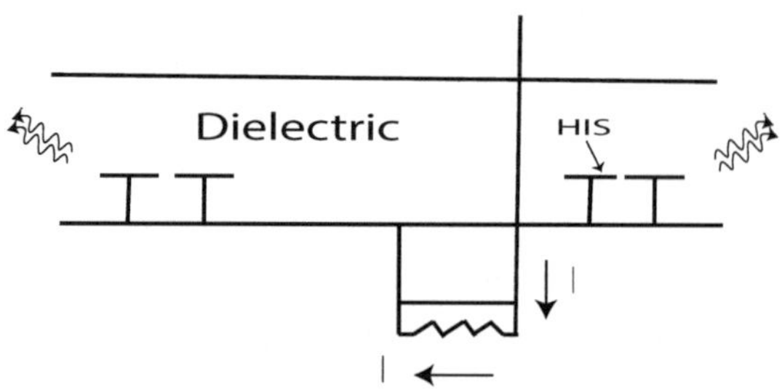

FIGURE 4.4 A multilayer PCB creating radiated noise through high-speed switching.

N is number of output switches (drivers), L_{eq}, is equivalent inductance when I current is passing through each switch during the simultaneous operation of multiple switches. This inductively coupled noise causes fluctuations in power supply voltage which consequently affect the output of different drivers and other signals, resulting in false switching or malfunctioning [8]. Moreover, higher clocking speeds in modern circuits also lead to switching noise. These EM noises propagate through the parallel plates and hence induce noise in power buses (vias) and exit from the board edges.

Faster and high-powered electronic circuitry is also a major source of EM waves [8]. Digital and electronic circuits emit at trace coupling, chip packages, heat sinks, power cables attached to the board, power buses and any low-impedance current pathway. Modern digital circuits consume high power for high-speed processing, and since power lines are embedded in multilayer boards, the high-speed charges moving in these power planes emit radiation like an antenna. Therefore, electronic devices themselves are a major source of EM pollution.

4.3 EMI MECHANISM

EMI is a complex process that occurs at various levels in an electronic device from chassis to board to component level to finally device level [8]. Due to involvement of multiple internal and external sources, and to reduce the emission from the electronic device itself, green EMI shielding of an electronic device is of great interest. Green EMI shielding ensures no or very little harm to device performance and to the external environment or devices [9]. The green EMI shielding is characterised using the "green index, g_s," proposed by Cao et al. [10], given as follows:

$$g_s = \frac{1}{S_{11}^2} - \frac{S_{21}^2}{S_{11}^2} - 1 \tag{4.2}$$

Here, S_{11} and S_{21} are reflection and transmission parameters, Figure 4.5. The green EMI index is a function of frequency (f) of the EM wave, temperature (T) and sample width (d):

$$g_s(f,T,d) = \frac{1}{\left|S_{11}(f,T,d)\right|^2} - \frac{\left|S_{21}(f,T,d)\right|^2}{\left|S_{11}(f,T,d)\right|^2} - 1 \tag{4.3}$$

A green EMI shielding material should have a green index value greater than 1, shielding effectiveness (SE) of more than 30 dB and should minimise reflection and maximise absorption of EM waves [10]. Therefore, higher impedance matching along with high SE and absorption losses leads to green EMI shielding.

Reflection losses are due to the impedance differences at the air–material interface, Figure 4.5, and these losses (SE_R) for a conductive material can be estimated using simplified Fresnel's equation as follows

$$SE_R(\text{dB}) = 20\log\frac{(\eta+\eta_0)^2}{4\eta\eta_0} = 39.5 + 10\log\frac{\sigma}{2\pi f\mu} \tag{4.4}$$

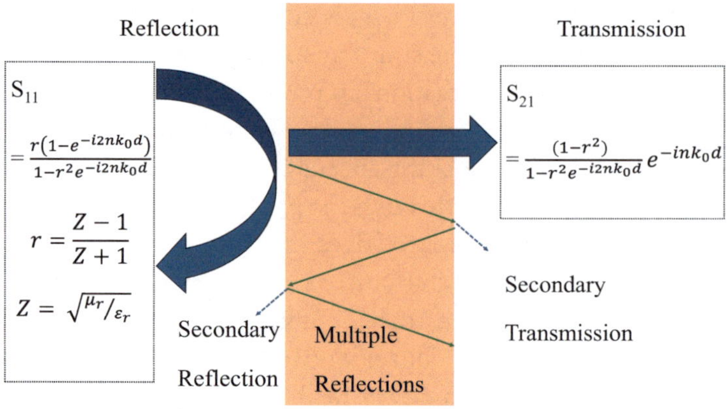

FIGURE 4.5 Mechanisms of EM wave attenuation. Here, Z, μ_r, ε_r k_0, n, and d, are the impedance mismatch, relative permeability, relative permittivity, vacuum wave number, refractive index and thickness of material.

η and η_0 represent impedances of shielding material and free space and σ is the conductivity of the shielding material. Equation 4.4 shows that by increasing conductivity of the sample, the reflection losses increase and these losses also depend on frequency (f) of the EM wave and the permeability (μ) of the material.

Once the EM wave is in the material, Figure 4.5, the absorption attenuation (SE_A) is governed by the combined effect of conduction, dielectric and magnetic losses and is expressed as follows:

$$SE_A\left(\text{dB}\right)=20\log e^{\alpha d}=20\left(\frac{d}{\delta}\right)\log e=8.68\left(\frac{d}{\delta}\right)=8.68d\sqrt{\pi f\mu\sigma} \tag{4.5}$$

Here, α and δ are attenuation coefficient and skin depth or optical length, respectively. Skin depth represents the depth of attenuation of EM into the material where the intensity of electric field is $1/e$ of the impinged EM wave.

The internal multiple reflection losses (SE_M) are due to metal–air interfaces, Figure 4.5, and are given by the following expression:

$$SE_M\left(\text{dB}\right)=20\log\left(1-e^{2\alpha d}\right)=20\log\left(1-d^{2d/\alpha}\right) \tag{4.6}$$

SE_M losses become significant for a sample with thickness greater than the skin depth or total attenuation losses (SE_T) of less than 15 dB [10]. Moreover, multiple interfaces within the material thickness (d) can further lead to internal scattering and absorption losses if the impedance differences at these internal interfaces are significant [11]. These internal scatterings therefore are important for a material to act as a green shielding material. The total shielding effectiveness (SE_T) is given by the following relation:

$$SE_T=SE_R+SE_A=-10\log\left(1-R\right)-10\log\left(\frac{T}{1-R}\right)=-10\log T \tag{4.7}$$

where

$$A + R + T = 1 \tag{4.8}$$

Here, A, $R\left(=|S_{11}|^2\right)$ and $T\left(=|S_{21}|^2\right)$ represent absorption, reflection and transmission energy coefficients, respectively. These coefficients depict the actual energy losses of the EM wave while SE_A (ratio of the energy absorbed to energy entering the material), SE_R (reflection capacity with respect to energy of the incident EM wave) and SE_T show the ability to absorb, reflect and totally shield the EM waves. Therefore, these two sets of indicators are different since a material with high SE_A does not necessarily absorb most of the energy of the incident wave [11].

4.4 EMI SHIELDING MATERIALS FOR ELECTRONIC APPLICATIONS

Various materials have been used for EMI shielding of electronic devices. Selection of the materials depends on the application and the required SE. Other factors also greatly affect the selection of the materials, and these include ease of fabrication, environmental and thermal stability, mechanical strength, stretchability, transparency and cost. In this section, conventional and smart materials will be considered. Also, materials with flexibility and transparency are discussed due to their use in future wearable and flexible electronic devices.

4.4.1 Conventional Materials

Conductive metallic materials are commonly used as shielding materials because they offer higher shielding efficiency, thermal conductivity and high strength and scratch resistance [12].

Conventional metallic EMI shields are, however, increasingly becoming unfit for future miniaturised ICs due to their poor corrosion resistance, high densities and limited flexibility [9, 13–15]. This miniaturisation of electronics is essential for wearable electronics and sensing applications. Moreover, the growth of these miniaturised electronics has accelerated due to increasing demand of interactions between different machines through electromagnetic waves, the IoTs. Furthermore, wearable electronics pose new challenges of stretchability and transparency which are difficult to address with conventional EMI shielding materials and technologies [16].

4.4.2 Electromagnetic Bandgap (EBG) Structures: Artificial Smart Materials

Periodic patterns that mitigate the propagation of EM waves of specific frequency for all states of polarisation and some or all incident angles are called EBGs [8, 17]. EBGs belong to a class of materials called metamaterials and these left-handed materials show negative index of refraction, compared to the positive refractive index governed by Snell's law in normal materials, Figure 4.6. The shift from radio wave to microwave communications requires more specialised materials like EBGs. These materials are used for antenna applications but also used in PCBs to control the propagation of EM waves. EBGs can be fabricated or designed in 1D, 2D or 3D with 3D designs being more suitable to tackle EM waves

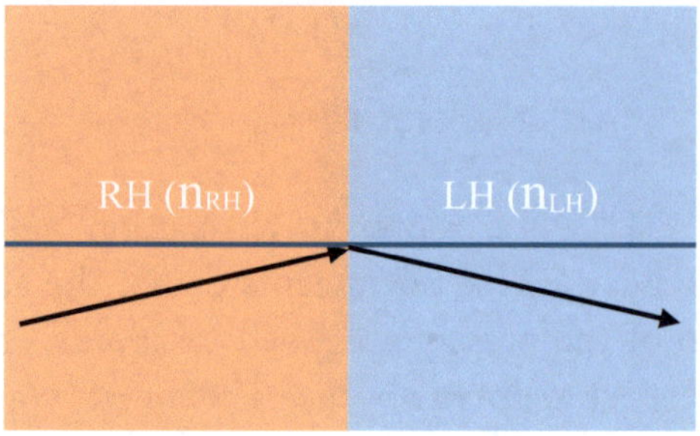

FIGURE 4.6 Demonstration of negative refraction.

propagating in all directions. EM waves with millimetre, micrometre and sub-micrometre ranges can be managed through EBGs.

When EBGs mitigate the propagation of EM waves of certain frequencies and at some angles, the frequencies are termed partial bandgap. Complete bandgap or global bandgap represents a frequency range of EM waves where EBGs inhibit the propagation of waves travelling in all directions. A single periodic structure can handle multiple bandgaps [18]. The bandgap formation is dictated through micro and macroscale resonances of the periodic pattern. Macroscopic resonance, or the Bragg resonance (also called lattice resonance), is controlled through periodicity of the pattern while the nature of each element controls the Mie resonance [19–21] or microscale resonance. The overlap of both resonance modes leads to maximum frequency bandwidth. One resonance phenomenon can dominate over the other based on structural characteristics and the state of EM polarisation, hence leading to blockage of certain frequencies. Hence EBG can be designed to either reflect the EM waves acting as a mirror or as a transparent medium [18], Figure 4.7.

A metallo-dielectric EBG (MDEBG) consists of an array of metallic patches of EBG embedded in a dielectric material which can effectively prevent a band of EM frequencies from propagation. However, the EM wave can propagate in space between the metal patches and the ground plane. To prevent this propagation of EM waves, metallic pins are inserted between the patches and the ground plane. The embedded EBG structures are expected to mitigate the EM waves in the 0.5–20 GHz range [18]. However, for high frequencies and larger bandwidth, more complex structures are needed. Figure 4.8 shows the efficacy of different embedded structures for EM mitigation. Therefore, for wideband radiation, a combination of different methods is employed to control the EM propagation.

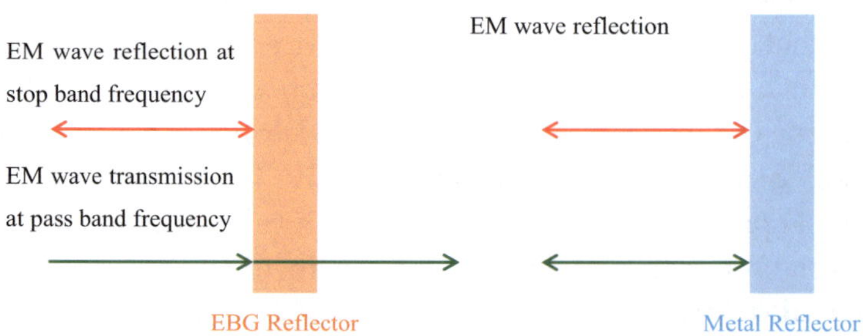

FIGURE 4.7 Comparison of EBG reflector with the conventional metal reflector.

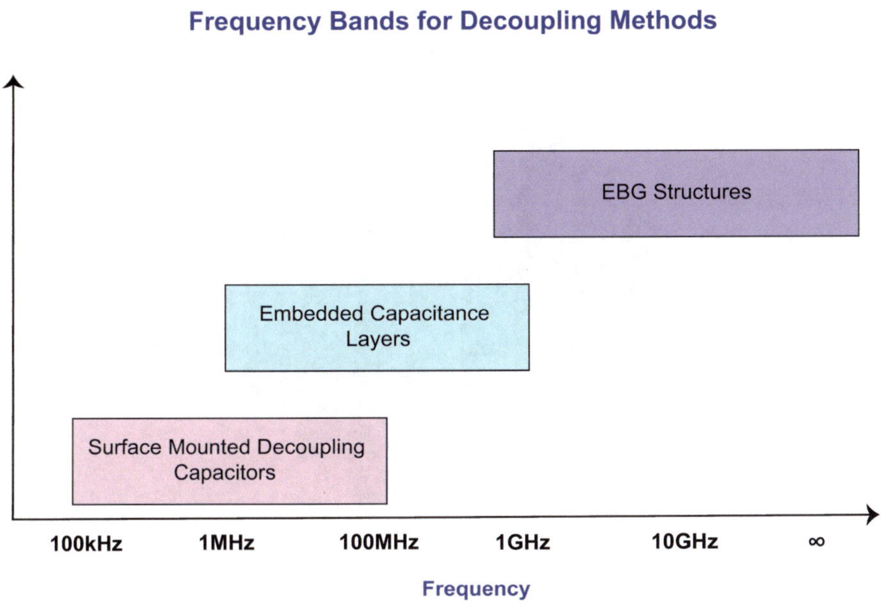

FIGURE 4.8 Efficacy of different EM reduction structures.

4.4.2.1 Types of EBGs

4.4.2.1.1 Defected Ground Structures (DGS)

These are easy to fabricate EBGs and are etched in the ground plane of a PCB or other devices which can suppress the EM waves. As the name suggests, defected ground means there is a resonance gap or slot etched in the ground plane of the electronic device which is directly below the signal transmission lines. Due to simplicity compared to photonic bandgap (PBG) type EBGs, DGS are being researched as microwave filters, oscillators and amplifiers [18]. However, the slot of DGS can also act as a radiation source.

4.4.2.1.2 Photonic Band-Gap Structures (PBGs)

Just like semiconductor devices operating on electrons, PBGs (photonic crystals) are periodic structures that operate on microwaves. These structures have a bandgap that inhibits the propagation of EM of specific frequencies lying in that band. PBGs work on the Bragg destructive interference phenomenon. There are various configurations of PBG available including 1D, 2D and 3D. Three D configurations are particularly important for EM waves travelling in multiple directions and a variety of 3D design elements are available.

PBG structures can be fabricated from dielectric or metallic materials. Metallic materials, however, play a significant role in EM wave energy losses at high frequencies. Dielectric constructions are important for frequency ranges from microwave to ultraviolet. There are various fabrication methods available for the fabrication of PBG structures. For 3D constructions, a 2D layer of defined thickness is sandwiched between two perfect electric conductors (PECs). This arrangement will confine the EM waves perpendicular to the periodic structures of PBG. For confinement of EM waves with the 2D PBG structure, the periodic structures are attached to a material of different dielectric constants [18].

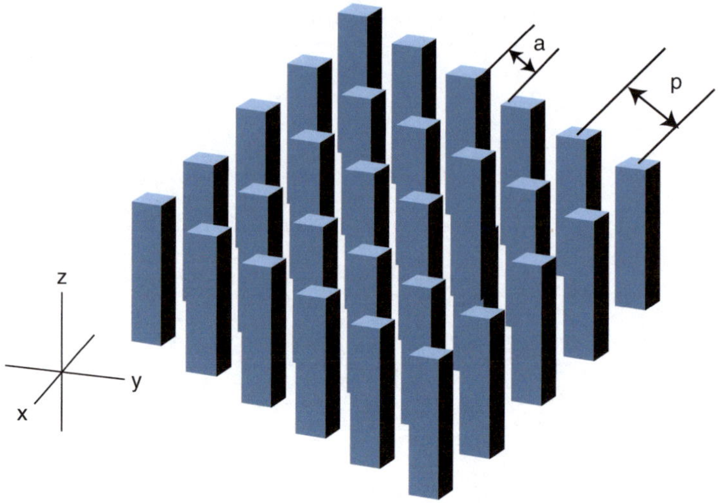

FIGURE 4.9 A 2D structure of PBG in air. Here a is patch size and P is periodicity of photonic crystal.

Like a metal or other crystal, the unit cell of a photonic crystal also has a specific lattice spacing (p), crystal geometry (shape or thickness etc.) and material properties such as magnetic, dielectric, conductive etc., Figure 4.9a. This unit cell controls the properties of the whole photonic crystal. When the lattice spacing a of this crystal becomes equal to the wavelength of EM waves propagating through the photonic crystal, the following expression approximates the bandgap of EBG,

$$\lambda_r = \frac{c}{fr\sqrt{\varepsilon_r}} = 2a \tag{4.9}$$

The energy of the propagating EM waves in this bandgap is attenuated and is given by S_{21} transmission loss usually measured using two-port network analysers, as illustrated in Figure 4.10.

The central wavelength or bandgap of a photonic crystal is based on the periodicity of the crystal and the ratio of the permittivity, ε_r (or refractive indices) of surrounding material and the motif. The periodic structure is characterised by (a) $\lambda_\tau = 2a$, (b) shape of the patches or motif and (c) filling factor, that is, ratio a / P (size of patch/periodicity). The filling factor is between 0.65 and 0.75 for 2D structures while 0.9–0.95 for 3D ones. Wide bandgaps are achieved using high dielectric material with air gap as motif giving rise to a ratio of ε_r 10/1, while a minimum of 2/1 is required to operate any EBG.

4.4.2.1.3 High Impedance Surfaces/Structures (HIS)

High impedance structures (HIS) are metallic structures characterised through their high surface impedance. HIS conduct DC currents while stopping the propagation of AC waves of certain bandwidth. HIS are 2D EBG with mushroom-like structures protruding from the surface, Figure 4.11. The properties of HIS are controlled using the equivalent LC

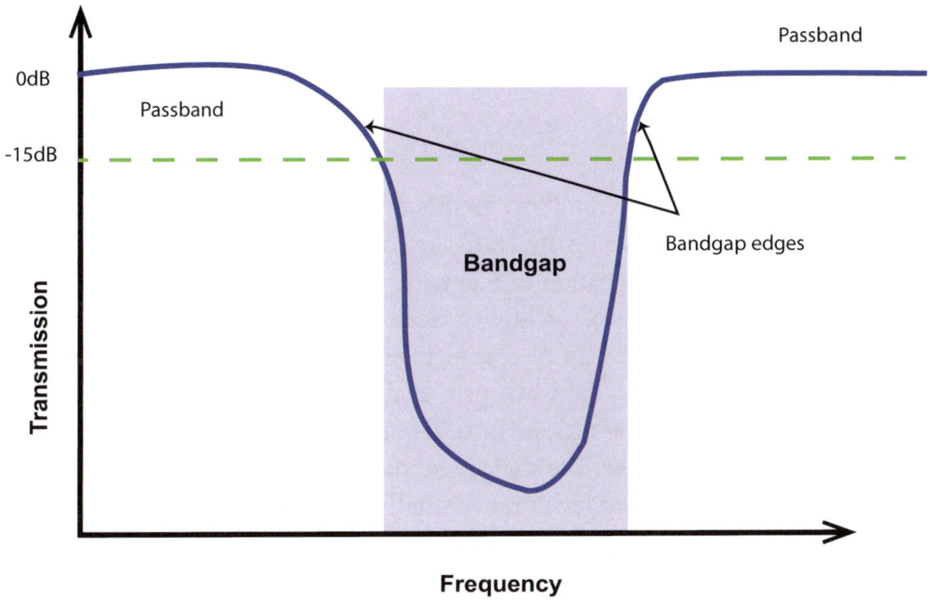

FIGURE 4.10 Illustration of stopband produced by a photonic crystal or EBG structure.

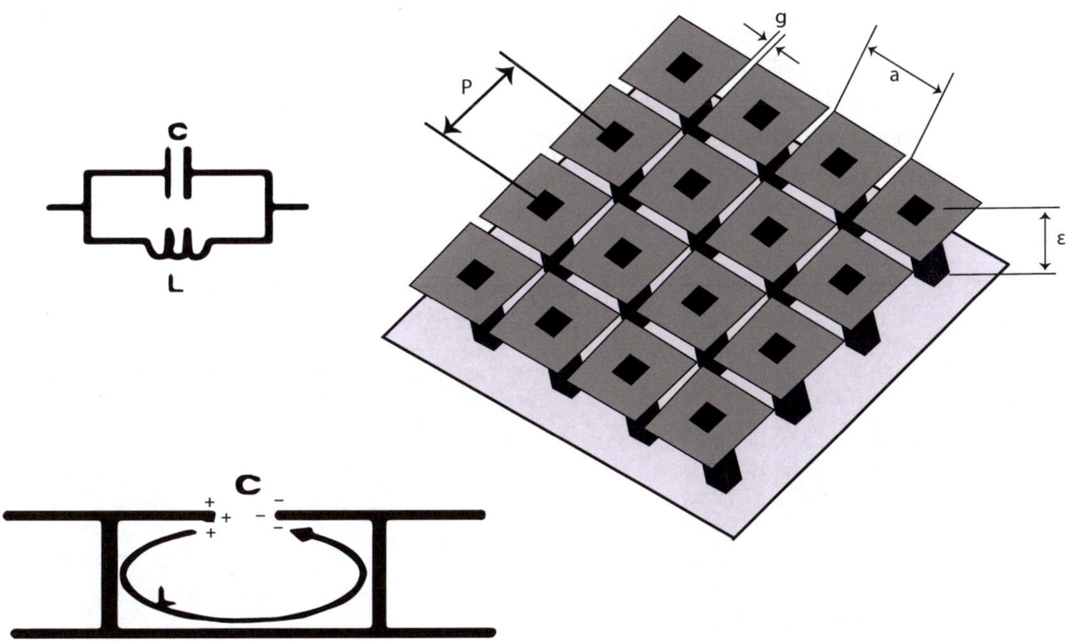

FIGURE 4.11 A high impedance surface characterised by patch size (a), gap size (g), periodicity P and dielectric constant ε_r (right image). The equivalent LC circuit is given at the left side.

circuit shown in Figure 4.11. All the metallic patches are folded into lumped elements containing capacitors and inductors. The closed current loops formed by metal patches that are connected to the ground through conducting poles result in the inductive component of the equivalent LC while proximity of metal patches, that is, gap between the metals, results in the capacitive part of this circuit, Figure 4.11. The impedance of HIS is given by the following equation:

$$Z_s = \frac{j\omega L}{1-\omega^2 LC} \tag{4.10}$$

At frequencies lower than the resonance frequency of *LC* circuit ($\omega < \omega_0$), the impedance is inductance dominated and as a result, transverse magnetic (TM) waves propagate along the surface of HIS. Due to propagation of magnetic waves, these HIS are also called artificial magnetic conductors (AMCs). At higher frequencies, the capacitance component of the *LC* circuit prevails and hence transverse electric (TE) waves travel along the HIS surface [22]. At frequencies $\omega = \omega_0$, the wave propagation along the HIS is diminished and there is no current flowing along this surface. The EM waves are reflected off the ground plane and when these waves meet the dielectric interface, they are again reflected. These EM waves propagating in zig-zag fashion eventually reach the end of HIS structures and diffract from the edges resulting in end-fire radiation [18]. These end-fired radiations, though attenuated during propagation, may function as a radiation source for nearby devices if not suppressed by HIS.

Shahrooz et al. reported the mitigation of switching noise produced in a fast speed PCB by parallel plate bus structures [6]. HIS EBGs were employed on a 6.5 cm × 10 cm PCB as a noise reducer, Figure 4.12. A ribbon of four rows of HIS structures were fabricated around the PCB board and these HIS structures was attached to one of the power planes. The source of noise was located at the centre of the board. The simulation results showed that the bandwidth of prohibited frequencies decreased from 6 to 0.4 GHz when patch size was varied from 4 to 40 mm while keeping via diameter (0.8 mm), gap size (0.4 mm), via length

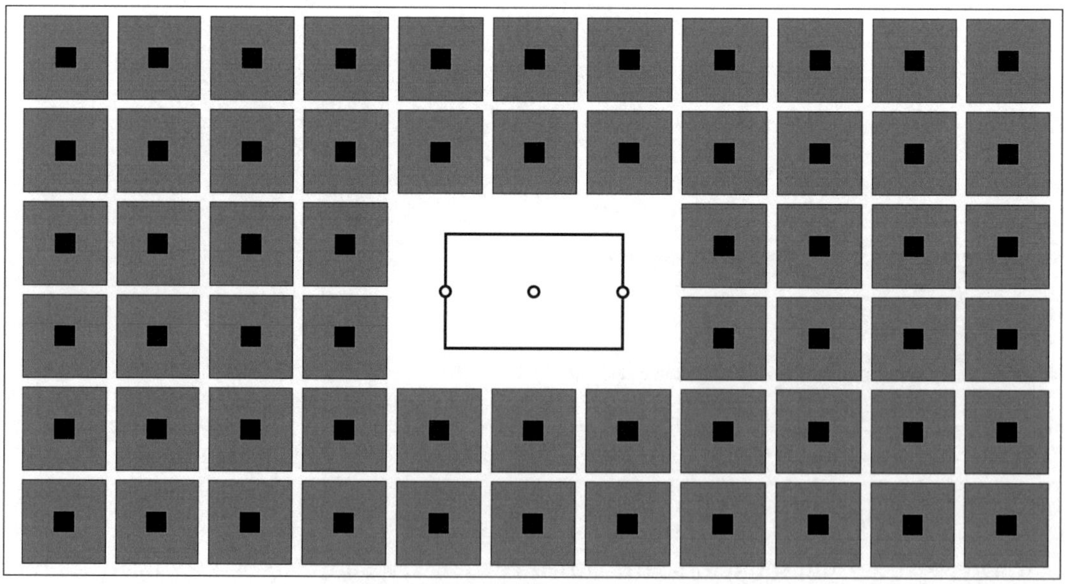

FIGURE 4.12 A 6.5 cm × 10 cm PCB board and a source of EMI located at the centre of board is surrounded with a ribbon of four rows of HIS structures.

(1.54 mm), distance between the parallel bus planes (3.08 mm) and permittivity constant. The experimental results with 5 mm patch size revealed suppression of EM waves in the 4–10 GHz range with average dB of 30 (in terms of S_{21} transmission parameter) and loss was higher than the predicted values at high frequencies. When a ribbon of 5 mm patch sizes was cascaded with 10 mm patch sizes at the outer boundaries of the board with a 20.6 cm × 16.6 cm PCB, additional forbidden bandgap of 2–4 GHz was achieved. Therefore, wideband EMI suppression was established with cascaded HIS structures. Due to unreasonably large patch size requirements for 0.5 MHz to 2 GHz, more complex designs of HIS or combinations of different mitigation methods are required (also see Figure 4.8). Moreover, miniaturisation of electronic devices requires miniaturised HIS structures to accommodate the space constraints. Also, cavity type HIS structures have been reported to mitigate the switching noise in PCBs [22].

4.4.3 Smart EMI Shielding Materials

Smart EMI shielding materials enable the tuning of shielding characteristics of a device under the mechanical (like tensile or compressive stress), chemical environment, temperature, light, phase change and crossover angle stimuli [23].

4.4.3.1 Mechanically Adaptive EMI Shielding Materials

Application of a force to a conductive material leads to mechanical strains in the materials which result in changes in electrical resistance of the material. Mechanical strain (i.e., change in length per unit original length, $\Delta L/L_0$) induced in a material is related to change in resistance (ΔR) as follows

$$\Delta R = 2R_0\left(\frac{\Delta L}{L_0}\right) \tag{4.11}$$

where R_0 is the resistance of the unstrained sample. The conductivity of the sample increases under compressive strains while the reverse happens under tensile strains. These conductivity changes hence cause the changes in EMI response of the material [23]. Higher conductivity under compressive strains demonstrates higher SE though may be counteracted by reduced sample thickness under the influences of these strains, Figure 4.13. Therefore, EMI shielding of the material can be tuned by the magnitude of the

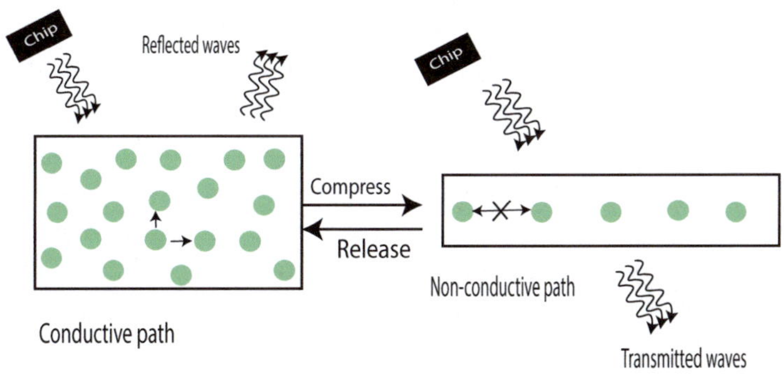

FIGURE 4.13 EMI shielding tuning mechanisms under mechanical stimuli.

applied stress. EMI SE tends to decrease under the tensile loads due to increased separation between the charge transport species, Figure 4.13 [23].

Cellular structures have been demonstrated to possess tunable EMI effectiveness under mechanical stimulus [24, 25]. MXene loaded polyaniline foam prepared using CO_2 at subcritical temperatures and these foams were encapsulated into polydimethylsiloxane PDMS [24]. These cellular structures showed SE from 24 to 40 dB with compression ratios 0.2–0.31. The shielding mechanism was based on absorption of EM waves manifesting their capability as a green EMI shielding material. Silver nanoparticle loaded ethylene–vinyl acetate copolymer based foam material with polypyrrole modifications has also demonstrated self-fixable shielding material [25]. Their SE was adjustable from 6 to 62 dB when compressive strains were reduced from 50% to 0%. These cellular structures have not only demonstrated a maximum SE of 108 dB in X-band region but also were hydrophobic in nature, a property required for electronic applications.

4.4.3.2 Shape Memory EMI Shielding Materials

Intelligent EMI shielding requires the material that can adjust its temporary shape to the original one under the influence of external stimulus such as temperature, light, pH, chemical reagent such as enzymes or simple water. Shape memory characteristics in polymers, for example, depend on hard and soft segments present in the material [26]. The soft segment of shape memory polymer controls the shape switching character while the hard segment is reinforced within the matrix of the soft segment. In contrast to shape changing materials, the shape switching behaviour of the shape memory polymers can be programmed. The mechanism of shape memory effect is explained in Figure 4.14, under the most common stimulus, temperature [27].

The dimensional changes during shape memory effect result in changes in SE. For example, if the thickness of the sample is reduced under the temporary stage of shape memory effect, the SE of the material is reduced [28]. Also, the reduced thickness results in multiple internal reflections and hence transmission of EM waves is more compared to their loss in absorption. However, when the material is switched back to its original dimensions, it behaves more like a lossy material or green shielding material. Examples of shape memory materials are melamine foam (MA)-coated MXene/silver nanowires/poly(ethylene glycol) composites, D-gluconic acid-chitosan/single-walled carbon nanotubes and crystalline paraffin layers on polyurethane foams (with tunable SE of 18–30 dB) and carbon foam coated with trans 1,4-polyisoprene (TPI) [23]. The tunability of SE of these shape memory materials depends on the shape fixity ratio ($R_f = 100\% \times \varepsilon/\varepsilon_{\text{load}}$, ε is strain) and shape recovery ratio ($R_r = 100\% \times \left(\varepsilon - \varepsilon_{\text{rec}}\right)/\varepsilon$). Higher values of R_f and R_r lead to better shape memory characteristics and hence better tunability of the EMI properties. The dimensional changes in shape memory materials during adjustment of SE, however, may induce stresses particularly in miniaturised devices. These dimensional changes limit the applicability of these materials in electronic devices due to reduced device reliability. The miniaturised devices need adaptive shielding characteristics with minimal large-scale deformation.

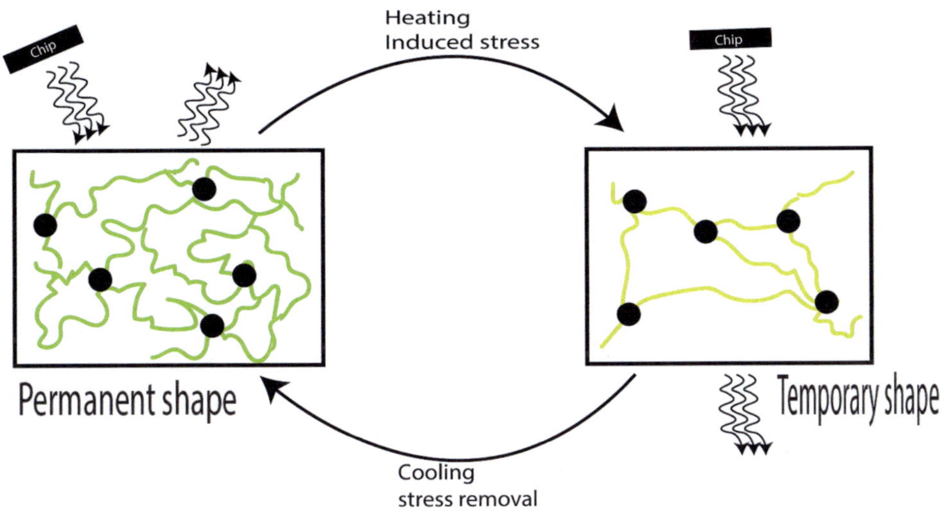

FIGURE 4.14 Mechanism of shape memory effect in polymers under thermal stimulus. At low temperature, the rigid chains (dark green lines) are connected using network points (black dots). When the temperature increases above the glass transition temperature (T_g), the chains become flexible. At this stage, if a load is applied at this elevated temperature (deformation temperature T_d), the network connections are changed. When the sample is cooled below T_g, under this load, the new network structure is preserved in temporary shape (light green lines). Now the load is removed. Later, if the sample is reheated under service conditions, the network structure is restored to the permanent initial position. Multiple programming can be performed using the same sequence of steps.

4.4.3.3 Phase Change EMI Shielding Materials

Phase change materials exhibit adaptive EMI shielding character to thermal or mechanical stimulus. Thermal excitation of phase change materials causes lattice distortion in vanadium dioxide (VO_2), for example, and hence changes its electron density, Figure 4.15. These lattice strains lead to transition of VO_2 from insulator to conductive state on heating (>68 °C) and vice versa upon cooling. Reversible phase change from insulating monoclinic to conductive tetragonal, the conductivity of the VO_2/cellulose nanofibres (CNF) composites varied between 0.3 and 445 S/m with adaptive shielding of 19–53 dB in the frequency range of 8.2–12.4 GHz [29]. At higher temperatures, poor impedance match between the impinging EM wave and the conductive phase of VO_2 resulted in reflection loss. Moreover, the free electrons in conductive tetragonal VO_2 further led to reflection losses. At elevated temperatures, the imaginary part of permittivity also increased due to increased electrical conductivity which in turn also increased the absorption loss in VO_2/CNF composites. Other phase change materials such as temperature responsive microspheres have also been reported which result in rearrangement of conductive charge carriers within, for example, a polymeric matrix and hence changes in conductivity and adaptive EMI under thermal excitation [23].

4.4.3.4 Chemically Tunable and Cross Angle-Responsive Materials

The EMI shielding performance of a material can also be tuned through material–chemical reagent interactions. A polar molecule such as water can change the conductivity

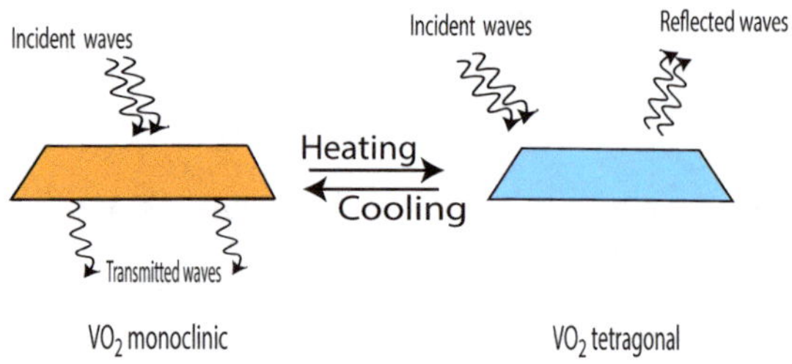

FIGURE 4.15 Illustration of the EMI shielding mechanism in phase change materials using VO_2 as an example. Phase change in VO_2 (from monoclinic to tetragonal) resulted in changes in conductivity and hence SE.

of cellular structure which in turn changes the EMI character of the material [30]. This change in conductivity due to strong polar molecules results in enhanced impedance mismatch between the material and EM wave and consequently more attenuation of EM waves due to reflection. Also, absorption losses occur within the material due to polarisation losses at the interface between polar water molecules and the cell walls. Therefore, changes in the dielectric properties of the material due to polar species result in self-adaptive EMI SE of the material that can be tuned through humidity control.

In conducting polymers, changes in oxidation states or doping due to the interaction between the polymer and the chemical reagent cause variations in conductivities of the polymer making them chemically responsive materials. During protonation, hydrogen ions (H+) are transported, for example, along the polyaniline chain which interacts with nitrogen atoms in the chain to produce polarons. These polarons increase the SE of the material. Under high pH values, the SE values are reduced due to changes in doping concentrations. Therefore, EMI shielding character of the material adapts to the environment. Li et al. reported that the PEBA/CNS cellular structure can regulate the SE between 17 and 38 dB by controlling the moisture contents [30].

4.4.4 Flexible and Transparent EMI Shielding Materials

The miniaturisation of electronic devices and the development of 4G and more recently 5G communications networks has increased the popularisation of portable devices such as laptops, smartphones and flexible wearable devices and sensors [9]. Portability of these devices not only demands light weightiness but also high flexibility along with high corrosion resistance and EMI efficiencies. Furthermore, these portable devices are also greatly slim which poses additional requirements of complex permittivity, permeability and electrical conductivity for the reliable performance of the device under multisource EMI environment. Therefore, advanced EMI shielding materials to meet these requirements have been extensively researched. Among these materials carbon-based materials such as graphene, carbon nanotubes, carbon dots and carbon fibres with polymer matrices are widely studied. Moreover, cellular structures, inherently conductive polymers, aerogels and cellulose also offer flexible substrates for EMI shielding [23].

Moreover, future wearable electronic devices also require transparency along with flexibility and typically conducting materials like indium oxide glasses with inherent brittle character are not suitable for flexible devices. Therefore, transparent stretchable polymers such as polydimethylsiloxane loaded with conductive fillers such as metal nanowires, carbon nanotubes and MXene are being explored for EMI shielding applications [16, 31].

4.5 MULTILEVEL EMI SHIELDING IN AN ELECTRONIC DEVICE

An electronic system is usually comprised of PCBs, system-in-chip (SiP) packages (SiP contains multiple ICs mounted on one package), interconnects and input/output (I/O) device cables, Figure 4.16. The reliable working of the device requires effective shielding of all its components. The selection of shielding module depends on the application, Table 4.1.

4.5.1 Package Level Shielding

The EMI shielding of electronic devices starts at the package level [32–34]. Various IC packages like SiP are essential for miniaturisation of electronic devices, Figure 4.16. Various components of SiP package are not only protected from the EM wave interference from

TABLE 4.1 Shielding Effectiveness Required For Various Electronic Devices [12]

Electronic Device	SE (dB)
Notebook computer	15–20
Desktop computer	15–20
Mobile phone	70–90
Cable tap	70–90
Workstation/server	30–40

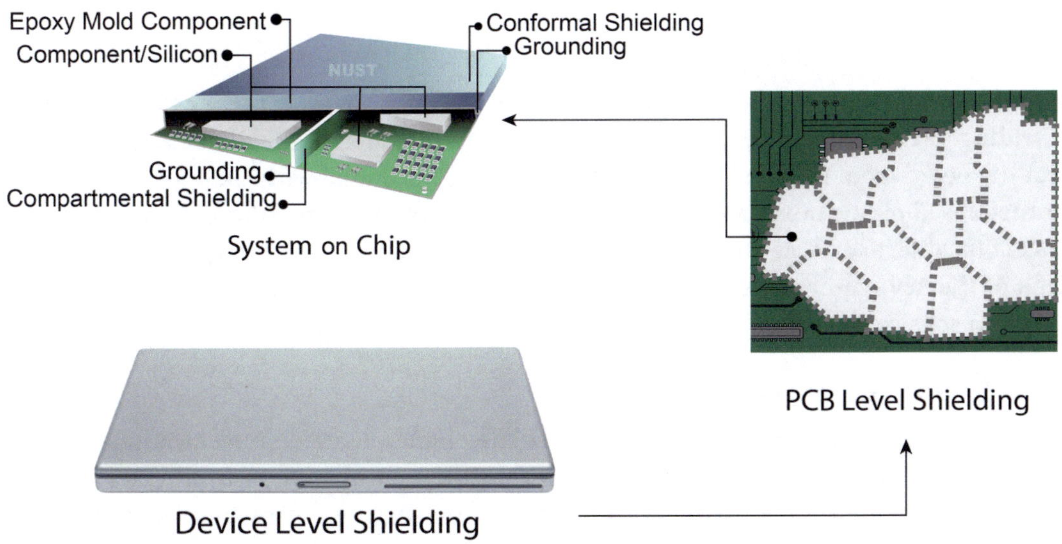

FIGURE 4.16 Multilevel EMI shielding of an electronic system/device.

external sources, but also isolation of various components of SiP is required. Conventionally, electronic packages are enclosed with metal lids or caps also called Faraday cages. These lids or cans are usually made of high conductivity material stainless steel, white copper, aluminium, nickel, silver and mumetal (nickel-iron alloy) and have very high shielding capability (100 dB with 0.1–1.6 mm thickness [12]). These metal enclosures are required to surround all the electronics and occupy a large space. This design approach, which requires large space, however, limits the miniaturisation and integration issues in multi-scenario EMI shielding. Additionally, voids between different EMI shielding elements pose problems related to the thermal management of the whole packaging [15]. Furthermore, the repair of electronic components becomes difficult with these metal cans since these are usually welded or soldered with PCB boards.

Recently, an advanced approach of "conformal shielding" is gaining importance. In this method, the packaging material is completely integrated with the shielding material. This integration of both materials minimises the space requirement of the whole electronic framework [13, 15, 35, 36]. The structure of the EMI shielding module is customised according to the electronic element [13, 35, 36]. Advanced fabrication techniques like layer-by-layer 3D printing, sputter deposition, electrospray and electrophoretic deposition, and spray deposition are utilised to integrate the EMI layer with the packaging material [15, 37–39]. 3D printing technology seems more promising for conformal coating of packaging material for high thermal management along with SE [15]. The print architecture can be designed for effective heat dissipation and can conform to the complex structure of miniaturised electronic devices. These conformal 3D printed structures also occupy less space and hence offer design flexibility.

For the isolation of components of SiP, laser beams are used to create the trenches between the components and subsequently filled with several types of materials to create fences. This compartmentalisation of SiP reduces cross talks between the components mounted on SiP.

4.5.2 PCB-Level Shielding

Board-level shielding involves shielding a part of the PCB board or the whole board. EMI shielding filters comprising inductors and capacitors attached to the PCB are used to protect the board from internal and external noise sources. Along with these filters, metallic or metallised plastic boxes are used to protect board components from EM wave interference [40]. Smart surfaces and materials are useful for adaptive shielding at the PCB level. One of the key requirements of board level shields is good thermal conductivity so that they can act as heat sinks for thermal management of the board.

4.5.3 System- or Device-Level Shielding

Enclosure of personal computers, laptops, mobile phones, large communication network systems and battery systems is required to protect the entire system from external EMI sources and to reduce the leakage of EM waves from these systems for electromagnetic compatibility (EMC) in multidevice environments. System- or device-level enclosure need to include penetrating conducting cables, air flow inlets/outlets and maintenance access areas, and these design features lead to void-related problems and hence EMI SE is compromised [7]. These voids or cavities need to be covered with conducting meshes that act like Faraday cages to avoid internal or external leakage fields. Furthermore, I/O cables need to be protected using coaxial cables or braided metal meshes that act like Faraday cages and/or conducting polymer cladding for full protection of electronic systems against EM waves.

4.6 CONCLUSIONS

This chapter has shown that the miniaturisation of electronic devices has popularised the use of electronic devices, and they are exposed to natural EM wave sources with frequency ranging from less than 30 MHz to 10 GHz. Furthermore, the latest telecommunication networks like 5G also transmit EM waves with frequency as high as 6 GHz in the midband range along with other low-frequency sources in the environment. The density of electronic components in an electronic system is continuously increasing along with high computational demands and as a result high powered electronic devices are in use. These high speed and high density electronics emit EM waves at ever higher energies. Moreover, to accommodate high density electronics in miniaturised devices, packaging technologies such as SiP are dominant in today's portable devices. The cross talk in these complex device structures and the external high-frequency emitters generate a complex EMI environment and demand development of advanced EMI mitigation technologies and protective materials. Conventional materials cannot cope with smaller dimensions and advanced manufacturing technologies like additive manufacturing are essential to produce conformal and compartment-level EMI systems along with board-level and device/system-level protection. Conformal EMI shielding solutions using advanced manufacturing technology not only result in space and weight saving but can also be optimised for the thermal management of high power electronics. Advanced artificial high impedance tunable surfaces are important to ensure the reliable functioning of the devices. The advancements in the field of smart materials along with the metamaterials have enabled the concepts of adaptive EMI shielding systems. However, current smart EMI shielding systems are usually adaptive to one or few stimuli. To deal with the complex EMI environment, advanced adaptive EMI shielding systems responsive to multiple stimuli are the need of the day. Future demands of flexible and transparent wearable electronics and sensors also require transparent elastomeric material with good shielding capabilities. Due to increased EMI pollution, green shielding materials with high absorption characteristics should be explored for device safety as well as human and other living beings' health.

ACKNOWLEDGEMENTS

I am indebted to my family and friends for their support. I am thankful to my master's student Syeda Mehak Ali Kazmi for her efforts to produce graphics used in various figures of this manuscript.

REFERENCES

1. Wang, Y.-H. and C.-C. Li, Highly effective EMI shielding composites for 5G Ka-band frequencies. *Applied Materials Today*, 2024. **36**: p. 102041.
2. Russell, C.L., 5G wireless telecommunications expansion: public health and environmental implications. *Environmental Research*, 2018. **165**: p. 484–495.
3. Weinstein, L.A., *Electromagnetic Waves*. 1988, Radio i svyaz: Moscow.
4. Morgan, D., *Handbook for EMC Testing and Measurement*. 2007, Institution of Engineering and Technology: Stevenage, England.
5. Isari, A.A., et al., Structural design for EMI shielding: from underlying mechanisms to common pitfalls. *Advanced Materials*, 2024. **36**(24): p. 2310683.
6. Violette, J.L.N., D.R.J. White, and M.F. Violette, Sources of electromagnetic interference, in *Electromagnetic Compatibility Handbook*, J.L.N. Violette, D.R.J. White, and M.F. Violette, Editors. 1987, Springer Netherlands: Dordrecht. p. 13–62.
7. Crawhall, R.J.H., EMI potential of multiple sources within a shielded enclosure. 1993, University of Ottawa.
8. Shahparnia, S. and O.M. Ramahi, Electromagnetic interference (EMI) reduction from printed circuit boards (PCB) using electromagnetic bandgap structures. *IEEE Transactions on Electromagnetic Compatibility*, 2004. **46**(4): p. 580–587.
9. Cheng, J., et al., Recent advances in design strategies and multifunctionality of flexible electromagnetic interference shielding materials. *Nano-Micro Letters*, 2022. **14**(1): p. 80.
10. Zhang, D.-Q., et al., Self-assembly construction of WS2–rGO architecture with green EMI shielding. *ACS Applied Materials & Interfaces*, 2019. **11**(30): p. 26807–26816.
11. Sajid, H.M., et al., Design of multilayered 2D nanomaterial composite structures for EMI shielding analysis. *ACS Omega*, 2022. **7**(40): p. 35586–35594.
12. Markham, D., Shielding: quantifying the shielding requirements for portable electronic design and providing new solutions by using a combination of materials and design. *Materials & Design*, 1999. **21**(1): p. 45–50.
13. Qu, Y., et al., Multifunctional AgNWs@MXene/AgNFs electromagnetic shielding composites for flexible and highly integrated advanced electronics. *Composites Science and Technology*, 2022. **230**: p. 109753.
14. Geetha, S., et al., EMI shielding: methods and materials – a review. *Journal of Applied Polymer Science*, 2009. **112**(4): p. 2073–2086.
15. Shi, S., et al., 3D-printed carbon-based conformal electromagnetic interference shielding module for integrated electronics. *Nano-Micro Letters*, 2024. **16**(1): p. 85.
16. Jung, J., et al., Highly stretchable and transparent electromagnetic interference shielding film based on silver nanowire percolation network for wearable electronics applications. *ACS Applied Materials & Interfaces*, 2017. **9**(51): p. 44609–44616.
17. Rahmat-Samii, Y. and H. Mosallaei. *Electromagnetic band-gap structures: classification, characterization, and applications*. in *2001 Eleventh International Conference on Antennas and Propagation, (IEE Conf. Publ. No. 480)*. 2001.
18. Dalia, M.N.E., A.E. Hala, and A.A. Esmat, Antenna designs with electromagnetic band gap structures, in *Metamaterial*, J. Xun-Ya, Editor. 2012, IntechOpen: Rijeka. p. Ch. 16.
19. Zhao, Q., et al., Mie resonance-based dielectric metamaterials. *Materials Today*, 2009. **12**(12): p. 60–69.
20. Kosulnikov, S., et al., Volumetric metamaterials versus impedance surfaces in scattering applications. *Scientific Reports*, 2021. **11**(1): p. 9571.
21. Butakov, N.A. and J.A. Schuller, Designing multipolar resonances in dielectric metamaterials. *Scientific Reports*, 2016. **6**(1): p. 38487.
22. Wei, X.C., et al., Applications of high impedance surfaces for surface wave elimination. in *2016 URSI Asia-Pacific Radio Science Conference (URSI AP-RASC)*. 2016.

23. Hou, X., et al., Recent progress in smart electromagnetic interference shielding materials. *Journal of Materials Science and Technology*, 2024. **186**: p. 256–271.
24. Jia, X., et al., Corrigendum to "Construction of compressible Polymer/MXene composite foams for high-performance absorption-dominated electromagnetic shielding with ultra-low reflectivity" [Carbon 173 (2021) 932–940]. *Carbon*, 2021. **176**: p. 660–661.
25. Chen, J., et al., Multifunctional shape memory foam composites integrated with tunable electromagnetic interference shielding and sensing. *Chemical Engineering Journal*, 2023. **466**: p. 143373.
26. Turan, D., G. Gunes, and F. Seniha Güner, Synthesis, characterization and O_2 permeability of shape memory polyurethane films for fresh produce packaging. *Packaging Technology and Science*, 2016. **29**(7): p. 415–427.
27. Zhao, Q., H.J. Qi, and T. Xie, Recent progress in shape memory polymer: new behavior, enabling materials, and mechanistic understanding. *Progress in Polymer Science*, 2015. **49–50**: p. 79–120.
28. Jia, X., et al., Construction of shape-memory carbon foam composites for adjustable EMI shielding under self-fixable mechanical deformation. *Chemical Engineering Journal*, 2021. **405**: p. 126927.
29. Liao, S.-Y., et al., Intelligent shielding material based on VO2 with tunable near-field and far-field electromagnetic response. *Chemical Engineering Journal*, 2023. **464**: p. 142596.
30. Li, X., et al., Multifunctional polyether block amides/carbon nanostructures piezoresistive foams with largely linear range, enhanced and humidity-regulated microwave shielding. *Chemical Engineering Journal*, 2023. **455**: p. 140860.
31. Feng, Y., et al., Transparent and stretchable electromagnetic interference shielding film with fence-like aligned silver nanowire conductive network. *Small Methods*, 2023. **7**(7): p. 2201490.
32. Xiong, W., et al., Analysis of electromagnetic shielding of IC package with thin absorbing material coating inside in two different configurations. in *2018 IEEE International Symposium on Electromagnetic Compatibility and 2018 IEEE Asia-Pacific Symposium on Electromagnetic Compatibility (EMC/APEMC)*. 2018.
33. Hoang, J.D.V., et al., Breakthrough packaging level shielding techniques and EMI effectiveness modeling and characterization. in *2016 IEEE 66th Electronic Components and Technology Conference (ECTC)*. 2016.
34. Al-Duhni, G.S.G., et al., Prefabricated and assembled electromagnetic shields inside electronic packages to reduce near-field capacitive and inductive coupling. *IEEE Transactions on Components, Packaging and Manufacturing Technology*, 2024. **14**(6): p. 1015–1024.
35. Kuo, H.-C., C.-W. Kuo, and C.-C. Wang, Effective low-frequency EMI conformal shielding for system-in-package (SiP) modules. *Microwave and Optical Technology Letters*, 2023. **65**(7): p. 1892–1897.
36. Liu, J., et al., Additive manufacturing of Ti3C2-MXene-functionalized conductive polymer hydrogels for electromagnetic-interference shielding. *Advanced Materials*, 2022. **34**(5): p. 2106253.
37. Li, R., et al., Mn-Co-Ni-O thin films prepared by sputtering with alloy target. *Journal of Advanced Ceramics*, 2020. **9**(1): p. 64–71.
38. Wang, Z., et al., Ultrahigh conductive copper/large flake size graphene heterostructure thin-film with remarkable electromagnetic interference shielding effectiveness. *Small*, 2018. **14**(20): p. 1704332.
39. Ning, F., et al., Integrated gas diffusion electrode with high conductivity obtained by skin electroplating for high specific power density fuel cell. *Small Methods*, 2023. **7**(2): p. 2201256.
40. Tong, C., EMI shielding materials and absorbers for 5G communications, in *Advanced Materials and Components for 5G and Beyond*, C. Tong, Editor. 2022, Springer Nature Switzerland: Cham. p. 143–172.

CHAPTER 5

Electromagnetic Interference Shielding for Defence Applications

Mohsin Ali Fakhar and Kamran Ahmed Khan

5.1 INTRODUCTION

In the prevailing global defence applications, the foremost consideration is being given to electronic systems, as they play a vital role in communication, surveillance, navigation and weaponry. However, reliance on these systems brings multiple consequences as they are extremely vulnerable to electromagnetic interference (EMI) that can downgrade performance and subsequently lead to malfunctions [1]. EMI can also compromise the security of military operations that can have a devastating effect on any defence operation. In parallel, the essential aspect to understand is that EMI is produced from both human-made sources and natural factors, that is, lightning, solar flares and electronic warfare measures. Hence, making EMI shielding a central constituent in the development, design and deployment of military hardware is essential [2]. This chapter intends to deliberate on the techniques, materials and most importantly principles used in EMI shielding purposely made for defence industry applications. It will also cover the important aspect of EMI shielding, the technologies implemented and practical applications in a variety of military defence systems.

5.1.1 Importance of EMI Shielding in Defence

EMI shielding holds great significance in defence applications such as operational integrity, security, confidentiality and compliance with important standards. The salient details in this regard are as follows.

5.1.1.1 Operational Integrity

EMI can easily interrupt the normal operation of electronic systems that will lead to downgraded performance or complete failure of the system. In defence applications such interferences can have stark consequences that encompass communication loss, inaccurate navigation and malfunctioning of weapon systems [3]. Therefore, ensuring operational integrity is of prime importance.

DOI: 10.1201/9781003512363-5

5.1.1.2 Security and Confidentiality

Military systems manage classified information and important strategic communications. Electromagnetic emissions from these systems can possibly be intercepted by hostile agencies leading to security breaches. EMI shielding supports containing these emissions and preventing unsanctioned access to sensitive data [4].

5.1.2 Compliance with Standards

Different military systems must observe rigorous EMI and EMC (electromagnetic compatibility) standards, that is, MIL-STD-461, which postulates the control of EMI features of subsystems and equipment. To effectively operate in the intended electromagnetic environment, it is mandatory to comply with such standards [5].

5.1.3 Different Types of EMI Shielding Materials

There are different types of materials used in EMI shielding. Details [6] regarding the same are as follows.

5.1.3.1 Copper

Copper is extensively utilised in EMI shielding owing to its excellent electrical conductivity and capability to absorb and reflect electromagnetic waves. It is normally employed in the form of sheets, foils and meshes.

5.1.3.2 Aluminium

Aluminium gives great stability between conductivity and weight, hence making it suitable for applications where weight carries significance especially in aerospace and mobile units. Furthermore, it is more cost-effective than copper.

5.1.3.3 Steel

Steel, specifically stainless steel, gives robust strength from a mechanical perspective in addition to EMI shielding. It is used in enclosures that often require both structural integrity and EMI protection.

5.1.3.4 Conductive Coatings

Conductive coatings made from nickel, silver or copper can be applied to non-conductive materials in order to impart EMI shielding properties. These coatings are useful in applications where lightweight and flexible shielding is required.

5.1.3.5 Conductive Polymers

Conductive polymers, which are plastics infused with conductive fillers such as metal or carbon particles, give flexibility and are lighter than metals. These materials are suitable for applications that do not require traditional metal shielding.

5.1.3.6 Metalised Fabrics

Metalised fabrics are fibres that are coated with a thin layer of metal and have the flexibility of textiles with the conductive properties of metals. These fabrics are utilised in tents and other flexible shielding solutions.

5.1.3.7 Composite Materials

Composite materials have different elements which can increase shielding effectiveness while maintaining structural integrity. They might include layers of conductive metal and insulating polymers, providing both EMI shielding and mechanical strength.

5.1.4 Methods of EMI Shielding

5.1.4.1 Enclosures and Cabinets

Electronic equipment is often encapsulated in shielded enclosures to prevent EMI. These enclosures are made from conductive materials and designed to block electromagnetic fields effectively. Proper design guarantees minimal gaps and seams that could let EMI penetrate [7].

5.1.4.2 Gaskets and Seals

To further uphold the integrity of shielded enclosures, conductive gaskets and seals are utilised to fill gaps and seams. These components ensure that enclosures remain tightly sealed, preventing EMI from entering or escaping.

5.1.4.3 Cable Shielding

Cables are a significant source of EMI. Shielded cables, especially those which have conductive braids or foil wraps, ensure signal integrity by preventing EMI.

5.1.4.4 Grounding and Bonding

Effective grounding and bonding are crucial for expelling electromagnetic energy securely. Appropriate grounding techniques ensure that absorbed electromagnetic energy is directed away from sensitive components which reduces the risk of interference.

5.1.4.5 Faraday Cages

A Faraday cage is an enclosure made of conductive materials that helps in blocking external static and non-static electric fields. These are used to protect highly sensitive equipment or secure facilities from EMI.

5.2 PRACTICAL APPLICATIONS OF EMI SHIELDING

5.2.1 Communication Systems

Defence communication systems have to operate in a very reliable manner, especially in the presence of possible interference from both hostile and friendly agencies. EMI shielding ensures safe and secure communication channels that are fundamental for coordinating military operations and making informed strategic decisions [8].

5.2.2 Radar and Navigation

Radar systems are very important for detection and tracking of objects while navigation systems help in guiding military assets accurately. EMI shielding is used to protect these systems from interference which ensures accurate detection and reliable navigation.

5.2.3 Computing and Data Storage

Computing and data storage carry great importance in military operations for various mission planning, execution of operations and gathering of intelligence information. Shielding these systems prevents corruption of data and unauthorised access to sensitive information.

5.2.4 Avionics and Vehicle Electronics

The electronics in aircraft and military vehicles have to operate reliably, especially under extreme conditions. EMI shielding ensures that avionics and other electronic systems function appropriately thus improving the performance and safety of these platforms.

5.2.5 Portable Devices

Portable electronic devices such as handhelds used by military personnel have to be shielded in order to operate correctly in various electromagnetic environments. EMI shielding ensures that these devices remain useful and steadfast in the field.

5.3 MILITARY KEYBOARD

EMI shielding has great significance and is crucial in military keyboards. Inappropriate EMI shielding in military keyboards can cause catastrophic malfunctioning like signal interference, data corruption and in some cases complete failure. A poorly designed keyboard can act as a conduit and EMI can circulate in the overall system which could disrupt other connected devices. Precisely, erratic cursor movements, unresponsive keys and failure to transmit important data during the execution of critical operations can occur. All defence-related operations demand high reliability from the technologies used for routine as well as sensitive activities. EMI can cause malfunction in electronic devices which can be devastating in critical situations. Therefore, it is very important to ensure that the keyboard and other associated electronic components function effectively without being affected by external electromagnetic fields. It is pertinent to mention that keyboard enclosures are made from conductive materials such as aluminium, copper or specialised alloys [9]. These materials establish a Faraday cage around the sensitive electronic components. The enclosure must be developed so that there are no gaps or seams that may allow penetration of EMI. Application of conductive coating to the internal and external surfaces of the keyboard housing can protect against EMI. Additionally, application of plating or spray can also ensure a uniform conductive layer that can give further protection.

In addition, EMI gaskets are used at joints and seams where two conductive surfaces join. This ensures continuous conductivity and prevents EMI leakage. Conductive sealants are also used around openings such as cable entry points to protect the integrity of the shielding. Shielded cables are also used in keyboards to protect them from EMI. Likewise, shielded connectors and proper grounding are significant to ensure shielding effectiveness. It extends through the entire cable system. For printed circuit boards (PCBs) of the keyboard, proper grounding techniques are used to minimise EMI susceptibility. Furthermore, in military keyboards, ferrite beads and chokes are also used on cables and traces to subdue high-frequency EMI. It is pertinent to highlight that it is imperative for military keyboards to comply with stringent standards (i.e., MIL-STD-461 and MIL-STD-464) to ensure their EMI shielding effectiveness.

EMI shielding in military keyboards encompasses elements from materials science, electrical engineering and testing and evaluation techniques to make sure that these devices can operate effectively in the challenging conditions during military operations. By employing coatings, gaskets, conductive enclosures and proper research and design, military keyboards can achieve the requisite steadfastness against EMI [7].

5.4 SUBMARINES

Submarines can be considered a pinnacle of technological advancement due to their complex and challenging operational requirements both above and under the sea surface. This is due to the conductive nature of seawater and the metallic vessel of the submarine. Besides this, high-power electrical systems, sensitive electronic equipment and wireless communication links susceptible to EMI. Any minute interference can degrade signal quality or disturb the overall electrical system and this may cause safety risks. Effective EMI shielding is needed to protect sensitive electronic systems of navigation, propulsion and weapon systems as shown in Figure 5.1. To ensure the operational integrity and security of the vessel's electronic system it is mandatory to have secure EMI shielding in submarines. Having said that, one must comprehend that submarines are complex and sensitive in nature, therefore effective EMI shielding is essential to maintain communication, navigation and functionality of all control systems. EMI shielding in submarines has a diverse and integrated approach that combines advanced materials and design techniques coupled with rigorous testing and evaluation to ensure the vessel's electronic systems operate effectively and securely. By use of shielded cabling, conductive enclosures, grounding and proper component design, submarines can ensure and protect their operational integrity and maintain stealth in challenging electromagnetic environments [10]. Details with respect to how EMI shielding is applied in submarines are laid as follows:

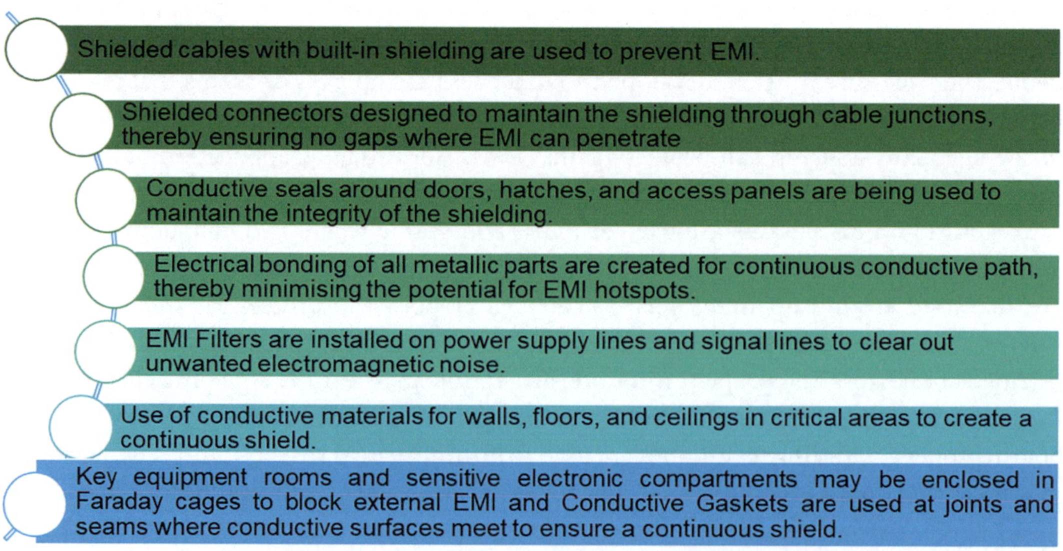

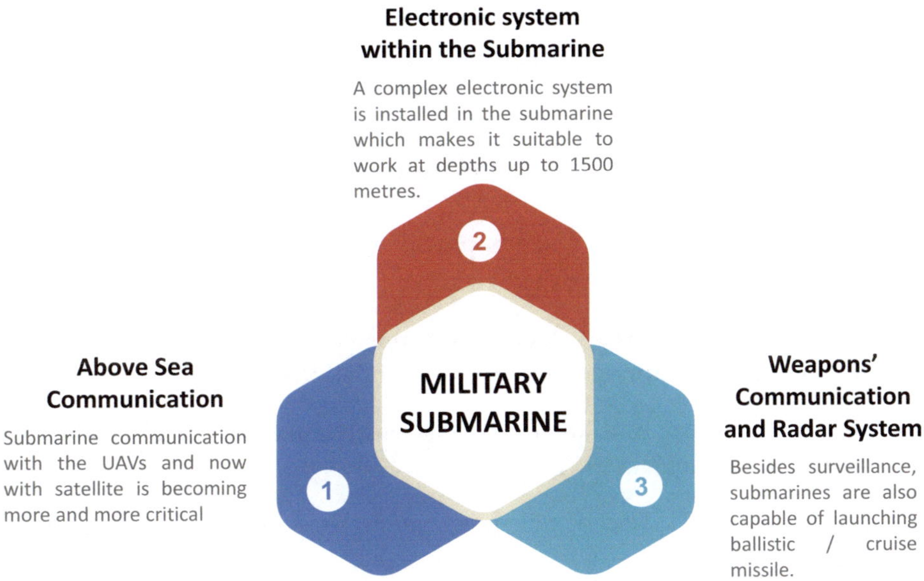

FIGURE 5.1 A schematic showing communication of submarine. © Copyright PresentationGO.com – Free Templates & Infographics for PowerPoint and Google Slides.

5.5 MILITARY HELMETS

EMI shielding in military helmets is very important to protect the advanced electronic systems that are integrated in modern helmets such as communication devices, sensors and displays [4]. EMI shielding in military helmets will ensure:

1. Communication Reliability
2. Protection of Electronic Systems
3. Safety

Military helmets incorporate very thin layers of conductive metals, that is, copper, aluminium or nickel. These materials act as coatings or are integrated into the helmet structure. Additionally, woven metal fibres are also used to line the inside of the helmet thus making a more flexible and lightweight shielding layer.

Additionally, carbon nanotubes and graphene-based coatings or composite structures offer lightweight and effective EMI shielding owing to their excellent electrical conductivity. In military helmets, various electronic modules within the helmet can also be encompassed in separate encapsulated compartments to prevent EMI emission or interference. Like military keyboards, military helmets must adhere to stringent standards (i.e., MIL-STD-461 and MIL-STD-810).

5.6 EMI SHIELDING IN ELECTROMAGNETIC WARFARE

Electromagnetic warfare (EW) comprises the electromagnetic spectrum that conducts military operations such as electronic attack (EA), electronic protection (EP) and electronic support (ES). EMI shielding plays a critical role in the effectiveness and steadfastness of systems used in these operations. EMI shielding protects the electronic systems, ensures data integrity, maintains consistent performance, prevents unintentional electromagnetic emissions, improves the ability of systems to resist jamming and protects from potential health hazards associated with exposure to high levels of electromagnetic radiation. All these activities are performed through the following measures [12].

1. **Faraday Cages**: Key facilities and equipment are often contained in Faraday cages to block external electromagnetic fields.
2. **Shielded Operation Centres**: Command and control centres are equipped with shielding to guarantee unremitting operation of electronic systems.
3. **Metallic Coatings**: Application of metallic coatings on equipment housings provides an effective barrier against EMI.
4. **Composite Materials**: Use of advanced composites for lightweight and effective shielding.
5. **Shielded Cables**: Cables with built-in shielding are used to prevent EMI from affecting signal integrity.
6. **EMI Gaskets and Seals**: Ensures all connectors and cable entry points are sealed with EMI gaskets to preserve continuous shielding.
7. **Grounding Systems**: Proper grounding of all electronic equipment is done to prevent potential differences that can cause EMI.
8. **Bonding and Electronic Filters and Suppressors**: Used to filter out unwanted electromagnetic noise.
9. **Ferrite Beads and Chokes**: Used on cables and circuits to subdue high-frequency EMI.

5.6.1 Applications in Electromagnetic Warfare

As shown in Figure 5.2, EMI shielding is essential in EW to guard and ensure the consistency of electronic systems used in defence operations. With the utility of shielded enclosures, conductive materials, shielded cabling, grounding techniques and rigorous testing,

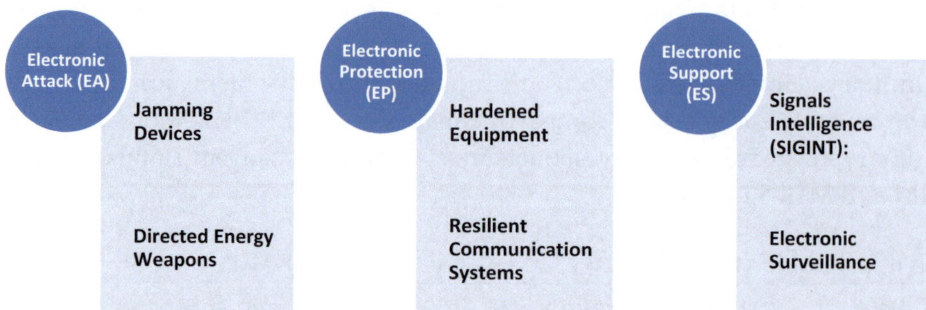

FIGURE 5.2 Electronic attack, protection and support.

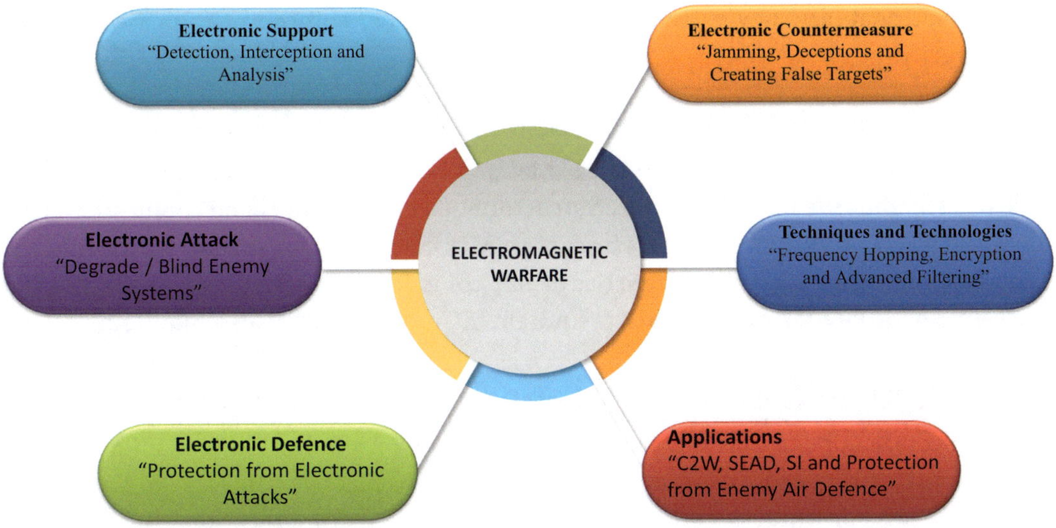

FIGURE 5.3 A schematic of electromagnetic warfare. © Copyright PresentationGO.com – The free PowerPoint and Google Slides template library.

defence forces can ensure the reliability, performance and safety of their electronic systems in complex and hostile electromagnetic environments.

A paradigm shift has been observed in war strategies, and from the last few decades EW has gained the most attention. Figure 5.3 is an illustration of the key aspects pertaining to EW. To protect territory, it is important to detect and intercept any threat. Similarly, the attacker will put efforts into degrading or blinding the enemy system and for this frequency hopping, encryption or advanced filtering will be used.

5.7 EMI SHIELDING IN BALLISTIC MISSILES

Ballistic missiles are considered one of the most critical aspects of modern warfare and therefore designing must cater for EMI shielding during the harsh environment of flight. Ballistic missiles are designed to deliver a heavy payload at an accurate position and any miscalculation can cause huge loss. Ballistic missiles are exposed to a range of electromagnetic phenomena such as lightning strikes, high-power electromagnetic pulses and electromagnetic radiations from different sources. One of the primary threats is the pulse attacks generated by nuclear detonations or high-power microwave sources which can induce high-intensity electric currents or voltages in the system. Additionally, ballistic missiles contain a complex electronic/electrical system which can also cause EM interference. The EM interference can interrupt guidance and control, navigation, communication links and the onboard electronic systems.

EMI shielding in ballistic missiles is vital for reliability and effectiveness of their electronic systems. It is highlighted that ballistic missiles have sensitive electronic components for navigation, guidance, communication and control [14]. Effective EMI shielding protects these components from all types of internal and external EMI which could otherwise lead to failure of the system. EMI shielding provides ballistic missiles operational reliability, navigation and guidance systems, communication systems and data integrity.

5.7.1 Methods of EMI Shielding in Ballistic Missiles

Ballistic missiles are protected with EMI shielding by using shielded enclosures, metallic enclosures, composite materials, layered structures, electronic filters and suppressors, EMI filters and so on. Additionally, compliance with MIL-STD-461 (for specifying the requirements for controlling EMI characteristics of electronic equipment) and MIL-STD-464 (for electromagnetic environmental effects requirements for systems) is also required.

5.8 EMI SHIELDING IN MILITARY COMMUNICATION SYSTEMS

Modern defence applications rely heavily on military communication to facilitate secure, dependable, real-time information. These systems are extremely vulnerable to EMI which can impair mission success and interfere with sensitive communication. Therefore, the system requires EMI shielding to be safe from electromagnetic threats. EMI shielding ensures communication integrity, security and confidentiality and compliance with standards. It is used for tactical radios, satellite communications, secure networks, electronic warfare systems and UAVs.

5.9 EMI SHIELDING IN MILITARY UAVS [13]

UAVs have become essential equipment integrated in today's military operations. They provide a range of capabilities including surveillance, reconnaissance and target acquisition. Keeping in view their sensitive electronic systems for communication, navigation and control, UAVs are particularly vulnerable to EMI. Effective EMI shielding is crucial to ensure their reliable operation, protect sensitive data and maintain mission success [15]. A schematic of EMI/EMC testing of a UAV is shown in Figure 5.4. It provides operational reliability, security, data integrity, and compliance with standards.

5.9.1 EMI Shielding Materials for UAVs

The materials used for EMI shielding of UAVs include conductive metals such as copper which, due to its excellent electrical conductivity, has the ability to absorb and reflect electromagnetic waves [17]. Likewise, aluminium is used as it offers a good balance between conductivity, weight and cost. Its lightweight nature is suitable for UAVs. Other materials being used for EMI shielding include conductive polymers, metalised fabrics, composite materials, gaskets and seals, cable shielding, grounding and bonding and Faraday cages.

Figure 5.5 is an illustration of the major electronic components needing EMI shielding the most and for this the foremost is the flight control computer (FCC) where all the details are stored related to any mission. UAVs also possess critical and intricate sensors for the power management system along with the communication sensors to contact ground stations. The power system and propulsion system are another vulnerable area of the UAV as EMI can affect this system and this is also considered a one-point failure of the system.

FIGURE 5.4 Schematic of EMI/EMC testing of a UAV. Reprinted from [16], licensed under CC BY 4.0.

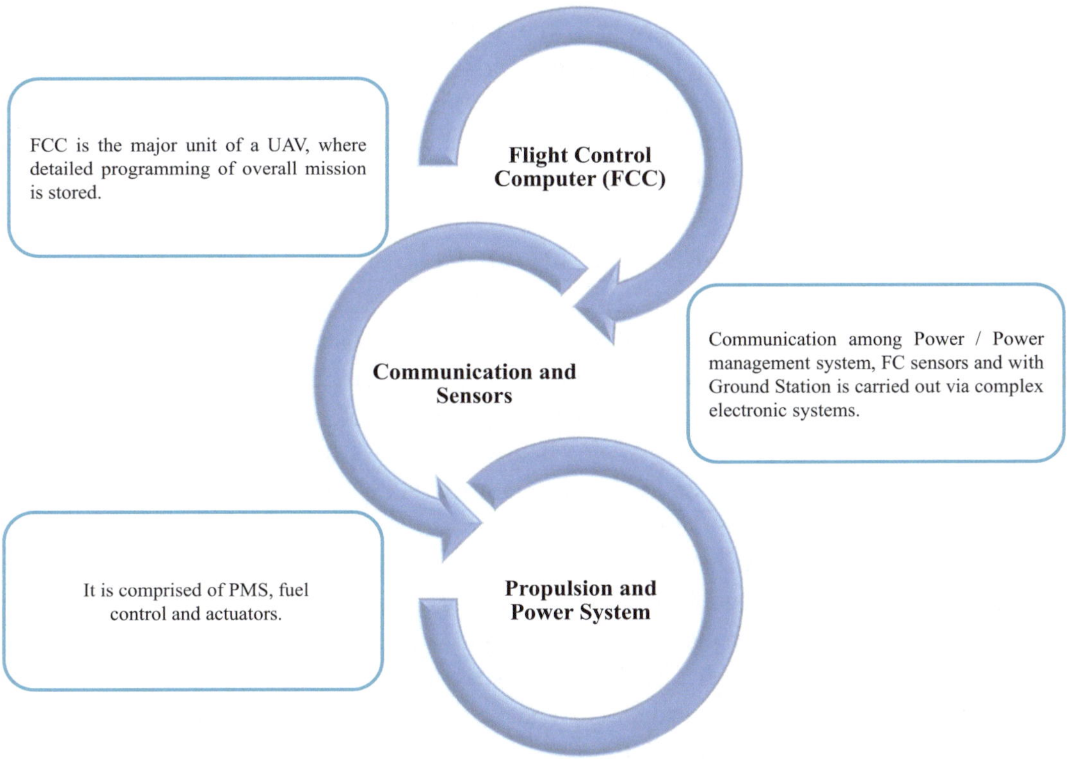

FIGURE 5.5 A schematic to understand complex data transference via electronic components.

5.10 EMI SHIELDING IN MILITARY DATA STORAGE

Military data storage systems are very critical for information essential for mission planning along with intelligence gathering purposes. These systems depend on sophisticated electronic components that are susceptible to EMI. EMI can damage system performance and compromise the security of classified information. Therefore, EMI shielding in military data storage ensures the integrity, reliability and security of military data storage. EMI shielding in military data storage can be achieved through conductive materials like copper by using its high electrical conductivity property as it can effectively absorb and reflect EMI waves. Likewise, the same can also be achieved through aluminium, conductive coatings and conductive barriers, composite materials and metal-polymer composites. Additionally, the employment of enclosures and cabinets for housing military data storage prevents electromagnetic fields from entering the system and internal emission from escaping. Likewise, gaskets, seals and cable shielding can also be used [18].

5.11 EMI SHIELDING IN MILITARY HELICOPTERS

Military helicopters are very important assets in warfare that perform a variety of missions including reconnaissance, close air support and evacuation. These helicopters rely heavily on advanced electronic systems for navigation, communication, weapons guidance and creating situational awareness. Despite their significance, these electronic systems are vulnerable to EMI which can worsen performance, interrupt communication and compromise mission effectiveness. EMI shielding in military helicopters is essential to ensure the reliability, security and safety of these aircraft in complex electromagnetic environments [2]. EMI shielding is thus very important in military helicopters as it ensures operational reliability, safety, security and compliance with standards.

EMI shielding materials for military helicopters include conductive metals such as copper and aluminium, conductive coatings (nickel, silver or copper coatings), conductive polymers and composite materials.

5.12 EMI SHIELDING IN ARMOURED VEHICLES

Like military helicopters, armoured vehicles are also critical assets in military operations, providing protection to personnel and equipment in combat situations. These vehicles are equipped with advanced electronic systems for communication, navigation, surveillance and weapon control. These electronic systems are susceptible to EMI which can disrupt their operation and compromise mission effectiveness. EMI shielding in armoured vehicles is critical to safeguard the operational reliability, safety, security and survivability of electronics in hostile environments.

EMI shielding materials for armoured vehicles include conductive materials such as copper that absorb and reflect electromagnetic waves. Likewise, steel, conductive coatings and conductive polymers can also be used. Composite materials are also used for ensuring EMI shielding in armoured vehicles. The various techniques being used for EMI shielding in such vehicles are enclosures and shields, gaskets and seals, cable shielding, grounding and bonding, and absorptive materials.

5.13 EMI SHIELDING IN MILITARY RADIO CONTROLLERS

Military radio controllers are devices used for conducting communication, command and control of various military assets including unmanned vehicles, artillery and infantry operations. These controllers are very important as they rely on electronic components and communication systems that are susceptible to EMI. EMI can disrupt communication links, sacrifice data integrity and affect operational effectiveness in critical military missions. Therefore, effective EMI shielding is mandatory to ensure reliable and secure communication in military radio controllers. Effective EMI shielding ensures communication reliability, data security and integrity, and compliance with standards [19].

EMI shielding materials for military radio controllers include conductive metals such as copper and aluminium, conductive coatings, nickel, silver and conductive polymers.

5.14 EMI SHIELDING IN FIGHTER AIRCRAFT

EMI shielding is very important for ensuring the operational integrity of fighter aircraft systems. Modern fighter jets have an array of sophisticated electronic systems including radar, communication, navigation and electronic warfare (EW) systems. To operate seamlessly in a highly electromagnetic environment these systems (both internally and externally) require effective EMI shielding.

EMI shielding in fighter aircraft ensures operational reliability, data security and integrity and compliance with standards. The different materials that are used for EMI shielding of fighter aircraft include conductive metals such as aluminium and copper, composite materials including carbon fibre composites and conductive coatings [1].

5.15 EM SPECTRUM IN EMI SHIELDING IN DEFENCE APPLICATIONS

The electromagnetic (EM) spectrum includes all forms of electromagnetic radiation which include low-frequency radio waves to high-energy gamma rays [20]. This spectrum is divided into various segments as in Table 5.1 each with unique properties and applications in defence and civilian technologies.

5.16 CHALLENGES IN EMI SHIELDING FOR DEFENCE APPLICATIONS

5.16.1 Multi-spectral Interference

Defence systems function across various segments of the EM spectrum concurrently. Effective shielding must therefore resist interference from a wide range of frequencies to assure system effectiveness, reliability and performance.

5.16.2 Harsh Environmental Conditions

Military operations that involve extreme environmental conditions must be properly EMI shielded with materials that are durable and ensure effectiveness in challenging environments.

TABLE 5.1 Relevant Spectrum, Applications and EMI Concerns

Sr No	Spectrum	Applications	EMI concerns
1	Radio Frequency (RF) Spectrum	Communication, radar systems, navigation (GPS) and electronic warfare (EW).	Can intercept communication signals, radar operation, and navigation systems, compromising mission success and safety.
2	Microwave Spectrum	Radar systems, satellite communication and electronic countermeasures.	Microwave interference can degrade radar resolution, disrupt satellite and communication links.
3	Infrared Spectrum	Thermal imaging for reconnaissance, target detection and missile guidance systems.	Interference can distort thermal images, affecting target identification and tracking accuracy.
4	Visible Light Spectrum	Target identification, signalling and secure communication using optical fibres.	Although less susceptible to EMI compared to other segments, interference can still affect optical communication systems.
5	Ultraviolet (UV) and X-ray Spectrum:	UV for sterilisation, X-rays for imaging and inspection in security and medical applications.	Interference can impact X-ray imaging quality and the effectiveness of UV sterilisation.

5.16.3 Weight and Size Constraints

Military equipment has to balance EMI protection with varying weight and size challenges. Lightweight shielding that does not compromise performance is therefore mandatory.

5.16.4 Integration with System Design

EMI shielding needs to be integrated while the system is in the design phase of defence systems to minimise interference without impacting system functionality or operational readiness.

5.17 FUTURE TRENDS AND DEVELOPMENTS

5.17.1 Advanced Materials

Sustained research and development into nanomaterials and meta-materials offer enhanced EMI shielding performance while reducing weight and complexity.

5.17.2 Adaptive Shielding Technologies

Research and development of adaptive materials and coatings which can maintain their shielding properties in response to changing electromagnetic environments.

5.17.3 Integration with Cyber-security

Junction of EMI shielding with cyber-security to create inclusive defences against electromagnetic and cyber threats.

TABLE 5.2 Types of Disasters Caused by Poor EMI Shielding

Sr No	Type of Disaster	Impact	Consequence
1	Communication Interference	This could result in losing communication between military units, command centres that can have worse consequences.	Miscommunication and delay in transmission of key information during critical operations.
2	Radar and Navigation Failures	Unable to detect or track threats, such as any lethal weapon missiles or aircraft.	Increased susceptibility to surprise attacks.
3	Weapon Systems Malfunction	Imprecise operation of guided missile systems.	Missed targets and failed mission objectives.
4	Avionics System Errors	Flawed readings that can result in system failure in avionics of aircraft.	Losing aircraft control, errors in navigation and potential or hazardous accidents.
5	Data Integrity Breaches	Threat to the security of classified information or sensitive data.	Vulnerability to cyber threats and potential for espionage.

5.18 POTENTIAL THREATS OR FAILURE OF EMI SHIELDING IN DEFENCE SECTOR

Although such cases being of sensitive nature are not documented, a few potential disasters solely attributed to failed EMI shielding and EMI in military and defence contexts are shown in Table 5.2.

5.19 EMI SHIELDING IN STEALTH TECHNOLOGY

EMI in stealth technology plays a critical role in maintaining the efficiency and stealth characteristics of military aircraft, ships and other systems [21]. Stealth vehicles are developed to reduce their radar cross-section (RCS) and electromagnetic emissions to avoid detection by radar or other electromagnetic sensors. Effective EMI shielding is mandatory to achieve and maintain these stealth capabilities.

5.19.1 Importance of EMI Shielding in Stealth Technology

Stealth technology has become an essential component of modern warfare, allowing military platforms to evade detection by radar systems and maintain operational superiority. A critical aspect of stealth technology is EMI shielding which plays a crucial role in reducing the radar signature of platforms and ensuring their detectability is minimised.

EMI shielding in stealth technology serves multiple purposes. Primarily, it ensures a reduced RCS making it more challenging for radar systems to detect the platform. This is particularly important in situations where the platform needs to operate in hostile environments or near enemy forces. In addition, EMI shielding helps prevent unintentional emissions that could increase the platform's detectability, further enhancing its stealth capabilities.

5.19.2 Methods of EMI Shielding in Stealth Technology

EMI shielding in stealth technology is ensured through structural design using composite materials, materials with embedded conductive fibres or coatings to provide EMI shielding, conductive coatings and materials such as RAM (radar absorbent material), shielded enclosures, Faraday cages, EMI filters and suppressors, grounding and bonding.

5.19.3 Standards and Testing

Stealth platforms must meet critical standards for EMI shielding to ensure their efficiency and compliance with operational requirements. It includes MIL-STD-461 that specifies requirements for the control of EMI and MIL-STD-188-125 that provides guidance on protecting military facilities and equipment.

5.19.4 Testing Methods

EMI shielding is a critical aspect of designing stealth anechoic chambers which are used to test the electromagnetic properties of various systems and materials. Two primary approaches to evaluating EMI shielding performance in such chambers are field testing and simulation and modelling techniques.

Field testing involves directly measuring the electromagnetic shielding effectiveness of materials and structures within the anechoic chamber environment. This approach provides empirical data on real-world performance but can be time-consuming and resource-intensive. In contrast, simulation and modelling techniques use computational methods to predict shielding behaviour without the need for physical testing.

5.19.5 Applications for Stealth Technology

5.19.5.1 Aircraft

Reduction in RCS: It ensures that stealth aircraft maintain their low RCS profile.

Avionics and Communication Systems: Protecting onboard systems from EMI to safeguard and maintain operations.

5.19.5.2 Naval Ships

Reduced Electromagnetic Signature: It reduces the ship's detectability by radar and other sensors.

EW Systems: Enhancing the effectiveness of electronic warfare systems.

5.19.5.3 Ground Vehicle

It ensures operational security through shielding internal communication and control systems and maintains survivability by maintaining functionality of mission-critical systems.

5.20 CASES OF FAILURE IN EMI SHIELDING IN DEFENCE

EMI can dramatically impact the safety and operations of fighter aircraft and other defence systems. Appended herein are a few incidents and concerns related to EMI shielding taken from practical historical examples. It is pertinent to mention that poor EMI shielding cannot only be declared as the prime reason for failure in these examples, but it played a role directly or indirectly.

5.20.1 EA-6B Prowler EMI Incidents

The EA-6B, an electronic warfare aircraft, faced a critical issue in its jamming system which caused interference with the onboard circuits leading to false navigation and inappropriate communication. After a thorough investigation, it was discovered that the primary reason behind this problem was poor EMI shielding.

Ensuring electromagnetic compatibility is a crucial aspect in the design and development of high-reliability systems such as spacecraft and launch vehicle electronic systems. The natural space environment is characterised by various complex and subtle phenomena that can have adverse effects on spacecraft, underscoring the importance of proper shielding and mitigation strategies. Similarly, the EMI challenges encountered in the EA-6B jamming system highlight the criticality of effective EMI shielding in maintaining the integrity and reliability of onboard systems.

The success of modern wireless communication systems is heavily dependent on the ability to operate effectively within complex electromagnetic environments. The special features of such systems such as diversified functions, use of a wide range of spectrum and high-density packaging, can contribute to interference problems that can result in mission delays or failures. Therefore, the design and implementation of robust EMI shielding solutions are essential to ensure reliable operation of these critical systems.

To address the EMI issue in the EA-6B jamming system, a comprehensive EMC programme should be established, starting with the definition of system-level EMC requirements. This would be followed by the generation of EMC specifications, the implementation of effective design techniques and the execution of rigorous testing requirements to validate the electromagnetic compatibility of the system.

5.20.2 F-35 Lightning II Development Challenges

The development of the F-35 Lightning II, a fifth-generation multirole fighter aircraft, has been a significant engineering feat, with the integration of complex electronic systems posing unique challenges. One of the primary areas of concern has been the potential for EMI affecting the sensor fusion capabilities of the aircraft.

To address these EMI-related challenges, extensive testing and modifications were undertaken to ensure the seamless operation of the F-35's electronic systems. The complexity of the aircraft's design, with its diverse range of electronic functions and high-density packaging, has necessitated a comprehensive approach to electromagnetic compatibility throughout the design and development process.

The F-35's sensor fusion capabilities, which integrate data from various onboard systems, are critical to its mission effectiveness. Ensuring that these systems can operate without interference from other electronic components has been a significant focus for the aircraft's engineers. The integration of advanced sensors, avionics and communication systems has required careful analysis and testing to mitigate potential EMI issues and maintain the aircraft's overall performance.

The challenges faced during the F-35's development are not unique to this programme, as similar concerns have arisen in the design of other modern launch vehicles and spacecraft. The high reliability and diverse functions of such systems can increase the risk of

EMI which can have severe consequences, from mission delays to total failure. Comprehensive EMC (electromagnetic compatibility) programmes are therefore essential in the design and development of space systems as well as military aircraft like the F-35.

5.20.3 USS Forrestal Fire (1967)

In 1967, a tragic incident occurred when a Zuni rocket was fired, leading to a disastrous fire and the loss of over 100 sailors' lives. The primary cause of this calamity was poor EMI shielding which led to the improper handling and malfunction of the electronic systems onboard the rocket. Inadequate EMI protection allowed electrical interference to disrupt the rocket's control systems, ultimately resulting in the catastrophic fire and loss of life.

5.20.4 Black Hawk Helicopter Incident (1993)

In 1993, two Black Hawk helicopters were shot down, and the official investigations suggested that this occurred due to EMI failures which increased communication and navigation errors among the defence systems. The lack of robust countermeasures against EMI in the missile defence network was a critical vulnerability that contributed to these tragic incidents.

5.20.5 Patriot Missile System Failures

During the Gulf War, the inability of the defence system to reliably intercept incoming Scud missiles was a significant challenge. One of the major contributing factors was EMI, which provided false targets and missed interceptions. The radar systems used were not optimised to handle the complex electromagnetic environment, resulting in false target information being relayed to the missile launchers.

5.21 CONCLUSION

EMI shielding has become increasingly crucial in the realm of military technology and defence systems. The rapid growth of technology and proliferation of electronic devices have led to a novel class of pollution known as EMI, which can be caused by various sources such as radio frequency interference, electrostatic discharge, electromagnetic coupling and electromagnetic conduction or induction. This electromagnetic pollution can have detrimental effects on the performance and reliability of military equipment, compromising their stealth, security and overall operational effectiveness.

To address this challenge, the use of advanced materials and structural design principles has become paramount. Polymer nanocomposites have emerged as a promising solution for EMI shielding, offering lightweight, flexible and high-performance characteristics. These materials can be engineered to exhibit high conductivity and efficient shielding capabilities, ensuring the protection of sensitive military systems from EMI.

In addition to the materials used, the structural design of EMI shielding systems is also crucial. Effective shielding requires the incorporation of conductive coatings, grounding techniques and strategic placement of shielding components to create a comprehensive and robust defence against electromagnetic threats.

Furthermore, rigorous testing methods are essential to ensure the reliability and performance of EMI shielding solutions in diverse electromagnetic environments. Military forces must employ comprehensive testing protocols to validate the shielding effectiveness, durability and compatibility of their EMI shielding systems, enabling them to maintain a strategic advantage and ensure mission success in modern warfare scenarios.

REFERENCES

1. Bijulin Greety, D.J., John Wessley, G.J., A study on EMI shielding in aircraft: introduction, methods and significance of using electrospun nanocomposites. *Journal of Space Safety Engineering*, 2024, 13, pp. 150–160.
2. Mishra, Raghvendra, George Thomas, Martin, Abraham, Jiji, Joseph, Kuruvilla, Thomas, Sabu. Electromagnetic interference shielding materials for aerospace application: a state of art. In Maciej Jaroszewski (ed.) *Advanced Materials for Electromagnetic Shielding*, 2018. Wiley.
3. Hamouda, Samir, Amneenah, Naji, Electromagnetic interference Impacts on electronic systems and regulations. *International Journal of Advanced Multidisciplinary Research and Studies*, 2024, 4, pp. 124–127.
4. Deepcoat. [Online] [Cited: August 14, 2024.] www.deepcoat.com/how-emi-shielding-helps-the-military/.
5. DoD. Requirements for the control of electromagnetic interference characteristics of subsystems and equipment. 2007.
6. Liu, Ji, Yu, Ming-Yuan, Yu, Zhong-Zhen, Nicolosi, Valeria, Design and advanced manufacturing of electromagnetic interference shielding materials. *Materials Today*, 2023, 66, pp. 245–272.
7. Geetha, S., Satheesh Kumar, K.K., Rao, Chepuri R.K., Vijayan, M., Trivedi, D.C., EMI shielding: methods and materials – a review. *Journal of Applied Polymer Science*, 2009, pp. 2073–2086.
8. Abdelal, Nisrin Rizek, *A Shield of Defense: Developing Ballistic Composite Panels with Effective Electromagnetic Interference Shielding Absorption*. 2024, Defence Technology, pp. 123–136.
9. Verma, Ritesh, Thakur, Preeti, Chauhan, Ankush, Jasrotia, Rohit, Thakur, Atul, A review on MXene and its composites for electromagnetic interference (EMI) shielding applications. *Carbon*, 2023, 170–190.
10. Yusof, Mohd, Kabir, Shahid, An overview of sonar and electromagnetic waves for underwater communication. *IETE Technical Review*, 2012, 29, pp. 307–315.
11. Mobility Engineer. *Mobility Engineer*. [Online] [Cited: August 6, 2024] www.mobilityengineeringtech.com/component/content/article/24599-nrl-0064.
12. Gordon, Don E., *Electronic Warfare Element of Strategy and Multiplier of Combat Power*. 1981, Elsevier Science.
13. Everything RF. [Online] [Cited: August 14, 2024] www.everythingrf.com/community/what-is-electronic-warfare.
14. Pai, Anand, Rodriguez-Millan, Marcos, Beppu, Masuhiro, Valverde-Marcos, Borja, Satish Shenoy, B., Experimental techniques for performance evaluation of shielding materials and configurations subjected to blast and ballistic impacts: a state-of-the-art review. *Thin Walled Structures*, 2023, 191, p. 111067.
15. Zhang, Z., Zhou, Y., Zhang, Y., Qian, B., Strong electromagnetic interference and protection in UAVs. *Electronics*, 2024, 13(2), p. 393.
16. Badeo, G.-P., et al., Development and testing of an AI-based specific sound detection system integrated on a fixed-wing VTOL UAV. In *Acoustics*, 2025, 7(3), p. 48.
17. Kim, S.-G., et al., *Review of Intentional Electromagnetic Interference on UAV Sensor Modules and Experimental Study*. 2022, UAV-Based Technology for IoT.

18. Nasreen Taj, M., Daruka Prasad, B., Ramarao, N., Nagabhushana, H., Reddy, A, Ashwini, K.R., Vinuta, Dhanalakshmi, K., Dielectric and structural properties of polyaniline-tungsten trioxide nanocomposites: for the packing of nano-electronic devices and EMI shielding. *Nano-Structures & Nano-Objects*, 2024, 39, p. 101219.
19. Bijulin Greety, D.J., Jims John Wessley, G., A study on EMI shielding in aircraft: introduction, methods and significance of using electrospun nanocomposites. *Journal of Space Safety Engineering*, 2024, 11, pp. 150–160.
20. Mhedi Hasan, Md., Moniruzzaman, Md., Kirawanich, Phumin, Alam, Touhidul, Yahya, Iskandar Bin, Alrashdi, Ayed M., Mobarak, Mahjabin, Soliman, Mohamed S., Islam, Mohammad Tariqul. Frequency and bandwidth modulation of a wide band-stop metamaterial for EMI shielding applications. *Optics & Laser Technology*, 2024, 172, p. 110515.
21. Hunjra, M., Fakhar, M., Naveed, K., Subhani, T., Polyurethane foam-based radar absorbing sandwich structures to evade detection. *Journal of Sandwich Structures and Materials*, 2017, 19, pp. 647–658.

CHAPTER 6

Electromagnetic Interference Shielding for Automotive Applications

Ramy Sadek and Jamal Chaouki

6.1 INTRODUCTION TO EMI IN AUTOMOTIVE APPLICATIONS

6.1.1 Electromagnetic Interference (EMI)

Electromagnetic interference (EMI) refers to the unwanted influence of electromagnetic fields on electronic circuits and devices. It can be radiated, conducted emissions or both. EMI can be categorised depending on its frequency range and the mechanisms by which it affects electronic systems; it also includes narrowband and broadband interference, continuous wave and impulsive noise [1, 2]. This phenomenon interrupts the normal function of low electromagnetic compatibility (EMC) electrical devices [3].

Recent technological advancements have introduced many EMI sources into daily life including satellite communications, automotive collision avoidance radar, railroad accident monitoring systems and millimetre-wave wireless LANs. These developments have led to a significant increase in electromagnetic energy emissions as depicted in Figure 6.1 [4].

Consequently, EMI problems are aggravated and become a genuine dilemma impacting the expected performance of sensitive electric devices which are numerous in modern automotive applications [5, 6].

6.1.2 Electromagnetic Compatibility (EMC)

Electromagnetic compatibility (EMC) signifies the capability of electronic parts and electrical devices to operate effectively under the effect of EMI without detrimental deterioration in their performance. EMC can be achieved by implementing EMI shielding materials and designs to attenuate EMI and protect devices from exposure to electromagnetic radiation [7].

6.1.3 Importance of EMI Shielding in Automotive Industry

The automotive industry has undergone remarkable growth owing to advanced electrical systems. Vehicles integrate several electronic control units (ECUs), sensors, actuators and

DOI: 10.1201/9781003512363-6

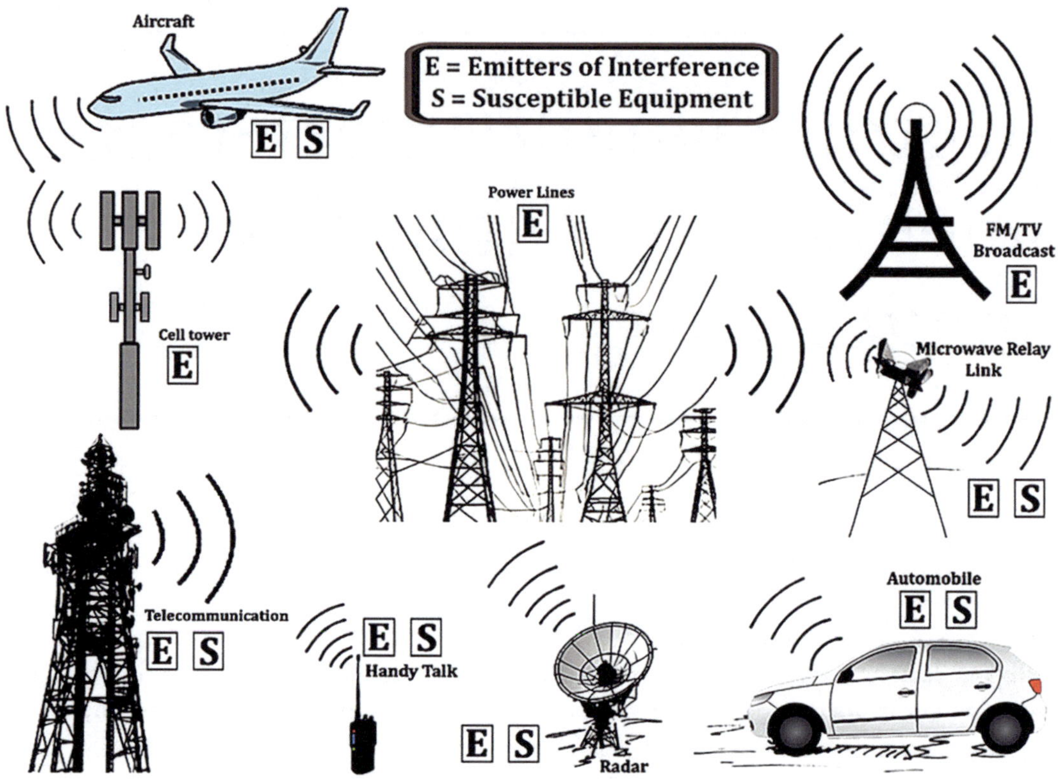

FIGURE 6.1 EMI sources and receivers, sub-images are licensed CC-BY-SA 4.0 International.

communication devices as demonstrated in Figure 6.2, which is pivotal to enhance user experience, functionality and safety. This requires EMI shielding to be applied to prevent electrical devices from malfunctioning [8].

EMI shielding imposes its importance as it provides safety, reliability, performance and compliance of automotive systems. EMI may cause catastrophic failures and disturb the stable operation of critical safety systems like anti-lock braking systems (ABS), airbags and stability control which require precise electronic control. As a result, EMI will induce unexpected vehicle behaviour or failures, rendering performance unpredictable and leading to an unsatisfactory user driving experience. The aforementioned criteria impact automotive manufacturers' regulatory standards for EMI emissions and susceptibility to determine compliance, leading to market disadvantages.

6.1.4 Overview of EMI Sources in Vehicles

Vehicles emit and receive EMI, creating a complicated electromagnetic environment. EMI is characterised in two primary types which require comprehensive shielding strategies.

1. Internal sources include ECUs, infotainment systems (e.g., GPS and Bluetooth), power electronics and electric motors.

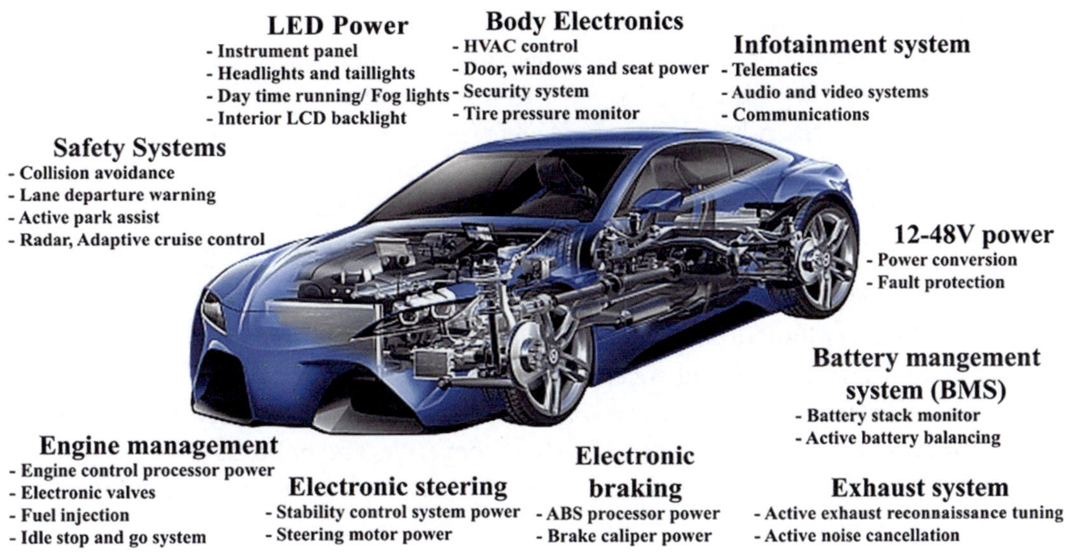

FIGURE 6.2 EMI sources in vehicles, image is licensed CC-BY-SA 4.0 International.

2. External sources include garage door openers, remotes, mobile phones, Bluetooth devices, MP3 players with Wi-Fi, third-party navigation and high-power transmitters like TV and radio towers.

In conclusion, as the automotive industry continues to innovate with more advanced electronic systems, the significance of adequate EMI shielding cannot be overstated. Shielding strategies must evolve in parallel with technology improvements to assure an efficient EMI shield and reduce radiation interference offering dependable electronic system performance.

6.2 TYPES AND SOURCES OF EMI IN VEHICLES

6.2.1 Internal Sources of EMI

The rapid pace of technological improvement opens the route for many electrical and sensitive devices which are internal sources of EMI in modern vehicles [9]. All these sources must be known for the development of EMI mitigation techniques.

1. **ECUs:** They are regarded as the brain of the vehicle and are responsible for controlling many automotive functions. Modern vehicles can have upwards of 100 ECUs, each managing different aspects of vehicle operation like engine management, air conditioning and transmission control as well as safety features such as airbags and ABS. These units generate EMI through their microprocessors and communication interfaces, particularly during data processing and communication operations. To illustrate, the high-speed switching of transistors and the data communication between ECUs over networks like Controller Area Network (CAN) and Local Interconnect Network (LIN) play a significant role in generating internal EMI.

2. **Infotainment Systems:** Also known as entertainment systems, they have evolved from simple radio receivers to full-fledged multimedia hubs featuring touchscreens, Global Positioning System (GPS) navigation, Bluetooth, Wi-Fi and cellular connectivity. Each of them has different frequencies to operate effectively. In turn, they become a potential source of EMI. For example, the GPS utilises the L1 and L2 frequencies of 1.57542 and 1.2276 GHz, respectively, while the frequencies used for Wi-Fi operate at 2.4 GHz and 5 GHz, which can interfere with other electronic systems if they are not shielded properly [10].

3. **Power Electronics:** Nowadays, hybrid and electric vehicles have gained high popularity to overcome the challenges of conventional cars and become more environmentally friendly [11, 12]. These vehicles necessitate the presence of high-power components like electric motors, inverters, converters and onboard chargers for converting DC to AC (and vice versa), which involve high-frequency switching in the range of 20 kHz to several hundred kHz. Consequently, electromagnetic waves are generated and may interfere with and disrupt the electronic circuits nearby.

4. **Wireless Charging Systems:** The wireless charging systems for EVs use high-frequency electromagnetic fields to effect energy transfer. These can generate significant EMI that could impact other nearby electronic systems. Effective shielding and filtering solutions are required to manage this interference and ensure that wireless charging operates safely and efficiently.

5. **Augmented Reality (AR) and Advanced Displays:** Introduction of AR into vehicles, both HUDs and digital dashboards, requires serious EMI management for these systems to be insensitive to interference. Proper operation of these systems is crucial to ensure driver information and safety. Display performance with shielding solutions should be clear without any interruption.

6.2.2 External Sources of EMI

Vehicles are unintentionally surrounded by multiple sources of EMI, which penetrate the vehicle's body and can reach and affect the regular operation of its sensitive electronic systems. This radiation is considered as an external source of EMI.

1. **Radio Frequency Interference (RFI):** Signals from radio and television broadcast towers, mobile phone base stations and other wireless communication devices generate electromagnetic radiation. The generated waves interfere with vehicles by inducing unwanted currents in vehicle wiring and systems, leading to interference in onboard electronics. For example, AM radio broadcasts range from 530 kHz to 1.7 MHz, while FM broadcasts occupy the band from 88 MHz to 108 MHz, which can affect the performance of sensitive devices.

2. **Electromagnetic Pulses (EMPs):** These can be generated from natural phenomena such as lightning or man-made nuclear detonations. They are transient electromagnetic energy, which propagates and represents a serious threat to vehicle

electronics. EMPs become a vital concern since an atmospheric discharge may strike and cause large-voltage surges, which would propagate throughout the vehicle's electrical system and could damage sensitive electronic components and cause malfunction.

3. **Wireless Communication Systems:** Another EMI challenge is brought about by the increased utilisation of wireless communication systems for vehicle-to-vehicle (V2V) and vehicle-to-infrastructure (V2I). In these systems, communications occur in the 5.9 GHz band for dedicated short-range communications (DSRC) as a key ingredient to autonomous and connected vehicles. This high-frequency signal employed in V2V and V2I communication has a high propensity to interfere with other electronic systems in the vehicle, for instance navigation and collision detection among others. Therefore, they should be accordingly managed and maintained to ensure proper EMI shielding.
4. **5G and Beyond:** Deployment of 5G networks in vehicles and beyond will increase the speed of data transfer and connectivity. But new challenges are expected to arise regarding EMI shielding because of the higher frequencies of the newer network systems. Materials for shielding and designs will have to change to allow them to block or absorb high-frequency electromagnetic waves while maintaining transparency to communication signals.

6.2.3 Conducted vs Radiated EMI

Conducted EMI is the type of EMI that is transmitted through electrical conductors. The usual source of this EMI arises from switching power supplies, electric motors and other components that introduce voltage and current fluctuations into the vehicle wiring. Conducted EMI can propagate through the vehicle's electrical system, affecting several sensitive components.

Radiated EMI is the other type of EMI which is emitted as electromagnetic waves and travels through the air. Such EMI sources can be internal or external, including wireless communication devices, radio frequency transmitters and many electronic circuits present within a vehicle. The radiated EMI can penetrate through the vehicle enclosures and couple inductively and capacitively into sensitive electronics. Figure 6.3 illustrates the complexity of EMI radiation between conducted and radiated emissions.

An electronic device can be affected by both kinds of emissions as demonstrated in Figure 6.4. In such a case, one device will play a role as a noise source while the other is called a victim. Usually, the noise source emits radiated emissions and, if connected to the victim through electrical wiring, a conducted emission will apply. This will create a coupling path between the noise source and the victim, leading to an increase in the EMI effect which will cause failure or malfunction of the device [13].

6.2.4 Frequency Ranges and EMI

The electromagnetic spectrum is the full range of electromagnetic radiation and consists of both electric and magnetic energies. They are arranged by frequency (f) or wavelength (λ)

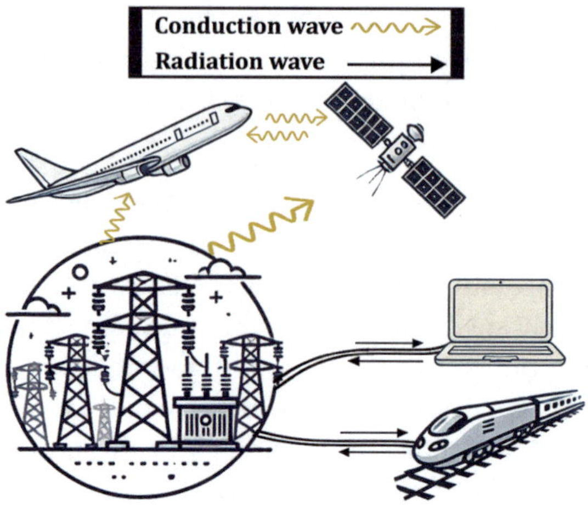

FIGURE 6.3 Complexity of EMI radiation: conducted and radiated emissions.

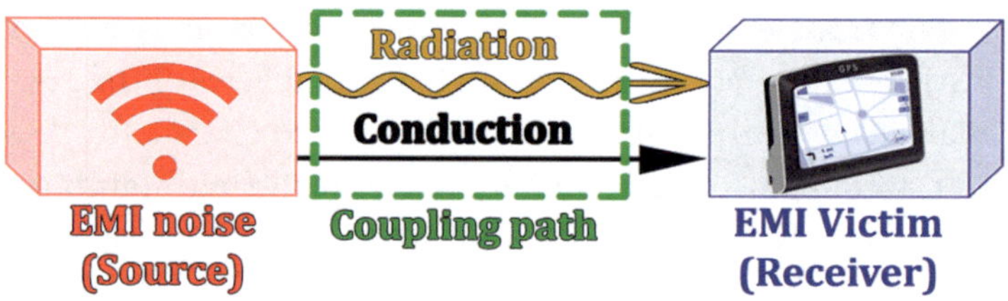

FIGURE 6.4 EMI impact between source and receiver.

which are related to each other (inversely proportional) according to the following equation [14]:

$$C = f\lambda \tag{6.1}$$

where

- C is the speed of light in a vacuum (299,792,458 m/s).
- f is the frequency, in hertz (Hz).
- λ is the wavelength (nm).

The spectrum is divided into several bands as demonstrated in Figure 6.5. They are divided into bands such as radio waves, microwaves, infrared, visible light, ultraviolet, X-rays and gamma rays, each having distinct characteristics such as how they are created, interaction with matter and practical uses [15].

Low-frequency EMI occurs below 150 kHz. It happens due to heavy machinery and power electronic parts in the vehicle and includes the noise from power inverters, battery management systems and electric motor drives. High-frequency EMI, which ranges from

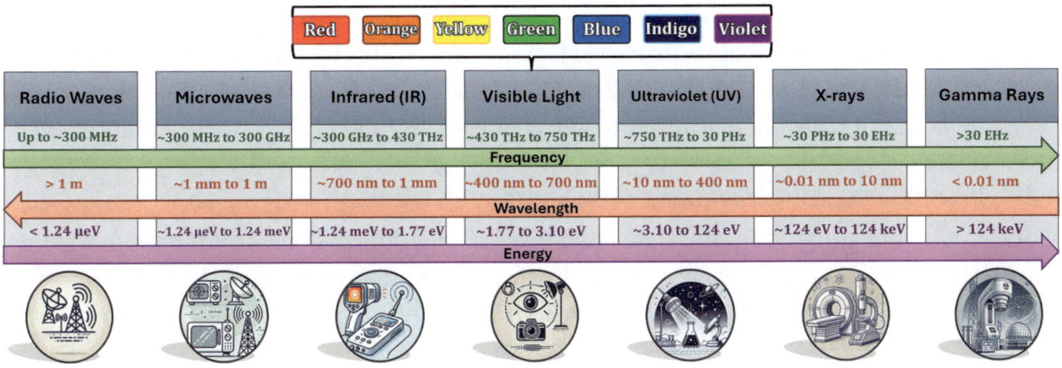

FIGURE 6.5 The electromagnetic wave spectrum.

150 kHz to several GHz, is produced by high-speed digital circuits, wireless communication systems, infotainment systems, GPS, Wi-Fi and mobile communications.

EMI types and their sources' investigation within a vehicle is very important in developing methods and designs to shield from wave interference. Identifying EMI sources and characterising them allows automotive engineers to design systems and select materials capable of minimising interference, hence ensuring the safe and reliable operation of modern vehicles.

6.3 EFFECTS OF EMI ON AUTOMOTIVE SYSTEMS

EMI can affect automotive systems at many different levels, potentially degrading the safety, reliability and performance of vehicles. As a car will continue to be electronically sophisticated, the necessity to understand and mitigate the effects of EMI will continue to become more vital. In this section, we explore some specific interference impacts on the various systems or components used within automotive applications.

6.3.1 Impact on Vehicle Safety Systems

The motivation is that vehicles have a strong dependence on electronic controls regarding various safety-critical systems. Such interference by EMI impacts their normal operations and leads to dangerous situations.

1. **ABS:** It prevents wheel lockup to retain steering control during braking. EMI might interfere with those sensors and the control unit which monitors wheel speed and then adjusts the brake pressure accordingly. This may have serious implications on the response by the system and could lead to late or incorrect operation of the ABS system, increasing the possibility of skidding or a catastrophic accident.
2. **Airbag Deployment Systems:** Airbags rely on crash sensors capable of deploying airbags in milliseconds when a collision is detected. EMI can spuriously trigger or delay the deployment system so that airbags either deploy when they should not or fail to deploy in time, both conditions being quite dangerous.

3. **Electronic Stability Control (ESC):** Also known as Electronic Stability Program, it acts to maintain vehicle control in extreme manoeuvres by modulating engine power and selectively applying the brakes. EMI in such a system would affect the sensors and actuators, probably leading to a loss of vehicle control during critical moments.

6.3.2 Interference with Navigation and Communication Systems

Modern cars are fully equipped with advanced navigation and communication systems. They have become essential for any driver by providing aid and comfort. EMI can interfere with these systems and prevent their stable and normal function.

1. **GPS:** GPS receivers can be a victim to interference easily since they rely on a very weak signal from the satellite. EMI radiated from internally mounted infotainment systems or from external radio frequency towers may cause a reduction in accuracy in the GPS data or loss of the signal, impacting navigation and route planning or services that rely on precise location data.
2. **Telematics and Infotainment Systems:** Telematics systems provide important data and communication services, including vehicle diagnostics, remote control features and emergency assistance. Infotainment systems integrate audio, video, navigation and communication services. EMI can be responsible for malfunctions such as distortion in audio, problems with displays or connectivity. It affects both the user experience and functionality of such systems.
3. **Vehicle-to-vehicle (V2V) and vehicle-to-infrastructure (V2I) communication systems:** They hold great importance for future development in both autonomous driving and advanced driver-assistance systems. In this system, there is wireless communication about the conditions of the road, traffic and hazards through data sharing. EMI can interfere with these communications, leading to delays or errors in received data and compromising vehicle safety and efficiency.

6.3.3 Disruption of Infotainment and Entertainment Systems

Infotainment and entertainment systems development and advancement found their way into car system luxury, adding to the amalgamation of technologies such as AM/FM radio, satellite radio, Bluetooth, Wi-Fi and digital display. They become a source and a victim for EMI interference.

1. **Audio Systems:** EMI easily causes noisy audio systems with the appearance of hissing, buzzing or static. It is very evident when one uses an AM/FM radio since the interference degrades the quality of the signal received. For digital audio, the result will be some dropouts or reduced clarity.
2. **Visual Displays:** Most of the state-of-the-art modern vehicles are equipped with all sorts of displays such as a touchscreen, instrument clusters and HUDs. EMI causes flickering, distortion or total loss of the display signal which influences driver interaction with vehicle systems.

3. **Wireless Connectivity:** It is represented by Bluetooth and Wi-Fi connections. EMI interferes with the signal and disrupts media streaming, audio phone calls or even the delivery of data. This is quite an issue with navigation systems when information is required in real time or if one relies on hands-free communications.

6.3.4 Effects on Battery Management Systems (BMS) in Electric Vehicles (EVs)

EVs have some unique EMI challenges, especially regarding BMS. The BMS will monitor and manage the state of charge of the battery pack, together with its temperature and general health. There are various ways in which EMI can cause serious damage.

1. **Data Corruption:** EMI can corrupt data normally emanating from sensors connected to a BMS. Thus, it provides wrong readings of the voltage, temperature and current of the batteries. This may result in improper charging or discharging of the batteries. It can reduce battery life or even cause safety issues.
2. **Control Signal Interference:** The BMS relies on control signals to manage battery operations and communicate with other vehicle systems. EMI can interfere with these signals, leading to issues like incorrect power distribution, inefficient energy use or even system shutdowns.
3. **Thermal Management:** It plays an important role in the operation of EV batteries to ensure their safety and durability. EMI can interfere with sensors and ECUs within a thermal management system, which may result in overheating or less effective cooling of cells and may lead to accidental fire.

6.3.5 System Reliability and Diagnostic Challenges

EMI can influence the overall reliability of vehicle electronics while also making diagnostic procedures more complex in general. Electronic component degradation will take place due to repeated EMI exposure over time, shortening life span and making components more prone to failure. Also, it can interfere with the onboard diagnostic system, creating false codes or even suppressing the real issues. This kind of issue creates big complications during maintenance processes and repair that may lead to misdiagnosis and unnecessary part replacements.

In general, EMI has been a big challenge in the automotive sector because it affects everything in a vehicle from systems to components. All these effects must be put into consideration when devising effective EMI shielding and mitigation strategies that will ensure the safety, reliability and performance of modern vehicles.

6.4 EMI SHIELDING MATERIALS AND TECHNOLOGIES

EMI shielding in automotive applications is basically a series of effective protections against unwanted electromagnetic radiation coming towards onboard electronic systems. It involves the suitable material and technology for attenuating EMI, hence ensuring high EMC that electronic systems work without interference. This section gives an overview of materials and technologies applied in EMI shielding for automotive applications [16].

6.4.1 Conductive Materials

Conductive materials are the most common types of EMI shielding material. They mainly depend on reflection shielding mechanisms, thus preventing electromagnetic waves from penetrating sensitive areas.

Metals such as copper, aluminium and steel are excellent EMI shielding materials since they have high electrical conductivity and reflective resonance to electromagnetic waves. Copper is very effective for a wide range of frequencies but is expensive in terms of cost. Aluminium is lighter and cheaper, though it provides a little lower shielding effectiveness. Steel materials, especially galvanised steel, are utilised for strength and resistance to corrosion, though far less effective than copper and aluminium.

Metal foams and meshes offer lighter versions of solid metals. They can be used in the shielding of heat-generating components like power electronics and electric motors because they serve as effective shields while allowing the flow of air for cooling. Common applications are found in enclosures and gaskets.

Conductive coatings are applied to non-conductive substrates to provide EMI shielding, which is predominantly made of copper, silver and nickel. These conductive coatings are usually applied by electroplating, sputtering and painting techniques. They take place when weight and space are critical concerns, such as in the shielding of ECUs and other sensitive electronics.

However, this type of shielding encountered numerous shortcomings, including high density, susceptibility to atmospheric corrosion for metal types, low mechanical flexibility and high processing costs [5, 17]. To address these drawbacks, advanced EMI shielding materials were favoured, developed and employed.

6.4.2 Advanced Materials

Advanced shielding materials have attracted the interest of current research. They can be classified as follows:

1. Magnetic materials or nanoferrites such as magnetite, strontium hexaferrites and barium hexaferrites.
2. Conductive polymers like polyaniline (PANI), polypyrrole (PPy) and polythiophene (PTh).
3. Nanocarbons such as fullerenes, nanocarbon black (CB), CNTs, reduced graphene oxide (RGO) and graphene.
4. Hybrid nanocomposites.

Most of these materials are distinguished by their high electrical conductivity, superior thermal stability and low density, providing a competitive candidate to the metals. They offer several advantages, including lightweight, flexibility, corrosion resistance, cost-effectiveness and durable EMI shielding [18–20].

6.4.2.1 *Magnetic Materials*

Magnetic materials are very effective in low-frequency magnetic fields commonly produced by electric motors, transformers and power lines.

From the magnetism perspective, materials can be classified into different categories based on their response to magnetic fields. Diamagnetic materials are those where the magnetic field in a material cancels each other. Examples include Pb, Cu, Si and NaCl, which are largely unaffected by magnetic fields, possessing a magnetic susceptibility on the order of -10^{-5} ($\chi_m < 0$, $\mu_r \leq 1$). While paramagnetic materials exhibit incomplete cancellation of internal magnetic fields. They are temperature dependent and their magnetic susceptibility ranges from $+10^{-5}$ to $+10^{-3}$ ($\chi_m > 0$, $\mu_r \geq 1$). Air, Pt, W and K are some examples of paramagnetic materials. Moreover, ferromagnetic materials derive their name from the element iron (Fe), which is one of the most prominent ferromagnetic elements. Other examples include cobalt, nickel and their alloys, all of which contain atoms with substantial permanent magnetic moments. These materials are characterised by large numbers of magnetic susceptibilities ($\chi_m >> 0$, $\mu_r >> 1$) [21].

Mu-metal is a type of nickel-iron alloy that has extremely high magnetic permeability; hence, it is very effective in shielding at low frequencies. Its applications vary from shielding sensitive components such as sensors and communication systems in electric and hybrid vehicles [22, 23].

Ferrites are basically ceramic materials composed of iron oxide combined with one or more other metals. They find applications in EMI suppression applications, such as cores in inductors and transformers and on cables as beads for the suppression of high-frequency noise. Some physical properties of ferrites depend upon the method of synthesis, the choice of the dopant, the concentration of the dopant and the site at which the doping is done [24].

In EMI shielding, ferromagnetic materials are very important to protect sensitive electric devices from the effect of external magnetic field that will cause disruption to their regular performance [25, 26].

6.4.2.2 *Conductive Polymers*

Based on the conduction nature of the polymer, they may be categorised into two main types [17]. The first type is the intrinsic conducting polymers (ICP), which refers to the polymers that could conduct electrical current intrinsically without the addition of any inorganic conductive substances. The other one is conductive polymer composites (CPC), referring to any non-conductive polymers that obtain conductivity upon addition of conductive filler into the polymeric matrix; hence, the polymer becomes electrically conductive.

Most recently, for the applications of EMI shielding, the attention has shifted to ICP benefiting from their high conductive property as well as the ability of its variation with simple oxidation or reduction treatments, hence they have secured an important place in developing commercially viable EMI shielding material [27, 28].

Generally, conductive polymers, such as polyaniline, polythiophene and polypyrrole, can be used as standalone materials or as coatings over other substrates. They offer some degree of flexibility, lightweight and corrosion resistance. Applications include any use where traditional metal shielding is impractical due to weight or design constraints.

6.4.2.3 Nanocarbons

Nanocarbons including fullerenes, CB, CNTs and graphene have emerged as a promising material for EMI shielding applications due to their outstanding electrical conductivity and light weight combined with mechanical strength [29]. Different carbon structures are shown in Figure 6.6.

Fullerenes are a form of carbon that consists of carbon atoms as molecules in a spherical structure. They are one of the first nanocarbons which drew attention for EMI shielding. In some applications, fullerenes may prove effective because they can absorb and dissipate electromagnetic energy. However, cost and complexity issues persist and several aspects of the integration of fullerenes into composite materials stand in the way of their large-scale usage in automotive applications [30].

CB is one of the amorphous kinds of carbon that has been used intensively as a filler material in conductive polymers. Due to its small particle size and high surface area, it enhances the electrical conductivity of composites and, thus, their performance in EMI shielding. Hence, in automotive applications, CB is usually added to the polymer matrices to form lightweight conductive composites with an easily mouldable capability into complex shapes and hence suitable for several vehicle components [31].

Carbon nanotubes (CNTs) are considered the most exciting new targets today because of their exceptional electrical, thermal and mechanical properties. One-dimensional CNTs are single-walled (SWCNTs) or multiwalled (MWCNTs), with the latter showing more effectiveness due to higher aspect ratio and electrical conductivity. When integrated into polymer matrices or other composite materials, CNTs form conductive networks that help strongly dampen electromagnetic waves. The high aspect ratio allows low loading levels so that the mechanical properties of the host material can still be retained while effective EMI

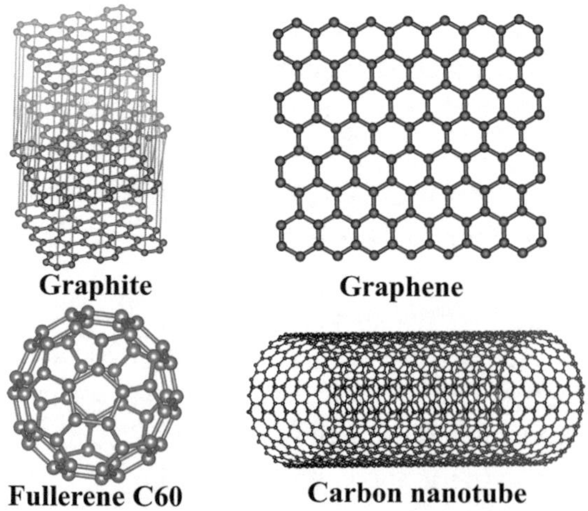

FIGURE 6.6 Different carbon structures.

shielding could be maintained. This makes them particularly attractive in automotive applications where weight reduction is of paramount importance [32, 33].

RGO is a form of graphene retaining some oxygen functional groups, which could be favourable in improving its dispersion in various matrices. RGO exhibits electrical conductivity that is very good and might be used in thin films or coatings in EMI shielding. Its structure allows effective absorption and reflection of electromagnetic waves, which makes it a promising material in light shielding applications [34]. The ability of being processed into large-area films or coatings enhances its applicability in automotive applications where the shielding should be uniform and effective over extended surfaces [35, 36].

However, nanocarbons are used for the shielding of critical electronic parts, including engine control units and advanced driver-assistance systems, from EMI. Such materials provide not only effective shielding but also contribute to the overall reduction of vehicle weight which has become highly relevant for fuel efficiency improvement and reduction of emissions.

6.4.2.4 Hybrid Nanocomposites

Recently, hybrid nanocomposites have attracted researchers' interest in the incorporation of ferrites, conductive polymers and nanocarbons to take advantage from each type for enhancing the EMI shielding efficiency covering different frequency ranges beyond their limitations and hence outperforming the performance of their counterpart [37–39].

Composite materials combine conductive and non-conductive materials to deliver tailored EMI shielding properties. They find widespread applications in the automotive industry, where weight, cost and mechanical properties are concerns [40].

Carbon fibre composite materials have very light and strong features that make them highly suitable for automobile applications. Further, they can be designed to offer EMI shielding with the help of conductive fillers or coatings. They are also applied to structural components, body panels and enclosures [41].

This new class of materials, or hybrid nanocomposites, presents a high surface area and tunable electrical properties, thereby representing an effective material for shielding over a wide range of frequencies. Currently, there is some research and development in this material group, but nanocomposites hold promising potential in future automotive applications.

6.4.3 Emerging Materials

The growth in the automotive segment paves the way for new materials and technologies such as metamaterials and liquid metal to address the increasing requirements for EMI shielding in today's modern vehicles.

6.4.3.1 Metamaterials

In general, metamaterials are a class of artificially fabricated materials engineered to possess electromagnetic properties uncommon in nature. Typically, artificially made metamaterials are composed of periodically structured units, called meta-atoms, which influence electromagnetic waves according to new principles. Such structures can be designed to possess negative permeability or permittivity, enabling exceptional possibilities for the control of electromagnetic wave propagation and absorption.

Metamaterials are promising a new frontier in the field of EMI shielding. Since metamaterials can have unique properties, this would serve well to block or redirect electromagnetic radiation in a manner that enhances shielding performance. Most of the research in metamaterials is experimental and involves the optimisation of properties and practical applications. With further development of metamaterials, important advances could be achieved in EMI shielding technologies, which offer better performance and versatility than traditional materials [42].

6.4.3.2 Liquid Metals

Due to their intriguing properties, liquid metals, especially gallium-based alloys, have emerged as one of the new promising material systems for EMI shielding applications. These materials are in a liquid state at room temperature or slightly above it; hence, they enable flexible and conformable shielding applications. Some of the main benefits of liquid metals include inherent EMI shielding, self-healing properties and easy application on various substrates. Gallium alloys, such as eutectic gallium-indium (EGaIn), have received much attention due to their high EMI shielding effectiveness [43]. These alloys have high electrical conductivity with low density. Self-healing properties involve the material recovering its shielding effectiveness after mechanical damage, making them particularly suitable for dynamic and flexible electronic applications. Besides, integrating liquid metals with flexible electronics holds immense potential for advanced design in automotive applications where rigid shielding solutions may be less feasible [44].

The combination of metamaterial and liquid metals will provide automotive EMI shielding solutions that will enable the industry to fully exploit these emerging technologies to ensure performance, flexibility and durability to meet the expanded needs of today's electronic systems.

6.4.4 Shielding Effectiveness and Design Considerations

The effectiveness of EMI shielding does not depend on the right choice of materials only, it depends equally on the design and implementation of the shielding solution [45]. Therefore, shielding effectiveness and EMI attenuation rely on the following parameters [46–49]:

- **Type of the Material Used for Shielding:** There is a variety of materials used to provide an effective shield against EM waves. They depend on several criteria and the targeted application.

- **Thickness and Coverage:** Shielding effectiveness depends directly on the material thickness used for shielding and the amount of coverage. Usually, shielding efficiency increases with the increase of material thickness but it is a point at which researchers compete to provide a thin shield with the highest possible efficacy. Full coverage, with appropriate thickness, is considered highly necessary to attain high performance regarding shielding.

- **Joints and Seams:** In automotive applications most of the shielding materials contain joints and seams. These areas sometimes show EMI leakages and weakness. A proper design and good seal are necessary in this area to maintain the integrity of the shield. Conductive gaskets and tapes are generally used for sealing gaps.
- **Grounding:** A proper grounding of the shielding material needs to be performed to prevent a build-up of static charges and ensure effectiveness in diverting electromagnetic waves away from the sensitive components.
- **Ventilation and Heat Management:** Shielding must not compromise the cooling of electronic components. In practice, most shielding designs also contain vents or heat sinks for handling thermal problems while maintaining EMI protection.
- **The Wave Frequency:** The shielding effectiveness may vary for the same material depending on the band frequency since each band has its own characteristics.
- **Distance Between the EM Source and the Shield:** shielding effectiveness is affected by the distance between the wave source and the shielding material, dividing measurement region into far field and near field regions. For far field region, plane wave exists, and the value of EM wave impedance is the same as the intrinsic impedance of free space. While in the near field region, the EM wave impedance will be different than the intrinsic impedance of free space.

This entails selecting the appropriate EMI shielding materials and methods to protect automobile electronic systems from EMI. Because car electronics continue to become more complex and integrated, inventive shielding solutions will be sought after, allowing for more research and development in this field.

6.5 DESIGN AND IMPLEMENTATION OF EMI SHIELDING IN VEHICLES

Vehicle EMI shielding design and implementation play an important role in ensuring reliable and safe operation of the automotive electronic system. This section describes the principles, strategies and practical considerations that are involved in designing and implementing effective EMI shielding in vehicles.

6.5.1 Principles of EMI Shielding Design

The primary philosophy of EMI shielding design is to ensure that external EMI does not affect sensitive electronic systems.

Electromagnetic waves are fundamentally composed of two essential components: the electric field (E) and the magnetic field (H). These two are mutually perpendicular with the electromagnetic wave propagating in a direction orthogonal to the plane formed by the E and H fields. Both have their characteristics which depend on their frequency and associated photon energies. Also, the ratio between these two fields' strength is known as wave impedance. In free space, the impedance is called intrinsic impedance and has a constant value of 377 Ω [17, 47].

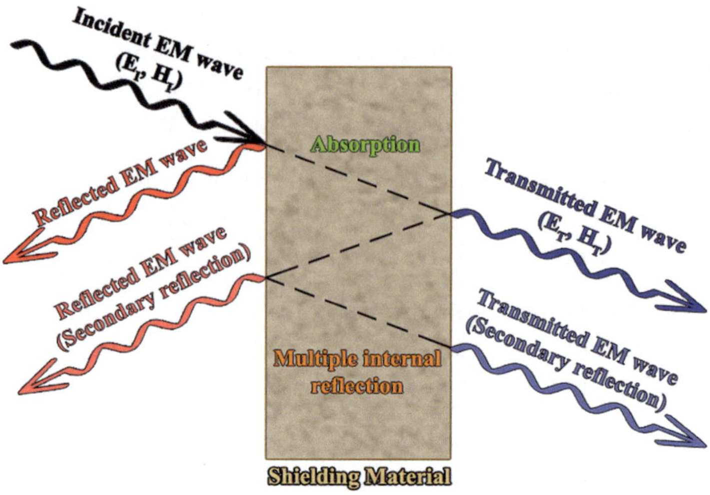

FIGURE 6.7 Simple schematic diagram for the EMI attenuation mechanisms.

The interaction of the shielding material and incoming EM waves is described and fully explained by a theory known as the theory of EM shielding [50]. The shielding efficiency of any material counts on absorption, reflection and multiple internal reflections [51, 52].

After the incident wave reaches the surface of the material, it is divided into several components, as illustrated in Figure 6.7. A portion of the wave is reflected from the front surface of the shielding material. A part of the wave is absorbed within the shielding material, contributing to energy dissipation. A fraction of the wave is reflected from the back internal surface of the material towards the front, where it possibly enhances the effectiveness, depending on the phase relation with the incident wave. The transmitted part passes through the material and emerges on the other side.

Shielding effectiveness is the attenuation of electromagnetic energy through a material or structure. It is usually measured in decibel units (dB). Its value depends on conductivity of the material, its thickness, the frequency of EMI and distance from the source [53]. It is mathematically represented by the ratio of the incident field to the transmitted field strengths [54].

$$SE_P = 20 \log_{10}\left(\frac{P_I}{P_T}\right) \tag{6.2}$$

$$SE_E = 20 \log_{10}\left(\frac{E_I}{E_T}\right) \tag{6.3}$$

$$SE_H = 20 \log_{10}\left(\frac{H_I}{H_T}\right) \tag{6.4}$$

The electromagnetic plane wave strength is noted as (P), electric field strength is (E) and magnetic field strength is (H). The subscript (I) stands for the incident magnitude of the field strength on EMI shielding material and the subscript (T) is for the transmitted magnitude.

Therefore, to judge the effectiveness of the shielding material, the total shielding efficiency (SE_T) needs to be considered. It is the sum of efficiency due to reflection, absorption and multiple internal reflections according to Equation (6.5).

$$SE_T = SE_A + SE_R + SE_M \tag{6.5}$$

SE_T is the total shielding efficiency while SE_A, SE_R and SE_M represent the absorption, reflection and multiple internal reflection shielding efficiency.

The multiple internal reflection losses are influenced by the thickness of the shielding material and take place multiple times inside the material between its two surfaces. These losses become negligible when SE_A value is high, typically exceeding 10 dB [55]. In addition, they are significant at low frequencies below 20 kHz which means they can be irrelevant for high frequency [5].

6.5.2 EMI Shielding Strategies in Vehicles

Diverse EMI shielding approaches may be employed in vehicle electronic systems according to the source and nature of the interference and system requirements. Figure 6.8 illustrates some applicable EMI shielding forms in vehicles.

A very common form of shielding is in the enclosures and housings used to shield the ECU, power electronics and other sensitive components. They take the form of conductive material enclosures that are designed to contain or exclude the electromagnetic fields.

Shielding cables can be helpful in preventing EMI from being conducted along wiring harnesses. It typically involves a conductive braid or a foil wrap that acts as a shield to prevent interference radiation inside or outside the cable.

For the effective implementation of EMI shielding, proper grounding and bonding are crucial. The safety of dissipating unwanted electromagnetic energy is possible by providing the path for the same through grounding, while bonding ensures electrical continuity between different parts of the vehicle. The potential differences are reduced, and hence the chances of EMI are minimised accordingly.

Conductive gaskets and seals are applied to joints and seams in shielded enclosures to avoid EMI leakage. Gaskets can be made of metal, conductive elastomers or even

FIGURE 6.8 EMI shielding applicable forms, sub-images are licensed CC-BY-SA 4.0 International.

conductive fabrics that make the shielded enclosure effective, even at the points where different materials meet.

Conductive paints and coatings can be applied to provide EMI shielding. These are often applied where adding a separate shielded enclosure is impractical. Coatings include materials such as nickel, silver or graphite, providing a conductive layer that can reflect and absorb electromagnetic waves.

6.5.3 Implementation Challenges

The main difficulties in applying an EMI shield are weight and space constraints. Vehicles are very strictly limited by weight and available space. Therefore, heavy or bulky shielding material cannot be easily fitted. It needs material selection and creative design solutions which balance EMI protection with these constraints in such applications.

Thermal management could be an important factor because most materials used for shielding provide a barrier to the dissipation of heat that may cause overheating in electronic components. This will demand the incorporation of thermal management solutions such as heat sinks, ventilation and thermally conductive materials in addition to EMI shielding.

Modern vehicles involve complex geometries that, in turn, cannot be totally shielded. The geometrical difficulties require flexible shielding materials, conductive fabrics or coatings that allow conformity to irregular shapes.

Material cost and process cost could be factors that may hugely affect the overall cost of EMI shielding solutions. Automakers are always challenged by the need to optimise budgets while looking for effective EMI protection, thus forcing them to use the most cost-effective materials and manufacturing processes.

6.5.4 Testing and Validation of EMI Shielding

Testing and validation techniques are necessary steps involved in every EMI shielding solution. Some of the major testing methods include shielding effectiveness testing. The shielding effectiveness testing is the measurement of attenuation of electromagnetic waves by the material used for shielding. Testing can be done using methodologies such as coaxial transmission line or reverberation chamber [56, 57]. Results obtained are converted into a shielding effectiveness value in decibels.

Electromagnetic compatibility testing offers confidence that a vehicle and its components are not emitting excessive EMI and can operate adequately in the presence of external EMI. The three primary categories into which EMC tests can be divided are emissions, immunity and susceptibility. Testing normally takes place in an appropriate EMC test facility, including an anechoic chamber as well as proper measurement equipment.

The thermal and environmental testing of EMI shielding materials is done to ensure that they can bear the extreme conditions a vehicle is usually subjected to. Tests for temperature cycling, humidity, vibration and mechanical shock are included.

6.5.5 Case Studies and Examples

Several case studies and examples can be enlisted to present the phenomenon of EMI shielding in a practical implementation perspective.

Recently, electric and hybrid vehicles attracted attention. They are designed with high-voltage power electronics and electric motors that are substantial EMI generators. Shielding strategies include shielded enclosures for inverters and converters, shielded cables for high-voltage wiring and special coatings on electric motor housings.

Another example is autonomous and connected vehicles. These vehicles rely mostly on V2V and V2I, Radar and LiDAR systems. EMI shielding majorly works to protect the sensitive communication and sensor systems against external and internal interference in these vehicles.

Infotainment systems' existence is not debatable to offer users' comfort. These systems are susceptible to both conducted and radiated EMI. The shielding strategies will include shielding enclosure, conductive coating on displays and audio and data cable routing with a guarded route.

Finally, EMI shielding in vehicles should be designed and implemented as a multidisciplinary process. Materials science, electrical engineering and configuration aspects play an important role in carrying out EMI shielding in an automobile. As cars keep changing and rising with advanced electronics, the need for effective and creative solutions to EMI shielding grows with each passing day.

6.6 REGULATORY STANDARDS AND TESTING FOR AUTOMOTIVE EMI SHIELDING

Regulatory standards and testing are extremely important in ensuring that automotive electronic systems have adequate EMI protection. This provides the basis on which vehicles are brought up to satisfy essential electromagnetic compatibility requirements, safety, reliability and performance in diverse environments. The section discusses key regulatory standards, test methodologies and compliance-related procedures related to EMI shielding for automotive applications.

6.6.1 Overview of Regulatory Bodies and Standards

Several international and national regulatory bodies regulate EMC and EMI shielding requirements for automotive applications. These ensure that vehicles and their components do not emit excessive levels of EMI and are immune to a certain level of EMI from external sources.

International Electrotechnical Commission (IEC) is the international standards organisation for electrical, electronic and related technologies. IEC 61967 and IEC 62132 are two major standards for measuring conducted and radiated emissions and immunity in integrated circuits, respectively, which have extensive applications in automobiles [58, 59].

The International Special Committee on Radio Interference (CISPR) establishes standards for RFI control. CISPR 12 and CISPR 25 are major standards relating to vehicles and their electrical/electronic subassemblies, including emission limits and test methods [60, 61].

The International Organization for Standardization (ISO) develops and publishes international standards for nearly every industry. The ISO norms related most to the subject include ISO 11452 and ISO 7637 which were created for the testing of immunity against radiated and conducted disturbances in vehicles and components, respectively [62, 63].

United Nations Economic Commission for Europe: The UNECE sets the EMC requirements through UNECE Regulation No. 10 for vehicles and electronic components. It places limits for both emissions and immunity so that vehicles will not interfere with other devices and have the ability to operate safely in an EMI-rich environment [64].

Federal Communications Commission of the United States of America and Industry Canada: In North America, FCC and IC set up the regulations for EMC compliance, especially regarding radio frequency emissions. Vehicles need to comply with such regulations in order to be sold in the respective markets [65].

6.6.2 Key Standards and Testing Methods

Automotive manufacturers should perform a series of tests to ensure their products comply with EMC regulations. There is a need for both emissions and immunity testing to demonstrate the capability of a vehicle or any of its components to coexist without causing or being subject to EMI.

Emission testing is where they measure electromagnetic emissions emitted from a vehicle or its component. These emissions are divided into two types of emissions. First, the electromagnetic waves emitted to the environment by the vehicle electronic systems known as radiated emission. The tests are done in an anechoic chamber, the vehicle is exposed to a number of frequencies to measure the intensity of the emissions. The second type is conducted emissions which are related to the electromagnetic disturbances that are transmitted along the power or data lines. Testing involves measuring the noise present on those lines to ensure it stays within acceptable limits.

Immunity testing characterises a vehicle's immunity against external EMI. The major types of immunity tests are based on radiated immunity tests where the vehicle must be exposed to external electromagnetic fields in order to realise whether the onboard electronic systems will be able to function correctly without degradation. Conducted immunity tests the capability of the vehicle to resist electromagnetic disturbances conducted along cables and wiring, including surges, transient and continuous wave phenomena.

Another important test is the electrostatic discharge (ESD) testing which gives the capability of resistance for a vehicle to an electrostatic discharge which may be induced either from human contact or through environmental conditions. ESD could cause malfunction or damage to electronic systems if not handled properly.

6.6.3 Testing Facilities and Equipment

Specialised facilities and equipment are required to carry out the tests mentioned earlier. Anechoic chambers, shown in Figure 6.9, are shielded rooms whose inner surfaces are covered with radiation-absorbing materials that dampen electromagnetic waves and provide an environment free from external EMI. They are used for radiated emissions and immunity testing.

Figure 6.10 demonstrates reverberation chambers used to test the immunity of various electronic devices to electromagnetic fields. Unlike anechoic chambers, which absorb electromagnetic waves, in reverberation chambers they are reflected to create a high-intensity environment for testing the strength of the device.

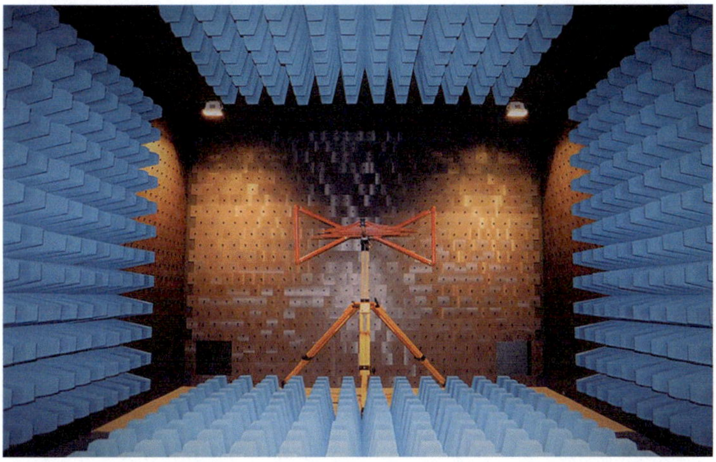

FIGURE 6.9 RF Fully Anechoic Room (FAR) host structure at INCESA Research Hub of Applied Science building, Craiova, Romania by Mihaelawojcik is licensed *CC-BY 4.0 International.*

FIGURE 6.10 Reverberation chamber, Supélec, France by Javalone is licensed *CC-BY-SA 4.0 International.*

Still, EMI receivers and spectrum analysers are used to measure the amplitude and frequency of electromagnetic emissions. These are very significant in the tests for emissions since they will indicate if a vehicle's emissions are within the limits of regulation. While signal generators and amplifiers are utilised in immunity tests in generating electromagnetic fields or conducted disturbances at varied frequencies and power levels.

6.6.4 Compliance and Certification Processes

The process of compliance with EMC standards is mandatory before the vehicles can be marketed and sold worldwide. In general, compliance and certification are completed through the following steps:

1. **Pre-compliance Tests:** It helps the manufacturer identify some of the major EMI issues early in the developmental stages so that costly redesigns and delays during the final certification process could be avoided.

2. **Third-Party Certification:** The official compliance tests are conducted very often by an independent, third-party testing laboratory. Laboratories obtain accreditation to make sure that the test procedures they use meet international standards.

3. **Documentation and Reporting:** Further to the testing, detailed reports are created to comprise test methods, results and any corrective actions taken. This documentation plays a critical role in regulatory submissions and is necessary for internal records.

4. **Labelling and Marking:** It includes the fact that when certification is effective, products are labelled to reflect compliance with applicable standards. This generally provides a means of entry into the market and assurance to the consumer.

6.7 CHALLENGES AND FUTURE TRENDS

With continuous innovation in the automotive industry, the demand for effective EMI shielding grows. Indeed, modern vehicles host a wide range of electronic systems that could be sensitive to EMI and hence call for modern ways of shielding. This section presents the current challenges in EMI shielding and future trends in developing more effective and efficient solutions.

6.7.1 Current Challenges in EMI Shielding

New vehicle designs, employment of new technologies and ever-tightening regulatory requirements continue to pose some challenges before the automotive industry in order to establish effective EMI shielding.

The development and complexity of electronic systems since the modern vehicle has dozens of ECUs, sensors, communication systems and infotainment devices. Together they form a matrix of electronic circuitry that increases the likelihood of EMI problems, because electronic devices are sometimes both emitters and susceptible to interference. Therefore, how to ensure real effectiveness of EMI shielding across all systems without excessive cost or added weight remains a big challenge.

Electrification and high-voltage systems take place with the increasing momentum towards EVs and HEVs. They bring onboard high-voltage systems and power electronics, thus introducing a high level of EMI. The mitigation of such interference is very critical in terms of ensuring safe and reliable operations of BMS, power inverters and electric motors. In the case of power electronics, high-frequency switching may be difficult to shield against.

Key features of autonomous and connected vehicles involve heavy reliance on sensors such as radar, LiDAR and cameras among others, and communication systems like V2V and vehicle-to-everything (V2X). These are highly prone to EMI; thus, potent shielding solutions to ensure functionality and safety are needed. In addition, the assembly of such technologies into one vehicle increases the possibility of mutual interference from one system to another.

Miniaturisation and weight reduction have become a vital concern in modern automotive sectors. With transportation becoming lighter and compact, the space for traditional shielding materials is further reduced. Additionally, manufacturers are under pressure to decrease vehicle weights to help improve fuel economy and increase range, especially for EVs. This requires novel, lightweight shielding materials without sacrificing effectiveness in shielding.

Cost is considered as a controlling factor. Effective EMI shielding can be expensive, as it concerns advanced materials and complicated designs. The automotive industry works within tight cost constraints, having to strike an optimal balance between the cost of solutions to EMI shielding and their effectiveness. The car manufacturers must find cost-effective ways of integrating shielding without making it much more costly to achieve the market demands.

6.7.2 Future Trends in EMI Shielding Materials and Technologies

Development of new materials and technologies is underway for EMI shielding as the automotive industry evolves to meet rapidly growing demands. These will address the challenges mentioned earlier, further enabling efficient, lightweight and cost-effective solutions.

Advanced composites such as carbon fibre, reinforced polymer and metal matrix composite usage is on the rise. These materials provide a good balance between lightweight and high shielding effectiveness. Current research is in progress to incorporate conductive fillers like CNTs or graphene in these composites seeking improved EMI shielding.

The unique electrical and mechanical properties of nanomaterials such as carbon nanotubes, graphene and metal nanoparticles are under investigation. In coatings, films and composite structures, these materials may provide lightweight, high-value EMI shielding. Another frontier of nanotechnology involves the development of materials that can provide multiple functionalities, thus enabling multifunctional shielding, thermal management and structural support.

Metamaterials are artificially manufactured materials that possess special electromagnetic properties not observed in naturally occurring materials. They can be designed to tune their permittivity or permeability for a specific response to the electromagnetic wave. The great potential of metamaterials for EMI shielding is their high selectivity and efficiency, especially in difficult environments or at specific frequencies where other natural materials are not effective.

It is important to develop flexible and stretchable EMI shielding material systems to achieve wearable electronics and flexible displays that will be embedded in vehicles. Such materials will be suitable and effective for use under geometrical complexity where rigid materials would be impracticable.

Conductive textiles and fabrics are also under development for automotive interior and exterior applications. These materials provide a lightweight, flexible alternative for EMI shielding in seat covers, headliners and other interior components. Conductive textiles can offer multifunctionality, such as heating and sensing.

6.7.3 Challenges and Future Trends in Regulatory Compliance

New frontiers in automotive technologies present a new challenge in EMI shielding and regulation compliance. These challenges concern the increased complexity of electronic systems, wireless communication and connectivity, global harmonisation of standards, emerging materials and advanced technologies.

Vehicles are increasingly equipped with sophisticated electronic systems such as ADAS and autonomous driving. The increasing complexity boosts the difficulty level for conducting EMC testing. These systems require tight EMI controls and strong test procedures. The wireless communication technologies proliferating inside the vehicle such as V2X communications raise the potential for EMI issues. Because these technologies present unique regulatory challenges, the standards are continuing to evolve.

In addition, the differences in regional EMC standards make the process of certification quite complicated for manufacturers targeting several regions globally. Efforts are underway to harmonise these standards and to make compliance procedures easier.

Newer materials and technologies, including especially metamaterials and graphene-based shielding, are very promising for EMI management. New testing methods and standards are required to exploit the full potential of such new developments.

Therefore, testing the shielding of automotive EMI and regulatory standards will play an essential role in ensuring that vehicles maintain certain standards of safety, performance and reliability. Moving on, with increasingly advanced automotive technologies, the industry continuously must update its testing methods and standards to cope with new challenges and opportunities.

6.7.4 Regulatory and Testing Evolution

The advancement of vehicle technologies will, in turn, require regulatory standards and testing methodologies to change to meet new challenges in EMI management.

There is a continuing effort towards harmonisation of EMC standards on a global scale for easy design and testing of vehicles intended for various markets. This should potentially help reduce the cost and complexity of compliance for manufacturers.

Dynamic and real-to-life testing becomes a necessary demand as the automotive industry evolves. The electromagnetic environment to which modern vehicles are exposed is dynamic and complex; traditional methods of testing cannot capture this crucial aspect. There is therefore an increasing interest in tests and simulations under real-world conditions offering a real-time and realistic test to understand and resolve EMI problems for a variety of operational scenarios.

Safety and cybersecurity present vital considerations, since the increased interconnectivity and automation in automobiles burden them with potential risks of cybersecurity and functional safety. EMI shielding in this regard helps protect the communication systems from interference and ensures the integrity of critical systems; thus, regulatory standards are evolving to address these new risks.

In a nutshell, EMI shielding for automotive applications moves at an intensive speed, considering the never-ending technological innovations and increased regulatory scrutiny that surrounds this field. The industry is undertaking huge challenges regarding EMI management, yet promising solutions are coming through with emerging materials and technologies. As increasingly connected, automated and electrified vehicles continue to make their mark, the need for effective and innovative EMI shielding methods will continue to drive into the future of design and engineering in automotive.

6.8 CONCLUSION

In this chapter, we have considered the vital role of EMI shielding within the automotive sector, where electronic content has become increasingly complicated and integral to vehicle safety and performance and connectivity. EMI can be internally generated as well as received from external sources that interfere with the operation of crucial systems such as safety controls, infotainment and communications. So, effective shielding becomes an essential requirement to overcome these risks. The materials commonly used to reduce EMI exposure include metal enclosures, conductive coatings and advanced composites.

Accordingly, the demand for more innovative EMI shielding solutions will continue to rise as vehicles electrify further and gradually move towards autonomous operation. Future challenges include controlling the development of EMI challenges with EVs, 5G and increasing electronics density. Meanwhile, engineers must balance the demands of effective shielding with cost, weight and design space constraints.

Further research and development are highly required from this perspective. Newly emerging materials such as nanomaterials, metamaterials and liquid metals have shown promise of substantial EMI shielding performance. They have added the advantage of being lightweight and multifunctional. Any development regarding new solutions would require collaboration among researchers, engineers and regulatory bodies because the technologies of vehicles keep evolving along with the regulatory standards.

EMI shielding is one of the crucial aspects in automotive design; surely, it's going to be of greater importance as time goes on in the industry. New materials and novelty design approaches will allow the sector of automotive to ensure that vehicles remain safe, reliable and free from the disrupting effects of EMI.

REFERENCES

1. C. Wang *et al.*, "Overview of carbon nanostructures and nanocomposites for electromagnetic wave shielding," *Carbon*, vol. 140, pp. 696–733, 2018, doi: 10.1016/j.carbon.2018.09.006.
2. N. Violette, *Electromagnetic Compatibility Handbook – Chapter 1: An introduction to electromagnetic compatibility*, Netherlands: Springer, 2013.
3. R. J. Perez, *Handbook of Aerospace Electromagnetic Compatibility – Chapter 10: EMI Effects in Flight Control Systems and Their Mitigations*. Wiley, 2018.
4. R. L. Yadava, *Microwave Engineering: Principle and Devices – Chapter 1: Basics of Microwave*. Notion Press, 2018.
5. X. C. Tong, *Advanced Materials and Design for Electromagnetic Interference Shielding – Chapter 1: Electromagnetic Interference Shielding Fundamentals and Design Guide*. CRC Press, 2016.
6. D. Weston, *Electromagnetic Compatibility: Principles and Applications, Second Edition, Revised and Expanded – Chapter 1: Introduction to EMI and the Electromagnetic Environment*. Taylor & Francis, 2001.
7. M. Feliziani, T. Campi, S. Cruciani, and F. Maradei, "Chapter 11 – Electromagnetic compatibility of automotive WPT systems," in *Wireless Power Transfer for E-Mobility*, M. Feliziani, T. Campi, S. Cruciani, and F. Maradei Eds.: Academic Press, 2024, pp. 361–395.
8. S. Sonko, C. D. Daudu, F. Osasona, A. M. Monebi, E. A. Etukudoh, and A. Atadoga, "The evolution of embedded systems in automotive industry: a global review," *World Journal of Advanced Research and Reviews*, vol. 21, no. 2, pp. 96–104, 2024.
9. F. Arena, G. Pau, and A. Severino, "An overview on the current status and future perspectives of smart cars," *Infrastructures*, vol. 5, no. 7, p. 53, 2020.

10. J. J. D'Costa and B. S. Ajith, "A review of autonomous vehicles," in *Self-Powered Cyber Physical Systems*, Rathishchandra R. Gatti, Chandra Singh, Rajeev Agrawal, Felcy Jyothi Serrao Eds., 2023, pp. 243–253.
11. N. Tiwari *et al.*, "Converting gasoline vehicle into an electric vehicle (EV)-A review," *Materials Today: Proceedings*, vol. 79, pp. 379–388, 2023, doi: 10.1016/j.matpr.2022.12.161.
12. F. Alanazi, "Electric vehicles: benefits, challenges, and potential solutions for widespread adaptation," *Applied Sciences*, vol. 13, no. 10, p. 6016, 2023.
13. X. Ruan, L. Xie, Q. Ji, and X. Yuan, "Introduction," in *Conducted Electromagnetic Interference in Power Converters: Modeling, Prediction and Reduction*, X. Ruan, L. Xie, Q. Ji, and X. Yuan Eds. Singapore: Springer Nature Singapore, 2024, pp. 1–24.
14. S. Sankaran and R. Ehsani, "Introduction to the electromagnetic spectrum," in *Imaging with Electromagnetic Spectrum: Applications in Food and Agriculture*, A. Manickavasagan and H. Jayasuriya Eds. Berlin, Heidelberg: Springer Berlin Heidelberg, 2014, pp. 1–15.
15. C. N. Stedwell and N. C. Polfer, "Spectroscopy and the electromagnetic spectrum," in *Laser Photodissociation and Spectroscopy of Mass-separated Biomolecular Ions*, N. C. Polfer and P. Dugourd Eds. Cham: Springer International Publishing, 2013, pp. 1–20.
16. R. Wilson, G. George, and K. Joseph, "Chapter 1 – An introduction to materials for potential EMI shielding applications: Status and future," in *Materials for Potential EMI Shielding Applications*, K. Joseph, R. Wilson, and G. George, Eds. Elsevier, 2020, pp. 1–8.
17. D. Jiang *et al.*, "Electromagnetic interference shielding polymers and nanocomposites – a review," *Polymer Reviews*, vol. 59, no. 2, pp. 280–337, 2019, doi: 10.1080/15583724.2018.1546737.
18. A. Kausar, "Chapter 14 – Hybrid polymeric nanocomposites with EMI shielding applications," in *Materials for Potential EMI Shielding Applications*, K. Joseph, R. Wilson, and G. George, Eds. Elsevier, 2020, pp. 227–236.
19. D. Saha, R. Cardinaels, A. Roc'h, T. A. Engels, and P. D. Anderson, "Designing multi-layer polymeric nanocomposites for EM shielding in the X-band," in *2018 International Symposium on Electromagnetic Compatibility (EMC EUROPE)*, 2018: IEEE, pp. 783–788.
20. K. R. Sahu and U. De, "Polymer composites for flexible electromagnetic shields," in *Macromolecular Symposia*, 2018, vol. 381, no. 1: Wiley Online Library, p. 1800097.
21. M. N. O. Sadiku, *Elements of Electromagnetics*. Oxford University Press, 2018.
22. S. R. Jog, I. Deshpande, M. S. Sutaone, and V. V. Badawe, "Analyzing and mitigating the near-field radiations from PCB of GPS-based vehicle tracking system," *IETE Journal of Research*, vol. 59, no. 4, pp. 392–399, 2013.
23. S. Jog and V. Bhatnagar, "Radiated immunity assessment of music infotainment system and recommending the enclosure design to mitigate effects," in *2016 International Conference on ElectroMagnetic Interference & Compatibility (INCEMIC)*, 8–9 Dec. 2016 2016, pp. 1–4, doi: 10.1109/INCEMIC.2016.7921490.
24. K. M. Srinivasamurthy *et al.*, "Effect of Ce3+ substitution on the structural, morphological, dielectric, and impedance spectroscopic studies of Co–Ni ferrites for automotive applications," *Indian Journal of Physics*, vol. 94, no. 5, pp. 593–604, 2020, doi: 10.1007/s12648-019-01495-7.
25. K. H. J. Buschow, *Handbook of Magnetic Materials* (vol. 13). Elsevier Science, 2001.
26. A. Goldman, *Chapter 8 – Metallic materials for magnetic shielding applications*, 1st ed. *Handbook of Modern Ferromagnetic Materials*. Springer US, 2012, p. 655.
27. G. Inzelt, *Conducting Polymers: A New Era in Electrochemistry*. Springer Berlin Heidelberg, 2012.
28. S. Ganguly, P. Bhawal, R. Ravindren, and N. C. Das, "Polymer nanocomposites for electromagnetic interference shielding: a review," *Journal of Nanoscience and Nanotechnology*, vol. 18, no. 11, pp. 7641–7669, 2018, doi: 10.1166/jnn.2018.15828.
29. H. Presting and U. König, "Future nanotechnology developments for automotive applications," *Materials Science and Engineering: C*, vol. 23, no. 6–8, pp. 737–741, 2003.

30. A. Kausar, "Fullerene nanofiller reinforced epoxy nanocomposites – developments, progress and challenges," *Materials Research Innovations*, vol. 25, no. 3, pp. 175–185, 2021.
31. J.-B. Donnet, *Carbon Black: Science and Technology*. Routledge, 2018.
32. A. Roy, C. Sreejith, S. Abhishek, G. Ragul, and I. Ghosh, "Effect of multi-walled carbon nanotubes on automotive and aerospace applications – case study," *International Journal of Emerging Trends in Science and Technology*, vol. 5, pp. 5102–5113, 2017.
33. V. K. Patel, N. Patel, M. D. Patel, H. A. Shukla, A. B. Patel, and N. J. Parmar, "Innovations in lightweight materials for automotive engineering," *Journal of Electrical Systems*, vol. 20, no. 10s, pp. 2121–2133, 2024.
34. R. Sadek, M. S. Sharawi, C. Dubois, H. Tantawy, and J. Chaouki, "Superior quality chemically reduced graphene oxide for high performance EMI shielding materials," *RSC Advances*, vol. 12, no. 35, pp. 22608–22622, 2022, doi: 10.1039/D2RA02678C.
35. K. Dericiler, N. Aliyeva, H. M. Sadeghi, H. S. Sas, Y. Z. Menceloglu, and B. S. Okan, "Graphene in automotive parts," in *Nanotechnology in the Automotive Industry*, Huaihe Song, Ghulam Yasin, Nakshatra Bahadur Singh, Ram K. Gupta, Tuan Anh Nguyen Eds.: Elsevier2022, pp. 623–651.
36. A. Elmarakbi and W. Azoti, "State of the art on graphene lightweighting nanocomposites for automotive applications," *Experimental Characterization, Predictive Mechanical and Thermal Modeling of Nanostructures and Their Polymer Composites*, pp. 1–23, 2018.
37. F. Qin and C. Brosseau, "A review and analysis of microwave absorption in polymer composites filled with carbonaceous particles," *Journal of Applied Physics*, vol. 111, no. 6, p. 4, 2012.
38. S. Kuester, N. R. Demarquette, J. C. Ferreira Jr, B. G. Soares, and G. M. Barra, "Hybrid nanocomposites of thermoplastic elastomer and carbon nanoadditives for electromagnetic shielding," *European Polymer Journal*, vol. 88, pp. 328–339, 2017.
39. R. Sadek, M. S. Sharawi, C. Dubois, H. Tantawy, and J. Chaouki, "Reduced graphene oxide/barium ferrite ceramic nanocomposite synergism for high EMI wave absorption," *ACS Omega*, vol. 8, no. 17, pp. 15099–15113, 2023, doi: https://doi.org/10.1021/acsomega.2c08168.
40. V. Volpe, S. Lanzillo, G. Affinita, B. Villacci, I. Macchiarolo, and R. Pantani, "Lightweight high-performance polymer composite for automotive applications," *Polymers*, vol. 11, no. 2, p. 326, 2019.
41. H. Ahmad, A. Markina, M. Porotnikov, and F. Ahmad, "A review of carbon fiber materials in automotive industry," in *IOP Conference Series: Materials Science and Engineering*, vol. 971, no. 3, 2020, p. 032011.
42. R. Kumar, M. Kumar, J. S. Chohan, and S. Kumar, "Overview on metamaterial: history, types and applications," *Materials Today: Proceedings*, vol. 56, pp. 3016–3024, 2022.
43. D. Mani *et al.*, "Elongated liquid metal based self-healing polyurethane composites for tunable thermal conductivity and electromagnetic interference shielding," *Composites Communications*, vol. 44, p. 101735, 2023.
44. G. Lu, E. Ni, Y. Jiang, W. Wu, and H. Li, "Room-temperature liquid metals for flexible electronic devices," *Small*, vol. 20, no. 9, p. 2304147, 2024.
45. B. Kim, H. Jeon, D. Park, G. Kim, N.-H. Cho, and J. Khim, "EMI shielding leadless package solution for automotive," *Journal of Advanced Joining Processes*, vol. 5, p. 100102, 2022, doi: https://doi.org/10.1016/j.jajp.2022.100102.
46. M. Jaroszewski, S. Thomas, and A. V. Rane, *Advanced Materials for Electromagnetic Shielding: Fundamentals, Properties, and Applications*. Wiley, 2018.
47. D. Weston, *Electromagnetic Compatibility: Principles and Applications, Second Edition, Revised and Expanded – Chapter 2: Introduction to E and H, Near and Far Fields, Radiators, Receptors, and Antennas*. CRC Press, 2017.
48. P. Saini and M. Arora, "Microwave absorption and EMI shielding behavior of nanocomposites based on intrinsically conducting polymers, graphene and carbon nanotubes," *New Polymers for Special Applications*, vol. 3, pp. 73–112, 2012.

49. W. Y. Chiang and Y. S. Chiang, "Effect of titanate coupling agent on electromagnetic interference shielding effectiveness and mechanical properties of PC–ABS–NCF composite," *Journal of Applied Polymer Science*, vol. 46, no. 4, pp. 673–681, 1992.
50. G. M. Kunkel, *Shielding of Electromagnetic Waves: Theory and Practice – Chapter 10: Shielding effectiveness Theory of Shielding*. Springer International Publishing, 2019.
51. S. Geetha, K. Satheesh Kumar, C. R. Rao, M. Vijayan, and D. Trivedi, "EMI shielding: Methods and materials – a review," *Journal of Applied Polymer Science*, vol. 112, no. 4, pp. 2073–2086, 2009.
52. R. Pandey, S. Tekumalla, and M. Gupta, "Chapter 21 – EMI shielding of metals, alloys, and composites," in *Materials for Potential EMI Shielding Applications*, K. Joseph, R. Wilson, and G. George, Eds. Elsevier, 2020, pp. 341–355.
53. J.-C. Huang, "EMI shielding plastics: a review," *Advances in Polymer Technology*, vol. 14, no. 2, pp. 137–150, 1995, doi: 10.1002/adv.1995.060140205.
54. A. Ohlan, K. Singh, A. Chandra, and S. K. Dhawan, "Microwave absorption behavior of core – shell structured poly(3,4-ethylenedioxy thiophene) – barium ferrite nanocomposites," *ACS Applied Materials & Interfaces*, vol. 2, no. 3, pp. 927–933, 2010.
55. K. Singh, A. Ohlan, and S. Dhawan, *Polymer-Graphene Nanocomposites: Preparation, Characterization, Properties, and Applications, Nanocomposites – New Trends and Developments*, Intech Open Science, 2012, pp. 37–71.
56. M. Magdowski, S. V. Tkachenko, and R. Vick, "Coupling of stochastic electromagnetic fields to a transmission line in a reverberation chamber," *IEEE Transactions on Electromagnetic Compatibility*, vol. 53, no. 2, pp. 308–317, 2011, doi: 10.1109/TEMC.2010.2097267.
57. J. R. Baker-Jarvis *et al.*, "Measuring the permittivity and permeability of lossy materials: Solids, liquids, metals, and negative-index materials," 2005.
58. *IEC 62132 – Integrated circuits – Measurement of electromagnetic immunity 150 kHz to 1 GHz – Part 4: Direct RF power injection method*, I. E. Commission, 2 February 2006, 2006. [Online]. Available: https://webstore.iec.ch/en/publication/6510
59. *IEC 61967 – Integrated circuits – Measurement of electromagnetic emissions, 150 kHz to 1 GHz – Part 4: Measurement of conducted emissions, 1 ohm/150 ohm direct coupling method*, I. E. Commission, 16 March 2021 2021. [Online]. Available: https://webstore.iec.ch/en/publication/68739
60. *CISPR 12 – Vehicles, boats and internal combustion engines – Radio disturbance characteristics – Limits and methods of measurement for the protection of off-board receivers*, T. I. S. C. o. R. Interference, 2 February 2006, 2007. [Online]. Available: https://webstore.iec.ch/en/publication/5
61. *CISPR 25 – Vehicles, boats and internal combustion engines – Radio disturbance characteristics – Limits and methods of measurement for the protection of on-board receivers*, T. I. S. C. o. R. Interference, 2 February 2006, 2021. [Online]. Available: https://webstore.iec.ch/en/publication/64645
62. *ISO 11452-4:2020(en) – Road vehicles – Component test methods for electrical disturbances from narrowband radiated electromagnetic energy – Part 4: Harness excitation methods*, I. O. f. Standardization, 2020. [Online]. Available: www.iso.org/obp/ui/#iso:std:iso:11452:-4:ed-5:v1:en
63. *ISO 7637-1:2023(en) – Road vehicles – Electrical disturbances from conduction and coupling – Part 1: Vocabulary and general considerations*, I. O. f. Standardization, 2023. [Online]. Available: www.iso.org/obp/ui/#iso:std:iso:7637:-1:ed-4:v1:en
64. *UNECE Regulation No. 10 – Uniform provisions concerning the approval of vehicles with regard to electromagnetic compatibility*, U. N. E. C. f. Europe, 2014. [Online]. Available: https://unece.org/fileadmin/DAM/trans/main/wp29/wp29regs/2015/R010r5e.pdf
65. *Radio Frequency Safety*, F. C. Commission, 2021. [Online]. Available: www.fcc.gov/general/radio-frequency-safety-0

CHAPTER 7

Electromagnetic Interference Shielding of Medical Equipment from a Mathematical Modelling Perspective

Immanuvel Paul and Shannen Chacko Rajan

7.1 INTRODUCTION

Electromagnetic interference, or EMI, remains a persistent issue in contemporary healthcare settings. Hospitals and clinics are equipped with numerous electronic devices that may either emit electromagnetic radiation unintentionally or be vulnerable to it. EMI refers to undesirable disturbances resulting from electromagnetic fields or currents, which can cause malfunctions or reduce the performance of sensitive equipment. When it comes to medical devices, this becomes particularly serious, as it may affect the functioning of critical life support systems or compromise the accuracy of diagnostic tools. For instance, exposure to background radio-frequency signals or sudden electromagnetic pulses can interfere with the normal functioning of monitors, ventilators or infusion pumps. Therefore, achieving electromagnetic compatibility, or EMC, which is the ability of devices to function reliably despite the presence of electromagnetic disturbances, is a fundamental consideration in both the design of medical devices and the planning of healthcare facilities.

Medical equipment can both emit EMI and be affected by it. For instance, devices such as electrosurgical instruments and radio-frequency ablation systems are designed to produce electromagnetic fields, which may unintentionally interfere with other nearby electronic equipment. At the same time, highly sensitive devices like electrocardiographs and imaging systems can easily pick up unwanted electromagnetic signals from the surrounding environment if they are not adequately shielded.

There are well-documented incidents where EMI has led to clinical complications. As early as the 1990s, reports highlighted that mobile phones could interfere with the

DOI: 10.1201/9781003512363-7

functioning of cardiac pacemakers [1]. This led to safety recommendations such as maintaining a distance of at least 15 cm between the phone and the implanted device [2]. While more recent pacemaker designs have improved resistance to such interference, studies suggest that modern smartphones pose minimal risk during regular use, and the basic concern remains valid. Strong or nearby electromagnetic fields have the potential to disrupt the performance of medical electronics [2]. Even in recent years, studies have shown that radio-frequency identification systems operating at certain frequencies can cause noticeable interference in pacemakers and implantable cardioverter defibrillators (ICDs) during laboratory testing. These observations clearly highlight the importance of developing and implementing effective strategies to reduce the impact of EMI in medical devices [2].

One of the key methods to ensure EMC is through the use of EMI shielding. This technique involves the application of conductive or magnetic materials to reduce the strength of electromagnetic fields, thereby either protecting sensitive components or limiting the emission of unwanted signals. In medical environments, shielding is implemented in various ways. For example, a magnetic resonance imaging (MRI) machine is placed inside a specially constructed room known as a Faraday cage, which is designed to prevent external radio signals from entering and also to contain the strong radio-frequency pulses produced by the scanner itself.

Implantable medical devices such as pacemakers are typically enclosed within metallic casings that help block incoming electromagnetic fields to some extent. Additionally, many hospital cables and connectors are designed with shielding, such as coaxial arrangements or braided covers, to prevent them from acting as unintended receivers of electromagnetic noise. These practices demonstrate a preventive approach to managing interference. Instead of depending solely on physical distance or limiting the power of nearby sources, important medical devices are engineered to tolerate a reasonable level of electromagnetic exposure without suffering from performance issues.

Given the critical nature of medical applications, considerable research and engineering efforts have been directed towards improving EMI shielding in this field [3]. One of the notable features of this area is the prominent role played by mathematical modelling. Through mathematical and computational approaches, engineers and researchers are able to predict how an electromagnetic field will behave in the presence of a shielding material or enclosure, even before any physical prototype is developed [4].

By applying Maxwell's equations under specific conditions, it becomes possible to estimate the shielding effectiveness (SE) of a particular material or design. This approach also helps in identifying weaknesses, such as the presence of resonant frequencies or the possibility of leakage through openings. In this way, the design can be improved step by step towards achieving optimal results. Modelling proves especially beneficial in the case of medical devices, where experimental verification can be expensive, time-consuming or potentially hazardous. For instance, testing an active implant by deliberately subjecting it to strong electromagnetic fields to observe failure is not practical within a living body. Accurate models allow for the simulation of worst-case interference scenarios, making it possible to design equipment that is more likely to meet EMC requirements in the very first round of regulatory testing.

This review concentrates on the EMI shielding of medical equipment, with emphasis on the mathematical modelling perspective. It begins by discussing the theoretical foundations of electromagnetic shielding, covering essential ideas such as reflection, absorption and skin depth, which influence the behaviour and effectiveness of shielding materials. Following this, the mathematical framework for analysing shielding problems is introduced, demonstrating how equations can be used to quantitatively assess the performance of a shield. The discussion then moves to numerical modelling methods, where techniques like the finite element method (FEM) and the finite difference time domain (FDTD) method are applied to represent complex situations involving medical devices. The review continues with an examination of specific applications across various categories of medical equipment. These include large imaging systems, with particular attention to MRI, as well as implantable devices such as pacemakers and ICDs, along with other biomedical instruments. In each case, the focus is on how modelling work has contributed to practical advancements in shielding design. The chapter concludes by identifying certain gaps in current understanding and practices, and by suggesting possible future directions for research and development that could lead to improved EMI protection in medical technologies. Throughout the discussion, important studies and regulatory standards are referenced, as they have played a significant role in shaping the current knowledge in this important area.

7.2 THEORETICAL BACKGROUND

The foundation of effective EMI shielding lies in classical electromagnetic theory. The interaction of electromagnetic fields with shielding materials is described by Maxwell's equations. In differential form and expressed in SI units, two of the curl equations for an isotropic medium are given as:

$$\nabla \times \mathbf{E} = -\mu \frac{\partial \mathbf{H}}{\partial t}, \tag{7.1}$$

$$\nabla \times \mathbf{H} = \mathbf{J} + \varepsilon \frac{\partial \mathbf{E}}{\partial t}, \tag{7.2}$$

where **E** and **H** represent electric and magnetic fields, respectively, **J** is the conduction current density, ε is the permittivity and μ is the permeability of the medium. In a conductive medium, such as a metallic shield, the current density **J** is related to the electric field through Ohm's law, written as $\mathbf{J} = \sigma \mathbf{E}$, where σ denotes the electrical conductivity.

By combining these equations in a region free of external sources (where free charges and currents are absent) and applying the curl operation to the first equation while substituting from the second, we arrive at the wave equation for the electric field. Eliminating the magnetic field leads to the following expression:

$$\nabla \times \left(\nabla \times \mathbf{E}\right) - \mu\varepsilon \frac{\partial^2 \mathbf{E}}{\partial t^2} - \mu\sigma \frac{\partial \mathbf{E}}{\partial t} = 0, \tag{7.3}$$

which, under the assumption of a homogeneous medium and time-harmonic fields varying as $e^{j\omega t}$, simplifies to the Helmholtz equation:

$$\nabla^2 \mathbf{E} + k^2 \mathbf{E} = 0, \tag{7.4}$$

where k represents the complex wavenumber of the medium. In free space, where the conductivity σ is zero, the wavenumber is purely real and given by $k_0 = \omega\sqrt{\mu_0 \varepsilon_0}$. However, within a conducting material, the presence of the term involving $\mu\sigma\partial\mathbf{E}/\partial t$ causes the wavenumber to become complex, which reflects the attenuation of the wave as it propagates through the medium.

An important quantity that characterises the attenuation of electromagnetic waves in conductive materials is the skin depth, denoted by δ. Skin depth refers to the distance within a conductor at which the amplitude of an incident electromagnetic wave falls to $1/e$ (approximately 37%) of its value at the surface. For a good conductor operating at an angular frequency ω, the skin depth is given by the following approximate relation:

$$\delta = \sqrt{\frac{2}{\omega\mu\sigma}}, \tag{7.5}$$

under the assumption that $\sigma >> \omega\varepsilon$, meaning the conduction current dominates over the displacement current in the material.

This expression indicates that an increase in electrical conductivity σ, frequency ω or magnetic permeability μ will result in a smaller skin depth. From a physical perspective, a smaller skin depth implies that the electromagnetic fields are confined to a very thin surface region of the conductor and decay rapidly as they travel inwards. For instance, stainless steel commonly used in medical applications, which is a moderate conductor, may exhibit a skin depth in the micrometre range when exposed to microwave frequencies. In contrast, at lower frequencies such as those associated with power lines (around 50–60 Hz), the skin depth may increase to the millimetre range. Materials like copper and aluminium, which are widely used for high-frequency shielding, exhibit extremely small skin depths at radio and microwave frequencies. This property makes them highly effective, since even a thin metallic layer can significantly attenuate the incoming electromagnetic energy.

The ability of a material or enclosure to reduce the transmission of electromagnetic energy is described by the parameter known as shielding effectiveness (SE). This quantity is typically expressed in decibels (dB) and is defined as the logarithmic ratio between the electromagnetic field (or power) before and after the application of the shield. Mathematically, it is given by:

$$SE_{(\mathrm{dB})} = 20\log_{10}\left(\frac{E_{\mathrm{incident}}}{E_{\mathrm{transmitted}}}\right) = 10\log_{10}\left(\frac{P_{\mathrm{incident}}}{P_{\mathrm{transmitted}}}\right), \tag{7.6}$$

where E_{incident} and $E_{\mathrm{transmitted}}$ represent the electric-field amplitudes before and after the shield, while P_{incident} and $P_{\mathrm{transmitted}}$ denote the corresponding power levels. The second expression follows from the fact that power is proportional to the square of the electric field in a uniform impedance setting. A larger value of SE in decibels implies greater attenuation and hence better shielding performance. For example, if the SE is 20 dB, the transmitted electric-field amplitude is reduced to one-tenth of the incident field, which corresponds to a power reduction by a factor of 100. Similarly, an SE of 60 dB indicates that the transmitted field is only 0.1% of the original amplitude, and the transmitted power becomes one millionth of the incident power.

In classical electromagnetic theory, the SE of a material is commonly expressed as the sum of three distinct components:

$$SE = R + A + B,$$

where R represents the loss due to reflection at the interfaces of the shield, A denotes the loss caused by absorption within the material and B accounts for any loss or gain arising from multiple internal reflections inside the shield structure.

The reflection component R is attributed to the impedance mismatch that occurs when an electromagnetic wave encounters a boundary between two media with differing impedances. In the case of a metallic shield, the impedance is typically much lower than that of free space, which is around 377 Ω. This difference results in a considerable portion of the incident wave being reflected back at the surface.

The absorption component A is related to the exponential attenuation of the wave amplitude as it propagates through the shield, a behaviour governed by the skin effect. The greater the thickness of the shielding material in comparison with its skin depth, the higher the absorption. For good conductors operating at high frequencies, the absorption loss can be approximated.

It is important to distinguish between near-field and far-field EMI when evaluating shielding approaches. The discussion so far assumes the presence of plane waves, which corresponds to far-field conditions. In contrast, within the near-field region, defined as being within a small fraction of the wavelength from the source, the electromagnetic field may exhibit a dominant electric or magnetic component. In such cases, the effectiveness of shielding depends on the physical nature of the dominant field.

For electric-field dominance, materials with good electrical conductivity are generally suitable. However, in the case of low-frequency magnetic fields, such as those produced by MRI gradient coils or power transformers, conventional conductive shields like thin copper or aluminium tend to be ineffective. Instead, materials with high magnetic permeability, such as mu-metal, are employed to guide and attenuate the magnetic flux.

This review concentrates on the more prevalent scenario observed in medical-device environments, where interference typically occurs in the radio-frequency range. In such cases, far-field considerations apply, and conductive shielding is generally effective. Nevertheless, the fundamental ideas of reflection and absorption continue to hold true across different field regimes, though the specific design considerations may vary depending on whether the interference is electric or magnetic in nature.

7.3 MATHEMATICAL FORMULATION

For the purpose of designing and quantitatively assessing shielding performance, mathematical models are employed to study standard or canonical scenarios. One widely studied case involves a plane electromagnetic wave incident on an infinite planar slab made of shielding material. Although this model is idealised, it captures the core physical behaviour and serves as the foundation for analysing more complex geometries.

The model typically assumes far-field conditions along with homogeneous material properties. These assumptions enable the application of analytical methods based on transmission line theory or wave propagation theory, thereby facilitating a clear understanding of the shielding mechanisms involved.

Let us consider a slab of material with thickness t, electrical conductivity σ, magnetic permeability μ and permittivity ε, placed between two semi-infinite regions, typically assumed to be free space on both sides in the context of an EMI shielding problem. When a plane electromagnetic wave with electric-field amplitude E_i is incident normally on the slab, a portion of the wave, denoted as E_r, is reflected back into the incident medium, while the remaining part enters the slab. As the wave propagates through the material, its amplitude diminishes exponentially due to energy loss within the medium. Upon reaching the second interface, where the slab meets air again, part of the attenuated wave exits as the transmitted field E_t, while the rest reflects back into the slab. By applying the appropriate boundary conditions, specifically the continuity of the tangential components of the electric and magnetic fields at both interfaces, and solving the wave equations in each region, one can obtain analytical expressions for the reflection and transmission coefficients. These are typically represented as R and T, or alternatively as Γ and τ.

For the sake of brevity, we present the outcome of such a derivation in a simplified manner. Under the assumption of a good conducting material and a shield of moderate thickness, where the effect of multiple internal reflections becomes negligible, the SE for normal incidence can be approximated as:

$$SE \approx 20 \log_{10} \left| \frac{Z_0 \mid Z_s}{2Z_0} \right|^2 + \frac{20t}{\delta} \text{ (dB)}, \tag{7.7}$$

where $Z_0 \approx 377\ \Omega$ denotes the intrinsic impedance of free space and Z_s represents the surface impedance of the shielding material.

In this expression, the first term corresponds to the reflection loss R, which arises due to the impedance mismatch between the incident medium and the shield. The second term represents the absorption loss A, where the ratio $20t/\delta$ (in decibels) is equivalent to $8.686\ t/\delta$ if expressed using the natural logarithmic base. It is important to note that this equation is an approximation valid for single-layer shielding, where the reflection component is estimated based on the mismatch between free space and the conductive material. For materials with very high conductivity, the surface impedance Z_s is much smaller than Z_0, resulting in a significant reflection loss, since the magnitude of the fraction

$\left|(Z_0 + Z_s)/(2Z_0)\right|^2$ becomes much less than unity. Meanwhile, the absorption term increases linearly with the thickness t of the shield (in the logarithmic decibel scale) and is inversely proportional to the skin depth δ. As a result, higher frequencies, which reduce δ, and greater material thickness both contribute to a higher absorption loss.

A more accurate analysis of shielding performance can be carried out using the full transmission line analogy, known as Schelkunoff's theory of shielding [5]. In this framework, the propagation constant in the shielding material is expressed as $\gamma = \alpha \mid j\beta$, where $\alpha \approx 1/\delta$ is the attenuation constant and β is the phase constant.

The wave impedance inside the material is given by:

$$Z_{\text{in}} = \sqrt{\frac{j\omega\mu}{\sigma + j\omega\varepsilon}}.$$

The reflection coefficient at the front surface of the shield is defined as:

$$\Gamma = \frac{Z_{\text{in}} - Z_0}{Z_{\text{in}} + Z_0},$$

where Z_0 is the intrinsic impedance of free space. The total transmission coefficient across the slab, accounting for multiple internal reflections, is given by:

$$T = \frac{\left(1 - \Gamma^2\right)e^{-\gamma t}}{1 - \Gamma^2 e^{-2\gamma t}}.$$

The SE, in terms of the transmission coefficient, can be written as:

$$SE = -20\ \log_{10}|T|.$$

In many practical situations, the magnitude of the reflection coefficient Γ is nearly equal to 1, indicating strong reflection, while the factor $e^{-2\alpha t}$ becomes very small, signifying substantial absorption through the material. Under such conditions, the dominant term in the denominator becomes negligible, and the transmission coefficient simplifies to $|T| \approx e^{-\alpha t}$. Consequently, the SE becomes:

$$SE \approx 20\ \log_{10}\left(e^{\alpha t}\right) = 8.686\ \alpha t \quad (\text{dB}),$$

which agrees with the absorption component derived earlier in the simplified formula.

This mathematical formulation reinforces the understanding that achieving high SE depends on two key factors: a significant impedance mismatch at the interface (to ensure strong reflection) and high attenuation within the material (achieved through sufficient thickness or high material losses).

When dealing with scenarios that involve multiple layers or composite materials, the mathematical modelling becomes more involved, though the underlying principles remain similar. In such cases, each individual layer contributes to the overall shielding by partially reflecting and absorbing the incident electromagnetic energy. To evaluate the combined effect, transfer matrix methods are often used to compute the net SE.

As an illustration, consider a composite shield comprising a non-conductive plastic substrate coated with a thin metallic film. The plastic layer by itself does not offer electrical conductivity, but the metallic coating contributes to both reflection and partial absorption. However, if the metal film is thinner than a few skin depths, a portion of the electromagnetic energy can penetrate into the plastic layer. Depending on the material properties, the energy may either dissipate, if the plastic exhibits losses at the given frequency, or pass through and reappear on the other side. By systematically applying the boundary conditions at each interface and solving the resulting system using a layered model, one can estimate the net shielding effectiveness (SE). Furthermore, this approach allows for the optimisation of individual layer thicknesses and material properties, enabling trade-offs between performance, weight and cost in the design of advanced shielding structures.

Mathematical models are also capable of capturing the frequency-dependent nature of shielding performance. In practice, the effectiveness of a shield can vary considerably with frequency. At higher frequencies, the skin depth δ becomes smaller, leading to increased absorption and therefore higher values of the absorption loss A. Additionally, finer structural features such as thin films or small meshes can still provide effective shielding at these frequencies. On the other hand, low-frequency magnetic fields tend to penetrate non-ferrous shielding materials with relative ease, making such fields more difficult to attenuate.

Understanding the frequency response of SE, denoted by $SE(f)$, is particularly important in the context of medical devices. These devices must comply with EMC standards across a wide frequency range. For example, the IEC 60601-1-2 standard prescribes immunity testing at multiple frequencies, spanning from kilohertz to gigahertz, to ensure safe operation of medical equipment.

Analytical models help in identifying critical behaviours, such as resonant frequencies at which shielding performance may degrade. This degradation can occur when the physical dimensions of the shield are on the order of half the wavelength, leading to unwanted transmission or resonance effects at apertures. Such phenomena, once predicted through theoretical analysis, can be further examined and validated using numerical simulation techniques. These analytical formulations are extremely useful for gaining insight into general trends and for establishing baseline expectations. However, in real-world scenarios, medical devices often exhibit complex geometries, include openings for ventilation or cabling, and incorporate materials with non-uniform properties. Due to these practical complexities, analytical methods alone are insufficient for accurate evaluation. As a result, we next consider numerical modelling techniques, which extend the mathematical principles discussed so far to accommodate realistic geometrical features and material distributions. These methods provide the flexibility needed to analyse detailed configurations that are typical in actual medical equipment.

7.4 NUMERICAL MODELLING

The analytical approaches to EMI shielding discussed earlier are generally applicable only to simple geometries such as planar layers, spherical shells or cylindrical enclosures, and often rely on idealised assumptions. In contrast, actual medical devices tend to have far more intricate structures. For instance, a pacemaker includes external leads, an MRI suite features doors and cable waveguides, and a ventilator may consist of several electronic modules along with ventilation slots.

To evaluate SE in such complex scenarios, numerical modelling techniques become essential. These computational approaches allow for the solution of Maxwell's equations in arbitrary geometries and material configurations. Typically, such methods are based on discretisation schemes that convert the continuous domain into a finite computational grid, enabling accurate analysis of electromagnetic field behaviour in realistic medical environments [6, 7].

Several widely used techniques are available for numerical electromagnetic modelling. Among them, one of the most prominent is the FEM. In this approach, the computational domain is divided into a mesh consisting of small subregions, such as tetrahedral elements in three dimensions. The solution is obtained using variational methods to determine the electromagnetic field distribution across the domain.

FEM is particularly suitable for problems involving complex geometries and material properties. It can be applied in both the frequency domain, by solving the Helmholtz equation, and directly in the time domain. For instance, Van Loock et al. [4] employed a three-dimensional FEM formulation of the vector wave equation, which is a frequency-domain representation of Maxwell's equations, to evaluate the SE of composite materials. Their work demonstrated how geometric modifications, such as introducing a saw-tooth pattern on the surface of the shield, could be used to adjust the balance between reflection and absorption in a predictable manner. Such design insights are difficult to obtain using analytical methods. To validate their simulation, they compared the FEM results with solutions derived from transfer matrix theory for flat-layered configurations. The good agreement between the two approaches added credibility to the numerical model.

The FDTD method solves Maxwell's curl equations directly in the time domain by advancing the solution in small steps across both space and time. It is widely adopted due to its relatively simple implementation and its ability to capture the system's response across a broad frequency range within a single simulation, typically through the Fourier transformation of time-domain signals.

FDTD is particularly effective for modelling intricate geometries, including complete medical devices with their outer enclosures, internal structures and cabling. In the context of shielding analysis, this method can be used to inject either transient pulses or continuous waves and then observe how the electromagnetic field propagates, how much enters the device and how the field is distributed internally. Historically, FDTD has also been used in radio-frequency safety studies, for example, to simulate heating effects in pacemaker leads when subjected to MRI fields. Such simulations involve modelling the human torso along with the implant to analyse induced currents. While this represents more of a safety assessment than a conventional SE analysis, it still demonstrates the versatility and applicability of the FDTD method in electromagnetic modelling of medical scenarios.

Commonly referred to as the boundary element method in the context of electromagnetic modelling, the Method of Moments (MoM) is particularly well suited for radiation and scattering problems. Unlike volume-based methods, MoM involves discretising only the surfaces of conductive objects and solving integral equations to determine the surface current distributions. MoM-based tools, such as those frequently used in antenna simulations, can be effectively applied to evaluate how external electromagnetic fields induce currents on the metal enclosure of a device. These surface currents are crucial in understanding both how a device may radiate and how it might admit external fields through structural openings. Such simulations can provide valuable guidance in practical design decisions, including the optimal placement of conductive gaskets and the careful shaping of apertures. For example, it is important to ensure that any required openings in the shielding structure do not unintentionally behave as slot antennas at certain frequencies, which could compromise EMC.

In addition to the standard approaches, several other numerical techniques are also used in electromagnetic modelling. These include the Transmission Line Matrix (TLM) method, finite integration techniques and a range of specialised solvers available in commercial software packages such as CST Microwave Studio, Ansys HFSS and COMSOL Multiphysics. Many of these commercial tools employ a hybrid approach, combining different numerical schemes or implementing proprietary algorithms to simulate electromagnetic fields in and around complex devices. A notable advantage of these platforms is the inclusion of built-in material libraries, which support frequency-dependent properties for permittivity ϵ, permeability μ and conductivity σ. This capability is particularly useful when simulating realistic shielding materials, such as carbon-loaded polymers or layered composite coatings, where material behaviour varies across frequency ranges.

7.5 APPLICATIONS OF EMI SHIELDING IN MEDICAL DEVICES

Different types of medical equipment encounter varying EMI challenges and, as a result, employ shielding strategies in different ways. In this section, we focus on three representative categories: (1) large-scale imaging systems such as MRI machines, which require shielding at the environmental level; (2) implantable medical devices like pacemakers, which depend on integrated shielding to ensure patient safety; and (3) general-purpose medical electronics commonly found in hospital environments, which typically use enclosure-based shielding to satisfy EMC requirements.

For each of these categories, we illustrate how mathematical modelling and numerical simulation have been applied to enhance the effectiveness of shielding solutions.

7.5.1 Imaging Equipment: MRI Shielding

MRI serves as a prominent example where shielding is both essential and clearly evident. MRI systems operate by transmitting and receiving radio-frequency signals, typically ranging from tens to hundreds of megahertz, depending on the strength of the magnetic field used. These systems are generally installed within a *shielded room*, which functions as a large-scale Faraday cage constructed using conductive panels that cover the walls, ceiling and floor.

The shielding in MRI installations serves two primary purposes. First, it prevents external radio-frequency interference from entering the room and corrupting the sensitive MRI signal measurements. Second, it confines the powerful RF pulses generated by the MRI system itself, thereby avoiding interference with nearby electronic equipment and ensuring compliance with regulatory limits on electromagnetic emissions.

In practical terms, MRI shielding is highly effective. It is common to achieve attenuation levels around 100 dB within the operational frequency range of the MRI system. This implies that only a minuscule fraction, approximately 10^{-10} of the incident power, penetrates the shielded enclosure when materials like copper or aluminium are used for lining.

From a modelling perspective, the design of an MRI shielded room requires careful consideration of all possible penetrations in the shielding enclosure. For instance, the door must incorporate RF-tight sealing mechanisms, often employing copper gaskets and RF fingerstock to maintain shielding continuity. Similarly, any openings for ventilation or cabling must be designed in such a way that they do not compromise the overall shielding performance.

Mathematical modelling plays a key role in assessing how small gaps or apertures may allow specific frequencies to leak through. A common analytical approach involves treating each aperture as a waveguide and determining its cutoff frequency, the threshold below which electromagnetic waves cannot propagate. To block MRI operating frequencies (e.g., 64 MHz for a 1.5 T system or 128 MHz for a 3 T system), the waveguides are typically constructed to be sufficiently long and narrow so that the RF signals remain well below the cutoff frequency. This allows for limited airflow while maintaining electromagnetic isolation.

Numerical simulations using methods like FEM or FDTD can be employed to evaluate whether the shielding remains effective even under non-ideal conditions, for example, if the door is slightly open or if a small aperture is introduced for plumbing access. Frequency sweep simulations are particularly valuable, as they help identify any resonant frequencies that might compromise shielding performance. Although the enclosed volume of an MRI room can support cavity resonances, the inclusion of RF absorbing materials and careful architectural design ensure that such modes do not coincide with the MRI system's operational frequencies. Another important consideration in MRI design is the magnetic shielding of the static magnetic field. While radio-frequency shielding addresses high-frequency electromagnetic waves, the static magnetic field generated by the MRI magnet, typically 1.5 T or 3 T, also needs to be controlled. To achieve this, passive shielding using iron laminates embedded in the room's walls or active shielding with specially designed compensation coils may be implemented to limit the spread of the magnetic field beyond the MRI suite.

Mathematical modelling, particularly through magnetostatic finite element simulations, is used to predict the extent of the static field in the surrounding area. One common goal is to identify the 5-gauss boundary, which defines the region around the magnet where the magnetic field strength falls to a level considered safe for the general public.

Although this issue pertains more to static or low-frequency magnetic fields rather than high-frequency EMI, it is still a critical component of the broader electromagnetic safety planning involved in MRI room design.

7.5.2 Implantable Devices: Pacemakers and ICDs

Implantable cardiac devices, such as pacemakers and ICDs, are critical for maintaining heart function and must operate reliably even in the presence of external electromagnetic fields. These devices are compact electronic systems implanted within a patient's body, comprising a pulse generator, which houses the circuitry and power source, typically enclosed in a titanium casing, and one or more leads that extend to the heart.

The titanium enclosure functions as an effective electromagnetic shield, owing to its conductivity, and provides considerable attenuation of external electric fields. However, it is not possible to completely isolate the device from its environment. The leads, which extend outside the shielded enclosure, can behave like antennas, capturing radio-frequency energy from the surroundings and conducting it back to the internal circuitry.

Therefore, electromagnetic protection for such implants involves more than just the shielding effect of the metallic case. Equal attention must be given to the internal design, particularly to filtering circuits that can suppress unwanted signals picked up by the leads. This combination of physical shielding and circuit-level filtering is essential to ensure the safe and uninterrupted operation of the device.

Mathematical modelling related to EMI in implantable devices has largely been motivated by concerns over interaction with external sources such as mobile phones, anti-theft detection systems, MRI scanners and, more recently, RFID readers and wireless charging systems. These interactions often involve near-field coupling scenarios. For example, a pacemaker implanted in a patient may lie only a few tens of centimetres away from a mobile phone antenna. In such a case, electromagnetic emissions from the phone, typically in the 900 MHz or 1.8 GHz frequency bands, may induce a voltage on the pacemaker's lead.

If this induced voltage is misinterpreted by the device's sensing circuitry as a physiological signal, it may lead to inappropriate device behaviour, such as inhibition of pacing or the delivery of unnecessary therapy. Early theoretical models approached this problem by representing the pacemaker lead as a simple wire antenna embedded in a lossy dielectric medium, representing the human body. Antenna theory was then used to estimate the magnitude of the induced currents.

To improve model accuracy, these studies often included the dielectric properties of body tissues and the fact that the pacemaker's metallic casing is electrically grounded through contact with the body and is connected to the lead. One important insight from such modelling was the identification of lead lengths that could resonate with particular frequencies, making devices more vulnerable in those frequency bands. These findings helped inform practical guidelines, including maintaining minimum separation distances and restricting exposure to specific frequency ranges.

One of the most intense electromagnetic environments that implantable devices may be exposed to occurs during MRI examinations. For a long time, MRI was considered contraindicated for patients with pacemakers, as the strong radio-frequency fields used in MRI, typically in the tens of megahertz, could induce currents in the device leads. These currents posed risks such as localised tissue heating or unintended electrical stimulation of the heart.

To address these concerns, computational modelling has been employed extensively. Using FDTD methods and anatomically detailed human body meshes, researchers have developed simulations to estimate the worst-case heating in pacemaker leads when subjected to MRI fields. These models solve for the distribution of electric fields within the body when an implant is present and exposed to the electromagnetic field generated by the MRI coil.

Such studies have been instrumental in the development of "MR-conditional" pacemakers. These are devices and lead systems that have been specially engineered with enhanced shielding, filtered lead designs and dedicated firmware to ensure safe operation under specified MRI conditions. In this context, shielding extends beyond the metallic casing to include radio-frequency filters placed on the leads, often integrated into the device header. These filters are designed to block MRI-frequency currents from reaching the internal circuitry while still allowing the device to detect genuine cardiac signals, which occur at much lower frequencies.

Modelling was essential in determining the electrical characteristics required of these filters to achieve effective attenuation of MRI-induced fields without impairing the core functionality of the device.

Regulatory guidelines set by bodies such as ISO and the FDA increasingly require implantable devices to demonstrate immunity to EMI. While these requirements are often fulfilled through experimental testing, such as exposing a pacemaker to controlled EMI fields inside a test chamber, mathematical modelling is frequently employed to interpret the results and extend the findings beyond the tested conditions.

For instance, Seidman et al. [8] observed that low-frequency RFID systems operating at 134 kHz were more likely to interfere with pacemaker function compared with higher-frequency RFID systems. A theoretical interpretation of this result pointed to the nature of electromagnetic coupling at different frequencies. At low frequencies, the pacemaker lead and the patient's body can together act as a large loop antenna, effectively coupling with the magnetic component of the field, which is dominant in the low-frequency regime. In contrast, at higher frequencies, the shorter wavelength and associated field cancellation effects result in weaker coupling.

Such insights are not evident from measurement alone and are better understood through analytical studies of inductive coupling and numerical simulations. These modelling efforts complement empirical data and contribute to a more robust understanding of EMI susceptibility in implantable medical devices.

In practical design, manufacturers of implantable devices implement several layers of EMI mitigation to ensure reliable operation. These include the conductive metallic enclosure that acts as a primary shield, feed-through filters located at the entry points where leads connect to the internal circuitry, robust internal circuit designs that suppress common-mode interference and software algorithms capable of distinguishing noise from physiological signals. For example, pacing systems often include logic that differentiates between genuine cardiac signals and transient noise.

Mathematical modelling plays a crucial role in refining each of these protective layers. For instance, a feed-through filter, typically realised as a small capacitor placed in the header to connect the lead to the device casing at radio frequencies, can be analysed using circuit-level equations to verify that it effectively redirects high-frequency signals to the case ground without compromising the integrity of the actual cardiac signal.

The device casing itself can be modelled as a miniature Faraday cage, allowing for estimation of the SE in decibels over a range of frequencies. However, in most implantable systems, the leads serve as the dominant coupling path for external fields, making them the limiting factor in overall electromagnetic protection rather than the shielding provided by the case.

7.5.3 Other Medical Equipment and Environments

Beyond MRI systems and implantable devices, a wide variety of medical equipment must be designed with EMC in mind and often incorporate shielding measures. This category includes diagnostic devices such as X-ray machines, ultrasound systems and EEG or ECG monitors; therapeutic equipment like infusion pumps, surgical lasers and defibrillators; and even auxiliary systems such as motorised hospital beds and lighting installations in operating theatres.

For most of these devices, shielding is typically applied at the level of the device enclosure or integrated into specific sub-components. Mathematical modelling is generally employed on a case-by-case basis, particularly when a specific EMI problem is identified during testing or deployment. Such targeted modelling helps engineers diagnose the issue and develop appropriate shielding or filtering strategies.

One illustrative example arises in hospital electrophysiology laboratories or operating rooms, where multiple electronic devices often operate in close proximity. A well-documented problem in such environments is the interference caused by electrosurgical units, which generate radio-frequency currents for tissue cutting or cauterisation. These RF emissions can unintentionally couple into other equipment, such as patient monitors or cardiac pacing systems, potentially leading to malfunction or incorrect readings.

Mathematical modelling can be employed to investigate how the RF current from an electrosurgical pencil may couple into monitoring leads, which can behave like receiving antennas under certain conditions. By simulating this coupling mechanism, engineers can explore effective mitigation strategies involving both shielding and filtering.

One proven solution involves the use of ferrite beads and braided shields on signal cables. These components add impedance and reduce high-frequency coupling, thereby improving electromagnetic immunity. The performance of such configurations can be predicted through simulations that model the cable and ferrite bead as a transmission line system with additional impedance elements, allowing optimisation of the protective measures before physical implementation.

Another important domain where shielding plays a significant role is in telemedicine and wireless medical devices. As modern healthcare increasingly adopts wireless technologies, such as Bluetooth-enabled vital sign monitors and Wi-Fi connected infusion pumps, it becomes essential to ensure that these devices do not emit excessive EMI and are sufficiently immune to external radio-frequency noise.

In such applications, shielding may take the form of a small RF enclosure surrounding the circuit board or a conductive coating applied to the interior surface of the device's plastic housing. The shielding design must be carefully validated, often through modelling of the device's antenna characteristics and electromagnetic field distributions.

Interestingly, in the context of wireless medical devices, the shielding strategy must strike a careful balance. While internal components, particularly the digital circuitry, must be shielded to prevent them from disturbing the antenna performance, over-shielding the antenna itself must be avoided, as it would impair wireless communication. Computational electromagnetic tools are especially valuable in these cases, as they allow designers to optimise the layout so that the internal electronics are well protected while still allowing effective radiation of the intended signals, such as those from a Bluetooth transmitter.

In some hospital settings, EMI is addressed at the level of an entire room or designated zone. For example, intensive care units may be assessed for background electromagnetic field levels to ensure that they remain below thresholds that could interfere with the operation of critical medical equipment. If a specific interference source is identified, such as an outdated diathermy unit emitting unwanted RF noise, the preferred approach may be to apply localised shielding around the source or the affected equipment rather than attempting to shield the entire room.

In such scenarios, modelling approaches can vary in complexity. In simple cases, basic analytical models based on the inverse-square law may suffice to estimate the decay of radiated fields with distance. For more complex environments, particularly where reflections and partial shielding are involved, advanced techniques such as ray tracing may be employed to predict the propagation of radio-frequency energy within the indoor space.

These modelling methods are often adapted from the field of telecommunications, where similar tools are used to predict wireless signal coverage in buildings. In the medical context, however, the focus shifts towards evaluating and mitigating worst-case EMI within sensitive clinical environments.

Finally, it is important to consider the role of emerging materials in various miscellaneous medical applications. Researchers have investigated the use of advanced materials such as conductive polymers, carbon-nanotube composites and MXene-based coatings as lightweight alternatives to traditional metallic shielding in medical devices.

To evaluate the suitability of these materials, it is essential to model their electromagnetic properties, particularly their frequency-dependent permittivity and permeability, and assess their shielding performance over the frequency range relevant to the application. For instance, a carbon-loaded polymer may provide effective absorption in the gigahertz range but exhibit reduced performance at lower frequencies such as in the megahertz range.

Mathematical modelling is instrumental in determining whether a hybrid shielding strategy is needed. One such approach could involve combining a thin metallic layer for low-frequency reflection with a composite material for high-frequency absorption, thus leveraging the advantages of both mechanisms.

As these novel materials transition from laboratory research to practical implementation in commercial medical devices, modelling will serve as the critical tool to demonstrate compliance with the stringent EMC requirements of the medical field.

Across the wide range of equipment types and operational scenarios discussed, the fundamental principles of electromagnetic shielding remain consistent: employ conductive or magnetic materials to block undesired electromagnetic fields and evaluate their effectiveness through a combination of theoretical analysis and numerical simulation.

The collective experience gained from different applications contributes to the formulation of general design practices and regulatory frameworks. For instance, standards established by organisations such as the FDA and IEC frequently emphasise the importance of shielding and include specific test protocols to verify a device's immunity to external electromagnetic fields. These tests serve as practical checks on the adequacy of the implemented shielding strategies.

Having reviewed the applications, we have seen that while much progress has been made, there are still gaps and challenges that ongoing research needs to address. We discuss these next, along with emerging trends and future directions in EMI shielding for medical devices.

7.6 GAPS IN CURRENT RESEARCH

Despite the significant progress and numerous successful implementations of shielding solutions, several challenges and unresolved issues continue to exist in the area of EMI shielding for medical equipment, particularly from the standpoint of modelling. Identifying and understanding these gaps is essential for steering future research efforts and enhancing both the safety and performance of medical technologies. Some of the key areas that merit further investigation include the following.

7.6.1 Complex Coupling Paths and Multi-Scale Modelling

Medical devices often exhibit intricate interactions with their surrounding environment, involving multiple coupling mechanisms. For example, in the case of an implantable device, the coupling pathway may include biological tissue, lead conductors and internal device electronics. Similarly, hospital equipment may be affected by both conducted and radiated interference, occurring through cables, power lines and free space.

Current modelling strategies often address these effects independently or rely on simplifying assumptions. Developing a comprehensive modelling approach that simultaneously accounts for various coupling mechanisms, such as solving for both conducted currents on cabling and radiated electromagnetic fields, is a significant challenge. Furthermore, the need for multi-scale modelling is apparent, where systems must be analysed across vastly different spatial scales, ranging from sub-millimetre electronic components to room-sized installations. Bridging this gap may require co-simulation strategies, where full-wave electromagnetic solvers are coupled with circuit-level simulators, or where detailed sub-models are embedded within broader coarse-scale frameworks. Some preliminary work in this direction exists, for instance combining an electromagnetic model for a pacemaker lead with a circuit model for the device's sensing logic, but further advances are needed to achieve robust integration.

This limitation is particularly apparent when attempting to predict EMC at the system level. Hospital environments may contain dozens of active electronic devices, and interference can be cumulative or arise from cross-talk between systems. At present, the capability to model an entire intensive care unit's electromagnetic environment remains beyond reach, highlighting the need for more advanced tools that support large-scale, multi-source simulations.

7.6.2 Material Characterisation at Relevant Frequencies

The effectiveness of electromagnetic shielding critically depends on the electrical properties of the materials involved, specifically their conductivity, permittivity and permeability. While traditional metals like copper and aluminium are well characterised, many modern shielding strategies rely on alternative materials such as conductive polymers, composite coatings and carbon-based materials, whose properties are less predictable and may vary significantly with frequency, temperature or fabrication conditions.

For example, a polymer loaded with carbon nanotubes might exhibit very different behaviour at 400 MHz compared with 2.4 GHz, and may also respond non-linearly under high field intensities. Accurate modelling of such materials requires precise knowledge of their frequency-dependent electromagnetic properties. However, there remains a substantial lack of standardised data for many of these newer materials in the frequency range relevant to medical EMC, typically between 0.1 and 1000 MHz.

In practice, researchers often resort to measuring material samples for input into simulations. However, discrepancies between measurement and modelled performance can arise, especially when the material is anisotropic or when the test sample does not exactly replicate the configuration used in the final device. The absence of comprehensive material libraries limits the reliability and accuracy of predictive models. Bridging this gap will require coordinated efforts between material scientists to characterise and potentially standardise advanced shielding materials, and electromagnetic modellers to integrate complex, frequency-dependent material behaviour into simulation tools.

7.6.3 Extreme Scenarios and Model Validation

A further limitation in current EMI modelling practices lies in the prediction of rare or worst-case scenarios. While most models provide deterministic outcomes based on specific input parameters, real-world interference phenomena often exhibit probabilistic characteristics, influenced by factors such as device orientation, user behaviour and environmental variability.

For example, a ventilator may only experience interference under unusual conditions such as when a nurse activates a walkie-talkie very close to a specific cable configuration. Such outlier events are difficult to replicate in simulations, especially since most modelling approaches are based on standardised test setups, such as applying a uniform field from a fixed direction at a defined intensity.

The challenge here is that non-standard field orientations, unusual polarisation states or unexpected couplings may expose vulnerabilities that remain hidden under conventional test conditions. Compounding this difficulty is the problem of experimental validation – reproducing these rare scenarios in a controlled laboratory setting is often impractical.

This gap points to the need for more advanced uncertainty quantification techniques within EMI modelling. Such methods involve running ensembles of simulations with varied parameters, such as randomised field directions, device positions or material properties, to generate a statistical distribution of possible outcomes. Although not yet widely adopted in shielding simulations, this approach holds promise for identifying edge cases and improving the overall robustness of EMI predictions.

7.6.4 Human Exposure and Coexistence with Shielding

In certain situations, electromagnetic shielding in medical equipment interacts with concerns related to human exposure to electromagnetic fields. For instance, introducing a shielding layer may unintentionally redirect electromagnetic fields towards the patient or clinical staff, thereby altering the local field distribution. Although safety limits for human exposure, such as those based on specific absorption rate (SAR), are regulated independently from functional EMC requirements, an ideal modelling framework would address both aspects simultaneously.

At present, models developed for shielding design and those used for assessing human exposure are typically created and analysed separately. Integrating these two domains into a unified simulation approach remains an open research area. The need for such integration becomes even more critical in the context of wearable medical technologies, such as insulin pumps or portable ECG monitors.

In these devices, shielding not only serves to protect internal circuitry but may also influence wireless communication performance. Since these devices are designed to transmit or receive telemetry data, the presence of shielding can affect antenna efficiency and overall communication reliability. This creates a coexistence problem, where achieving EMC must be balanced with maintaining robust communication links.

Solutions may involve the use of partial-shielding structures or frequency-selective surfaces that allow transmission at certain frequencies while blocking unwanted signals. However, modelling these complex interactions, particularly in dynamic human-worn scenarios, remains challenging and is not yet routine in design workflows.

7.6.5 Standardisation of Modelling Approaches

Although not a research gap in the traditional sense, there exists a notable lack of standardisation in modelling practices for EMC assessment in medical devices. Currently, different research groups and industrial teams may adopt varying assumptions, boundary conditions and numerical methods, making direct comparison between modelling results difficult.

A simulation model's credibility heavily depends on its validation, and regulatory authorities are often cautious about accepting simulation-based evidence without accompanying experimental data. The establishment of standardised benchmarks, such as a canonical pacemaker model exposed to well-defined interference scenarios, would help to align modelling practices and build trust in simulation as a reliable engineering tool.

Such benchmarks could serve as reference problems that different modelling tools and teams are expected to simulate with consistent results. While the community has begun taking steps in this direction through inter-laboratory comparison studies and efforts to correlate models with experimental measurements, these efforts remain limited.

For simulation-based evidence to gain broader acceptance, akin to the way clinical trials are trusted for therapeutic device evaluation, further progress is needed. The goal is not to elevate modelling to the level of clinical trials in terms of evidentiary weight but to establish its reliability as a trustworthy complement to physical testing.

7.7 FUTURE DIRECTIONS

Looking ahead, the field of EMI shielding for medical equipment is poised to benefit from several emerging trends and innovative approaches. Here we outline some promising future directions that can enhance SE, make modelling more powerful and improve the overall electromagnetic safety of medical devices.

7.7.1 Advanced Shielding Materials and Metamaterials

Traditional EMI shielding has relied heavily on metallic materials due to their high electrical conductivity and well-established properties. However, recent research has started to explore a wide array of alternative materials such as conductive polymers, carbon-based nanomaterials (e.g., graphene and carbon nanotubes), MXenes and other two-dimensional (2D) materials, as well as metamaterials that offer engineered electromagnetic responses [9].

In future shielding architectures, it is likely that composite material systems will play a key role. These may combine high conductivity for reflection with intrinsic absorption properties. For instance, a thin metamaterial layer could act as a frequency-selective surface, specifically reflecting undesirable frequency components, while an adjacent layer might absorb broadband electromagnetic noise. Such combinations promise lightweight and mechanically flexible shielding options, enabling, for example, patch-like shields that can conform to irregular device shapes or even integrate into garments for protecting wearable medical electronics.

One particularly promising direction is the development of absorption-dominant shielding materials [10]. These are designed to suppress reflection (which can lead to secondary interference) while maximising electromagnetic energy dissipation through conversion to heat. The design of these materials often involves computational optimisation techniques, and it is anticipated that AI-assisted design strategies may further aid in creating novel micro- and nano-structured geometries for broadband, multi-band and application-specific shielding.

Another emerging frontier is the incorporation of *biocompatible shielding materials*, especially for implantable medical devices. While metals like titanium are already widely used due to their biocompatibility, polymer-based conductive composites offer the potential for new form factors, lighter weight or selective permeability to support both shielding and wireless telemetry. The characterisation and modelling of such advanced materials remain a challenging task due to their frequency-dependent and sometimes nonlinear electromagnetic properties. Nevertheless, ongoing progress in materials science, coupled with enhanced electromagnetic simulation capabilities, is expected to make such novel shielding materials viable for future medical-device applications.

7.7.2 Integration of Machine Learning in Modelling

Recent advances in artificial intelligence (AI) and machine learning (ML) are beginning to make a notable impact in the field of EMC, including the domain of EMI shielding. One promising direction is the use of ML algorithms to develop surrogate models that approximate the results of complex electromagnetic simulations. For instance, a neural network can be trained on a dataset comprising numerous simulation results, learning the mapping between input parameters such as material properties, layer thickness, frequency and geometric features, and the resulting SE [11, 12].

A noteworthy example is provided by Muthusamy et al. [13], who used an artificial neural network (ANN) to predict the EMI SE of coated fabric materials. The ANN demonstrated a high level of agreement with experimental data, indicating the potential of such models to replace or supplement traditional simulation approaches [13].

In future applications, these ML-based models could be embedded into design workflows to offer instantaneous predictions of shielding performance, thereby avoiding repeated full-wave electromagnetic simulations. Such tools would be especially beneficial in the iterative design of medical devices, where small modifications (e.g., in shielding layer or enclosure geometry) often require extensive re-analysis. Furthermore, ML techniques can assist in optimisation tasks, such as identifying the most efficient multi-layer shield structure to meet a specified SE target while minimising weight and cost.

ML may also find application in post-processing experimental or test data. For example, anomaly-detection algorithms could be employed to automatically identify patterns of interference or failure modes that are not easily discernible through manual inspection.

As simulation and measurement data accumulate over time, especially through collaborative or open-source efforts, ML models are expected to become more robust and reliable. In the long term, one can envision a practical design environment where the engineer inputs the intended medical application, operational frequencies and regulatory requirements, and the ML-driven system recommends suitable shielding configurations, complete with uncertainty bounds and compliance predictions.

7.7.3 Personalised and Adaptive Shielding Solutions

Looking towards the future, the concept of adaptive or even active shielding is emerging as a potential direction, especially in high-risk medical environments. Active EMI cancellation, which involves sensing incoming interference and generating an equal-but-opposite signal to suppress it, has already found application in fields like acoustic noise cancellation (e.g., headphones). Extending this concept to electromagnetic domains, particularly in medical devices, poses several technical challenges, but it may be viable for repetitive or narrowband sources of interference, such as power-line harmonics or fixed-frequency RF sources.

Incorporating such active cancellation circuits into medical equipment could provide a dynamic layer of EMI mitigation, operating in parallel with conventional passive shielding methods. For example, a wearable medical monitor might employ an onboard sensing-and-cancelling unit to reduce its vulnerability to interference in real time.

As wearable medical electronics continue to evolve, the notion of *personalised shielding* becomes relevant. Imagine a scenario where a patient with an implantable cardiac device dons a smart vest embedded with conductive textiles and sensors. This vest could continuously monitor the ambient electromagnetic environment and activate additional shielding circuits or reconfigure shielded zones in response to rising EMI levels. Though speculative at present, such systems could become practical with advancements in real-time sensing, materials and embedded processing.

From a modelling perspective, active shielding systems require a multidisciplinary approach. Alongside traditional electromagnetic simulations (e.g., FDTD or FEM), one must incorporate aspects of control theory and signal processing to design and tune the feedback and cancellation mechanisms. This creates a rich area for future research that intersects electromagnetics, electronics and cyber-physical systems.

Overall, the synergy between passive and adaptive shielding methods could eventually lead to medical devices that are not only robust but also responsive, capable of adjusting their shielding performance based on dynamic environmental conditions.

7.7.4 Enhanced Standards and Simulation-Led Certification

As numerical modelling techniques mature, their role in regulatory and certification processes is expected to grow significantly. Currently, most EMI compliance for medical devices relies heavily on standardised physical tests. However, future standards may begin to formally recognise *simulation-based evidence* as part of the certification process, especially when the models are well validated and have shown predictive reliability in previous applications.

This anticipated shift will require the development of standardised modelling frameworks and protocols, as briefly discussed in the earlier section on existing gaps. A natural extension of this concept is the emergence of *digital twins* for medical devices – high-fidelity virtual replicas that can be tested in simulated electromagnetic environments to evaluate their SE and susceptibility. This approach is already practised in aerospace and automotive sectors, where simulation data complements physical testing to save cost and design time.

In the medical context, such digital twins might include not just the device but also surrounding environments and human anatomical phantoms. For instance, a standard digital human torso model with implantable devices embedded at different locations could be developed and shared across the research community. Researchers and manufacturers could then use this as a benchmark platform to test their shielding or filtering strategies against various EMI sources.

The value of this approach lies not only in cost reduction but also in the ability to run multiple *what-if* scenarios that would be difficult to achieve in physical tests. For example, one could simulate the effect of EMI from a nearby radio transmitter in various patient postures or operating room layouts.

However, for simulation-led certification to gain wide acceptance, the community must establish strong evidence of model fidelity. This involves extensive model-to-experiment validation, reproducibility across software tools and transparency in assumptions. Organisations such as the IEC or FDA may eventually develop simulation validation benchmarks, just as they currently provide physical testing standards.

In summary, while not a new technological advancement, the integration of advanced modelling into the regulatory framework represents a major future direction – one that could make EMI certification more flexible, insightful and cost-effective.

7.7.5 Cross-Disciplinary Approaches: EMC Meets IoT and 5G

With the rapid expansion of healthcare technologies into the Internet of Things (IoT) and telemedicine domains, the EMC landscape is evolving. Medical devices are now expected to operate reliably in increasingly dense wireless environments, especially with the advent of 5G and the future deployment of 6G communication systems. These networks introduce higher operating frequencies, such as 3.5 GHz for mid-band 5G and millimetre-wave bands around 28 GHz, which pose new challenges for traditional EMI shielding strategies.

At such high frequencies, even minute discontinuities or apertures in a shielding enclosure can behave like slot antennas, significantly degrading the SE. Furthermore, the human body itself begins to act as a scattering or blocking element, which alters the local electromagnetic field distribution in unpredictable ways. This makes the modelling and prediction of EMI effects in hospital settings highly complex.

As a result, future modelling strategies will need to incorporate realistic and cluttered indoor environments, possibly using ray-tracing methods or hybrid field solvers. The hospital of the future may benefit from environmental control of electromagnetic fields – not just at the device level but also at the infrastructural level. One proposed idea is the use of *electromagnetic metamaterial coatings*, engineered surfaces applied on walls or equipment that can absorb or redirect EM waves with high precision.

For example, certain walls in sensitive units (such as ICUs or imaging rooms) could be painted with EM-absorbing coatings. While currently such materials are confined to laboratory use (e.g., in anechoic chambers), advancements may make them viable for clinical settings. Additionally, switchable or tunable surfaces (using varactor-loaded metasurfaces, for instance) could dynamically adjust their properties to suppress EMI only when needed, maintaining wireless communication performance the rest of the time.

Striking the right balance between effective shielding and sufficient wireless connectivity will be a critical design objective. Over-shielding could disrupt medical telemetry and wearable sensors, while under-shielding may expose sensitive electronics to unwanted noise. This field, therefore, represents a highly interdisciplinary research opportunity, linking EMC, RF communications and smart materials. It is expected that close collaboration between communication engineers, biomedical device designers and materials scientists will be essential for progress in this domain.

7.8 CONCLUSION

Effective EMI shielding in medical devices is fundamentally a multidisciplinary challenge, encompassing physics, electrical and materials engineering, computational modelling and biomedical constraints. Within this ecosystem, mathematical modelling emerges as a key enabler, offering a structured and predictive framework to design, evaluate and refine shielding strategies across a wide spectrum of device types and operational scenarios. As healthcare infrastructure becomes increasingly reliant on compact, interconnected and wireless electronic systems, the importance of ensuring EMC becomes paramount.

Robust shielding, guided by accurate modelling, will play a central role in this effort. From analytical formulations that provide foundational insights to sophisticated numerical simulations capable of capturing real-world complexities, modelling empowers engineers and researchers to make informed decisions early in the design cycle. Moreover, emerging tools such as AI-driven surrogate modelling, adaptive shielding materials and integrated exposure analysis offer exciting avenues for further enhancement.

Ultimately, the goal is straightforward but non-trivial: to ensure that medical devices perform reliably, without interruption or error, regardless of the surrounding electromagnetic environment. This demands continuous advancement in both theory and practice and a collaborative approach that bridges domains. With such concerted efforts, it is indeed possible to envision a healthcare future where EMI-related risks are proactively mitigated, thereby upholding patient safety and reinforcing the integrity of modern medical care.

REFERENCES

1. D.L. Hayes, P.J. Wang, D.W. Reynolds, M. Estes, J.L. Griffith, R.A. Steffens, W.A. Carlo, G.K. Findlay, and C.M. Johnson. Interference with cardiac pacemakers by cellular telephones. *New England Journal of Medicine*, 336(21):1473–1479, 1997. doi: 10.1056/NEJM199705223362101.
2. Fahd Nadeem, Cao Thach Tran, Estelle Torbey, Daniel Philbin, Carlos Morales, and Michael Wu. Interference by modern smartphones and accessories with cardiac pacemakers and defibrillators. *Current Cardiology Reports*, 24(4):347–353, 2022.
3. Shufang Zheng, Yuyin Wang, Xuesheng Wang, and Heng Lu. Research progress on high-performance electromagnetic interference shielding materials with well-organized multilayered structures. *Materials Today Physics*, 40:101330, 2024.
4. F. Van Loock, P.D. Anderson, and R. Cardinaels. A monolithic numerical model to predict the EMI shielding performance of lossy dielectric polymer nanocomposite shields in a rectangular waveguide: Design of an absorption-based sawtooth-shaped layer. *Applied Mathematical Modelling*, 134:108–125, 2024.
5. Sergei A. Schelkunoff. The impedance concept and its application to problems of reflection, refraction, shielding and power absorption. *Bell System Technical Journal*, 17(1):17–48, 1938.
6. V. Pulletikurthi, I. Paul, K.A. Prakash, and B. Prasad. On the development of low frequency structures in near and far laminar wakes. *Physics of Fluids*, 31(2), 2019.
7. I. Paul. On the isotropy of temperature fluctuations in passive scalar turbulence. *Physics of Fluids*, 36(10), 2024.

8. S.J. Seidman, R. Brockman, B.M. Lewis, J.W. Guag, M.J. Shein, W.J. Clement, J. Kippola, D. Digby, C. Barber, and D. Huntwork. In vitro tests reveal sample radiofrequency identification readers inducing clinically significant electromagnetic interference to implantable pacemakers and implantable cardioverter-defibrillators. *Heart Rhythm*, 7(1):99–107, 2010. doi: 10.1016/j.hrthm.2009.09.071.
9. Reza Peymanfar, Haniyeh Dogari, Elnaz Selseleh-Zakerin, Mohammad Hossein Hedayatzadeh, Sara Daneshvar, Nasim Amiri-Ramsheh, Hossein Ghafuri, Ali Mirkhan, Guangbin Ji, and Bagher Aslibeiki. Recent advances in microwave-absorbing materials fabricated using organic conductive polymers. *Frontiers in Materials*, 10:1133287, 2023.
10. Qiang Peng, Meng Ma, Qindan Chu, Hao Lin, Wenting Tao, Wenqin Shao, Si Chen, Yanqin Shi, Huiwen He, and Xu Wang. Absorption-dominated electromagnetic interference shielding composite foam based on porous and bi-conductive network structures. *Journal of Materials Chemistry A*, 11(20):10857–10866, 2023.
11. Aravin Prince Periyasamy, Lekha Priya Muthusamy, and Jiri Militký. Neural network model applied to electromagnetic shielding effectiveness of ultra-light Ni/Cu coated polyester fibrous materials. *Scientific Reports*, 12(1):8609, 2022.
12. Lakhdar Sidi Salah, Mohamed Chouai, Yann Danlée, Isabelle Huynen, and Nassira Ouslimani. Simulation and optimization of electromagnetic absorption of polycarbonate/CNT composites using machine learning. *Micromachines*, 11(8):778, 2020.
13. L.P. Muthusamy, J. Militký, S. Jamal, and A. Anand. Neural network model applied to electromagnetic shielding effectiveness of ultra-light Ni/Cu coated polyester fibrous materials. *Scientific Reports*, 12(1):10021, 2022.

CHAPTER 8

Recent Progress in Polymer-Based EMI Shielding Materials

Jonathan Tersur Orasugh, Lesego Tabea Temane, and Suprakash Sinha Ray

8.1 INTRODUCTION

With the use of current electronic gadgets and wireless communication systems, the issue of electromagnetic interference (EMI) has become more of a concern, as is evident from the frequent impact on the functionality of electronics and the consequent loss of data or system failure [1]. They are deemed to offer protection to the devices as well as the reliability of the electronic system from the effects of electromagnetic waves (EMWs). Standard or conventional EMI shields, which are mostly metals, provide very good shielding. Still, their drawbacks include high reflection of the incident EMI waves, high density, easy corrosion and difficulty in shaping, especially for applications requiring lightweight and flexible characteristics [2]. This has resulted in significant interest in polymer-based EMI shields because they are lightweight, do not corrode easily and are easily processed into desired consumer shapes.

Polymer systems that incorporate CNTs, Gr, magnetic metal NPs, clay-based materials (recently proposed) and/or MXenes are also well suited for the development of functional EMW shields that also exhibit desirable mechanical properties [3, 4]. The enhancement of *SE* in these composites is best achieved by optimising the dispersion and distribution of fillers that possess conductivity higher than that of the polymer matrix [3]. Recent studies have been centred on optimising the interface betwixt the filler and polymeric matrix, achieving improved filler distribution and creating new composite structures that would ideally increase the EMI *SE* of polymer-based materials [3].

The distinctive value of this chapter is that it presents the latest trends in the field of polymer-based EMI shields, as well as new approaches and methods for the development of *SE*. Recent developments in the different conductive fillers in polymer matrices and the latest techniques that are used in solution processing, melt mixing and in-situ

DOI: 10.1201/9781003512363-8

polymerisation are thus discussed. Besides, advanced research in the fabrication of hybrid materials by incorporating multiple types of fillers to stimulate mutual enhancement and obtain high EMI *SE* is elaborated. To this purpose, this chapter provides valuable insights into recent advancements in polymer-based EMI shields to give a vision of the present/future and the possible uses of these nanocomposites in diverse state-of-the-art materials.

8.2 METHODOLOGY

This chapter reviews recent advancements in polymer-based EMI shielding materials by systematically examining the latest research articles, review papers and relevant patents published over the past decade. An all-inclusive literature search was performed using major scientific databases like Scopus, Web of Science and Google Scholar to identify significant studies related to the synthesis, characterisation and utilisation of polymer-based EMI shields. Keywords such as "polymer-based EMI shielding," "conductive fillers," "CNTs," "Gr," "metal NPs," "3D printing and additive manufacturing polymers EMI shielding mechanism," "smart and responsive materials polymers EMI shielding mechanism" and "nanocomposites" were employed to ensure an inclusive search. Two hundred works of literature were afterwards selected, sorted out and then 95 of the selected works were finally chosen and considered in this chapter. The selection criteria for inclusion in this review were based on the novelty of the research, the significance of the findings and the practical applicability of the results. Studies that provided insights into the interaction mechanisms between polymers and conductive fillers, as well as those that proposed novel fabrication techniques like solution processing, melt blending and in-situ polymerisation, were given particular emphasis. Additionally, recent review articles and their analyses were examined to contextualise findings and identify emerging trends in the field. This systematic approach ensures a comprehensive and up-to-date understanding of the current state and future directions of polymer-based EMI shielding materials.

8.3 FUNDAMENTALS AND THEORY OF EMI SHIELDS

EMI shielding is a critical aspect of protecting electronic devices from external EM fields that can cause malfunctions, degrade performance or disrupt functionality, as earlier highlighted. When designing and fabricating polymer-based EMI shields, effective EMI shielding is achieved by understanding the interaction of EMWs with shields and developing materials that reflect, absorb or attenuate these waves based on the desired application niche [3]. This section discusses the fundamental theories behind EMI shielding, including transmission line theory, reflection loss (SE_R) absorption loss (SE_A) multiple scattering (SE_M) and impedance matching design, each crucial for designing efficient shields.

EMI shielding is classified as magnetic field shielding, electrostatic field shielding or EM field shielding based on the source of EMW interference. In most practical scenarios, the focus is on EM field shielding. EMWs consist of electric and magnetic fields travelling in perpendicular $(\perp)$ directions to each other and propagating in the direction of the electric field. EMI shields can effectively obstruct the propagation of EMWs by reflecting and absorbing them, thus providing EMI shielding capabilities [3]. EM shielding refers explicitly to using conductive or magnetic materials to contain EMWs within a defined spatial area, preventing potential damage and interference to nearby living organisms and equipment. As per EMW theory, the electric and magnetic fields of high-frequency EMWs are mutually dependent and support each other. Therefore, in practical applications, it is typically sufficient to shield either the electric or magnetic field, as doing so will also mitigate the unshielded component of the other field, leading to effective EM radiation shielding. The transmission line theory currently offers a detailed comprehension of the EMI shielding mechanisms and is widely accepted [3].

8.3.1 The Transmission Line Model/Theory

Transmission line theory indicates that EMWs interacting with EMI shields are influenced by three primary mechanisms: SE_R, SE_A and SE_R $\left(SE_M\right)$ [2, 3]. The effectiveness of a material in shielding and attenuating EMWs is quantitatively assessed using the parameter EMI *SE*, measured in decibels (dB). This evaluation involves estimating the field strength with respect to the magnetic field (H) and electric field (E) both before and after the shielding analysis [2, 3]. Typically, when EMWs encounter a shielding material, a significant portion is reflected as a result of the impedance mismatch betwixt the shield and free space, as illustrated in Figure 8.1. The waves that are not reflected penetrate the shield and are progressively attenuated via multiple internal reflections or dielectric or magnetic losses till entirely absorbed and transformed into heat, as illustrated in Figure 8.1. Any remaining EMWs that pass through the shield are the transmitted waves. The degree of attenuation or loss of EMWs is a crucial factor in assessing a material's *SE* [2, 3]. The overall *SE* (SE_T) of EMI shields is the aggregate of SE_R, SE_A and SE_M contributions and is calculated using the Schelkunoff theory [3].

$$SE_T = SE_R + SE_A + SE_M \tag{8.1}$$

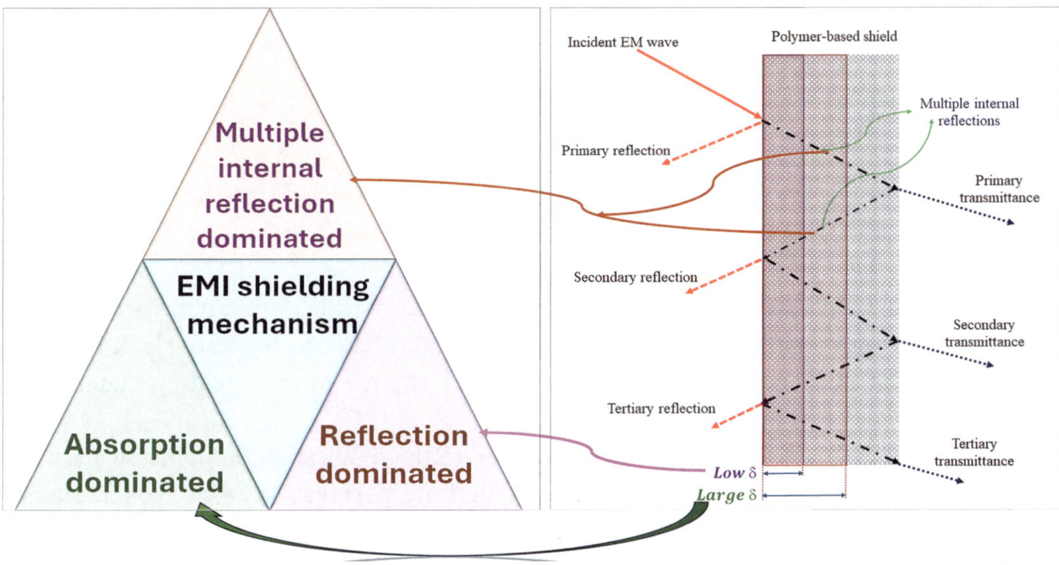

FIGURE 8.1 Schematic demonstration of EMI shielding mechanism of polymer-based materials.

SE denotes the shielding effectiveness (SE) in decibels (dB), R is the SE_R (dB), A is the SE_A (dB) and M is the multiple SE_R (dB). From this expression SE_M can also be written in the form:

$$SE_M = 10\log\left[1 - 2\times10^{-0.1SE_A} + \cos 0.23SE_A + 10^{-0.2SE_A}\right] \tag{8.2}$$

Assuming the EMI shields were uniform, the Schelkunoff model will serve as an effectual model for systematically simulating EMI SE [3]. Based on Schelkunoff's model and transmission line theory, the expression of multiple SE_M can be further detailed using Equation (8.2). SE_M is closely linked to the non-uniformity of the polymeric shield and its magnitude is influenced by SE_A. Research indicates that SE_M can be considered negligible when SE_A exceeds 15 dB. Additionally, according to Schelkunoff's theory, SE_M becomes insignificant if the shields' thickness exceeds the skin depth (δ) [5]. The skin depth δ is referred to as the distance at which the intensity of the incident EMWs reduces to $1/e$ of its original value within the polymeric shield. High-frequency EM radiation only penetrates the shields' surface layer of an electrical conductor [3, 5]. Empirical evidence shows that the intensity, power or strength of the transmitted EMWs decreases exponentially as δ increases. The skin depth "δ" can be expressed using Equation (8.3) [3, 5].

$$\delta = \sqrt{\frac{2}{\omega\sigma\mu}} = \sqrt{\frac{1}{\pi\sigma f}} \tag{8.3}$$

With respect to δ, the SE_R, SE_A and SE_M are also represented thus:

$$SE_R = 39.5 + 10\log\left(\frac{\sigma}{2\pi f\mu}\right) \tag{8.4}$$

$$SE_A = 8.68d\left(\frac{\sigma f\mu}{2}\right) \tag{8.5}$$

$$\text{and } SE_M = 20\log\left(1 - e\frac{-2d}{\delta}\right) \tag{8.6}$$

The variables f, μ, σ, and d correspond to the frequency of EMWs, magnetic permeability (μ), electrical conductivity (EC) and shielding material thickness, respectively. It has been found that the primary parameters affecting the performance of the shielding material are σ, μ, f, and d.

According to Equation (8.3), δ is closely related to the material's conductivity, permeability and frequency. Specifically, increased conductivity, permeability as well as frequency lead to a smaller δ [3]. When the shields' thickness is significantly greater than δ, fewer EMWs are able to penetrate the material. A lower relative permeability, higher conductivity plus lower frequency of EMWs result in greater SE_R. As described in Equation (8.4), higher conductivity tends to cause a larger percentage of incident EMWs to be reflected at the material's surface, reducing penetration into the shielding material. For materials dominated by SE_A, increasing thickness can enhance the material's ability to absorb EMWs, particularly when conductivity is low, as shown in Equation (8.5). A large δ and effective impedance matching help minimise significant surface reflection triggered by superior conductivity, allowing most incident EMWs to enter and be further attenuated via absorption [3, 5].

Defective or non-uniform interfaces within the shield can cause recurring reflections and scattering of EMWs, resulting in losses of multiple reflected waves. Equation (8.6) indicates that a material with a lower thickness can still attain high SE_M. Thus, the contribution of SE_M to the overall SE (SE_T) should not be overlooked. However, when SE_A exceeds 15 dB, multiple SE_M is generally included in SE_A. Studies have shown that shields typically exhibit higher SE_A values compared to SE_R [3]. It is important to note that for conductive shields, SE_R often occurs before SE_A, meaning that SE_R only accounts for surface reflection losses, excluding secondary or tertiary reflections. Comparing SE_R and SE_A alone is inadequate to fully capture the complex attenuation route of EMWs.

The mechanism of EMI shields may also be evaluated by means of reflection coefficient (R), absorption coefficient (A), and transmission coefficient (T). These parameters (coefficients) represent the EMWs' ratios absorbed, reflected and transmitted by the EMI shield. The scattering parameters $\left(S_{11} \text{ and } S_{21}\right)$ are used to calculate the power coefficients A, R, and T, which are determined by employing a vector network analyser (VNA) [3]. Variations in testing instruments may result in different calculation equations for "R" and "T." These shielding coefficients can be used to determine SE_A and SE_R, with the following equations providing the relevant equations [1, 3, 5, 6].

$$R = \left|S_{11}\right|^2 \tag{8.7}$$

$$A = 1 - R - T \tag{8.8}$$

$$T = \left|S_{21}\right|^2 \tag{8.9}$$

$$SE_R = -10\log\left(1 - R\right) \tag{8.10}$$

$$SE_A = -10\log\left(\frac{T}{1 - R}\right) \tag{8.11}$$

$$SE_T \approx SE_R + SE_A, \text{ where } SE_A > 15 \text{ dB} \tag{8.12}$$

The SE components in terms of SE_T, SE_R and SE_A quantify SE in terms of energy whereas the power coefficients T, A and R does so from a power perspective. According to Equation (8.11) for SE_A calculation, a higher reflection coefficient (R) for EMWs leads to a substantial increase in the SE_A value due to the smaller denominator in the Equation [5]. Notably, a high SE_A value simply indicates that EMWs entering the shield are successfully attenuated because of substantial absorption losses. Hence, when evaluating a material's shielding mechanism, it is important to differentiate clearly between SE $\left(SE_A \text{ and } SE_R\right)$ and power coefficients (A and R).

8.3.2 Reflection Loss (SE_R)

SE_R (reflection loss) occurs at the interface betwixt two media with different EM impedances. A significant impedance mismatch causes a fraction of the EMW to be reflected. Reflection is a major mechanism in materials with high EC, which present a substantial impedance mismatch with air or vacuum. The SE_R can be expressed as:

$$R = 20\log\left(\frac{Z_0 + Z_S}{Z_0 - Z_S}\right) \tag{8.13}$$

Z_0 and Z_S are the impedance of free space (~377 Ω) and the surface impedance of the shielding material.

According to classical EM theory, the main factor influencing how EMWs reflect is the difference in the wave impedance where the shielding material and air meet. Greater EC increases the impedance difference, resulting in a more significant EMW reflection [1, 3, 5]. It is well known that having a sufficient number of mobile charge carriers in EMI shields is essential for effectively reflecting and attenuating EMWs. Thus, enhancing conductivity and improving impedance matching have proven to be effective methods for reducing SE_R [1, 3, 5]. However, it is important to note that the high reflectivity of EMWs can lead to significant secondary contamination. As a result, there has been growing interest in attenuating EMWs through SE_A.

High-conductivity materials like metals (e.g., copper, aluminium) have low surface impedance, leading to high SE_R and making them effective EMI shields by reflecting incident EMWs [1, 3, 5, 7].

8.3.3 Absorption Loss (SE_A)

SE_A (absorption loss) refers to the EM energy dissipation inside the shielding material, usually converted into heat. This phenomenon is more prominent in materials with high μ and EC. SE_A mainly consists of dielectric loss but also magnetic loss, which are influenced by the shielding material's permeability (μ) along with dielectric constant (ε) [1, 3, 5, 7]. Dielectric loss refers to the loss of EMWs due to interactions with the electric fields within the shielding material, which primarily involves polarisation loss of dipoles under circumstances of high frequency and conduction loss [3, 5, 8]. It is important to remember that the extremely high range of 10^3–10^6 GHz frequency is where the majority of ionic and electric dipole polarisation occurs. Thus, the leading cause of interfacial polarisation is the polarisation loss of polymer-based EMI shields in the 8.2–12.4 GHz X-band range [3, 9]. Eddy current loss, dimensional resonance loss as well as natural resonance loss are typical instances of magnetic loss [3]. The SE_A is determined by the material's thickness, EC, μ and the incident wave's frequency. The SE_A is calculated using Equation (8.14):

$$A = 8.68t\sqrt{\pi\sigma fu} \tag{8.14}$$

μ is the material's magnetic permeability, σ is its EC, t is the shielding material's thickness and f is the incident EMW frequency. Absorption becomes a dominant mechanism at higher frequencies and is crucial for non-metallic, magnetic and composite materials designed for EMI shielding.

Furthermore, the complex dielectric constant $\left(\varepsilon_r = \varepsilon' - j\varepsilon''\right)$ is the primary determinant of the dielectric loss intensity; it is split into the actual part (ε') as well as the imaginary part (ε''), as represented by the following expressions [3].

$$\varepsilon_r = \varepsilon' - j\varepsilon'' \tag{8.15}$$

$$\varepsilon'' = \frac{\sigma}{2\pi f \varepsilon_0} \tag{8.16}$$

$$\tan\,\delta_a = \frac{\varepsilon''}{\varepsilon'} \tag{8.17}$$

Here, σ denotes conductivity, f is frequency, δ_d is the dielectric loss angle and ε_0 is the dielectric constant of free space. The material's capacity to store microwaves (MWs) is indicated by the real component of the dielectric constant (ε'), whereas the imaginary part (ε'') represents the capacity for dissipating energy. Maxwell's principle states that polarisation and charge accumulation at the interfaces of two adjacent materials might result from differences in their EC [3]. Consequently, a high dielectric constant enhances interfacial polarisation, leading to greater energy consumption of EMWs.

8.3.4 Multiple Reflection Loss (SE_M)

The energy dissipated as a result of EMWs being reflected and scattered due to variations in EC among different surfaces within a material is referred to as SE_M. When an EMW reflects multiple times within the shielding material, SE_M occurs [1, 3, 5]. When the shielding material is thin in comparison to the input wave's wavelength, this effect becomes significant since it permits the wave to reflect internally and, if not properly absorbed, potentially pass through the shield. When the thickness or absorption qualities of the material are high enough to adequately attenuate the EMW, SE_M becomes minimal. Equation (8.3) provides an approximation for it. The contact between the wave and the material increases with multiple reflections, increasing the amount of energy dissipated and enhancing the effectiveness of the shield. But when SE_A is predominant, the effect is diminished.

8.3.5 Design of Polymer-Based EMI Shields: Impedance Matching for Minimal Reflection and High Absorption

EMI shielding materials (EMI shields) can be divided into two kinds according to the main shielding mechanism that they use: absorption-dominated and reflection-dominated materials, as pictorially presented in Figure 8.1 [3]. For materials where SE_A is the main mechanism, the ability to absorb EMWs is crucial for their attenuation. The MW absorption capability of MW absorbing materials is closely associated with their complex permeability $\left(\mu_r = \mu' - \mu\varepsilon''\right)$ and complex permittivity $\left(\varepsilon_r = \varepsilon' - j\varepsilon''\right)$. To achieve optimal EMW absorption, MW absorbing materials typically need to meet two requirements. To guarantee that the majority of the incident EMWs can enter the absorber with the least amount of reflection, they must first achieve impedance matching (Z). In order to efficiently attenuate the incoming EMWs as much as feasible, the MW absorbing materials should also possess substantial dielectric or magnetic loss capabilities. It is possible to express the requirements for impedance matching using Equations (8.18) and (8.19) [3, 6].

$$Z = \frac{Z_{\text{in}}}{Z_0} = \sqrt{\frac{\mu_r}{\varepsilon_r}} \tag{8.18}$$

$$Z_{\text{in}} = Z_0 \sqrt{\left|\frac{\mu_r}{\varepsilon_r}\right|} \tanh\left[j\left(\frac{2\pi fd}{c}\right)\sqrt{\mu_r \varepsilon_r} \right] \tag{8.19}$$

$$\alpha = \frac{\sqrt{2\pi f}}{c} \times \sqrt{\mu''\varepsilon'' - \mu'\varepsilon' + \sqrt{\left(\mu''\varepsilon'' + \mu'\varepsilon'\right)^2 + \left(\mu'\varepsilon'' + \mu''\varepsilon'\right)^2}} \tag{8.20}$$

$$RL = 20\log\left(\frac{Z_{\text{in}} - Z_0}{Z_{\text{in}} + Z_0}\right) \tag{8.21}$$

In Equation (8.19), Z_{in} and Z_0 represent the input impedance of the EMW absorbing material and the impedance of free space (377 Ω), respectively, while c denotes the speed of light in a vacuum. As indicated by expression (8.18), perfect impedance matching $(Z = 1)$ can be attained by making sure that the permeability of the material is approximately equal to its dielectric constant. When EMWs transition from free space (with impedance Z_0) to the interface of a shielding material (with impedance Z_{in}), minimal reflection occurs only if Z_0 equals Z_{in}. However, in reality, Z_0 is typically greater than Z_{in} because the hyperbolic tangent function $(\tan h)$ usually has values less than 1, and the complex permeability (μ_r) is generally lower than the complex permittivity (ε_r) [3]. Achieving optimal impedance matching is crucial for excellent absorption performance in the development of MW absorbing materials. The attenuation capability of MW absorbing materials can be assessed by computing the attenuation constant "α" as expressed in expression (8.20) [3].

Larger μ'' and ε'' were helpful in improving the absorption as well as the attenuation capacity of EMWs, as shown in Equation (8.20). Transmission line theory states that SE_R (otherwise known as RL) is an indirect way to represent the SE_A capacity of MW absorbing materials to EMI shielding. The SE_A to EMWs peaked as the RL value decreased to its minimum. The equation displayed the expression (8.18). Z_{in} has to be as close to Z_0 as feasible in order to minimise the SE_R (lowest RL value). This might be accomplished by raising the μ_r and decreasing the surface's ε_r. Both the shielding material's ε_r and μ_r exhibit frequency-dependent characteristics (non-constant values) within the MW frequency band, and their values rapidly decrease with an increase in frequency. Achieving $\mu_r = \varepsilon_r$ in the wide band is challenging because the ε_r varies greatly, ranging from several thousand to extremely small, and is often less than 20.

Selecting materials with low ε_r can improve impedance matching when building shielding components [10]. EMWs could, for instance, penetrate a low-dielectric, well-insulated wave-transmitting material from the air without reflecting, but such as material has negligible EMW absorption functionality. To guarantee that EMW energy is effectively attenuated throughout its journey, the wave absorber needs to be placed evenly all around the channel of propagation [3, 10]. A wave absorber's absorbing performance was traditionally measured using the optimum SE_R $\left(R_{\mathrm{Lo}}p\right)$ and maximum effective absorption bandwidth (EAB). The frequency region exhibiting an *RL* value of less than −10 dB was designated as the EAB. The shielding material met practical criteria by effectively absorbing 90% of EMWs while the *RL* fell below −10 dB. To evaluate EMWs' absorbing characteristics, three more criteria should be considered: the minimal *RL* value (RL_{min}), matrix filling ratio and absorption thickness [3, 10].

For polymer-based EMI shields to function as desired, impedance matching is an essential design strategy, especially when it comes to ensuring high absorption and minimal reflection [3, 10]. For situations where internal absorption of EM energy is required, proper impedance matching permits more of the incident EMWs to enter the shielding material instead of being reflected at the surface. Impedance matching is achieved by designing materials with an impedance that gradually transitions from free space to the material's interior, reducing reflection and enhancing absorption. Strategies for impedance matching include the careful design of layered structures, advanced composite materials and the adoption of surface treatment and/or coating [6]. Using multilayered materials with gradually varying electrical and magnetic properties can provide an impedance gradient, improving absorption [1, 3, 6], where each layer's impedance is designed to progressively match the impedance of adjacent layers and free space. Moreover, incorporating different types of fillers, such as conductive (CNTs, Gr) and magnetic (ferrites, iron oxides) fillers, into a polymer matrix has been proven to help tune the composite's overall impedance [1–3, 5, 6]. These materials provide a balance of conductivity and μ that aids in impedance matching.

Surface treatments and coatings have also been adopted as an effective tool that aids impedance matching [3, 5, 6, 8, 11]. Applying surface treatments or coatings that alter the surface impedance can improve impedance matching. Thin conductive coatings on insulating or conductive substrates, depending on the predetermined design of the composites' structural architecture, have reportedly been adopted to achieve the desired impedance properties [3, 5, 6]. For instance, Luo et al. [11] coated Co NPs with PPy conducting polymer in order to achieve excellent impedance matching in the composite. The authors produced Co NPs using $CoCl_2 \cdot 6H_2O$ and $NaBH_4$ as a reductant, with oleic acid as a dispersant, in a water bath at 40 °C. According to the report, the Co NPs generated had an average size of 30 nm. Secondly, PPy/Co nanocomposites were created via in-situ polymerisation. A core-shell structure was formed by wrapping Co NPs in PPy. PPy/Co nanocomposites outperformed separate components in EMW absorption, indicating their potential for MW absorption. Figure 8.2 shows that when the thickness increased, the absorption peak shifted towards the low-frequency direction. At 13.8 GHz and 7.2 GHz bandwidth, a matching thickness of 3.0 mm resulted in a maximum *RL* of −20.0 dB [11].

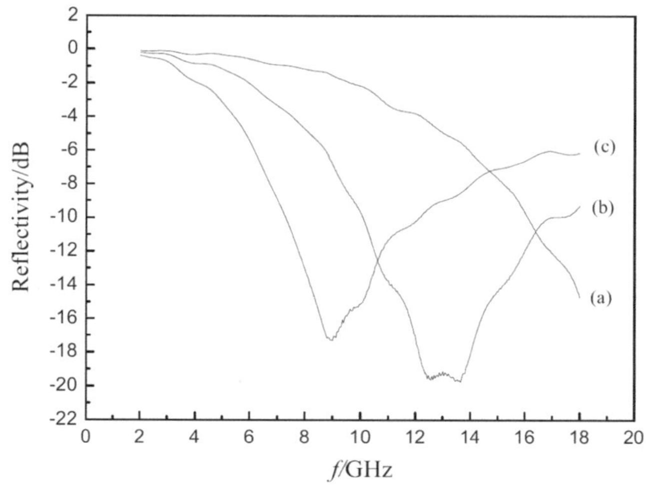

FIGURE 8.2 The reflectance curves of PPy/Co nanocomposites with different thicknesses: (a) 2.0 mm, (b) 3.0 mm and (c) 4.0 mm. Reproduced with permission from Luo et al. [11], Copyright 2014, Elsevier Science Ltd.

Impedance matching is critical for applications that require regulated energy absorption, such as EMI shielding in electronic equipment, where the absorbed energy must be managed to avoid heat accumulation.

Understanding the principles and theory of EMI shielding is essential for developing materials that successfully shield electronic equipment from EMI. The interaction of SE_R, SE_A, multiple SE_R and impedance matching determines a material's SE. Using these concepts, researchers and engineers may create improved EMI shielding solutions that fulfil the stringent requirements of modern electronic applications, balancing high absorption with minimal reflection while maintaining device reliability and performance.

8.4 POLYMER MATRICES FOR EMI SHIELDING

Polymer-based materials are found to have great potential in applications for EMI shielding owing to their features, including being lightweight, flexible and easy to process. These polymer matrices, when incorporated with active filler materials such as conductive fillers, produce efficient EMI shielding composites that are significant in guarding electronic devices against EMI waves. The choice of the right polymer matrix that will enable the desired attributes or performance to be attained in the fabricated EMI shield is important. The different kinds of polymers most often applied in EMI shielding composites and their characteristic features are briefly presented here.

8.4.1 Types of Polymers Used in EMI Shielding Composites

Some of the polymers used to produce polymeric EMI shields include thermoset polymers and thermoplastic polymers such as polyethylene (PE) [12, 13], polypropylene (PP) [12], polystyrene (PS), polycarbonate (PC), polyether ether ketone (PEEK) [14] and polyvinyl chloride (PVC), since they are easily processed and recyclable. These are versatile materials in the sense that they can be formed into different shapes depending on the required kind of product. For instance, polypropylene is preferred because of its low density and superb electrical insulation ability, while polycarbonate has high impact strength and transparency, which are useful for electrical parts such as capacitors [2–4, 15]. Other kinds of polymeric materials that are widely utilised in this respect are thermosetting polymers, for example, cellulose [16], epoxy (EP) resins, phenolic resins and polyester resins, as these polymers possess high thermal stability as well as good mechanical strength. When cured, these polymers have a high degree of cross-linking and are thermostable, which is why they are used in high-temperature applications. EP resins are particularly popular in EMI shielding due to their excellent adhesive properties, chemical resistance and ability to form strong bonds with conductive fillers [17]. Moreover, conductive polymers like polyaniline (PANI), polypyrrole (PPy) and polythiophene (PTh), which are inherently conductive materials used either as the primary matrix or as coatings, have also been reportedly used in EMI shield fabrication. These polymers offer the advantage of combining both conductive and matrix properties, providing flexibility in design and application [8]. Last but not least among this category are the elastomers. Among the elastomers adopted in this niche, polydimethylsiloxane (PDMS) and other silicone-based elastomers have reportedly been used for their flexibility, elasticity and excellent environmental stability [18–22]. These elastomers are particularly suitable for EMI shielding applications where mechanical flexibility and stretchability are required, such as in wearable electronics or flexible devices [7].

8.4.2 Polymers Properties Relevant to EMI Shielding Applications

The foremost property desired of polymer-based EMI shields is EC. When mixed with conductive and/or magnetic fillers or their hybrid forms, the polymer matrix should facilitate the creation of conductive pathways for effective EMI shielding. The inherent electrical property of the utilised polymeric matrix influences the percolation threshold (the minimum filler content needed to achieve conductivity in the polymer-based composite shield). Moreover, polymers with higher dielectric constants can enhance the conductivity of the composite at lower filler loadings [6]. The mechanical properties of these shields are also extremely important in view of structural robustness for the intended use. Of particular interest are the mechanical strength, flexibility and durability of the polymeric matrix, which determine the structural properties of the EMI shields. Elastomers have also been used to provide flexibility in systems that need to be stretchable and deformable, in contrast to the high strength and rigidity of polycarbonate and EP polymers that are suitable for use in structural members [23]. Another important desirable property with respect to the application of polymer-based EMI shields is thermal stability. Low operating temperatures may be required in certain applications, while the shielding material also needs low moisture absorption and good thermal stability at high operating temperatures. Epoxies and phenolics are examples of thermosetting polymers that possess good heat resistance, thus making them suitable for high-temperature conditions. Thermal stability enables the material to retain the *SE* as well as structural capacity over a broad temperature range [6].

Once more, processability remains a crucial factor when it comes to polymers used for polymer-based EMI shields. Processability in general, and particularly formability, coatability and extrudability, are essential factors for the practical applicability of polymer-incorporated EMI shielding materials. The use of thermoplastics provides the ability to recycle them for reuse and is cost-effective in processing. On the other hand, thermosetting polymers, while more difficult to process, provide robust performance once cured [17]. Also, the ability of these materials to withstand stripping by chemical agents, moisture and other environmental factors is a *conditio sine qua non* to enable these shields to serve the purpose of EMI shielding in demanding environments. Talking about polymers with suitable characteristics like chemical resistance, EP and fluoropolymers are notable, as they help predict the shields' performance in different settings [17, 24].

Therefore, the selection of the polymeric matrix plays a crucial role in determining EMI *SE*, as well as in the mechanical characteristics, thermal stability and service life of the composite. Through the proper choice of polymers and the regulation of their behaviour in the presence of the conductive filler, it is possible to create new efficient EMI shields to meet the prerequisites of today's communication and electronic gadgets.

8.5 MAJOR AND NOTABLE ADVANCES IN POLYMER-BASED EMI SHIELDS

EMI shields based on polymers have improved significantly in the recent past, particularly where polymer EM shields are concerned. These materials are essential in shielding electronic equipment from EM interference that causes device malfunction or data loss.

8.5.1 Development of Lightweight and/or Flexible Composites

Some of the earliest developments include the use of ultralight and, more importantly, flexible polymer composites for EMI shielding [22, 25]. Scientists have recently loaded polymer matrices with conductive fillers, such as CNTs, Gr, metal NPs and MXenes, to obtain materials with good SE_s, low density and flexibility. These are most effective in portable electronic devices, wearable electronics and aerospace applications because metallic shields of similar quality are heavy and inflexible [26, 27].

With a focus on creating lightweight and efficient polymer systems, Zhang et al. [26] introduced a porous polyimide (PI) composite that incorporates PANI-modified Gr oxide (GO) and multiwalled CNT (MWCNT) hybrid fillers for heat-resistant EMI shielding. Utilising a PI matrix, PANI-modified GO (PANI-g-GO) and MWCNT as hybrid fillers, porous PI composites for EMI shields were logically developed and manufactured using an inventive combination of in-situ polymerisation and thermally assisted non-solvent induced phase separation (TNIPS) processes, as shown in Figure 8.3(a). The composites were lightweight while exhibiting excellent heat resistance due to their porous structure and PI matrix. Additionally, hybrid fillers allow for efficient conducting networks in the PI matrix, as demonstrated in Figure 8.3(b), which improves the efficacy of EMI shielding. The porous PI composite, known as PPIC12, has a low density of 0.56 g/cm^3, a high T_{10} of 554.81 °C and exceptional EMI shielding performance up to 28.2 dB at a thickness of 2 mm, as shown in Figure 8.3(c). It also has a hybrid filler content of 12 wt.% [26].

Another illustration can be found in the work of Shen et al. [28]: in their report, a low-density (0.28–0.4 g/cm^3) flexible high-performance polyetherimide (PEI)/Gr@Fe_3O_4 (G@Fe_3O_4) composite foams were created using a phase separation method, as shown in

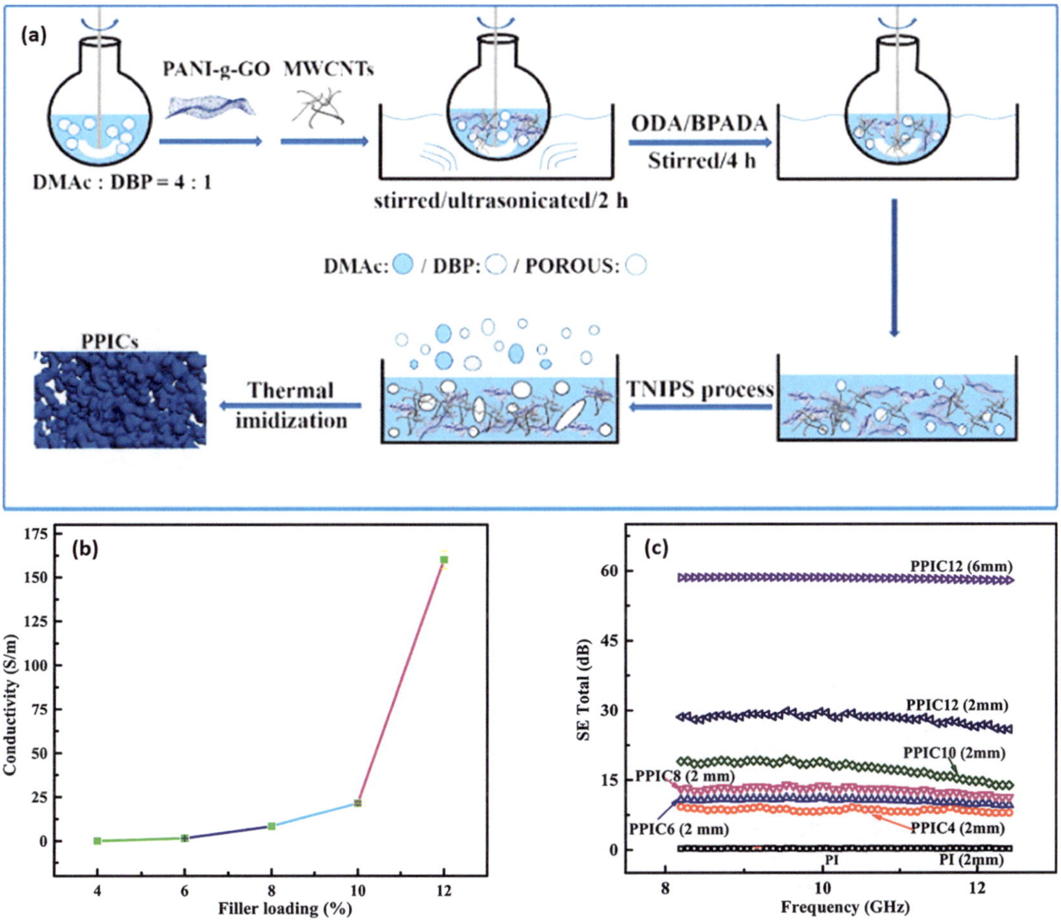

FIGURE 8.3 (a) Schematic of the facile fabrication of the porous polyimide composites (PPICs) through in-situ polymerisation and thermally assisted non-solvent induced phase separation (TNIPS) process; (b) Conductivity of the PPICs; and (c) *SE* total, *SE* reflection and *SE* absorption of PI (2 mm) and PPICs (2 mm or 6 mm). Reproduced with permission from Zhang et al. [26], Copyright 2021, Elsevier Science Ltd.

Figure 8.4(a). Figures 8.4(b) and (c) show that the PEI/Gr@Fe_3O_4 foam with a Gr@Fe_3O_4 loading of 10 wt.% has a good specific EMI *SE* of ~41.5 dB/(g/cm^3) in the X-band. Moreover, it was discovered that most of the applied MW was absorbed rather than reflected back, which was caused by multiple reflections, improved impedance matching and EMW attenuation. Meanwhile, the resultant foams demonstrated superparamagnetic character and a low thermal conductivity (TC) of 0.042–0.071 W/m·K. This method is quick, scalable and highly reproducible, which could aid in commercialising these composite foams and their widespread application as EMI shields in aeroplanes and spacecraft [28].

8.5.2 Enhanced Filler Dispersion and/or Distribution

The efficiency of EMI shielding is primarily determined by the qualities of the polymer matrix, functional fillers such as conductive or magnetic fillers, and the uniformity with which they are loaded [27, 29–31]. Enhanced filler dispersion and distribution are essential

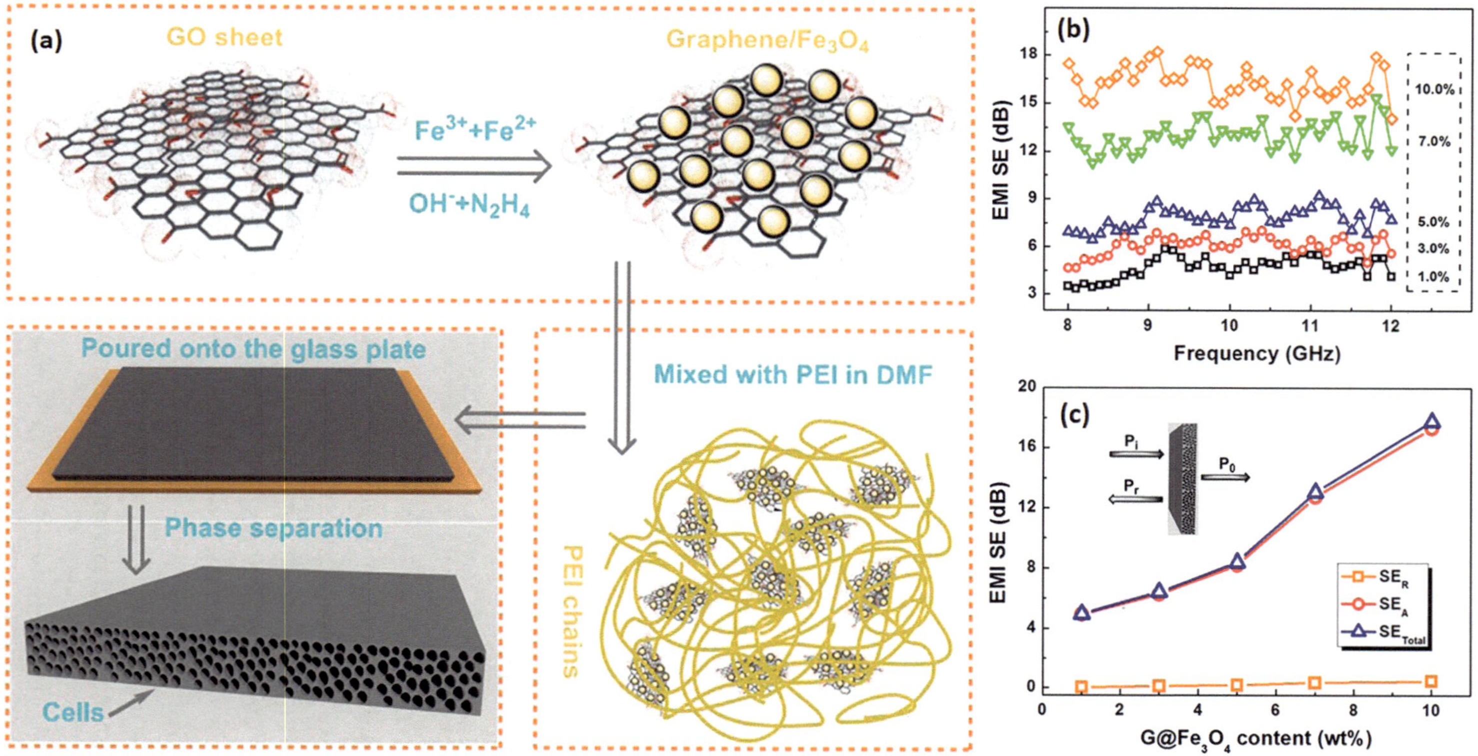

FIGURE 8.4 (a) Schematic of the preparation of $G@Fe_3O_4$ hybrid and PEI/$G@Fe_3O_4$ composite foams by a WVIPS process; (b) EMI shielding effectiveness as a function of frequency measured in the 8–12 GHz range of the PEI/$G@Fe_3O_4$ foams with various $G@Fe_3O_4$ concentrations; and (c) SE_{total}, SE_R and SE_A of microcellular foams at 9.6 GHz. Reproduced with permission from Shen et al. [28], Copyright 2013, ACS Publications.

methods for developing high-performance polymer-based EMI shields. This method aims to evenly distribute conductive fillers, such as CNTs, Gr or metal NPs, throughout the polymer matrix. By establishing uniform distribution, the interconnected network of fillers can effectively block and absorb EMWs, resulting in enhanced EMI *SE*. Several processing technologies, such as solution processing, melt blending and in-situ polymerisation, have advanced and improved, resulting in better filler dispersion [29]. These techniques help prevent filler agglomeration and ensure that the fillers are well integrated within the polymer, maximising the material's EC and shielding capabilities.

Moreover, the treatment of fillers with various agents, such as surfactants or compatibilisers, and the use of surface treatments of fillers have also been used to improve interfacial adhesion and proper dispersion to avoid the formation of conductive networks [29]. Furthermore, the uniform distribution of fillers improves the mechanical properties and thermal stability of the shields, making them more durable and practical in a variety of applications, including electronics, aerospace and telecommunications, where EMI shielding is critical for device performance and security [29]. Once more, bio-based Eucommia ulmoides gum (EUG) composites with segregated network architectures were reportedly made by straightforward mechanical mixing in the work of Han et al. [32], utilising the characteristics of highly cross-linked rubbers with limited molecular chains. Highly cross-linked EUG (HCE) successfully separated the conductive fillers into the EUG phase, whereas EUG crystallisation encouraged filler dispersion within the EUG's amorphous area. A very effective 3D conductive network was produced by the segregated architecture and crystallisation, which enabled the EUG/HCE/CNT composite to achieve an exceptional EMI *SE* of 83.0 dB and a notable SE_A spanning from 8.2 to 12.4 GHz [32].

A study carried out by a team of experts, for instance, established that the blend composition, the kind of filler in each blend component, and the dispersion and distribution of these fillers all affect the performance of EMI shielding. The fillers employed in this work include multiwalled CNTs, short MWCNTs produced through ball milling and Fe_3O_4 coated over rGO (rGO-Fe_3O_4) [30]. Additionally, they found that the absorption/reflection component of EMI material shielding is controlled by the frequency under consideration as well as the dispersion and distribution of fillers in the matrix (which is further influenced by manufacturing, matrix selection and filler/polymer interaction). These elements and the crucial filler-polymer compositions needed to enhance absorption-based shielding efficacy will be the subjects of future research [30]. Furthermore, the kind and concentration of fillers, their dispersion and distribution and the filler-polymer interaction were found to all have an effect on SE_T. PMMA (10 wt.%) also worked well as a compatibiliser but did not improve SE_T significantly.

Moreover, in their study, Ravindren et al. [31] developed a conductive polymer composite that absorbs electromagnetic radiation by utilising the concepts of preferred distribution and double percolation. The polymer matrix was a 50/50 wt.% binary combination of EMA and EOC, with MWCNT serving as the functional filler. Adding 3 wt.% MWCNT resulted in a *SE* of −18.66 dB, while adding 15 wt.% MWCNT resulted in a total *SE* of −34.06 dB, in addition to strong thermal and mechanical qualities. They believe their work provides insight into the development of a scalable and economical EMI shield, which has potential for academic and industrial applications [31].

Choudhury et al. [33] conducted an in-depth analysis of a thermoplastic elastomeric blend containing EPDM and LLDPE. Among the various compositions, the 60/40 ethylene propylene diene rubber (EPDM)/linear low-density polyethylene (LLDPE) mix demonstrated better thermoplastic elastomeric properties. To that end, they thoroughly investigated the effect of different carbon black (CB) concentrations in EPDM-LLDPE-VCB (ELV) composites in order to create a cost-effective, resilient and flexible material for EMI shielding in industrial applications. Their analysis highlights the importance of polymer and filler selection, filler quantity and filler distribution method on the final properties of ELV composites. Their work shows that making ELV composites using melt blending is an excellent method for developing an inexpensive, robust, lightweight and flexible EMI shielding material. In particular, achieving a *SE* of –33.5 dB in samples as thin as 2 mm and a CB concentration as high as 50 phr indicated that the method was suitable for large-scale implementation and satisfied the requirements for commercial viability [33].

8.5.3 Hybrid and/or Multilayered Composites

It has been reported that the application of more than one type of conductive filler will yield additive effects and produce improved EMI shielding properties. For example, combining Gr with other functional fillers such as CNTs or metal NPs has been shown to enhance both electrical and sometimes mechanical performance of the host polymeric matrix [16, 27]. There has also been advancement in the designs that use multilayers of materials with different compositions to achieve high *SE* over an extended range of frequencies [34]. These approaches have been used to optimise the electrical, mechanical, topographical and temperature characteristics of the structure depending on the particular application and microstructure [34]. This method is primarily used to improve the *SE* by reducing reflectivity, although there are two distinct circumstances in which it is applied. The first method simply stacks several layers of polymeric composites having the same active filler concentration, such as Gr, MXene, carbon and/or Fe_2O_3 NPs [27]. Figure 8.5 shows that increasing the multi-reflection method leads to improvement. While multi-reflection enhances overall reflection, the effect on perceived absorption is more pronounced, resulting in a larger *SE*-to-reflectivity ratio. This technique is effective for nanocomposites made of poly(butylene adipate-co-terephthalate) (PBAT) [35], cellulose (buckypaper) [36], PDMS [37], polylactic acid (PLA) [38] and similar materials.

An interesting example is presented by Zhang et al. [38]. The authors offered alternate building methods for conductive and magnetic layers. Combining copper-plated BF (Cu@BF) and iron-plated BF (Fe@BF) with PLA in a raw matrix of sustainable bamboo fibres (BF) and biodegradable PLA produced the magnetic and conductive layers. After alternately stacking magnetic and conductive layers and applying heat to press them together, a low *R*-value biomass multilayer composite with high EMI *SE* and "multi-(absorption-reflection-reabsorption)" structures was produced. According to Figure 8.6, the authors developed a shielding mechanism that highlighted the attenuation process of EMWs. The "multi-(absorption-reflection-reabsorption)" structures are responsible for multilayer materials' strong EMI *SE* and low *R*-values. Based on the findings, adding more alternating layers decreased R_{min} from 0.40 to 0.25 and increased *SE* within the X-band from 45.02 dB (3-layer)

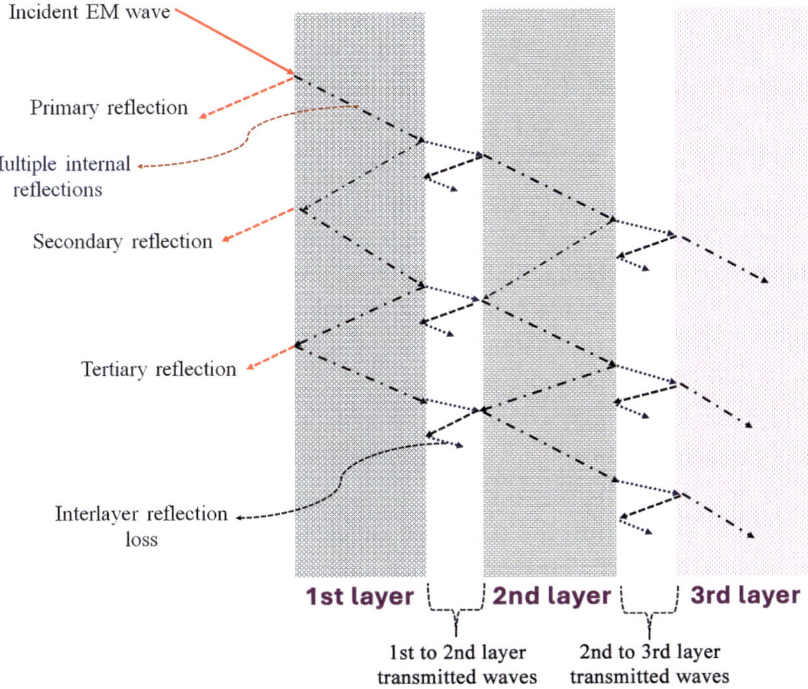

FIGURE 8.5 A pictorial illustration of multilayered systems' effectiveness against EMI waves.

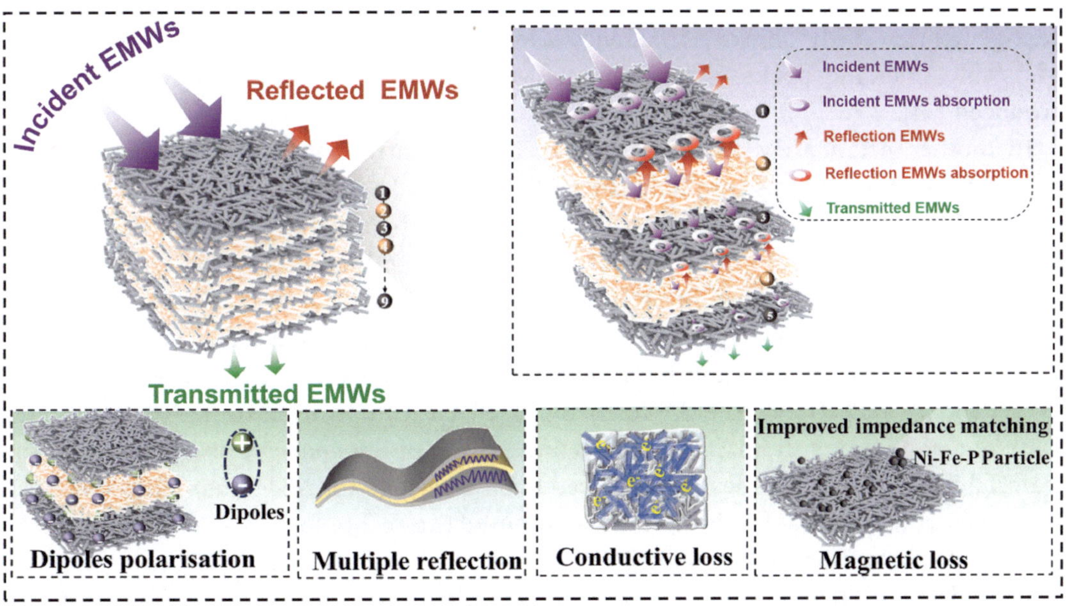

FIGURE 8.6 The possible electromagnetic shielding mechanism for alternating composite film material. Reproduced with permission from Zhang et al. [38], Copyright 2024, Elsevier Science Ltd.

to 80.2 dB in the 9-layer system. With a 75 dB *SE* from 1 to 18 GHz, the 9-layer composite exhibits good efficiency, low reflectivity and broadband shielding. Because of its improved impedance matching, the magnetic Fe@BF/PLA layer efficiently absorbs incident EMWs. The remaining EMWs are reflected back for additional absorption owing to natural resonance brought on by impedance mismatches, while some of them experience conductive loss when they interact with electrons as they approach the Cu@BF/PLA layer. It is inevitable that some EMWs pass through the Cu@BF/PLA layer and may be absorbed by the magnetic layer that follows. The repeated absorption and reflection that EMWs experience as the number of layers rises significantly lengthens their loss route. Moreover, a great deal of effective shielding is also attributed to interfacial polarisation, multiple reflections and the combined conductive and magnetic losses [38].

The second method uses a multilayered construction where the amount of carbon filler increases gradually as one layer is added to the next. The primary goal is to prevent the existence of an interface where two media have a significant variation in their dielectric constants ε_r, which would limit the reflectivity at each contact [27].

With conductive (MWCNT) filler concentrated in the surface layers and the inner layer selectively foamed using the CO_2 foaming method, a sandwich-structured TPU-MWCNT foam was proposed and constructed by Song et al. [34]. Their results showed that EMI *SE* and absorption percentage are higher in sandwich-structured foams with a low expansion ratio. The maximal *SE* and absorptivity were 53.3 dB and 0.66 dB, respectively, at a 5% MWCNT concentration. Sandwich structures with specified middle-layer thickness can attenuate EMWs with specific wavelengths, resulting in frequency-selective EMI shielding. This is due to the Fabry-Pérot resonance phenomenon. Furthermore, by prolonging the foaming time, the middle-layer thickness increased, causing the frequency of resonant waves to fluctuate, resulting in a peak frequency range of 8.7–12.4 GHz. Lastly, to improve next-generation EMI shields, frequency-selective TPU-based conductive foams with adjustable shielding peaks and superior EM attenuation capability were successfully produced [34].

In another instance, a group of researchers demonstrated that a hybrid system based on Bi_2Se_3-PVDF flexible films exhibited robust EMI *SE* due to dielectric dissipation as well as multiple internal reflections created by inorganic Bi_2Se_3 NPs included in the PVDF matrix, as depicted in Figure 8.7 [39]. A simple tape-casting method and multiple folding were used to make numerous folded Bi_2Se_3-PVDF composite films that could be easily expanded for industrial applications. When the EMI *SE* of the composite materials was analysed, it showed typical absorption-dominant behaviour and had a sizeable suppression capacity for secondary reflections. The 35 vol.% Bi_2Se_3-PVDF composite film had a single-layer thickness of 60 μm with an SE_A of 20 dB and an SE_R of less than 5 dB. After folding the material into eight layers, the SE_A increased to 50 dB, while the SE_R remained below 5 dB, as shown in Figure 8.8(a–f). These authors proposed this technology as an effective strategy to fabricate absorption-dominant EMW absorbers and/or EMI shields [39].

Hybrid systems have established themselves as highly effective EMI shielding materials, though researchers are still working towards achieving full commercial implementation [16, 39, 40].

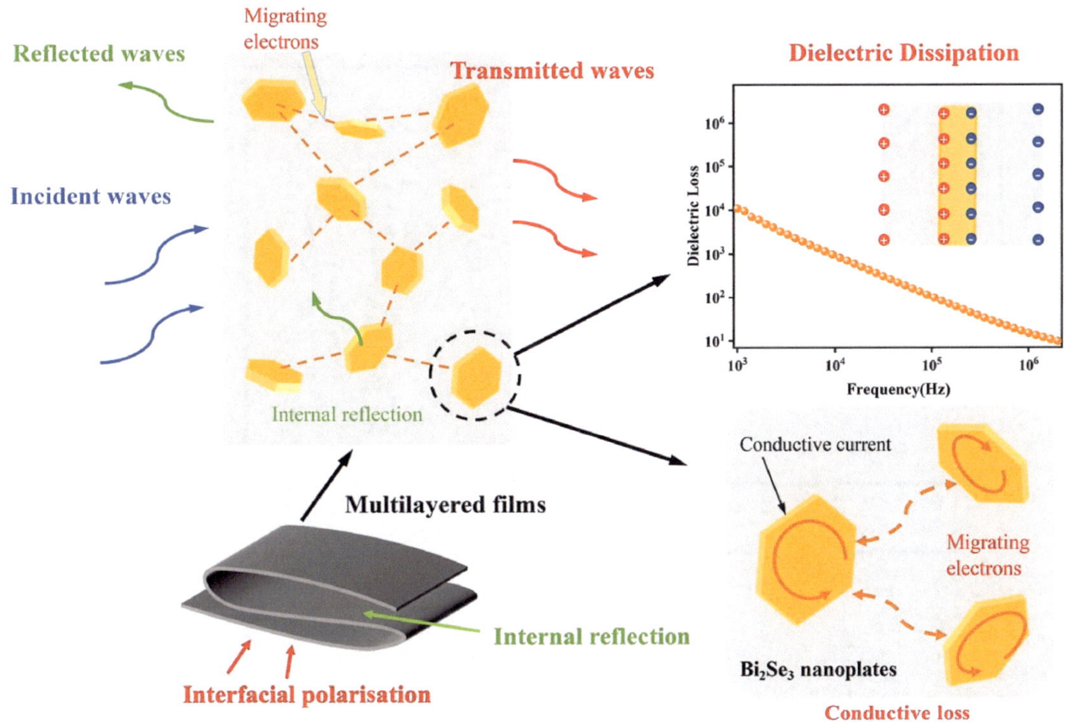

FIGURE 8.7 Schematic diagram of the EMI shielding mechanism of Bi_2Se_3/PVDF nanocomposite films. Reproduced with permission from Pu et al. [39], Copyright 2024, Elsevier Science Ltd.

8.5.4 Incorporation/Use of Conducting Polymers

With growing interest in incorporating conducting polymers into EMI shielding composites, PANI, PPy and PTh have attracted increasing attention [27, 41]. These polymers possess inherent EC and are usable in an extensive variety of applications such as coatings, films and fibres. Incorporating conductive polymers either as primary fillers or in conjunction with other fillers has paved the way for the development of innovative shields that can provide additional features such as sensing or self-healing mechanisms [41].

Nguyen et al. [41] presented a flexible, lightweight yet freestanding polymer composite film made of SWCNTs and PEDOT:PSS. The films were produced using a simple tabletop technique involving vacuum filtration followed by dip-coating. The 2.2 μm composite/hybrid EMI shield displayed excellent EC and outstanding *SE*, with EMI *SE* of up to 55.53 dB and SSE/t of up to 2,230,000 dB cm^2/g in the X-band. The composite's exceptional *SE* is attributed to the high conductivities of PEDOT:PSS and SWCNT, which work synergistically within the composite. Furthermore, the films showed outstanding stability, retaining performance even in harsh environments. The SWCNT film shows potential for large-scale applications. Thus, the scientists believe their composites provide a lightweight, highly efficient EMI shield suitable for a wide range of future [41].

Kumari et al. [42] developed PANI and PEDOT conducting polymer and Gr composites for EMW attenuation using an in-situ chemical-oxidative polymerisation process. Changes in IR band strength and presence/shifting indicate interactions and variations in "σ_s" with

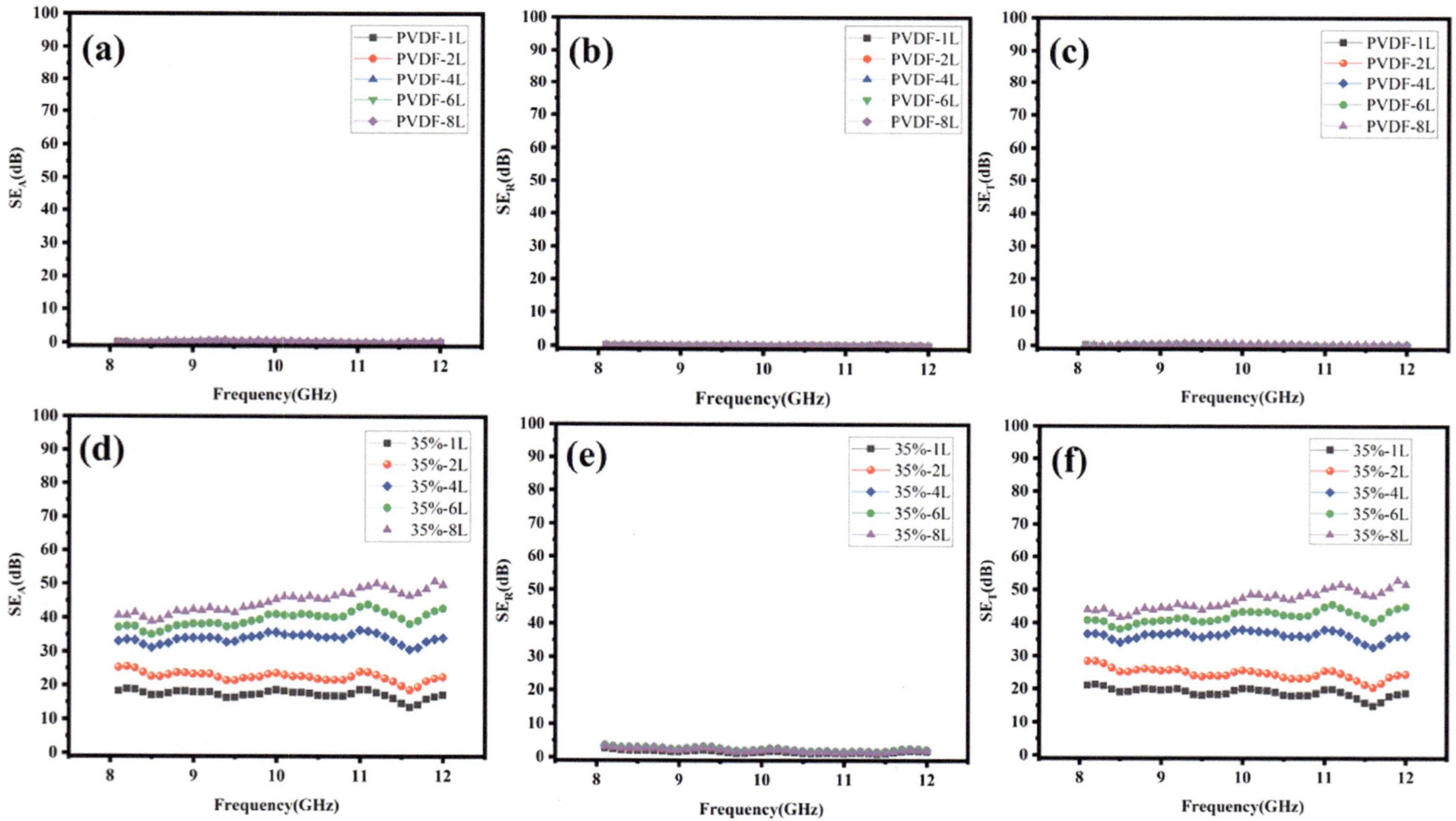

FIGURE 8.8 (a) SE_A, (b) SE_R and (c) SE_T of PVDF polymer multilayer films vary with frequency; (d) SE_A, (e) SE_R and (f) SE_T of 35% Bi_2Se_3/PVDF nanocomposite multilayer films vary with frequency. Reproduced with permission from Pu et al. [39], Copyright 2024, Elsevier Science Ltd.

filler addition. The virgin polymer@Gr composite samples showed good *SE* in the Ku-band frequency range. The PG nanocomposite's SE_T was measured at 18 GHz for a sample with a thickness of 2 mm. This value was 46.62 dB, which was much higher than the *SE* value required for industrial applications. At higher frequencies, the composite specimen effectively addressed the issue of electromagnetic contamination, shielding more than 99% of incident EMWs, as illustrated in Figure 8.9(a–d). The composite's *SE* results were attributable to greater synergistic dielectric losses and enhanced specimen conductivity. The Cole-Cole plot analysis revealed the existence of Maxwell-Wagner, dielectric and electronic polarisations in the samples. Increasing "σ_s" while decreasing skin depth resulted in an increase in total *SE*. A thorough examination of the samples' EM shielding properties in connection to various parameters revealed that they are outstanding shields appropriate for commercial applications in the telecommunications, electronic and defence industries [42].

Wang et al. [43] utilised the styrene-butadiene-styrene (SBS) copolymer to integrate the advantages of polymers with magnetic minerals. PS-S@SBS/Pyr is a formable block material made by combining pyrrhotite (Pyr) tails and waste PS. This material excels in temperature management, electrical insulation and EMI protection simultaneously. The bidirectional

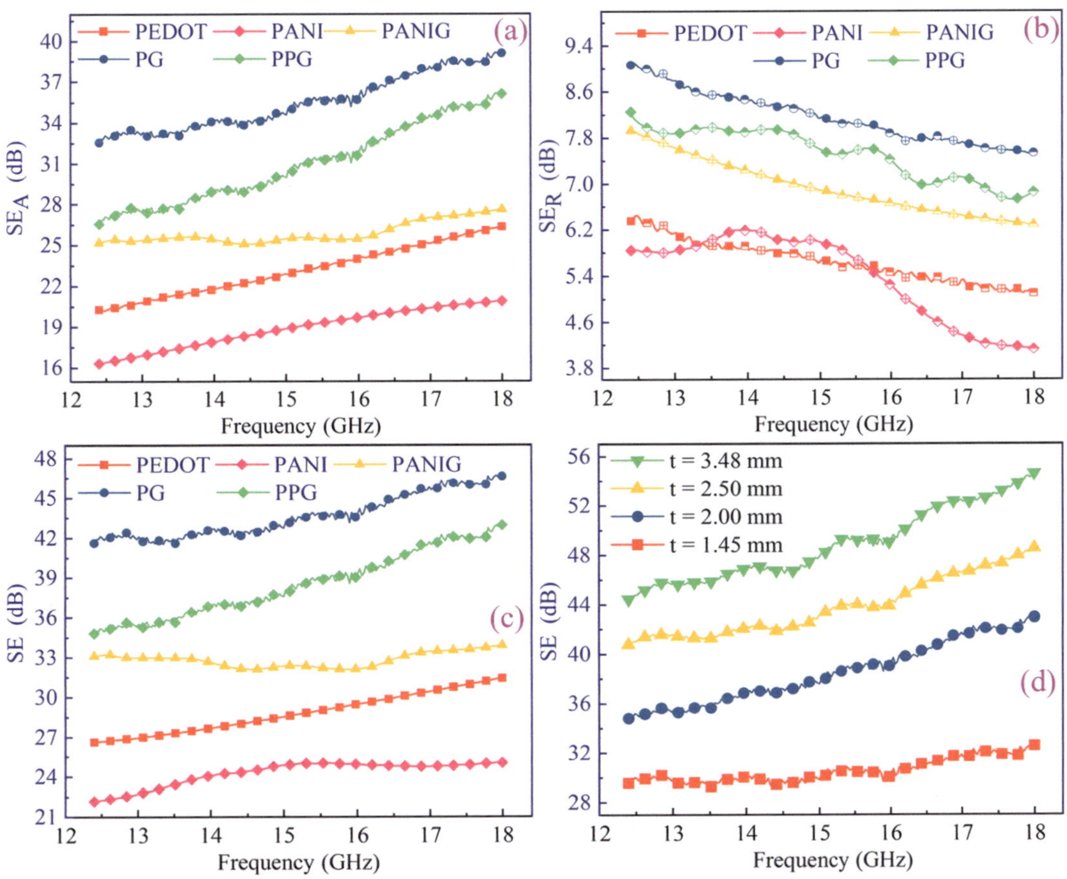

FIGURE 8.9 Plots of (a) SE_A, (b) SE_R and (c) SE_T as a function of frequency for all samples. (d) Plot of SE_T versus thickness (*t*) for PPG sample. Reproduced with permission from Kumari et al. [42], Copyright 2024, RSC.

bridging effect of SBS reduced the interfacial thermal resistance between PS and Pyr. The TC of PS-S@SBS/Pyr can reach 1.36 W/m·K with a resistivity of over 108 Ω. Furthermore, PS-S@SBS/Pyr maintains a consistent EMI *SE* of 33.1 dB in the 8–12 GHz frequency region, with absorption shielding acting as the dominant mechanism [43].

Chopra et al. [44] successfully developed PANI-coated barium hexaferrite (BaM) composites/hybrids and their incorporation or coating onto cotton fabric-based textile substrates as flexible and effective EMI-protective textile materials using sol-gel auto-combustion and self-assembly coating methods. Adding a thin layer of BaM (900 °C)/PANI NCs on cotton fabrics dramatically improved the hybrid's EMI *SE*. The treated cotton fabric attained a total EMI *SE* of –34.9 dB with only a 1 wt.% loading of BaM 900 °C/PANI NCs. The outstanding *SE* was attributable to the three-dimensional nanonetwork of conductive PANI and ferromagnetic BaM 900 °C NHFs, with their combined electrical and magnetic properties. The findings indicated that shielding performance could be further improved at this low weight percentage of nanomaterials, making functionalised cotton fabrics promising candidates for a variety of commercial applications, including high-performance EMI-protective textiles for advanced flexible devices [44].

8.5.5 Green and Sustainable Materials

As awareness of environmental impact increases, there has been increased emphasis on the generation of green and sustainable EMI shields [35]. There are studies and developments in biodegradable polymers, especially the use of bio-based polymers, biodegradable materials and eco-friendly synthesis methods to develop EMI shielding composites. The application of conductive fillers, which can be obtained in the process of reclamation of plants or in the process of obtaining products from green waste, such as carbon obtained from biomass, has been previously studied to develop environmentally friendly shielding solutions. This, aligned with the global inclination towards sustainability, gives a roadmap to the development of sustainable EMI shields [35].

Taking into consideration the significance of using eco-friendly, lightweight, high-efficiency and high-absorbing EMI shields in light of the global promotion of sustainable manufacturing, Yang et al. [35] reportedly investigated layered structural PBAT/Fe_3O_4@MWCNTs/Ni/Ag composite foams with high *SE* and low reflection. Isolated network and $scCO_2$ foaming were integrated to create the innovative system. Overlapping nickel particles on top of phase-separated PBAT/Fe_3O_4@MWCNTs microspheres formed the isolated conductive network, and the silver (Ag) layer attached to the top surface of the composite adds to its unique layered structure. By promoting magnetic loss and repetitive reflecting absorption of EMWs, the Fe_3O_4@MWCNTs NPs and microcellular structure increased MW dissipation. Simultaneously, the Ag particle layer was mostly responsible for the dielectric loss. The actions stated earlier naturally integrate several loss modes, resulting in an EMI shielding system known as "absorption-reflection-reabsorption" as per Figure 8.10. Their tests also revealed that the fabricated composite could achieve an EMI *SE* of 68.0 dB while maintaining an exceptionally low reflectance of only 23%. Furthermore, these foam composites showed strong EMI shielding even after 500 cycles of adhesion trials [35].

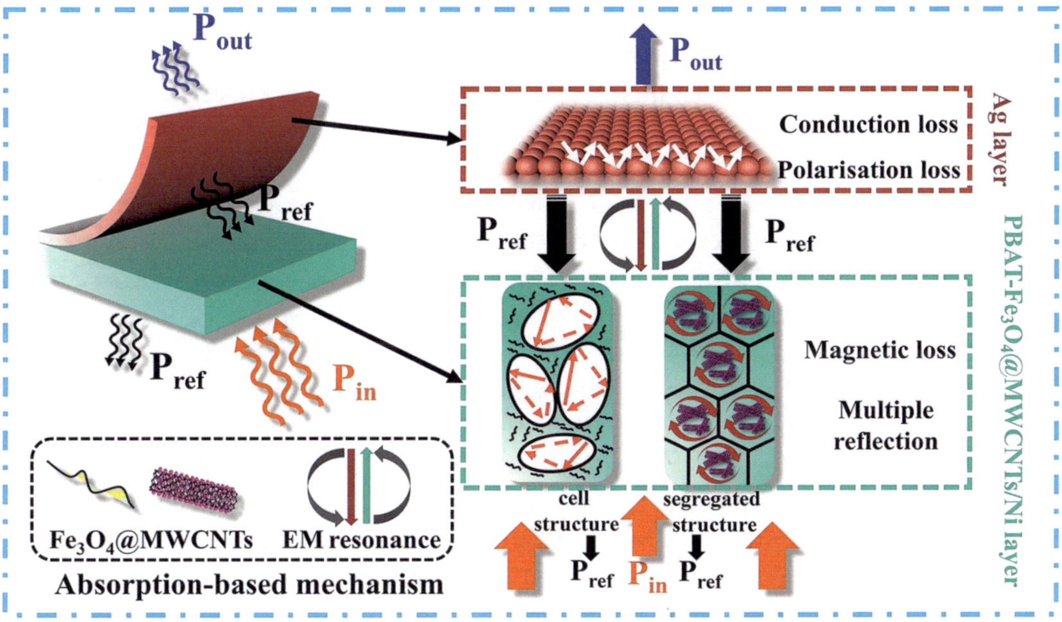

FIGURE 8.10 Scheme of EM wave dissipation mechanism within the composite foams. Reproduced with permission from Yang et al. [35], Copyright 2023, Springer Nature.

Xue et al. [45] created a multilayered composite containing Fe^{+2} and Fe^{+3} oxide (Fe_2O_3 and Fe_3O_4) NPs, PTh nanofibre (NF) arrangements and a gold nanometre layer. When modest amounts of Fe_3O_4 were utilised, the amalgamation of conductive polymer NF arrays and magnetic NPs produced an EMI *SE* of ≥30 dB within X-band frequency. This composite outperformed a similar composite without Fe_3O_4 by about 10 dB in terms of EMI *SE*. Magnetic NPs were responsible for the better EMI *SE* by introducing magnetic loss to attenuate EMWs and improving the impedance match betwixt the arrays and EP resin layers. Moreover, adding NPs allowed the material to demonstrate an absorption-dominant EMI shielding mechanism, dramatically reducing the secondary reflection of EMWs. As a result, this unique environmentally friendly EMI shielding compound shows potential for use in high-power electronic equipment [45]. Renewable wood as the raw material offers substantial structural benefits in the study of EMI systems because of its unique 3D pore structure. Thus, Feng et al. [46] developed magneto-electric lignocellulose matrix composites made from delignified wood that provide adequate EMI shielding. A PEDOT conductive covering and magnetic Fe_3O_4 were polymerised in-situ in delignified wood to create the composite. Significant internal reflections and dielectric loss may result from Fe_3O_4-PEDOT@wood's conductive three-dimensional pore structure. Combining the magnetic loss of Fe_3O_4 with the dielectric loss, the substrate showed significant EMI *SE*, which may have been due to the combined action of the two losses. High EMI *SE* up to 59.8 dB, magnetic (26.7 emu/g), conductivity (10^3 S/m) and SE_A/SE_T ratios (84.2% to 95.7%) at 2 mm in the X-band were all demonstrated by the Fe_3O_4-PEDOT@wood. A strong tensile and compressive strength of 100.8 MPa was also displayed by the substance [46].

Liu et al. [47] created segregated honeycomb-based structures within the biopolymer composites by using a "two-step melting" technique in conjunction with selectively supercritical CO_2 (Sc-CO_2) foaming. A segregated structural composite was created by melt-blending a high-melting-point PLA-CNTs composite material first, then crushing and grinding it, and then combining it with low-melting-point PBAT. Sc-CO_2 foaming was then applied to this composite, using varying foaming temperatures to produce highly selective cell topologies that capitalised on their unique foaming windows. About 3.98×10^9 Ωcm of good electrical insulation, excellent EMI shielding of ~30.1 dB and low density of 0.69 g/cm^3 are the characteristics of the composite foams. Moreover, it was demonstrated that the addition of carbon as a filler and a porous structure did not compromise the composite foams' "green" credentials because they maintained remarkable biocompatibility and degradability. Their work helps to design healthcare biomaterials and devices with controlled electrical properties and EMI shielding performances by effectively balancing lightweight characteristics, stable electrical insulation, as well as absorption-dominant EMI shielding – structurally green and biopolymer composites [47]. In a different study, the combination of expanded graphite and thermally expandable microspheres (TEMs) was shown to be a method for producing absorption-dominated EMI shields, also known as green EMI materials [48]. The impedance matching between the composite and the surrounding air was enhanced by the TEMs present in the EG-waterborne PU mixture. Therefore, as Figure 8.11(a) illustrates, the sample's EMI *SE* reached 38.4 dB when the TEM content was 5 wt.% and the EG content was 15 wt.%. Figure 8.11(b) shows that the composite's average reflection power coefficient was 0.27, indicating that the material primarily shields EMWs through absorption. The resultant material had a low reflection power coefficient, outstanding EMI *SE* and remained stable after undergoing the compression-recovery test [48].

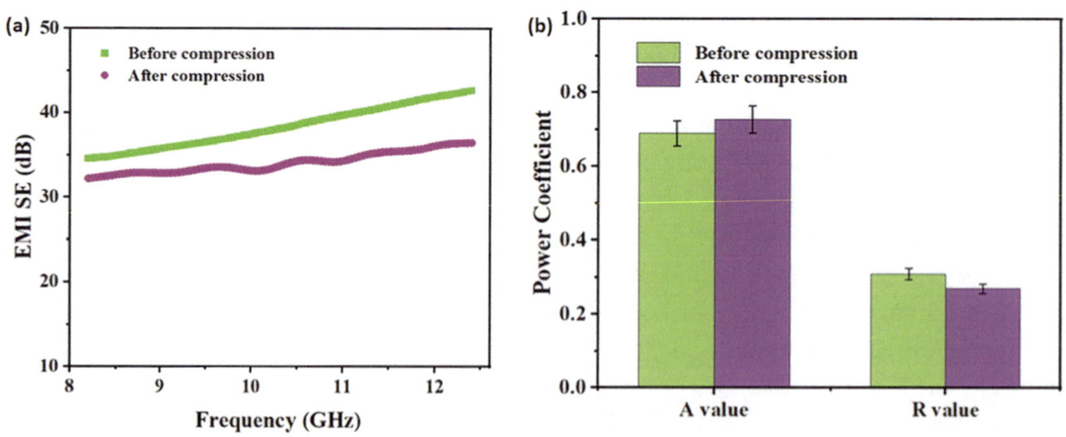

FIGURE 8.11 (a) EMI *SE* and (b) average power coefficients of A and R of GEFs-15/5 before and after 40 cycles of compression. Reproduced with permission from Zhou et al. [48], Copyright 2024, ASC Publications.

8.5.6 Improvement in Thermal Management via Enhanced Thermal Conductivity (TC)

Recent developments have also aimed at increasing the TC of polymeric EMI shields. Thermal management is an essential parameter for the reliability and performance of electronic devices, particularly in high-power applications. Fillers that function well in enhancing electrical and TC in polymers, including Gr and metal particles, has been used to develop composites that provide both EMI shielding and thermal dissipation [24, 49–51]. Table 8.1 provides a summative overview of diverse polymer-based EMI shields that have improved thermal management so that readers can grasp them better.

Enhancement in TC with respect to the content of Gr nanoplatelets resulting in a corresponding increase in the EMI *SE* as well as the EC have been demonstrated in the work by Łapińska et al. [24]. In their study, Gr nanoplatelets (GNPs), MWCNTs and MWCNTs-GNP blends were utilised to create flexible nanocomposites based on THV (a type of fluoropolymer). The top-performing samples were chosen for further study of EC and TC. The main conclusions of their investigation are thus: the hot-pressing technique promoted longitudinal and impulsive orientation of the filler, while tetrafluoroethylene, hexafluoropropylene as well as vinylidene fluoride monomeric units (THV) were used as the matrix; the fillers did not limit the mobility of the polymer chains; THV easily crystallises, which makes it essential for temperature-dependent mechanical analysis; the GNP/MWCNT samples showed no synergistic effect; EMI shielding for the whole set was mainly controlled by the absorption mechanism, with the saturation point noted at 10 wt.% GNP inclusion; the optimum EMI *SE* was attained with 10 wt.% GNP loading, resulting in a total *SE* (SE_{TOT}) of 23 dB at 5 GHz for a specimen thickness of 1 mm, as shown in Figure 8.12. The EC of GNP/THV composites followed a power law, reaching a maximum σ of 1.49 S/cm; TC showed a linear trend without a percolation threshold, reaching a maximum of $\kappa = 1.65$ W/mK at 15 wt.% GNP loading, as indicated in Figure 8.12 [24].

Lu et al. [73] used template carbonisation along with concurrent etching to create a $Ti_3C_2T_x$ MXene hybrid carbon foam (HCF) having outstanding performance and variable EMW absorption. The combined impact of porous as well as multilayered structures enhanced the number of EMW reflections in HCFs. In contrast, increasing interfaces increased polarisation along with conduction losses, as illustrated in Figure 8.13. Changing the MXene content can improve impedance matching by adjusting the ratio of polarisation loss to conduction loss. The HCF-2 achieved an RL_{min} of −72.24 dB at an average thickness of 2.84 mm and the HCF-3 attained an EAB_{max} of 6.55 GHz at 2.41 mm thickness, both with a 10% filling. HCF's absorption band can be modified between 8 and 14 GHz by altering MXene concentration. $Ti_3C_2T_x$ hybrid carbon foams perform well in EM-thermal conversion within a ramp rate of 85.81 °C/s from RT to 200 °C and radar wave absorption, making them suitable for MW heating, communication systems and radar technology applications. These authors also revealed that numerical simulations of the EM field were undertaken to evaluate the performance of the HCF in practical applications, as well as the radar cross-section (RCS) calculation. The 180 mm × 180 mm metal plate covered with HCF was used as the model for the simulation's reflection of the 11.49 GHz beam. Figure 8.13(a–c) shows the HCFs' RCS. At $\theta = 0^\circ$, pure CF exhibited −10 dB RCS, indicating poor absorption

TABLE 8.1 Recent Developments Aimed at Increasing the TC of Polymer-Based EMI Shielding Materials

Development Approach	Formulation	Active Filler or Component Amount	Fabrication Approach	Impact on Thermal Conductivity	Thermal Conductivity (TC) (W/mK)	EC	EMI Shielding (dB)	Thickness (mm)	Frequency (GHz)	Reference
Incorporation of high TC/ hybrid fillers										
	Transition metal carbide (MXene) @polystyrene (PS)	0.26 vol.%	Electrostatic assembly (self-assembly)	Significantly increases TC due to the high intrinsic TC of fillers.	High (not specified)	1081 S/m	>54	2	8.2–12.4	[49]
	MWCNT-Fe_3O_4@ Ag-EP	15 wt.%	SI + LbL	The utilised hybrid filler led to enhanced EMI shielding and TC.	0.46	0.280 S/cm	35	2	8.2–12.4	[50]
	Fe_3O_4/MWCNTs/ wax	20	SI + co-precipitating + in-situ polymerisation	The incorporation of MWCNTs led to improved conductivity and EMI shielding	–	Enhanced (not stated)	75 (*RL*)	≤5	7–18	[52]
	PS/sulphur-(S) SBS/ pyrrhotite (Pyr)	85 wt.%	Sulphuration + powder mixing + hot compression moulding	The use of hybrid filler resulted in better TC and EMI shielding.	1.36	10^{-5}	33.1		8–12	[43]
Surface modification of fillers, composite or hybrid										
	PANI/Fe_3O_4/ MWCNTs/wax	20	SI + co-precipitating	The incorporation of PANI led to decreased conductivity and EMI shielding	–	Enhanced (not stated)	16 (*RL*)	≤5	8–15	[52]
	Aramid NF/PPy	–	SM + in-situ polymerisation + vacuum-assisted filtration + hot compression moulding	The coating with PPy led to increased conductivity and EMI shielding	High (not specified)	3546 S/m	≥ 35	46 μm	6.0–26.5	[53]

Graphene (Gr)-based composites										
	Cu/rGO-EP	6.0–1.2 wt.%	Resin impregnation	The inclusion of rGO and Cu nanowires provided excellent TC, aiding in heat dissipation and EMI *SE*.	0.51	120.8 S/m	47	2	8.5–12.5	[51]
	3D porous Gr nanoplatelets (GNPs)/rGO-EP	0.1 wt.% rGO and 20.4 wt.% GNPs	SI + freeze-drying + resin impregnation	The addition of rGO and GNPs improved TC, which aids in heat dissipation and EMI *SE*.	1.56	179.2 S/m	15	3	8.2–12.4	[54]
	Gr/PMMA	1.8 vol.%	Solution mixing + super critical CO_2-assisted foaming	The inclusion of Gr led to enhanced conductivity and, consequently, EMI *SE*	–	3.11 S/m	13–19	4	8–12	[55]
	GNPs/PU	50 wt.%	SI/SM + brush coating	The inclusion of Gr resulted in improved conductivity and, as a result, EMI *SE*	41.60	2546.5 S/m	67.6	0.4	8.2–12.4	[56]
Engineered 3D network structures or foams										
	Thermoplastic urethane (TPU)/CNTs	4 wt.%	Melt mixing + foaming	CNTs inclusion facilitated efficient heat transfer across the shield, significantly increasing TC.	High (not specified)	2.74×10^{-6} S/m	−30.4	1.5	5.9–18.0	[57]
	$CoFe_2O_4$@CNTs/PDMS-expandable polymer microsphere foams	6.78 wt.%	SM + vacuum-assisted filtration	$CoFe_2O_4$@CNTs inclusion led to enhanced EMI *SE* and conductivity (thermal and electrical)	0.106	7.12 to 20.30S/m (with variation in density)	48.85	–	8.2–12.4	[58]

(*Continued*)

TABLE 8.1 Recent Developments Aimed at Increasing the TC of Polymer-Based EMI Shielding Materials (Continued)

Development Approach	Formulation	Active Filler or Component Amount	Fabrication Approach	Impact on Thermal Conductivity	Thermal Conductivity (TC) (W/mK)	EC	EMI Shielding (dB)	Thickness (mm)	Frequency (GHz)	Reference
	MXene/Co/C-waterborne polyurethane	21.8 wt.% MXene and 14.5 wt.% Co/C	SM + freeze-drying + hot pressing	Enhance electrical and thermal conductivities and EMI *SE* achieved via MXene/Co/C inclusion	3.29	46 S/cm	42	1.9	8.2–12.4	[59]
Alignment of anisotropic structure										
	rGO/cellulose NF (CNF)	50 wt.% rGO/CNF	SM + vacuum-assisted filtration + hydroiodic acid in-situ reduction	Enhances directional TC, improving heat dissipation in specific orientations (high in-plane TC) and EMI *SE*	56 (anisotropic TC)	4057.3 S/m	26.2	23 μm	8.2–12.4	[60]
Use of metal-polymer composites										
	MXene/AgNW-PVA	MXene (1 mg/mL) and AgNW (3 mg/mL)	SI (solution intercalation) + co-precipitation SPC (spin coating) + spray coating	Enhanced electro-photo-TC while maintaining adequate EMI shielding properties.	Great (not mentioned)	5.46×10^{-3} S/cm	32.0–36.4	<10 μm	18–27	[61]
	MXene/AgNW-PC	MXene (1 mg/mL) and AgNW (3 mg/mL)	SI (solution intercalation) + co-precipitation SPC (spin coating) + spray coating	The spray coating of hybrid NPs led to EMI shielding even as the electric properties improved.	Excellent (not mentioned)	–	≥35.0	<10 μm	8–12	[61]
	Poly(vinylidene fluoride) (PVDF)-$SnBi_{58}$	50 wt.%	Melt mixing	In the low-melting-point alloy ($SnBi_{58}$), inclusion resulted in enhanced EMI *SE* and TC.	6.38	10,500 S/cm	−68.79	2	8.2–12.5	[62]

	Carbon Fabric(CF)/ carbonyl iron powder (CIP)/ PDMS	87 wt.% CIP in 13 wt.% PDMS	Ball milling + solvent casting in a mould (moulding)	The CF and CIP synergistic interface resulted in outstanding TC, EC and EMI *SE*.	5	≤45 S/cm	120	3	26–40	[63]
Integration of phase change materials (PCMs)	PI/MXene/PVA/ Fe_2O_3	36.86 wt.%	SM + dip-coating + spray coating	MXene provides thermal management via excellent TC.	0.28 and 0.31	564.65 S/m	59.42	20 µm	8.2–12.2	[64]
	Poly(arylene ether nitrile) (PEN)/ MWCNTs	20 wt.%	SM + vacuum filtration + hot-pressing	MWCNT inclusion results in improved EMI *SE*, EC and heat insulation of the hybrid	0.12	–	43	1.5	8.2–12.4	[65]
	PEG/ rGO-C/ MXene-PVA		SI + lyophilisation + annealing + impregnation-hot-pressing or moulding	The introduction of rGO-C resulted in a combination of ultra-high in-plane TC, outstanding EMI *SE* and anti-leakage capabilities.	20.96	915–4716 S/m	70.5	1	8–12.5	[66]
	MXene/ Poly(3,4-ethylene-dioxythiophene)-polystyrene sulfonate (PEDOT:PSS)/ polyacrylamide (PAM)	1.8 wt.%	SM + SC	MXene inclusion results in improved EMI *SE*, EC and TC of the hybrid	0.64	0.9 S/m	38.7	3	8–12	[67]
	Lignin carbon aerogel/nickel/ PEG (PLGN-3)	8 mg/mL Gr	Vacuum-assisted adsorption + freeze-drying + thermal annealing + impregnation	Dual-network encapsulated PCM prevents leakage and provides excellent TC, EC and EMI *SE* of hybrids.	0.8288 ± 0.0134	1597.5 S/cm	69.9	–	8.2–12.4	[68]

(*Continued*)

TABLE 8.1 Recent Developments Aimed at Increasing the TC of Polymer-Based EMI Shielding Materials (Continued)

Development Approach	Formulation	Active Filler or Component Amount	Fabrication Approach	Impact on Thermal Conductivity	Thermal Conductivity (TC) (W/mK)	EC	EMI Shielding (dB)	Thickness (mm)	Frequency (GHz)	Reference
	MXene/delignified wood-supported phase change composites (PMPCMs) MXene/cellulose	16 mg/mL	Delignification + impregnation	MXene greatly enhanced the EM shielding effect.	–	–	34.12	1.1	8.2–12.4	[69]
	Wood aerogel/ PEG/phytic acid@ MXene	10 wt.% MXene	Delignification + self-assembly + encapsulation	MXene inclusion enhanced the TC, EC and EMI *SE* of the hybrid	0.82	–	44.45	3	8.2–12.4	[70]
	PEG/Ni–F/ PPy@ cellulose NFs (CNF)- PVA (PPCN) (PPCN-3)	10 mg/mL CNF and 30 mg/mL PPy	In-situ polymerisation + vacuum impregnation + freeze-drying + vacuum impregnation	Introducing hybrid fillers improved the TC, EC and EMI *SE*.	0.8653	186680 S/m	110.37	–	8.2–12.4	[71]
	Ni@C-r-GO/ AgNW–PI@PEG	50 wt.%	In-situ polymerisation + freeze-drying + deep coating	The inclusion of hybrid filler enhanced the TC, EC and EMI *SE*.	82.5	10^5 S/m	62.3	3.5	7.8–12.6	[72]

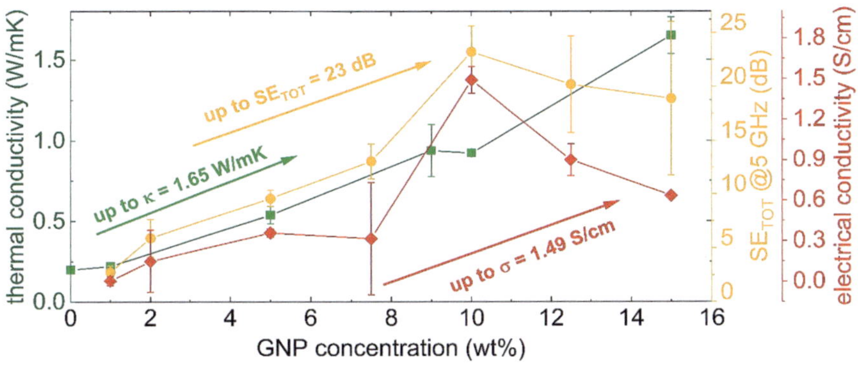

FIGURE 8.12 Thermal conductivity, EC and total EMI *SE* obtained for 1–15 wt.% graphene nanoplatelets (GNPs). Reproduced with permission from Łapińska et al. [24], Copyright 2023, SPE.

performance. However, at $\theta \approx 60^{\circ}$, absorption improved. Interference cancellation is the name given to these phenomena. The standard component of the wave balances out once an EMW has an incidence angle that is transmitted into the medium because the thickness of the material matches the wave's quarter wavelength. All HCF samples showed good EMW absorption, but the HCF-2 stood out with an RCS of –52.25 dB at $\theta = 0^{\circ}$, which was in line with the *RL* findings and demonstrated outstanding EMW absorption capabilities [73].

In another study, MWCNT-Fe_3O_4@Ag/EP composites were created using the solution casting (SC) technique [50]. When the mass ratio of MWCNTs-NH_2 to Fe_3O_4@Ag-COOH was 9:1 (MF-10), the associated EP nanocomposites had the best EC as per Figure 8.14(a) and EMI *SE* as shown in Figure 8.14(b–d). The MF-10/EP nanocomposites with 15 wt.% MF-10 had a satisfactory EMI *SE* of 35 dB, high EC of 0.280 S/cm, satisfactory TC having a thermally conductive coefficient λ of 0.46 W/mK as depicted in Figure 8.14(e), excellent Young's modulus of 4.60 GPa and hardness value of 0.26 GPa, besides exceptional thermal stability supported by a THRI of 183.4 °C. The researchers found that adding Fe_3O_4@Ag NPs improved the interaction between MF-10, resulting in higher λ and EMI *SE* values. Additionally, it reduced hysteresis loss of EMWs and provided more interfaces for reflection and reabsorption, leading to improved attenuation [50]. They proposed that the produced MWCNT-Fe_3O_4@Ag-EP nanocomposites have good comprehensive features, making them appropriate for a wide range of applications such as aviation equipment, medical equipment, industrial electronic equipment as well as wireless base stations.

Wang et al. [43] established a simple method for fabricating a thermally conductive nanocomposite having EMI shielding along with insulating properties using two solid waste resources, namely Pyr tailings in addition to waste PS. The PS/S@SBS5/Pyr85 has a filling limit of 1.31 W/mK at 85 wt.% filler content, indicating outstanding performance. TC's augmentation efficiency relative to rejected PS exceeded 1800%. PS-S@SBS5/Pyr85 showed an increased TC while maintaining an EC of less than 10^{-9} S/m owing to Pyr's peculiar semi-conductivity and the interface enhancement given by S@SBS. A TC of 1.36 W/mK in specific applications might be achieved by increasing the filler content PS-S@SBS/Pyr to

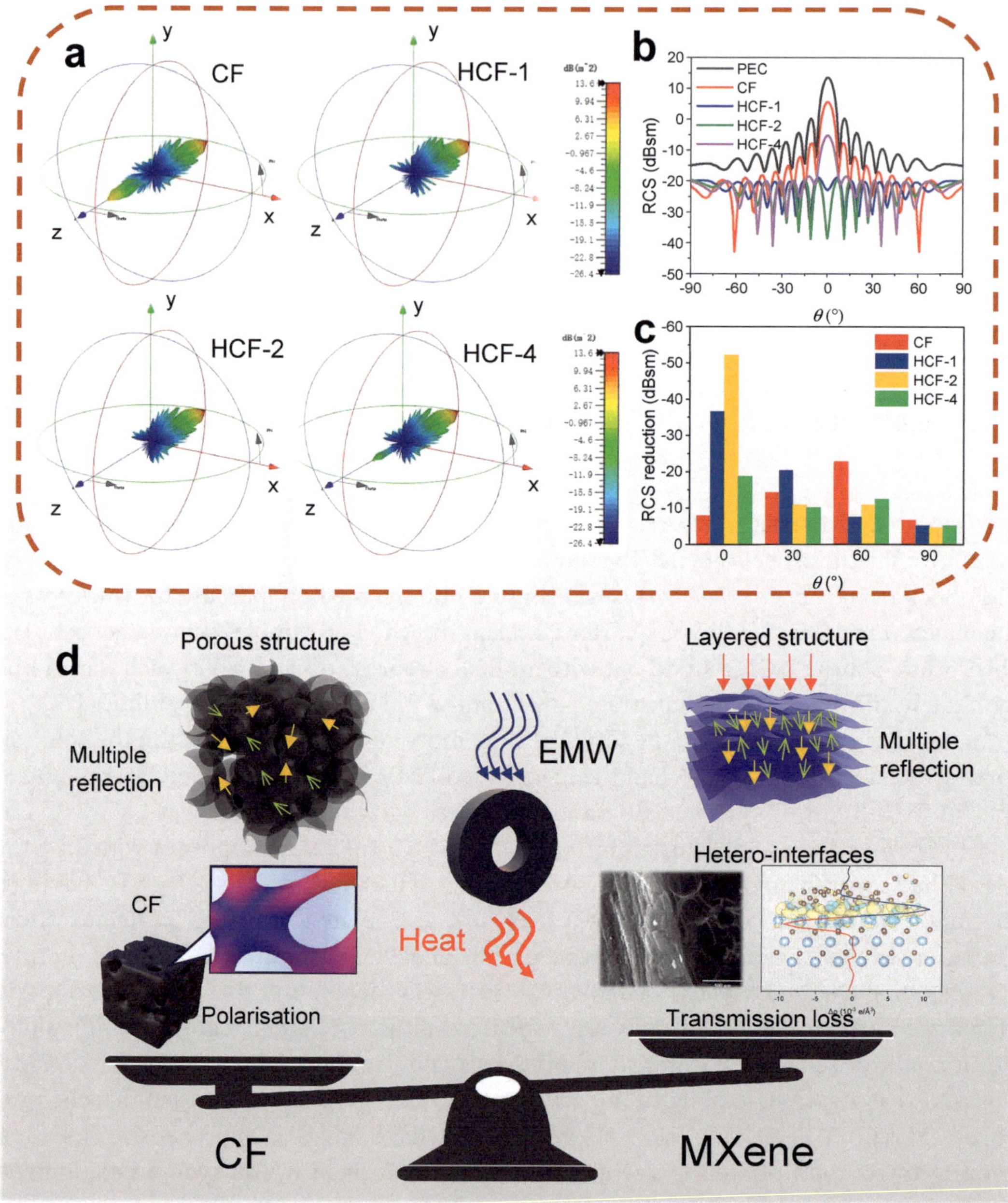

FIGURE 8.13 Electromagnetic wave response simulation and electromagnetic wave absorption mechanisms. (a) the far-field reflections of HCFs; (b) the RCS of HCFs as a function of θ; (c) the RCS losses of HCFs; (d) a schematic diagram of the electromagnetic wave absorption mechanism. Reproduced with permission from Lu et al. [73], Copyright 2024, Elsevier Science Ltd.

95 wt.%. They also revealed that PS/S@SBS5/Pyr85 has the potential to be an efficient thermal management material in applications requiring electrical insulation, as shown by the experimental findings of LED heat dissipation investigations. Notably, the surface working temperature of LED modules using PS/S@SBS5/Pyr85 was measured at 26.8 °C, suggesting

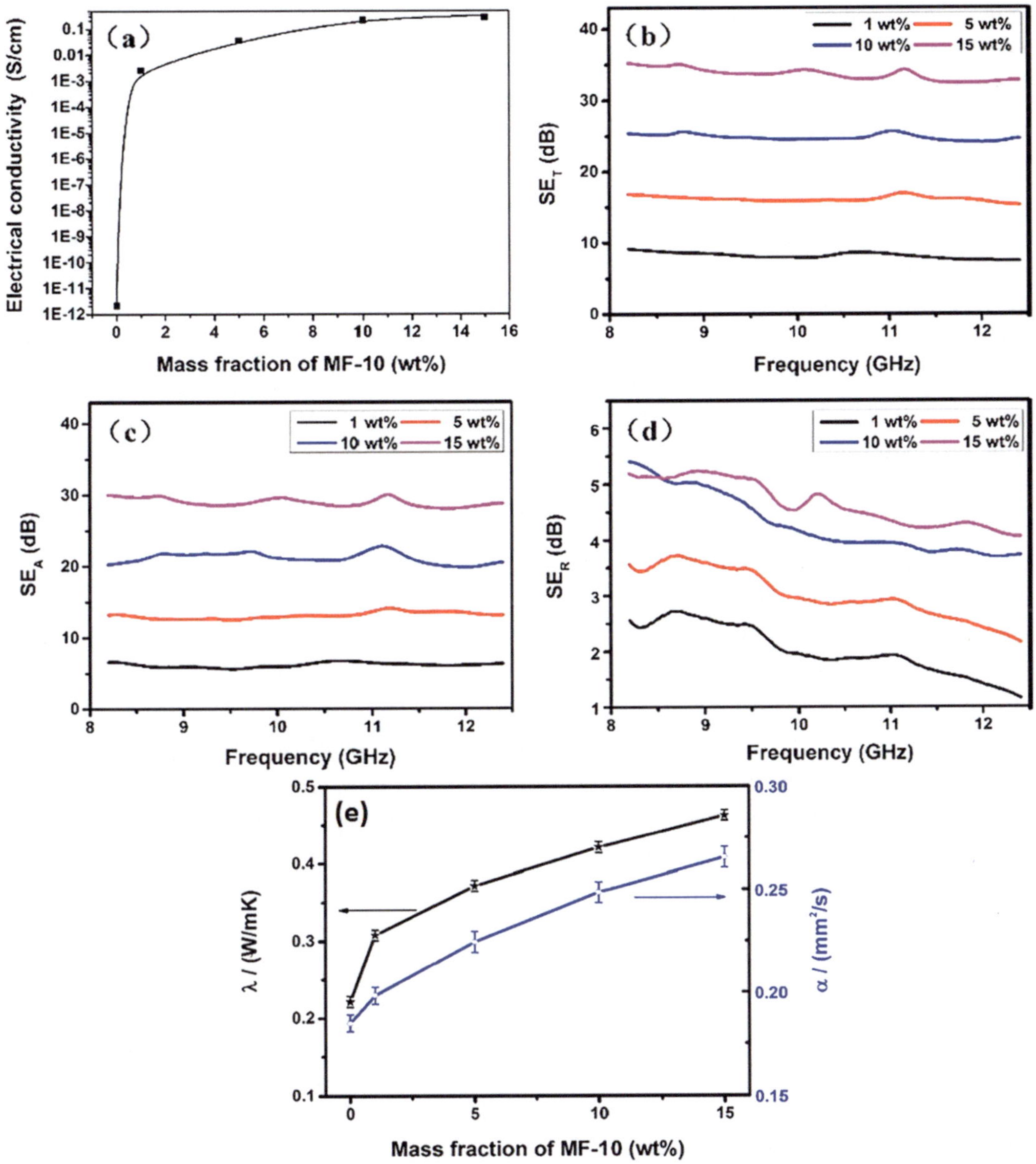

FIGURE 8.14 (a) Log EC vs mass fraction of MF-10; EMI *SE* for MF-10/epoxy nanocomposites at X-band: (b) SE_T, (c) SE_A, (d) SE_R (A colour version of this figure can be viewed online.) and (e) mass fraction of MF-10 affecting the λ and α value of the MF-10/epoxy nanocomposites. (A colour version of this figure can be viewed online.) Reproduced with permission from Wang et al. [50], Copyright 2019, Elsevier Science Ltd.

a temperature reduction of 20.8 °C compared to standard plastics. Furthermore, PS/S@SBS5/Pyr85 expertly utilised Pyr's magnetism and semi-conductivity properties, displaying continuous EMI *SE* within the 8–12 GHz frequency band and obtaining a maximum SE_{Total} of 33.1 dB. Further analysis revealed that the EMI shielding performance of PS/S@SBS5/Pyr85 was primarily due to Pyr's magnetic loss, emphasising absorption shielding. Surface analysis was used to confirm the production of C-S and C-S-S-C bonds, which served as the foundation for developing a theoretical model that uses S@SBS as an interface enhancer. Their model revealed the integrated mechanism responsible for PS/S@SBS/Pyr's thermal conduction, electrical insulation and EMI shielding properties. Finally, the authors proposed that the PS/S@SBS/Pyr preparation technique offers a significant advantage in addressing the unavoidable cost-related issues associated with supporting and framing heat dissipation materials. At the same time, it demonstrated remarkable EMI shielding capability, indicating considerable likely applications in temperature control of electrical and electronic equipment [43].

Last but not least, Su et al. [64] demonstrated that the MXene coating could be effectively deposited by dip-coating onto the PI aerogel's inner wall. According to reports, the produced MXene/Fe_3O_4-PVA (MFP) coating, which was applied by spraying upon the exterior of composite phase change material (CPCMS-4), demonstrated exceptional thermal stability, a high latent enthalpy (154.74 J/g, maintaining enthalpy efficiency of 91.31%) as well as dependability for thermal energy storage. It also had a multilayer stacked structure. Moreover, CPCMS conductivity in addition to dielectric losses for adjustable EMI shielding, photo-thermal and electric-thermal along with MW-thermal conversions and storage, were significantly boosted by the use of high-performance MXene coatings. CPCMS-4, with high MXene content, has a remarkable electric conductivity of 164.08 S/m, increasing its electro-thermal conversion efficiency to 72.37%. The EMI *SE* of CPCMS-4 reaches 59.42 dB. Furthermore, exhibiting energy conversion efficiencies of up to 89.09 and 12.47%, respectively, the photo- and MW-heating experimental results show that CPCMS-4 has remarkable photo- and MW-to-thermal conversion capabilities. The finite-element simulation's findings show that, when employed as an MW receiver, the porous skeleton of polyimide (PI)/MXene is effective in converting MWs into thermal energy. According to the authors' hypothesis, these characteristics, in addition to its strong solar, electric and MW-to-thermal conversion and storage capabilities, make CPCMS a competitive option for a range of sustainable energy gathering and EMI shielding [64].

8.5.7 Smart and Responsive Materials

Another advancement is the use of smart EMI shielding materials that change their effectiveness with changes in stimuli such as temperature, light, pH or EM fields [72, 74, 75]. This kind of material may offer shielding properties that are sensitive to a specific environment and may thus offer a level of flexibility in protection that other shields cannot offer. Thus, smart material is highly appreciated for next-generation electronics where the adaptability and tunability of EMI shields is much demanded [64, 69, 70, 72, 74].

Hu et al. [74] have created a special smart EMI material. To achieve this, highly transmissive, self-adjusting smart EMI hydrogels had to be developed with changing EMI *SE* values. Layer-by-layer (LbL) polymerisation as well as solvent displacement were used to create a pH-responsive pliable hydrogel and a thermochromic PNIPAM microgel to create the material. Notably, the designability of the high-concentration microgel and the metal salt solution's selectivity allowed for the accurate fabrication of a three-dimensional pattern and the modification of the specimen's original light transparency. The chloride conjunction inside the polyconic hydrogel matrix produced an EC of 0.810 S/m, as shown in Figure 8.15(a), and an EMI performance that was praiseworthy, as shown in Figure 8.15(b and c). The hydrogel substance displayed absorption-based EMI shielding ($A > 0.7$) because of its porous structure along with the unimpeded movement of metallic ions within it. As shown in Figure 8.15(d and e), its EMI *SE* value around 0–27.6 dB was also intelligently variable in response to changes in external pH. The outcome of these efforts was the development of a smart EMI protective window with a configurable EMI *SE* value, a morphable form and tunable transmission of light (~7–80%) [74].

In a different study, scientists developed Ni@C/rGO/AgNW/PI@PEG green EMI shielding and energy-storing material by systematically integrating the layers of impedance matching layer, dissipating layer along with reflection layer in accordance with the impedance matching principle [72]. The impedance matching layer Ni@C/rGO-300 minimised EMW reflection from the material's surface; the dissipation layer Ni@C/rGO-600 provided outstanding EMW dissipation capabilities; and the reflection layer AgNW-PI ensured that the EMWs were reflected back into the material. The Ni@C/rGO/AgNW/PI@PEG composite shield purportedly has an EMI shielding performance of 62.3 dB and an ultra-low *R*-value of 0.04. The energy loss density cloud maps in CST simulations indicated the material's intrinsic process of EM energy loss. Rather, heat was transferred to the Ni@C/rGO backbone, which had long vertical channels for energy storage as well as dissipation, from bottom to top via the extremely thermally conductive AgNW/PI membrane. Ni@C/r-GO/AgNW/PI@PEG exhibited good stability after 50 cycles of heating and cooling with a η value of ≥96.29%. Consequently, the impedance matching-based composites were suggested to hasten the application of EMI shields for battery heat dissipation packaging and algorithms plus portable electronic device thermal management [72]. Another group investigated PEG/Ni-F/PPy@CNF/PVA, so-called "PPCN" phase transition composites with diverse energy-storage capacities and outstanding EMI shielding qualities [71]. They discovered that PPCNs had a notable energy-storage density with a corresponding enthalpy of melting of 126.18 J/g, a relative enthalpy efficiency of up to 99.98% and outstanding TC with an improvement of 182% compared to PEG. Furthermore, because of the synergistic impact of the nickel foam along with the conductive network, the PPCNs exhibited an extraordinarily high EC of 186,680 S/m and great light-absorbing capabilities, allowing them to perform highly efficient photothermal, electro-thermal and quick magneto-thermal conversions. Furthermore, the PPCNs had a good EMI *SE* of 110.37 dB, much exceeding industry standards. Thus, the new multi-source-driven, morphologically stable PPCN phase change composites with good EMI shielding performance were proposed for use in the development of new energy sources and devices. However, due to nickel foam's large pore size, interfacial thermal resistance persisted even after refinement, lowering heat transfer efficiency. The authors said that in the near future, researchers would be able to improve the TC of the composites and enhance the heat transfer by employing thermally conductive fillers and so on, which is crucial for the practical application of CPCM [71].

Singh et al. [76] have reportedly developed a bio-mimetic CEC for thermally adaptive EMI shielding using a water-entrapped SZTO/SWCNT/PDMS composite with an active guest in a passive host system. Their research revealed that besides enhancing SE, polar

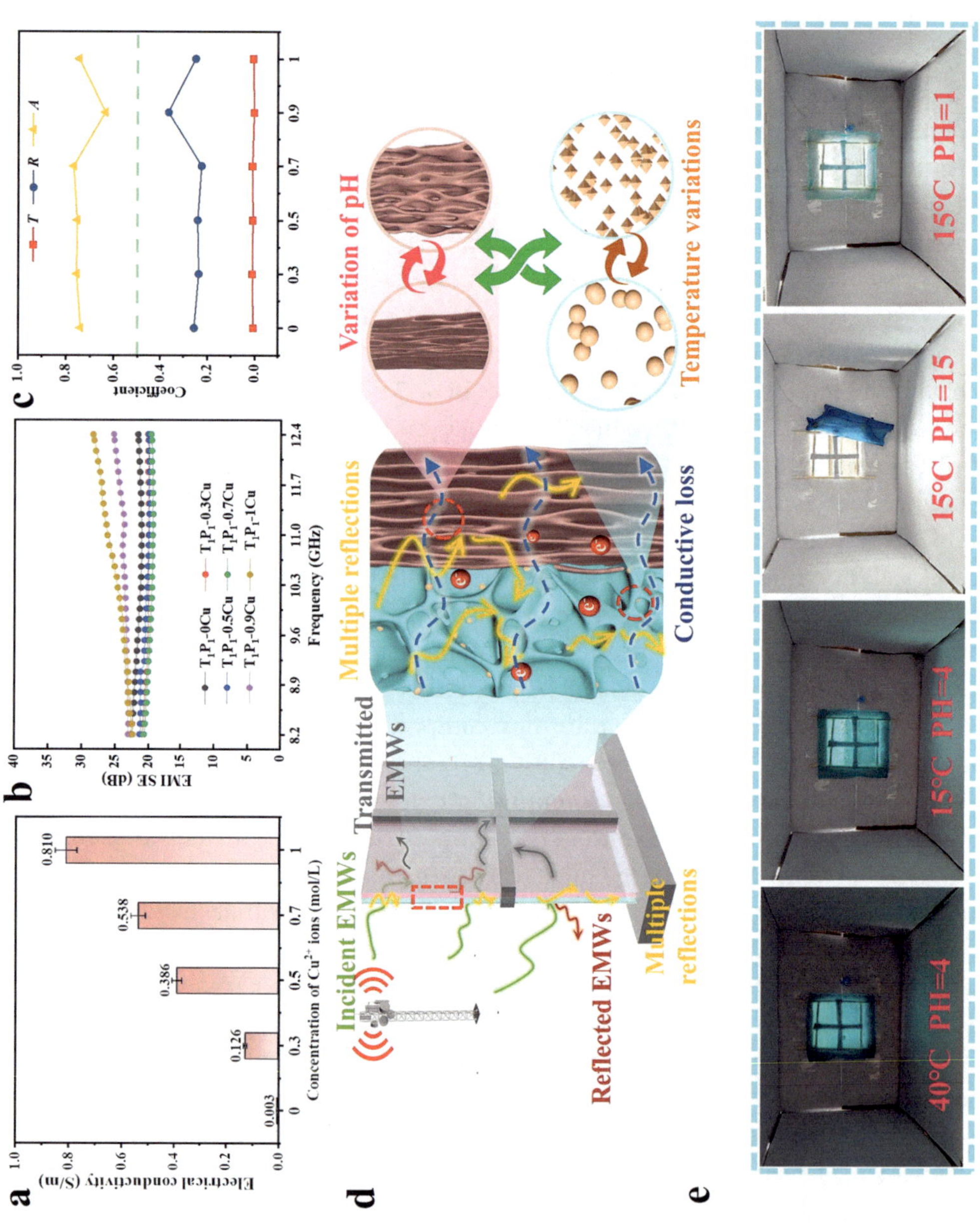

FIGURE 8.15 (a) Electrical conductivity of bilayer hydrogels with different $CuCl_2$ contents; (b) and (c) EMI *SE* values of hydrogels at different Cu^{2+} ion concentrations and their power coefficients; (d) Diagram showing EMI mechanism for controlled smart windows; (e) T1P1-0.5Cu hydrogel smart windows for practical applications. Reproduced with permission from Hu et al. [74], Copyright 2024, Elsevier Science Ltd.

water acted as an active agent, modifying the shielding response through thermal phase transitions that led to microstructural changes. As the temperature increased from 24 °C to 100 °C, the composite exhibited a consistent and reversible improvement in shielding performance, reaching up to 50.4 dB from 38.6 dB. Additionally, the composite achieved a total absorption-dominant shielding improvement of 32.6 dB (20.8 dB due to water content and 11.8 dB due to thermal modifications) while maintaining controlled reflection shielding without the need for extra fillers. Consequently, the integration of thermally adaptive smart green shielding performance along with the structural design principles and outcomes established a foundation for creating shields suitable for advanced and future electronic applications [76]. Wang et al. [75] have also created PANI/sodium alginate (SA) gel EMI shielding with switchable characteristics, allowing for effective tuning of its EMI *SE*. This dynamic modulation was attributed to the effect of solvent polarity on the molecular structures of PANI and SA. The variations in EMW attenuation were influenced not only by the solvent's polarity but also by the specific interactions within the PANI/SA hydrogel system. In this system, the carboxyl groups on the surface of SA acted as proton reservoirs, facilitating additional PANI doping through ionised H+ in water, which led to enhanced EMI shielding properties. The PANI/SA gel was reported to undergo structural changes, with stacked and coiled molecular chains depending on the solvent's polarity. These structural modifications affected the conjugation of PANI and electron mobility, triggering a switch from an "on mode" to an "off mode" in terms of EMI SE. The reverse shift from "off mode" to "on mode" was easily achieved by solvent replacement. As a result, the EMI *SE* of the PANI/SA gel could be precisely adjusted and reversibly changed without degradation using a straightforward solvent exchange method. The researchers suggested that this solvent-controlled approach combining durability and reliability holds great potential for the development of advanced smart and multifunctional EMI shielding materials [75].

8.5.8 Advanced Processing Techniques

8.5.8.1 Layer-by-Layer (LbL) Assembly

Polymer-based EMI shielding materials are being increasingly fabricated using LbL assembly techniques such as LbL polymerisation, LbL coating and/or LbL vacuum-assisted filtration [74]. It fulfils the advantage of providing close controlled tolerance for the alternating layer and shielding thickness of alternating layers for the production of multilayer composite shields. Hence, through layering, arranging the layers in a different manner by using conductive materials and dielectric or through the use of nanomaterials and other composite patterns, the electrical as well as mechanical properties of the composite can be aligned to create an inclination [77]. LbL assembly is also beneficial to fabricate thin, lightweight and flexible EMI shielding films.

Liu et al. [77] effectively created multifunctional C-MWCNTs/PEDOT:PSS)/polyimide (PI) composite films (CPFs) with flexible layered architectures utilising a simple LbL deposition process. Initially, they noncovalently modified the CNTs to increase their dispersion and homogeneity in the films. After adding PEDOT:PSS, the CPFs had a high EC of 2458.08 S/m at 72.69 MPa, as per Figure 8.16(a). They also displayed acceptable electrical heating

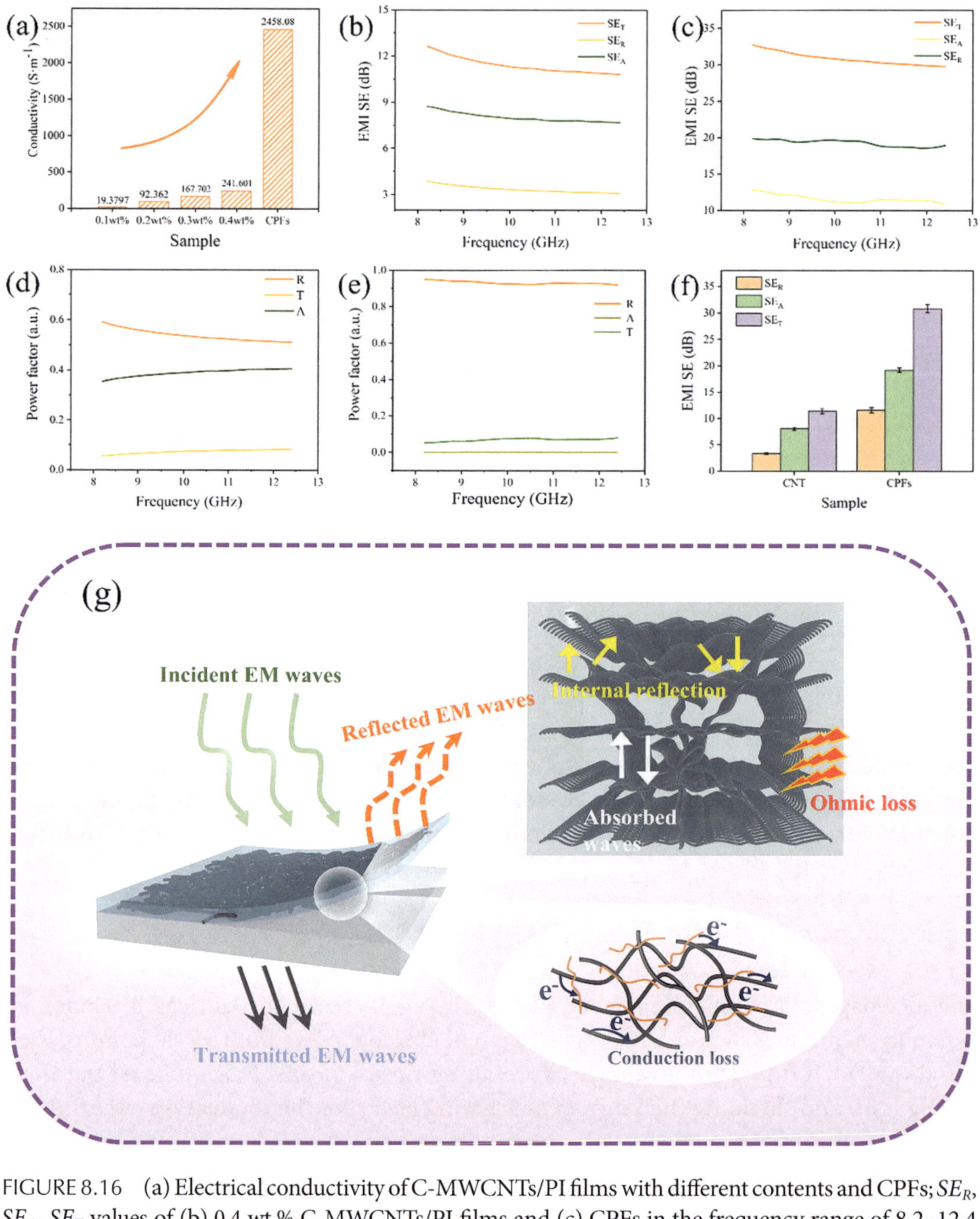

FIGURE 8.16 (a) Electrical conductivity of C-MWCNTs/PI films with different contents and CPFs; SE_R, SE_A, SE_T values of (b) 0.4 wt.% C-MWCNTs/PI films and (c) CPFs in the frequency range of 8.2–12.4 GHz; (d) R, A, T power coefficients of 0.4 wt.% C-MWCNTs/PI film; (e) R, A and T power coefficients of CPFs in the frequency range of 8.2–12.4 GHz; (f) average SE_T, SE_R and SE_A values of the two films; and (g) schematic of the EMI shielding mechanism of CPFs. Reproduced with permission from Liu et al. [77], Copyright 2024, Elsevier Science Ltd.

capabilities, reaching the maximum steady-state temperature after about 45 seconds of applying voltage. Furthermore, the CPFs demonstrated a consistent heating effect at 125 °C at 5 V. The addition of the PI layer gave the CPFs excellent flexibility, ensuring that the film retained optimal electrical heating performance even under deformation. The film's high conductivity improves its EMI shielding performance, achieving a good EMI *SE* value of 32.69 dB at a low thickness of 47 μm, resulting in lightweight material. Figure 8.16(b–e) depicts the EMI SE_T and A, T and R values of CPFs and 0.4 wt.% C-MWCNTs/PI films in the frequency range of 8.2–12.4 GHz. The 75-μm-thick C-MWCNTs/PI film has a low EMI *SE* of 12.62 dB. The EMI SE_T of CPFs with a thickness of around 47 μm climbs to 32.69 dB when the film conductivity increases. Because conductivity has a considerable influence on the value of R, comparing Figure 8.16(d and e) indicates that the value of the R curve grows dramatically as film conductivity increases. Figure 8.16(f) shows that, similar to the trend of conductivity change, the shielding effect of the layer increases with conductivity. The chart indicates that the value of SE_R is always less than the value of SE_A, indicating that increased film conductivity has an effect on the film's EMI shielding mechanism. According to Figure 8.16(g), when the EMW first reached the film's surface, there was an impedance difference between the film's conductive C-MWCNT/PEDOT:PSS layer and the open space, resulting in an impedance mismatch that caused the majority of the incident EMWs to be reflected. The film absorbed the remaining EMWs, which interacted with electrons in the conductive network within it. This interaction is thought to generate current via ohmic loss, resulting in EMW attenuation. The conductive layer's two-layer structure resulted in many reflections of the EMWs within the film, increasing EMW loss. Interfacial polarisation losses were also predicted to occur at layer interfaces, causing electrons to migrate in order to reduce EMWs. Thus, these benefits are thought to make the film suitable for wearable multifunctional electronic devices that match market expectations for thermal management and EMI shielding, and it has potential for future applications [77].

Zhang et al. [78] explored flexible multifunctional Janus nanocomposite films with magnetic and conductive properties. These films consist of magnetic cobalt carbide nanowires combined with bacterial cellulose (Co@C NW/BC) on the top layer and conductive $Ti_3C_2T_x$ MXene on the bottom layer, fabricated using a vacuum-assisted LbL filtration technique followed by hot pressing. The dual-sided structure results in distinct EMI shielding and thermal management characteristics. When exposed to EMWs from the Co@C NW/BC side, the films demonstrate a high EMI SE of 49.8 dB, with an enhanced microwave absorption (SE_A) of 33.9 dB at a thickness of just 43 μm. On the $Ti_3C_2T_x$ side, improved electrical heating is observed, reaching 120 °C at 6 V, with a photo-thermal conversion temperature of 77.8 °C under a light intensity of 2.0 kW/m^2 and effective thermal stealth properties with a low radiation temperature of 88.4 °C on a 240 °C hot stage. The films also exhibit strong mechanical properties with a tensile strength of 80 MPa, highlighting their potential for EMI shielding and thermal control applications [78]. Ji et al. [79] proposed a co-extrusion method to precisely control the alignment and distribution of graphite-CNT and hexagonal boron nitride (h-BN) layers in alternating multilayer silicone rubber (SR) composites. These structures address the needs of highly integrated multifunctional electronic devices, offering polymer composites with anisotropic thermal and EC as well as EMI shielding. The alternating layers form efficient conductive networks and help separate electrical and thermal conduction pathways in the through-plane direction. Additionally, destructive interference of EMWs between graphite-CNT layers enhances overall SE, with noticeable shifts in the shielding peak as the thickness of the h-BN layer is reduced. Four-layer SR composites showed anisotropic thermal and electrical conductivities along with frequency-selective EMI shielding. With a thickness ratio of 4:1 for the graphite-CNT and h-BN layers, the SR composites achieved an in-plane TC of 8.71 W/mK, a breakdown voltage of 8.0 kV/mm and an EMI *SE* ranging from 31 to 43.2 dB. The ordered hierarchical structure of the composites can be modified for various heat dissipation and electrical conduction applications, offering new approaches for designing multifunctional polymer composites for advanced electronic devices [79]. Sha et al. [80] developed an all-in-one lightweight, ultra-thin and flexible PE@PET/MXene film, utilising a unique PE@PET nonwoven fabric as a support structure. A polydopamine coating inspired by mussel adhesion allows for uniform MXene loading, enhancing the film's EC. The hot-pressing process melts the polyethylene (PE) sheath, which acts as an adhesive to secure the MXene nanosheets while maintaining the polyethylene terephthalate (PET) core fibres as an internal "steel" framework. The resulting PE@PET/MXene films demonstrate multiple enhanced properties including high mechanical strength, EC, stability and excellent EMI shielding. The films achieved an EMI *SE* of 50.44 dB and an ultra-high specific shielding effectiveness (SSE/t) of 10,606 dB cm^2/g making them highly effective for EMI shielding applications [80].

Several other reports exist on the utilisation of LbL for the fabrication of efficient EMI shields for advanced applications, such as the works by Sha et al. [80], Cheng et al. [81], Rezvan et al. [82] etc.

8.5.8.2 3D Printing and/or Additive Manufacturing (AM)

Recent advancements in 3D printing and/or AM techniques have presented further opportunities for the fabrication of polymer-based EMI shielding materials for high-tech equipment or devices [83–89]. These techniques enable one to design rather complex shapes and to produce very specific designs that cannot be obtained using more conventional means. Composites for EMI shielding that are printed in 3D can be modelled to fit the required electronic components better, and hence improved performance can also be achieved. Furthermore, due to the possibility of additive manufacturing (AM), it is also possible to implement a combination of various types of fillers and develop multiple levels of structures, which will also improve the EMI *SE* [83–85]. Ribeiro dos Anjos et al. [90] investigated the manufacturing and performance of acrylonitrile butadiene styrene (ABS)-based carbon ternary hybrid composites combining CNT, GNPs and CB for EMI shielding applications. The influence of nanocomposite processing on EM characteristics was investigated by altering the mixing strategy, which included either direct extrusion with simultaneous inclusion of all elements or the preparation of a master batch followed by dilution. The effects of nanofiller shape and processing techniques on nanocomposites' behaviour were thoroughly explored. These nanocomposites' filaments were created by AM using material extrusion, and the resulting components were tested for EMI *SE* in the X-band frequency region. Their findings demonstrated that sample morphology, which is controlled by the processing approach, has a considerable impact on the EMI *SE*. Particularly, ternary hybrids 3/3/3 wt.% "CNT/GNP/CB" nanocomposites showed promising EC of 0.003 S/cm, SE_T of 29 dB and mechanical performance with an elastic modulus of 3080 MPa. They discovered that shields prepared or processed via direct extrusion had a clear advantage, as the nanocomposites were validated as feedstock filaments for 3D printing, and the printed sample outperformed the injection-moulded behaviour for the composition 3/3/3 wt.% "CNT/GNP/CB", achieving ET of 40 dB at 11.8 GHz. Furthermore, their findings supplied useful expertise in designing nanocomposite formulations for AM applications in EMI shielding, offering a nuanced understanding of the link between processing techniques, nanocomposite morphology and the final material properties [90].

Recognising that creating segregated structures in conductive polymer composites is a promising method for enhancing EMI SE, Zou et al. [83] examined 3D-printed carbon fibre/PLA composites with different filling ratios, patterns, thicknesses and carbon fibre contents. Their results showed that the carbon fibre was evenly distributed within the PLA matrix. While the printed carbon fibre/PLA composite exhibited a higher modulus than the injected counterpart, its strength decreased. The incorporation of carbon fibre into 3D-printed PLA enabled the formation of an interconnected conductive network. At a 50% filling ratio combined with a grid filling pattern, 5 mm thickness and 20% carbon fibre content, the composite achieved a superior overall performance with an SE of 99.67%. Moreover, the main shielding mechanism was absorption. This research demonstrated that 3D printing offers an efficient way to produce carbon fibre/PLA composites with outstanding EMI shielding properties suitable for electronic devices and radiation sources [83]. In another study [84], carbon-based inks consisting of various amounts of graphene and CNT nanoparticles were optimised for 3D printing by adjusting their rheological properties. The use of 3D printing allowed for the creation of customised architectures with adjustable structures and multifunctionality. The resulting printed structures exhibited an *SE* of 61.4 dB while maintaining a lightweight density of 0.076 g/cm^3 and an exceptional specific shielding of 802.4 dB cm^3/g. As a proof of concept, the 3D-printed c-*SE* module was integrated into core electronics, replacing traditional metal modules and offering enhanced electromagnetic compatibility (EMC) and thermal dissipation. This innovation filled a gap in the development of carbon-based c-*SE* modules and opened doors for the next generation of high-performance customisable shields for integrated electronics [84].

Another approach involved the development of lightweight honeycomb structures using 3D printing, followed by a conductive Ag nanowire (AgNW) spray coating, to achieve both effective EMI shielding and air venting for thermal cooling [91]. The use of AgNWs with a high aspect ratio resulted in a remarkable surface conductivity of 1 Ω/square. These samples demonstrated up to 35 dB EMI attenuation, corresponding to about 99.97% reduction, with the shielding primarily driven by an absorption mechanism accounting for up to 75% of power absorption. These findings present significant benefits for industries requiring both air venting and EMI shielding. The combination of 3D printing and Ag nanowire coatings allows for the fabrication of complex customisable structures with effective EMI shielding. Future research could explore new materials and production methods to develop innovative shielding solutions for industries like electronics and telecommunications [91]. Thus, the integration of 3D printing technology along with a conductive Ag nanowire spray enabled the creation of complex and customisable structures with efficient EMI *SE*. Future studies in this area could look at new materials and manufacturing techniques to help produce unique EMI shielding solutions in a range of industries, notably electronics, telecommunications and other purposes.

8.5.8.3 Others

It is true that a vast amount of literature has focused on improving conventionally known approaches or their amalgamation as earlier discussed. However, limited literature is also present on new approaches that are not well known; these approaches are thus discussed here.

Lee et al. [23] introduced a novel parasitic reaction-driven method for fabricating segregated MXene/polycarbonate (PC) EMI shields. Figure 8.17 illustrates the synthesis process for creating the MXene/PC nanocomposites. The method involved miscible mixing and precipitation, beginning with the preparation of positively charged PC suspensions facilitated by a cationic surfactant. At 35 °C, PC and CTAB were dissolved in DCM and DI water, respectively. The two solutions were then mixed and homogenised at 16,000 rpm for five minutes. To precipitate PC microspheres, the PC emulsion was added to EtOH under vigorous stirring. The resulting PC suspension was stirred for two hours as MXene suspension (approximately 4 mg/g) was gradually added. Vacuum-assisted filtration was used to create MXene/PC microspheres, which were then dried in a vacuum oven at 80 °C for 12 hours. Finally, the materials were hot pressed at 200 °C for 15 minutes under 45 MPa to form square samples. MXene/PC nanocomposites were produced with varying MXene concentrations (0.19, 0.38, 0.76, 1.15, 1.94 and 2.75 vol.%) [23].

Another unique approach that has been reportedly utilised is the amalgamation of ultrasonication-assisted SC followed by phase separation along with in-situ reduction in

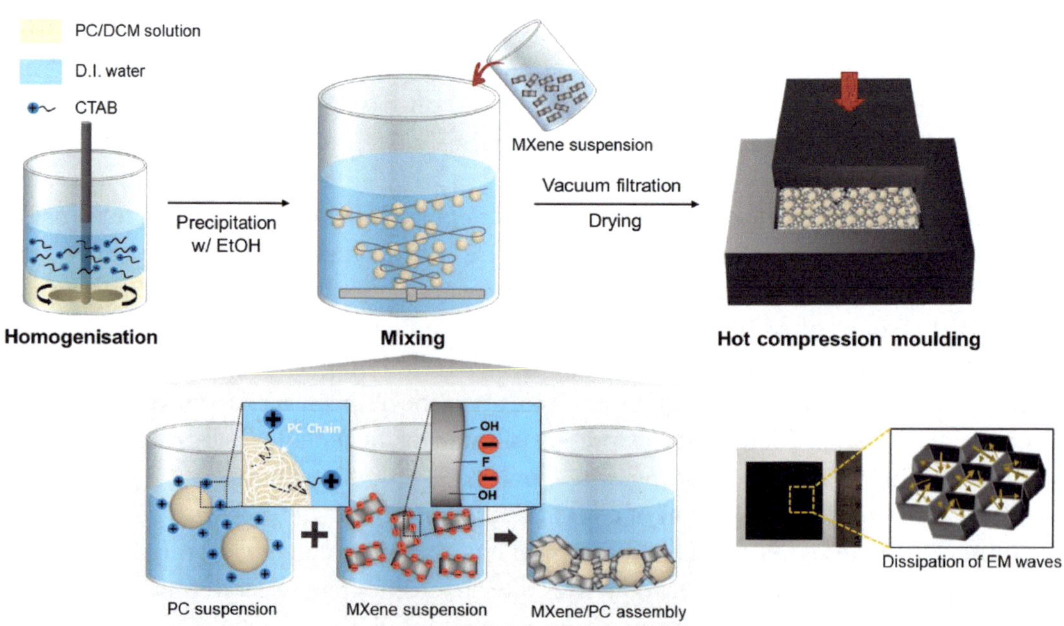

FIGURE 8.17 Schematic image of fabrication of MXene/PC nanocomposites with a segregated structure. Reproduced with permission from Lee et al. [23], Copyright 2023, Elsevier Science Ltd.

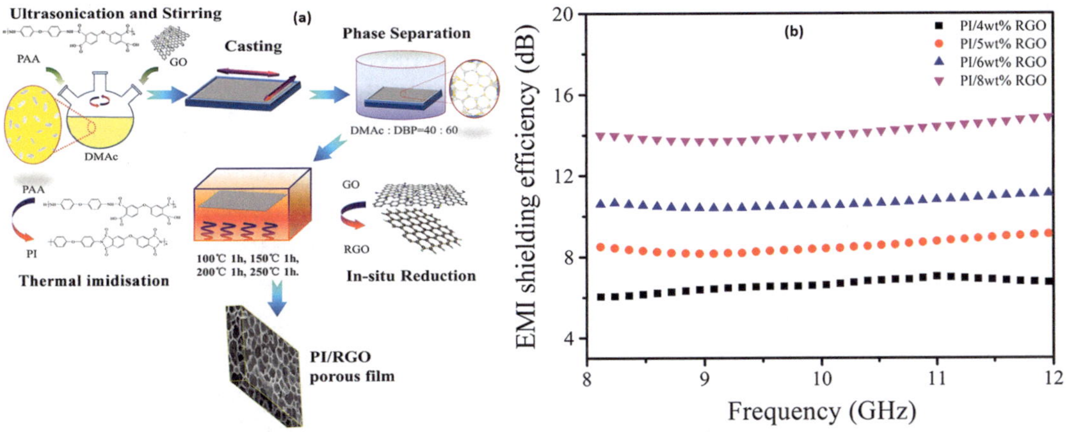

FIGURE 8.18 (a) Schematic for the fabrication of PI/RGO porous films and (b) EMI *SE* of PI/RGO porous films. Reproduced with permission from Yang et al. [92], Copyright 2017, Wiley.

the work by Yang et al. [92], as depicted in Figure 8.18(a) [92]. They first fabricated porous poly(amic acid) (PAA)/GO composite films using a non-solvent induced phase separation (NIPS) method. Subsequently, the PAA was thermally imidised into PI and the GO dispersed within the PAA matrix was thermally reduced to RGO in-situ. The resulting porous PI/RGO composite film, with a thickness of 500 μm and made from an original PAA/GO composite containing 8 wt.% GO, exhibited an effective EC of 0.015 S/m and an excellent specific SE of 693 dB cm^2/g. As shown in Figure 8.18(b), the EMI *SE* values of the samples demonstrated minimal frequency dependence across the tested frequency range. The EMI *SE* of the PI/RGO porous films significantly increased from 6.1 to 7.1 dB for PI/RGO with 4 wt.% GO to 13.7–14.9 dB for PI/RGO with 8 wt.% GO. Additionally, the thermal stability of the porous PI/RGO composite films improved considerably. The 5% weight loss temperature rose from 525 °C for the porous PI film to 538 °C for the composite film. Tensile testing revealed that the composite film had a tensile strength of 6.97 MPa and a Young's modulus of 545 MPa [92].

The major advances overviewed here demonstrate the shifts in numerous directions within the last few decades to develop polymer-based EMI shields. Scientists are gradually expanding the use of these EMI shields due to the development of innovative materials, creative approaches to working with those materials and intricate and sophisticated designs. Since the use of shielded electronic devices has become a trend worldwide, the shielding requirements against EMI will increase, so developers will invent even more efficient, lighter and multifunctional shielding solutions.

8.6 RECENT TRENDS

Innovations in polymer-based EMI shielding materials have been influenced by current trends focusing on lightweight, flexible and high-efficiency materials [3]. No doubt, one of the major trends, as per the literature overviewed in this chapter, is the integration of new conductive fillers, including CNTs, Gr and MXenes, into the polymer matrix [3]. These nanomaterials provide high TC, EC, dielectric properties and/or special light weightiness character, and a high extent of surface area that sharply increases the protective ability of these polymer composites/hybrids [3, 6]. Scientists have also conducted investigations into what are called filler composites/hybrids, systems where a blend of different types of fillers is used, such as Gr and metal NPs, which are also overviewed herewith. Table 8.1 presents some of these instances. These kinds of composite systems take advantage of the electrical characteristics, TC and/or dielectric constant of each filler type to enhance EC, TC as well as increase mechanical properties, thus enhancing EMI *SE* [3, 5, 6]. Another trend that has recently been presented relates to the use of green and eco-friendly polymer-based EMI shields [5, 48, 60]. This entails developing polymeric materials with biodegradable or recycled polymers with the incorporation of conductive and magnetic fillers in order to address the need for EMI shielding while at the same time minimising the impact of these materials on the environment. Further, the development of processing technologies for the fabrication of these polymer composites includes 3D printing and electrospinning, which have great potential to develop intricate structures and architectures within the composite material, which will help improve its EMI shielding efficacy. These techniques enable one to exert optimal control over the placement and alignment of the fillers, thus enabling efficient and effective shielding [3, 48, 60]. Additionally, flexible and wearable electronics have also kindled the search for polymer-based EMI shields that are efficient for shielding while being lightweight and conformable [22, 25–28, 31–33]. This has resulted in the search for materials to incorporate polymeric materials like textile fabrics or coatings, therefore adapting them for use as clothing or wearable devices [93–95]. Therefore, the most recent trends in polymer-based EMI shielding materials are expected towards the improvement of performance, eco-friendliness and adaptability in the current applications of electronics [3, 5].

8.7 CHALLENGES AND FUTURE DIRECTIONS

Despite the considerable advancement made in polymer-based EMI shielding materials, there are certain shortcomings that need to be overcome in order to improve the efficiency of these materials and also to explore the areas of their application. The first of these is the problem of uniform dispersion and distribution of conductive fillers in the polymer matrix. Weak dispersion results in the agglomeration of conductive particles that impairs the conductive network and the overall EMI *SE* [15]. A higher level of filler dispersion or a single-dispersed structure can be achieved by an advanced processing technique or the application of surfactants or compatibilisers, which makes the final EMI shielding product more expensive and complicated [15]. Additionally, the filler content needs to be balanced as high filler loadings are observed to result in higher density, poor mechanical flexibility as well as poor processability, which can be disadvantageous for applications where lightweight and flexible material is desired [3, 15]. Another challenge is achieving an optimal combination of EC and mechanical performance characteristics. High EC is desirable for EMI shielding and is generally achieved at the cost of mechanical strength and flexibility. However, the creation of composites with desirable mechanical characteristics and improved *SE* has been one of the ongoing research areas [2, 3]. Further, the thermal stability, UV resistance and chemical resistance of these materials are important for their long-term use in all environments.

As for future trends, there is increasing emphasis on preparing multifunctional polymer-based EMI shields that, in addition to offering shielding properties, would also act as thermal conductors, self-healing or smart materials with sensing ability. The application of various types of fillers to obtain composite materials with synergistic effects is considered one of the perspectives for obtaining enhanced performance [2, 3]. Also, there is an increased focus on the use of bio-based polymers and the development of green synthesis approaches that can effectively meet the rising demand for environmentally friendly products and processes [3]. Addressing these challenges through innovative material design, advanced processing techniques and sustainable practices is essential for the future development of polymer-based EMI shields. Continued research and collaboration across diverse disciplines would pave the way for next-generation materials that meet the evolving needs of modern electronic and communication technologies.

8.8 CONCLUSION

The development of polymer-based EMI shielding materials represents a significant advancement in materials science, driven by the need for lightweight, flexible and efficient shielding solutions in a world increasingly reliant on electronic devices. Through the integration of various conductive fillers such as CNTs, Gr, MXene and metal NPs, significant progress has been made in enhancing the *SE* of polymer composites. These materials offer several advantages over traditional metal-based shielding, including reduced weight, corrosion resistance and the ability to be processed into complex shapes, which is crucial for modern electronic applications. Recent innovations have focused on optimising the dispersion of fillers within the polymer matrix, improving filler-polymer interactions and developing hybrid materials that leverage the synergistic effects of multiple fillers to achieve superior shielding performance. The ongoing research and technological advancements in this field are expected to lead to even more efficient and multifunctional EMI shields, with potential applications spanning consumer electronics, automotive, aerospace and telecommunications industries. As the demand for robust EMI shielding solutions continues to grow, polymer-based composites are poised to play a pivotal role in meeting these challenges, offering a versatile and effective approach to mitigating electromagnetic interference.

ACKNOWLEDGEMENT

The authors would like to acknowledge financial support from the Department of Science and Innovation (C6ACH77), the Council for Scientific and Industrial Research (086ADMIN) and the University of Johannesburg (086310), South Africa.

REFERENCES

1. Mamatha, G.M. et al., Polymer based composites for electromagnetic interference (EMI) shielding: the role of magnetic fillers in effective attenuation of microwaves, a review. *Hybrid Advances*, 2024. **6**: p. 100200.
2. Ray, S.S., L.T. Temane, and J.T. Orasugh, *Graphene-bearing Polymer Composites: Applications to Electromagnetic Interference Shielding and Flame-retardant Materials*. 2024, Springer Nature.
3. Zheng, S., et al., Research progress on high-performance electromagnetic interference shielding materials with well-organized multilayered structures. *Materials Today Physics*, 2024. **40**: p. 101330.

4. Ruiz-Hitzky, E., C. Ruiz-Garcia, and X. Wang, *MXenes and clay minerals in the framework of the 2D organic–inorganic hybrid nanomaterials. Chemistry of Materials*, 2023. **35**(24): p. 10295–10315.
5. Orasugh, J.T., L.T. Temane, and S.S. Ray, Nanocellulose-based conductive composites: a review of systems for electromagnetic interference shielding applications. *International Journal of Biological Macromolecules*, 2024: p. 133891.
6. Kruželák, J., et al., Progress in polymers and polymer composites used as efficient materials for EMI shielding. *Nanoscale Advances*, 2021. **3**(1): p. 123–172.
7. Saritha, A., et al., Elastomer-based materials for EMI shielding applications, in *Materials for Potential EMI Shielding Applications*. 2020, Elsevier. p. 121–143.
8. Zheng, S., et al., Recent advances in structural design of conductive polymer composites for electromagnetic interference shielding. *Polymer Composites*, 2024. **45**(1): p. 43–76.
9. Cai, Z., et al., Ti3C2Tx MXene/graphene oxide/Co3O4 nanorods aerogels with tunable and broadband electromagnetic wave absorption. *Chemical Engineering Journal*, 2023. **462**: p. 142042.
10. Liu, J.R., et al., Electromagnetic wave absorption properties of ε-Fe3N/Y2O3 nanocomposites derived from Y2Fe17 intermetallic compound. *Journal of Magnetism and Magnetic Materials*, 2004. **277**(3): p. 251–256.
11. Luo, J. and D. Gao, Synthesis and microwave absorption properties of PPy/Co nanocomposites. *Journal of Magnetism and Magnetic Materials*, 2014. **368**: p. 82–86.
12. Mamunya, Y., et al., Influence of conductive nano- and microfiller distribution on electrical conductivity and EMI shielding properties of polymer/carbon composites. *Composites Science and Technology*, 2019. **170**: p. 51–59.
13. Temane, L.T., S.S. Ray, and J.T. Orasugh, Review on processing, flame-retardant properties, and applications of polyethylene composites with graphene-based nanomaterials. *Macromolecular Materials and Engineering*, 2024: p. 2400104.
14. Patadia, M., et al., Enhanced multifunctionality in carbon fiber/carbon nanotube reinforced PEEK hybrid composites: superior combination of mechanical properties, electrical conductivity, and EMI shielding. *Composites Part B: Engineering*, 2024. **284**: p. 111674.
15. Zeng, Z., et al., Thin and flexible multi-walled carbon nanotube/waterborne polyurethane composites with high-performance electromagnetic interference shielding. *Carbon*, 2016. **96**: p. 768–777.
16. Zhou, Y., et al., Asymmetric multilayered cellulose nanofiber composite membranes with electrical-magnetic dual-gradient architectures towards excellent electromagnetic interference shielding performance. *Composites Science and Technology*, 2024. **255**: p. 110729.
17. Sankaran, S., et al. *Thermosetting polymer composites for EMI shielding applications*. in *2006 9th International Conference on Electromagnetic Interference and Compatibility (INCEMIC 2006)*. 2006. IEEE.
18. Yu, D., et al., A super-stretchable liquid metal foamed elastomer for tunable control of electromagnetic waves and thermal transport. *Advanced Science*, 2020. **7**(12): p. 2000177.
19. Yao, B., et al., Highly stretchable polymer composite with strain-enhanced electromagnetic interference shielding effectiveness. *Advanced Materials*, 2020. **32**(14): p. 1907499.
20. Zhang, M., et al., Stretchable liquid metal electromagnetic interference shielding coating materials with superior effectiveness. *Journal of Materials Chemistry C*, 2019. **7**(33): p. 10331–10337.
21. Huang, K., et al., Stretchable microwave absorbing and electromagnetic interference shielding foam with hierarchical buckling induced by solvent swelling. *Carbon*, 2020. **157**: p. 466–477.
22. Nan, Z., et al., Flexible electromagnetic interference shields: materials, structure and multifunctionalization. *Materials Science and Engineering: R: Reports*, 2024. **160**: p. 100823.
23. Lee, J., et al., Parasitic reaction driven facile preparation of segregated-MXene/polycarbonate nanocomposites for efficient electromagnetic interference shielding. *Surfaces and Interfaces*, 2023. **40**: p. 103101.

24. Łapińska, A., et al., Flexible carbon-based fluoropolymer composites for effective EMI shielding and heat dissipation. *Polymer Composites*, 2024. **45**(5): p. 4319–4337.
25. Mondal, S., et al., Highly conductive and flexible nano-structured carbon-based polymer nanocomposites with improved electromagnetic-interference-shielding performance. *Materials Research Express*, 2017. **4**(10): p. 105039.
26. Zhang, C., et al., Rational design and fabrication of lightweight porous polyimide composites containing polyaniline modified graphene oxide and multiwalled carbon nanotube hybrid fillers for heat-resistant electromagnetic interference shielding. *Polymer*, 2021. **224**: p. 123742.
27. Yao, Y., et al., Polymer-based lightweight materials for electromagnetic interference shielding: a review. *Journal of Materials Science*, 2021. **56**(11): p. 6549–6580.
28. Shen, B., et al., Lightweight, multifunctional polyetherimide/graphene@Fe_3O_4 composite foams for shielding of electromagnetic pollution. *ACS Applied Materials & Interfaces*, 2013. **5**(21): p. 11383–11391.
29. Thomassin, J.-M., et al., Polymer/carbon based composites as electromagnetic interference (EMI) shielding materials. *Materials Science and Engineering: R: Reports*, 2013. **74**(7): p. 211–232.
30. Sushmita, K., et al., Distribution of carbon nanotubes in polycarbonate-based blends for electromagnetic interference shielding. *ACS Applied Nano Materials*, 2022. **5**(1): p. 662–677.
31. Ravindren, R., et al., Investigation of electrical conductivity and electromagnetic interference shielding effectiveness of preferentially distributed conductive filler in highly flexible polymer blends nanocomposites. *Composites Part A: Applied Science and Manufacturing*, 2019. **118**: p. 75–89.
32. Han, L., et al., Enhanced EMI shielding effectiveness of green Eucommia ulmoides gum composites through heterogeneous and crystalline structures. *Industrial Crops and Products*, 2024. **219**: p. 119033.
33. Choudhury, S.N., et al., Enhancing electrical characteristics and electromagnetic interference shielding effectiveness in thermoplastic elastomeric polymer blends by utilizing the selective distribution of conductive black. *Functional Composite Materials*, 2024. **5**(1): p. 4.
34. Song, P., et al., Frequency-adjustable electromagnetic interference shielding performance of sandwich-structured conductive polymer composites by selective foaming and tunable filler dispersion. *Composites Communications*, 2022. **34**: p. 101264.
35. Yang, J., et al., Layered structural PBAT composite foams for efficient electromagnetic interference shielding. *Nano-Micro Letters*, 2023. **16**(1): p. 31.
36. Jing, L., et al., Reduced graphene oxide porous films containing SiC whiskers for constructing multilayer electromagnetic shields. *New Carbon Materials*, 2024. **39**(6): p. 1191–1201.
37. Shi, Y., et al., Multi-interface assembled N-doped MXene/HCFG/AgNW films for wearable electromagnetic shielding devices with multimodal energy conversion and healthcare monitoring performances. *ACS Nano*, 2022. **16**(5): p. 7816–7833.
38. Zhang, Q., et al., Millefeuille-inspired biomass alternate multilayer composite, for excellent absorption-dominated, broadband EMI shielding and joule heating. *Composites Part B: Engineering*, 2024. **283**: p. 111620.
39. Pu, X., et al., Absorption-dominant EMI shielding performance of multiple folded Bi2Se3/PVDF nanocomposite films. *Journal of Materiomics*, 2024. **10**(1): p. 27–36.
40. Xu, J., et al., Competitively assembled aramid-MXene Janus aerogel film exhibiting concurrently robust shielding and effective anti-reflection performance. *Advanced Functional Materials*, 2024. **34**(34): p. 2400732.
41. Nguyen, T.T.L., et al., Exceptional electromagnetic interference shielding using single-walled carbon nanotube/conductive polymer composites films with ultrathin, lightweight properties. Carbon, 2024: p. 119567.
42. Kumari, S., et al., Enhanced microwave absorption properties of conducting polymer@graphene composite to counteract electromagnetic radiation pollution: green EMI shielding. *RSC Advances*, 2024. **14**(1): p. 662–676.

43. Wang, Z., et al., Unification of thermal conductivity, electrical insulation, and electromagnetic shielding: a bulk thermal conductive polymer from solid waste. *Chemical Engineering Journal*, 2024. **498**: p. 155065.
44. Chopra, A., et al., Leveraging BaFe12O19/PANI nanocomposite incorporated cotton fabric as an effective EMI shielding flexible material for X-band frequency. *ACS Applied Electronic Materials*, 2024. **6**(4): p. 2242–2257.
45. Lingchong, X.U.E., et al., A magnetically augmented eco-friendly conductive polymer composite for X-band electromagnetic interference shielding. *Journal of Advanced Manufacturing Science and Technology*, 2024: p. 2025004.
46. Feng, W., et al., Electromagnetic porous lignocellulosic matrix composites: a green electromagnetic shielding material with high absorption efficient electromagnetic interference. *International Journal of Biological Macromolecules*, 2024. **275**: p. 133505.
47. Liu, T., et al., Green segregated honeycomb biopolymer composites for electromagnetic interference shielding biomedical devices. *Chemical Engineering Journal*, 2024. **493**: p. 152438.
48. Zhou, Y., et al., Green electromagnetic interference shielding material based on thermally expandable microspheres. *ACS Applied Materials & Interfaces*, 2024. **16**(35): p. 46789–46797
49. Sun, R., et al., Highly conductive transition metal carbide/carbonitride (MXene)@polystyrene nanocomposites fabricated by electrostatic assembly for highly efficient electromagnetic interference shielding. *Advanced Functional Materials*, 2017. **27**(45): p. 1702807.
50. Wang, L., et al., Electromagnetic interference shielding MWCNT-Fe_3O_4@Ag/epoxy nanocomposites with satisfactory thermal conductivity and high thermal stability. *Carbon*, 2019. **141**: p. 506–514.
51. Yang, X., et al., Synchronously improved electromagnetic interference shielding and thermal conductivity for epoxy nanocomposites by constructing 3D copper nanowires/thermally annealed graphene aerogel framework. *Composites Part A: Applied Science and Manufacturing*, 2020. **128**: p. 105670.
52. Cao, M.-S., et al., Ferroferric oxide/multiwalled carbon nanotube vs polyaniline/ferroferric oxide/multiwalled carbon nanotube multiheterostructures for highly effective microwave absorption. *ACS Applied Materials & Interfaces*, 2012. **4**(12): p. 6949–6956.
53. Guo, D., et al., Aramid nanofiber/polypyrrole composite films for broadband EMI shielding, wearable electronics, joule heating, and photothermal conversion. *ACS Applied Nano Materials*, 2023. **6**(16): p. 15108–15118.
54. Liang, C., et al., Superior electromagnetic interference shielding 3D graphene nanoplatelets/reduced graphene oxide foam/epoxy nanocomposites with high thermal conductivity. *Journal of Materials Chemistry C*, 2019. **7**(9): p. 2725–2733.
55. Zhang, H.-B., et al., Tough graphene–polymer microcellular foams for electromagnetic interference shielding. *ACS Applied Materials & Interfaces*, 2011. **3**(3): p. 918–924.
56. Li, Y.-K., et al., High-efficiency electromagnetic interference shielding and thermal management of high-graphene nanoplate-loaded composites enabled by polymer-infiltrated technique. *Carbon*, 2023. **211**: p. 118096.
57. Zhao, B., et al., Highly compressible polymer composite foams with thermal heating-boosted electromagnetic wave absorption abilities. *ACS Applied Materials & Interfaces*, 2020. **12**(45): p. 50793–50802.
58. Xie, Z., et al., Robust CoFe2O4@Carbon nanotube/polydimethylsiloxane foams with low thermal conductivity for electromagnetic interference shielding. *ACS Applied Nano Materials*, 2023. **6**(23): p. 21733–21740.
59. Ran, L., et al., Fabrication of MXene based sandwich-like films for excellent flexibility, electromagnetic interference shielding and thermal management. *Composites Part A: Applied Science and Manufacturing*, 2023. **173**: p. 107672.
60. Yang, W., et al., Ultrathin flexible reduced graphene oxide/cellulose nanofiber composite films with strongly anisotropic thermal conductivity and efficient electromagnetic interference shielding. *Journal of Materials Chemistry C*, 2017. **5**(15): p. 3748–3756.

61. Zhou, B., et al., Flexible MXene/silver nanowire-based transparent conductive film with electromagnetic interference shielding and electro-photo-thermal performance. *ACS Applied Materials & Interfaces*, 2020. **12**(36): p. 40859–40869.
62. Zhang, P., et al., Low-melting-point alloy continuous network construction in a polymer matrix for thermal conductivity and electromagnetic shielding enhancement. *ACS Applied Polymer Materials*, 2019. **1**(8): p. 2006–2014.
63. Wang, Y.-H. and C.-C. Li, Highly effective EMI shielding composites for 5G Ka-band frequencies. *Applied Materials Today*, 2024. **36**: p. 102041.
64. Su, J., et al., Experimental investigation and numerical simulation on microwave thermal conversion storage properties of multi-level conductive porous phase change materials and its multifunctional applications. *Applied Thermal Engineering*, 2024. **253**: p. 123774.
65. You, J., et al., Aromatic phase change microspheres constructed nanocomposite films for fluorine-free self-cleaning, absorption-dominated EMI shielding, and high-temperature thermal camouflage. *Composites Part B: Engineering*, 2024. **283**: p. 111650.
66. Hu, B., et al., Millefeuille-like phase change composite with multi-level oriented architecture for efficient thermal management and electromagnetic interference shielding. *Chemical Engineering Journal*, 2024. **483**: p. 149240.
67. Zhu, T.-Y., et al., Multifunctional MXene/PEDOT:PSS-based phase change organohydrogels for electromagnetic interference shielding and medium-low temperature infrared stealth. *ACS Applied Materials & Interfaces*, 2024. **16**(12): p. 15372–15382.
68. Yan, R., et al., Phase change composite based on lignin carbon aerogel/nickel foam dual-network for multisource energy harvesting and superb EMI shielding. *International Journal of Biological Macromolecules*, 2024. **277**: p. 134233.
69. Yue, H., et al., Ti3C2Tx MXene/delignified wood supported flame-retardant phase-change composites with superior solar-thermal conversion efficiency and highly electromagnetic interference shielding for efficient thermal management. *Energy*, 2024. **286**: p. 129441.
70. Chen, Y., et al., Leakage proof, flame-retardant, and electromagnetic shield wood morphology genetic composite phase change materials for solar thermal energy harvesting. *Nano-Micro Letters*, 2024. **16**(1): p. 196.
71. Yan, R., et al., Cellulose-reinforced foam-based phase change composites for multi-source driven energy storage and EMI shielding. *Composites Communications*, 2024. **51**: p. 102047.
72. He, J., et al., Multifunctional phase change composites for green electromagnetic interference shielding and thermal response prepared under the guidance of impedance matching strategy. *Nanoscale*, 2024. **16**(35): p. 16622–16631.
73. Lu, J., et al., Porous-multilayered Ti3C2Tx MXene hybrid carbon foams for tunable and efficient electromagnetic wave absorption. *Carbon*, 2024. **229**: p. 119477.
74. Hu, X., et al., Smart windows with adjustable electromagnetic interference shielding using hydrogel technology. *Chemical Engineering Journal*, 2024. **493**: p. 152533.
75. Wang, Y.-Q., et al., Controllable proton-reservoir ordered gel towards reversible switching and reliable electromagnetic interference shielding. *Materials Horizons*, 2024. **11**(4): p. 978–987.
76. Singh, P.P., A. De, and B.B. Khatua, Bio-inspired smart composite architecture for thermally tunable green EMI shielding. *Applied Surface Science*, 2024. **643**: p. 158643.
77. Liu, X.-C., et al., Caffeic-acid-functionalized MWCNTs and PEDOT:PSS formed composite flexible films with "reinforced concrete" structure for electrical heating and EMI shielding. *ACS Applied Materials & Interfaces*, 2024. **16**(17): p. 22391–22402.
78. Zhang, Y., et al., Flexible multifunctional magnetic-conductive Janus nanocomposite films towards highly efficient electromagnetic interference shielding and thermal management. *Composites Science and Technology*, 2024. **256**: p. 110756.
79. Ji, Y., et al., Thermally and electrically anisotropic silicone rubber composites with frequency-selective EMI shielding. *Composites Science and Technology*, 2024. **256**: p. 110780.

80. Sha, Z., et al., All-in-one integrated flexible PE@PET/MXene films for high-performance electromagnetic shields with self-reinforced conductivity and mechanical properties. *Carbon*, 2024. **216**: p. 118595.
81. Cheng, R., et al., Sandwich-structured polyacrylate composite films for electromagnetic interference shielding with low reflection characteristics. *Progress in Organic Coatings*, 2024. **187**: p. 108155.
82. Rezvan, G., et al., Hybrid functional membranes through layer-by-layer assembly of Ti3C2Tx MXene and gelatin-stabilized calcium phosphate nanospheres. *Applied Materials Today*, 2024. **37**: p. 102144.
83. Zou, L., et al., 3D printing of carbon fiber powder/polylactic acid with enhanced electromagnetic interference shielding. *Diamond and Related Materials*, 2024. **141**: p. 110583.
84. Shi, S., et al., 3D-printed carbon-based conformal electromagnetic interference shielding module for integrated electronics. *Nano-Micro Letters*, 2024. **16**(1): p. 85.
85. Qureshi, N., et al., Exploring conductive filler-embedded polymer nanocomposite for electrical percolation via electromagnetic shielding-based additive manufacturing. *Advanced Materials Technologies*, 2024: p. 2400250.
86. Sanandiya, N.D., et al., Chitosan-based electroconductive inks without chemical reaction for cost-effective and versatile 3D printing for electromagnetic interference (EMI) shielding and strain-sensing applications. *Carbohydrate Polymers*, 2024. **337**: p. 122161.
87. Subramanian, J., et al., 3D-printed honeycomb structure filled with nanofillers for efficient electromagnetic interference shielding performance. *Proceedings of the Institution of Mechanical Engineers, Part E: Journal of Process Mechanical Engineering*, 2024: p. 09544089231221677.
88. Zhang, G., et al., Advancements in 3D-printed architectures for electromagnetic interference shields. *Journal of Materials Chemistry A*, 2024. **12**(10): p. 5581–5605.
89. Temane, L.T., J.T. Orasugh, and S.S. Ray, Polymer additive manufacturing: an overview, in *Reference Module in Materials Science and Materials Engineering*. 2024, Elsevier.
90. Ribeiro dos Anjos, E.G., et al., Advanced ternary carbon-based hybrid nanocomposites for electromagnetic functional behavior in additive manufacturing. *Applied Materials Today*, 2024. **40**: p. 102362.
91. Abbas, R., et al., 3D printed lightweight honeycomb vent structures with subsequent coating of silver nanowires for efficient electromagnetic interference (EMI) shielding. *Heliyon*, 2024. **10**(9): p. e30429
92. Yang, H., et al., Preparation of porous polyimide/in-situ reduced graphene oxide composite films for electromagnetic interference shielding. *Polymers for Advanced Technologies*, 2017. **28**(2): p. 233–242.
93. Orasugh, J.T., et al., Progress in polymer nonwoven textile materials in electromagnetic interference shielding applications. *Functional Composite Materials*, 2024. **5**(1): p. 5.
94. Orasugh, J.T. and S.S. Ray, Graphene-based electrospun fibrous materials with enhanced EMI shielding: recent developments and future perspectives *ACS Omega*, 2022. **7**(38): p. 33699–33718.
95. Orasugh, J.T. and S.S. Ray, Functional and structural facts of effective electromagnetic interference shielding materials: a review. *ACS Omega*, 2023. **8**(9): p. 8134–8158.

CHAPTER 9

Recent Developments of Polymer Nanocomposites with Carbon Nanostructures and MXene Fillers in EMI Shielding

Bedriye Ucpinar and Ayse Aytac

9.1 INTRODUCTION

The use of electromagnetic (EM) waves is increasing day by day due to rapid technological developments and widespread applications in aerospace, telecommunications, health, broadcasting and military sectors. The escalating demand for electronic and communication technologies entails several associated challenges. Digitalisation-induced electromagnetic interference (EMI) results in electromagnetic pollution, posing significant risks. EMI not only deteriorates the overall performance of electronics, causing malfunctions and security vulnerabilities, but also contributes to human health problems such as insomnia, headaches, asthma and cancer [1, 2]. The prevention of adverse health effects and the protection of electronic equipment from EMI have become the subject of considerable concern and rigorous research. Among the various strategies developed to overcome these challenges, the use of shielding is a critical method [3].

EMI shielding is the placement of materials around devices that block or attenuate electromagnetic interference to prevent its propagation. The shield can act as a barrier, absorbing or reflecting EMI and thus preventing this interference from entering the protected area [1, 3]. The performance of a shielding material is determined by the losses that occur because of absorption, reflection and multiple reflection mechanisms in the EMI arriving at the shield (Figure 9.1). The total shielding effectiveness (SE) of the material, with the contribution of each of these mechanisms, is measured in decibels (dB) [1, 4]. The reflection mechanism, which is generally dominant in EMI shielding, is based on the

DOI: 10.1201/9781003512363-9

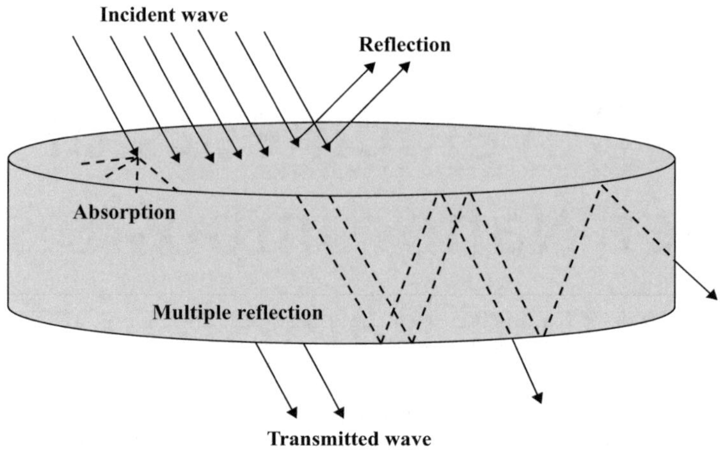

FIGURE 9.1 Schematic representation of EMI shielding mechanisms.

interaction of the EM field with the free electrons in the shielding material. When EMI strikes the shield, the impedance mismatch between the environment and the surface causes reflection. The shielding material must therefore be electrically conductive. Absorption depends mainly on the permeability and dielectric constant of the material, so a high dielectric constant is another desirable property of shields. Multiple reflection occurs as a result of non-uniformity of electron paths in the material. For example, filler-matrix interfaces in composites can cause multiple reflections [4, 5].

Metals have traditionally been preferred for EMI shielding for many years due to their excellent electrical conductivity. However, they have several drawbacks, including high density, poor corrosion resistance, limited design flexibility and high cost [6]. Polymer materials are often the focus of significant research and development due to their many advantages over traditional metal-based shielding materials. They offer key benefits such as light weight, corrosion resistance, design flexibility and cost-effectiveness [2, 7]. On the other hand, the inherently non-conductive character of many polymers poses a challenge to their direct use in EMI shielding. To circumvent this limitation and make polymers capable of EMI shielding, several strategies are being employed, including the incorporation of conductive fillers and the development of polymer composites (PCs) and nanocomposites (PNCs) [4, 7].

9.2 PNCs FOR EMI SHIELDING APPLICATIONS

Composites are materials made by combining two or more components to form a new material with features that are relatively different from the individual components. In polymeric composites, a polymer serves as the matrix, and fibres, tubes, particles, flakes or platelets can serve as the reinforcing material. PNCs are a special type of composite in which the reinforcement material is at the nanoscale [8, 9]. Nanofillers play a critical role in tuning the properties of PNCs, including their EMI SE. The performance of PNCs is strongly controlled by various factors related to the nanofillers, including their aspect ratio, size, physical geometry, intrinsic conductivity and magnetic properties [9, 10].

In addition to nanofillers, the manufacturing process is another important parameter that influences the properties of nanocomposites. In situ polymerisation, solvent casting and melt compounding are the basic nanocomposite production techniques [9].

As discussed earlier, the intrinsic electrical and magnetic properties (electrical conductivity, permittivity, permeability) of the filler and matrix, the amount and geometry of the filler, the matrix-filler interactions, and the production process type and conditions all affect PNCs' electrical conductivity and thus their capacity to function as EMI shields. The key factor for electrical conductivity in composites is the formation of a continuous network of conductive fillers dispersed in the polymer matrix. The filler concentration at which the conductive network is formed in the matrix, allowing the movement of electrons, is called the percolation threshold. At the percolation threshold, the composite turns from an insulator to a conductor, and an increase in the EMI SE value is also achieved [9, 11]. The high conductivity of the shield will ensure reflection of the incident EM waves. The unreflected EM wave propagates through the shield. Another shielding mechanism, absorption, depends on the presence of electric and magnetic dipoles in the shielding structure. These dipoles interact with the EM wave and undergo polarisation. Permeability (μ) and permittivity (ε), which are physical properties formed by polarisation, are critical factors affecting absorption and EMI SE [8]. Permeability and permittivity are complex properties; their real parts represent the storage of energy in magnetic and electric fields, respectively. The imaginary parts represent the dissipation of energy within the material. Accordingly, materials with high permeability and dielectric permittivity can improve shielding performance by increasing absorption efficiency [2, 8, 12].

Nanofillers are selected according to the desired purpose within the material. This allows the properties to be precisely tuned to meet specific requirements. Since electrical conductivity is the primary criterion for EMI shielding, electrically conductive nanoparticles are incorporated into the polymer matrix, yielding conductive polymer composites (CPCs) [7]. Conductive nanofillers in the insulating polymer matrix boost the material's conductivity while also affecting its dielectric characteristics. These nanofillers act as polarisable dipoles within the matrix. As the amount of nanofiller increases, the number of polarisable dipoles also increases, enhancing its dielectric permittivity. Furthermore, as the amount of nanofiller increases, more matrix-filler interfaces emerge, resulting in charge accumulation at these interfaces. This leads to increased interfacial polarisation and dielectric permittivity (Maxwell-Wagner-Sillars, MWS effect) [8, 11, 12]. Additionally, nanocapacitors can be generated in CPCs. Nanofillers act as conductive electrodes, while the polymer matrix serves as a dielectric medium. Increasing the number of nanocapacitors improves the matrix's energy storage ability (the real part of the dielectric permittivity) and hence the EMI SE [11].

Commonly used nanofillers in CPCs are metallic and carbon nanostructures due to their excellent conductivity [7, 11, 13]. Carbon nanostructures are especially noteworthy due to their lightweight nature, high surface area, enhanced thermal and mechanical properties and the diversity of allotropes present. These properties make them highly advantageous for use in CPCs for EMI shielding [1, 3]. Alongside the widely preferred carbon-based fillers, two-dimensional (2D) transition metal carbides/nitrides/carbonitrides, collectively known as MXenes, have gained considerable attention over the past few years due to their high electrical conductivity [14, 15]. This chapter addresses the EMI shielding applications of CPCs reinforced with carbon nanostructures and MXenes.

9.2.1 Carbon Nanostructures-Reinforced Nanocomposites

Carbon nanostructures or carbon-based nanofillers are commonly used in CPCs due to their outstanding electrical conductivity and ability to absorb EMI over a wide frequency range. The main carbon nanostructures include carbon black (CB), carbon nanotubes (CNTs), graphite/graphene and carbon nanofibre (CNF). Their unique properties such as high conductivity and permittivity, low density, high aspect ratio and excellent thermal and mechanical stability make them ideal for polymeric EMI shielding materials [1, 9, 16]. One of the major advantages of utilising carbon nanostructures in CPCs is that many of their unique properties depend on their structure. The size, geometry and distribution of the filler in the matrix are critical factors in producing better EMI shielding materials. Furthermore, high aspect ratios allow enhanced properties to be achieved at lower nanofiller levels. This reduces density and cost while facilitating processability [1, 9, 17]. The properties of some carbon nanostructures and the studies of EMI shielding materials based on them are highlighted as follows.

9.2.1.1 Carbon Black-Reinforced PNCs for EMI Shielding

CB consisting of >95% carbon is obtained by thermal decomposition of hydrocarbons. CB, which has a distorted graphite structure, can be produced in numerous types by changing the production method and process parameters according to the usage area [9, 18]. It is widely used in many areas such as paints, fillers, coatings and ink toners, 90% of which are in the rubber industry. CB is attractive because of its abundance, large surface area, resistance to chemicals and corrosion, cost-effectiveness and electrical conductivity [19, 20]. Thanks to its good electrical conductivity, CB is used as a filler in CPCs and is being investigated specifically for EMI shielding applications.

It has already been discussed that obtaining effective conductive networks in an insulating matrix depends on the proper dispersion of the filler in the matrix. However, CB has strong van der Waals forces and tends to aggregate into large clumps. This makes uniform dispersion challenging and leads to a high percolation threshold and weak mechanical properties [19, 21]. Therefore, research focuses on the uniform distribution of CB in the matrix [22, 23], optimising the percolation threshold [19, 21] and developing various composite structures [24–26] to obtain high EMI SE.

The effect of the amount of CB on the percolation threshold for EMI SE of CB-reinforced epoxy resin nanocomposites was investigated by Maryam et al. [21]. The EMI SE increased with increasing CB concentration, reaching a high of 20 dB at 25 wt% CB concentration. A sharp increase was detected at 20 wt% concentration, which is CB's percolation threshold. At the percolation threshold, the CB formed a conductive network within the matrix, and the conductivity of the nanocomposite reached its peak. They concluded that epoxy/CB nanocomposites exhibited high reflection and low absorption. The electrical conductivity of CB alters with surface area and chemisorbed oxygen complexes. Although CB with a large surface area and a small number of surface functional groups is more conductive, it raises production costs [27]. Rahaman studied the characteristics of an EPDM nanocomposite containing two types of CBs: conductive (Printex) and non-conductive (HAF). He discovered that the EMI SE value of the Printex-filled nanocomposite at 40 phr was approximately seven times higher than that of the HAF-filled nanocomposite [27]. As observed in the study by Maryam et al. [21], the percolation threshold of CB is also high, hence numerous strategies are being investigated to minimise it. One of these is the incorporation of CB into a blend matrix composed of two or more polymers. Depending on the polymer, CB is selectively localised in one of the two phases or at the interface. This yields a lower percolation threshold [19, 28]. Nath et al. prepared conductive CB (VCB) reinforced thermoplastic polyurethane (TPU)/polybutylene adipate-co-terephthalate (PBAT) blends via the solvent casting process. It is reported that VCB is located in the low-viscosity PBAT phase, resulting in a reduced percolation threshold. At a percolation threshold of 12.5 wt%, the total shielding effectiveness (SE_T) of the nanocomposite shifted from 14 to 20 dB. At X-band, the 1-mm-thick sample with 30 wt% VCB had an EMI SE of 25 dB [28]. As can be seen, the nanocomposite exceeded the commercial EMI SE limit of 20 dB. Lowering the percolation threshold and enhancing the electrical conductivity of nanocomposites appear to be useful methods for improving EMI SE. On the other hand, as conductivity increases, shielding is achieved through reflection, resulting in secondary radiation pollution. At this stage, research focuses on increasing shielding through absorption. Cell structure in CPC foams can improve absorption efficiency by improving polarisation loss and multiple reflections [24–26]. Wu and co-workers [26] prepared CB-reinforced polypropylene (PP) foams by the core-back foam injection moulding (FIM) method. The effect of CB content and expansion ratio was studied. While the study found that filler amount and expansion ratio had varied effects on EMI SE, it also presented an effective approach for controlling the EMI performance of CPC foams without affecting material weight. The data of some nanocomposites including CB are given in Table 9.1.

TABLE 9.1 Summary of PNCs Including Carbon Black

Nanocomposite	CB Concentration (wt%)	Preparation Method	Thickness (mm)	Frequency (GHz)	EMI SE (dB)	Reference
Epoxy/CB	25	Solvent casting	2.5	0.01–13.6	21	[21]
XNBR/EMA/VCB	45	Melt blending/compression moulding	2	8.2–12.4	33.5	[19]
EPDM/LLDPE/VCB	50	Melt blending/compression moulding	1	8.2–12.4	27.5	[23]
PS/EVA/CB	15	Melt blending/compression moulding	1	8.2–12.4	20.8	[29]
EMA/EOC/CB	30	Solvent casting/compression moulding	2	8.2–12.4	30.4	[30]
EMA/TPO/VCB	30	Solvent casting	1	14.5–20	29	[22]
PP/CB	30	Foam-injection moulding	–	5.38–8.17	31	[26]
EPDM/CB	28.57	Two-roll mixing mill	1	8.2–12.4	49.9	[27]

Note: EMA: poly(ethylene methyl acrylate), EOC: ethylene octane copolymer, EPDM: ethylene propylene diene rubber, LLDPE: linear low-density polyethylene, PP: polypropylene, TPO: thermoplastic polyolefin, VCB: Vulcan XC72, XNBR: carboxylated acrylonitrile butadiene rubber and XNBR: carboxylated acrylonitrile butadiene rubber.

9.2.1.2 Carbon Nanotube-Reinforced PNCs for EMI Shielding

Carbon nanotubes (CNTs), another type of carbon allotrope used in CPC manufacturing, are tubular nanomaterials that are rolled graphene sheets with sp2 bonds between carbon molecules [17]. Their diameters are typically on the scale of a few nanometres (nm), while their lengths can be many times longer. Depending on the configuration of the carbon network, different CNTs can be generated. CNTs are further characterised as single-walled (SWCNT), double-walled or multi-walled carbon nanotubes (MWCNT) based on the number of carbon layers on their side walls [17, 20]. CNTs have unique properties due to their high aspect ratio, large surface area and special structure. They are characterised by high modulus (1.2–1.8 TPa) and tensile strength (50–200 GPa), as well as excellent electrical (10^2–10^6 S/cm) and thermal conductivity (about 6000 W/m·K) [9]. CNTs are highly effective in improving the electrical conductivity and EMI SE of CPCs at low filler concentrations. Their high intrinsic conductivity, high aspect ratio and thin long structure facilitate the formation of effective conductive pathways, yielding high conductivity and EMI SE at low percolation thresholds. These properties make CNTs ideal for a broad range of applications, from military to aerospace and from electronics to automotive, where effective EMI shielding is critical [8, 17]. Many studies on conductivity and EMI shielding performance have been published in the literature, with variables including CNT type and characteristics, polymer matrix type and processing method.

Lee and co-workers investigated the effect of the aspect ratio of MWCNTs on the EMI SE of polydimethylsiloxane (PDMS) nanocomposites. Samples were prepared by mixing three types of MWCNTs with PDMS and moulding them with a thickness of 1 mm. An increase in electrical conductivity and EMI SE performance was obtained as the aspect ratio of the MWCNTs increased. It was reported that at the highest aspect ratio, an EMI SE of around 30 dB was achieved with a 10 wt% addition of MWCNTs [31]. A similar study was conducted by Shin et al. for TPU/MWCNT nanocomposites. TPU nanocomposites containing MWCNTs of various lengths were fabricated using the non-solvent induced phase separation (NIPS) strategy. Long MWCNTs provided high electrical conductivity (1.9×10^{-3} S/cm) by forming a more efficient conductive network within the matrix. The long MWCNT/TPU and single-walled CNT/TPU nanocomposites exhibited EMI SE of 37 and 30 dB respectively, while this value was measured to be 18 dB for short MWCNTs. The authors suggested that the films they produced could be used in applications such as military, wearable electronics and artificial intelligence, thanks to their superior EMI shielding performance [32].

Solution casting and melt blending are the most frequently used simple methods for the preparation of polymer/CNT nanocomposites. In their 2024 study, Rani and Malik prepared polyamideimide (PAI)/MWCNT nanocomposites through solution casting and assessed their electromagnetic interference (EMI) shielding performance. The nanocomposite containing 5.0 wt% MWCNT exhibited a conductivity of 0.145×10^{-4} S/cm and an EMI SE of 17 dB in the X-band [33]. Poly(3-hydroxybutyrate-co-3-hydroxyvalerate) (PHBV)/CNT bionanocomposites were prepared by the melt blending method and the effect of CNT amount on properties was explored. The eco-friendly PHBV/CNT nanocomposites showed a percolation threshold of 0.66%. Increasing the CNT ratio improved both the dielectric and EMI SE performance. The 3 mm thick sample with 5 wt% CNT was found to have an EMI SE of approximately 43 dB [34]. The effects of the polylactic acid/high-density polyethylene (PLA/HDPE) volume ratio on the properties of 1.0 vol% CNT/PLA/HDPE nanocomposites and the effects of CNT ratio were determined by Al-Saleh and El-Methaly. In PLA/HDPE blends, CNTs were selectively localised in HDPE. Electrical conductivity and EMI SE increased with increasing PLA content in the blend. The EMI shielding performance was reported to be a considerable function of the microstructure of the nanocomposite [35].

As with other carbon nanostructures, the dispersion of CNTs in the matrix is vital to their EMI shielding performance. CNTs, which can become entangled due to their high aspect ratios and 1D structures, tend to aggregate due to van der Waals forces [8]. Optimisation of various criteria such as process method and conditions, CNT surface treatments and so on are frequently studied to improve the dispersibility of CNTs and thus the EMI SE of nanocomposites. An EMI SE of 37.38 dB in the X-band range was attained by incorporating 20 phr of MWCNT into a thermoplastic polyurethane (TPU) matrix. The electrical conductivity of the TPU/MWCNT nanocomposite enhanced from 10^{-13} to 10^{-3} S/cm. These improved properties were attributed to the superior dispersion achieved by double extrusion followed by the solution casting process [36]. Functionalisation of CNTs is an effective approach for this purpose. Huang and co-workers [37] used imidazolium functionalised polyurethane (IPU) to improve the distribution and compatibility of MWCNTs in a poly(lactic acid)/polycaprolactone (PLA/PCL) matrix. CNTs were functionalised with IPU and nanocomposites were prepared using a torque rheometer. It was observed that IPU improved the dispersion of CNTs, ensuring their selective localisation at the interface and in the PCL phase. Thus, a considerable reduction in the percolation threshold was attained. The EMI SE for PLA/PCL/8CNT/0.8IPU nanocomposites enhanced to 35.6 dB. Furthermore, the mechanical properties of this nanocomposite were significantly enhanced compared with the nanocomposite without the compatibiliser. In another study, to improve the dispersion of nanotubes in polyamide 12 (PA12)/MWCNT composites, Lee et al. modified MWCNTs with flavin mononucleotides (MCNT) and annealed (MCNT(H)) and prepared nanocomposites by melt blending/hot pressing methods. The annealed nanotubes exhibited good dispersion even with increasing loading. The best EMI SE performance of 29.4 dB was obtained in the PA12/MCNT(H) (10 wt%) composite [38]. In a different paper, polyketone-grafted (CNT-g-PK) and hydroxylated MWCNT-reinforced polyketone (PK) nanocomposites were prepared by the reactive blending method and their properties were investigated to ensure effective dispersion of MWCNTs in the matrix. Electron microscopy has proven that PK grafting improves the interfacial interactions between filler and matrix. The mechanical properties of the PK/CNT-g-PK nanocomposites were observed to improve with an increase in the MWCNT content. An electrical conductivity of 6.98×10^{-4} S/cm and an EMI SE of 22 dB at X-band were reported for a 1 mm thick PK/CNT-g-PK sample containing 5.0 wt% MWCNT [39].

In addition to various strategies applied to enhance the dispersion of CNTs, certain PNC structures such as double percolation, sandwich, foam and segregated structures have been studied in recent years to achieve high EMI SE at low CNT concentration [40–43].

A segregated composite structure was designed with CNT-coated polydimethylsiloxane (PDMS) microspheres. In the X-band region, the nanocomposite showed 47 dB EMI SE at 2.2 vol% CNT content. Moreover, absorption was obtained as the dominant shielding mechanism [40]. Wang et al. developed an EMI shielding material reinforced with CNTs using dynamically crosslinking polyurethane Diels-Alder (PUDA) bonds as the matrix. The PUDA granules and CNTs were mechanically blended in a planetary ball mill and then hot-pressed to produce segregated CNT/PUDA nanocomposites. Using this technique, conducting networks with selectively distributed CNTs were established at the interfaces of the PUDA regions, resulting in an EMI SE of 41 dB in the X-band with a CNT loading of 7 wt% [44]. The segregated PLA/CNT nanocomposites were reported to exhibit a high electrical conductivity of 12.0 S/m at 0.6 vol% CNT loading. This result is due to the conductive network structure obtained at a low percolation threshold of 0.04 vol% in the segregated structure. This high conductivity value resulted in a high EMI SE of 35.5 dB [45]. The segregated structures provide a large number of interfaces for absorption and especially multiple reflection of the incident EM waves, and thus an absorption-dominated shielding mechanism is achieved.

Another favourable structure for the absorption and multiple reflections of EMI is the foam structure. The light weight, high porosity and high EMI SE values of foam composites are especially prominent in areas such as stealth technology in fighter jets where light weight is important [14]. Chen et al. fabricated co-continuous polystyrene (PS)/polymethyl methacrylate (PMMA)/MWCNT nanocomposites by melt compounding and supercritical carbon dioxide ($scCO_2$) foaming. A porous and double-percolated structure resulted in strong conductivity (1.6 S/m) and 23 dB EMI SE at a low MWCNT concentration of 1.61 vol% [41]. Wang et al. prepared CNT/polyimide (PI) foam and investigated its properties. During foam preparation, they used poly(vinyl pyrrolidone) (PVP), a surfactant that can be removed in situ during polymerisation, resulting in a very light foam with a density of 32.1 mg/cm^3. It was emphasised that with this method, CNTs formed excellent conductive networks and thus the EMI SE reached 41 dB. As expected, the dominant shielding mechanism was absorption [46]. Besides these structures, in a novel study, CNT/polytetrafluoroethylene (PTFE) double nanofibrous membranes were fabricated using a shear-induced in situ fibrillation method and their EMI SE performance was evaluated. With this technique, a spider web-like structure was obtained from CNTs and PTFE fibrils, resulting in high conductivity and EMI SE of 25–42 dB in the thickness range of 100–520 μm. The obtained films maintained their EMI shielding performance even under harsh conditions, with high hydrophobicity and chemical resistance. It was also mentioned that the films exhibited fast and effective thermal conversion abilities, thus imparting self-heating and de-icing performance. These multifunctional materials offer great potential in sectors such as wearable electronics and aerospace [47].

For the preparation of EMI shielding materials, 3D printing technology is being researched as an additive manufacturing (AM) method where the structure can be modified as needed [48, 49]. Xu et al. [49] prepared PLA/MWCNT nanocomposites by solution mixing followed by filament extrusion and 3D printing. EMI SE increased with increasing MWCNT content and reached 47 dB at 17 wt% MWCNT. The data of some nanocomposites including CNT are given in Table 9.2.

TABLE 9.2 Summary of PNCs Including Carbon Nanotubes

Nanocomposite	CNT Concentration	Preparation Method	Thickness (mm)	Frequency (GHz)	EMI SE (dB)	Reference
PVDF-HFP/ MWCNT	15 wt%	Solvent casting	0.15	8.2–12.4	36	[50]
PAI/MWCNT	5 wt%	Solvent casting	1.2	8.2–12.4	17	[33]
TPU/MWCNT-L	10 wt%	Solution blending	0.8	0–18	37	[32]
TPU/MWCNT	20 phr	Extrusion + solvent casting	0.15	8–12	37	[36]
PC/ABS/ MWCNT	10 wt%	Extrusion	2	2	12	[51]
PP/MWCNT	5 wt%	Extrusion	2	8.2–12.4	21	[52]
PA11/PLA/ MWCNT	5 wt%	Melt blending	2	8.2–12.4	5	[53]
PA12/ MWCNT(H)	10 wt%	Melt blending	–	8–18	29.4	[38]
PK/CNT-g-PK	5 wt%	Melt blending	1	8–12	22.4	[39]
PLA/PCL/CNT	8 wt%	Melt blending	1	8.2–12.4	35.6	[37]
PHBV/CNT	5 wt%	Melt blending	3	8.2–12.4	42	[34]
PLA/HDPE/CNT	10 vol%	Melt blending	1	8.2–12.4	43	[35]
PDMS/MWCNT	10 wt%	Paste mixing/ milling	1	1–2	30	[31]
ABS/CNT	5 wt%	Foaming	2	8.2–12.4	28	[43]
PS/PMMA/ MWCNT	1.6 vol%	Foaming	2	8.2–12.4	23.1	[41]
ABS/CNT	6 wt%	3D printing	2.5	8.2–12.4	4	[48]
PLA/CNT	17 wt%	3D printing	3	8.2–12.4	47	[49]
TPU/CNT	4 wt%	3D printing/ compression	1	8.2–12.4	43	[54]

Note: ABS: acrylonitrile butadiene styrene, HDPE: high density polyethylene, PA11: polyamide 11, PAI: poly (amide imide), PC: polycarbonate, PCL: polycaprolactone, PK: polyketone, PMMA: polymethyl methacrylate, PVDF-HFP: poly(vinylidene fluoride-co-hexafluoropropylene).

9.2.1.3 Carbon Nanofibre-Reinforced PNCs for EMI Shielding

Another electrically conductive carbon nanostructure is the CNF. CNFs are cylindrical structures formed by stacking graphene layers in different variations. CNF, whose structure resembles CNT and carbon fibre (CF), has a dimension varying between them. Thanks to its nano-size and high aspect ratio, it exhibits superior electrical, chemical and mechanical properties compared with CF [9, 55]. On the other hand, there are significant differences in size and structure between CNT and CNF. There are more edges thanks to the graphene layers stacked on the outer wall of the CNF. This facilitates the bonding of functional groups to the CNF and the transfer of electrons [56]. CNFs, which stand out for their properties such as high electrical and thermal conductivity, thermal stability and lightness, are used in EMI shielding, sensor and energy applications. Numerous studies use it as a filler in PCs and evaluate its EMI performance.

Mohan et al. produced epoxy nanocomposites containing CNF in various ratios and studied their properties using the solution mixing method. DC conductivity measurements revealed that the percolation threshold of the epoxy/CNF nanocomposite was 1.5 wt%. Dielectric measurements indicated the presence of dipolar and interfacial polarisations. EMI SE was measured in the X and Ku bands using the waveguide method. A 0.5-mm-thick sample containing 8 wt% CNF exhibited 30 and 35 dB SE in the X and Ku bands, respectively [56]. Another study evaluated the electrical and EMI shielding performance of 5–20 wt% CNF/TPU nanocomposites produced by solution casting. The 15 wt% CNF nanocomposite demonstrated the optimum performance in terms of morphological and electrical properties, with an EMI SE in the range of 30–42 dB in the X-band. It was emphasised that flexible films produced with a thickness of 0.7 mm can be used as shields in the coating and packaging of electronic devices [57]. Another epoxy/CNF nanocomposite was fabricated and studied by Chakradhary and Akhtar. A 4 wt% CNF provided 12 dB EMI SE to a 2-mm-thick sample [58]. In these two epoxy/CNF nanocomposites produced with similar techniques, a low amount of CNF provided low EMI SE. Serrato and co-workers reported on the performance of PVDF/CNF nanocomposites fabricated using a Banbury mixer and compression moulding. The electrical conductivity enhanced with the amount of CNF, reaching a percolation threshold at 5 wt%. The EMI SE value over the range of 30 kHz to 1.5 GHz was reported to vary from 6 to 16 dB, with some inconsistencies due to agglomeration [59].

As can be seen, it is noteworthy that very high EMI SE values cannot generally be achieved with the addition of CNF. As with other carbon nanostructures, the distribution and interaction with the matrix play a critical role in the electrical properties of CNF-reinforced nanocomposites. In fact, it is expected to provide strong interaction with the matrix due to the stacked-cup structure, which is mentioned as a difference from CNT and causes more edge formation on the surface. However, it is known that CNF tends to aggregate due to its nano-size and large surface area and has weak interaction with the polymer matrix [60, 61]. The distribution of CNFs in the matrix varies depending on the CNF surface treatments, process techniques and conditions. In techniques such as solution casting and melt blending, many factors such as temperature, pressure, mixing speed and time affect the distribution. Mondal et al. used two different manufacturing techniques to improve the dispersion of CNF in chlorinated polyethylene (CPE) and compared the properties of the final material. In the first technique, CPE and CNF were directly melt mixed and then compression moulded. In the second technique, CPE/TPU was first solution cast into film and then cut into small pieces. These pieces were then melted blended and compression moulded. The second technique provided better CNF distribution, resulting in higher electrical conductivity, EMI SE and tensile strength. Solution-cum-melt processed samples exhibited EMI SE of 24 dB at 10 wt% CNF, whereas melt-processed samples exhibited 22 dB [61]. The data of some nanocomposites including CNF are summarised in Table 9.3.

9.2.1.4 Graphene- and Derivatives-Reinforced PNCs for EMI Shielding

Graphene is a two-dimensional single-layer graphite sheet consisting of sp2-bonded carbon atoms arranged in a honeycomb structure. It can be considered the building block of all sp2-structured carbon allotropes. 1D CNT can be obtained by rolling it into a cylinder,

TABLE 9.3 Summary of PNCs Including Carbon Nanofibres

Nanocomposite	CNF Concentration (wt%)	Preparation Method	Thickness (mm)	Frequency (GHz)	EMI SE (dB)	Reference
Epoxy/CNF	15	Solvent casting	0.7	12.4	42	[57]
Epoxy/CNF	4	Solvent casting	2	8.2–12.4	12	[58]
PVDF/CNF	15	Melt mixing	0.82	30 kHz–1.5 GHz	16.4	[59]
CPE/CNF	10	Melt mixing	1	8.2–12.4	22	[61]
CPE/CNF	10	Solution + melt mixing	1	8.2–12.4	24	[61]
PVDF/CNF	8	Melt spinning	0.23	0.015–3	9	[62]
PBS/CNF	7	Melt mixing	2.5	8.2–12.4	25	[60]
PBS/CNF	7	Melt + foaming	2.5	8.2–12.4	14	[60]

Note: CPE: chlorinated polyethylene, PBS: poly(butylene succinate), PVDF: poly(vinylidene fluoride).

and 3D graphite can be obtained by depositing multiple layers of graphene [9, 63]. Graphene is one of the strongest materials known, with a tensile strength of 130 GPa and a Young's modulus of 1 TPa. It has a high specific surface area of 2600 m^2/g, and electrical and thermal conductivity of ~6000 S/cm and ~5000 W/m·K, respectively. Thanks to its extraordinary charge carrier mobility and other excellent properties, graphene-based studies are increasingly intensifying in energy storage, sensors and EMI shielding materials. While monolayer graphene cannot be used much due to the difficulty of producing large quantities and good quality, derivatives such as graphene nanosheets (GNPs), graphene oxide (GO) and reduced graphene oxide (rGO) are preferred in CPC manufacturing [20, 63]. GO has oxygenated functional groups such as hydroxyl, ether, diol and epoxy. This makes it remarkable for obtaining uniform PNCs. On the other hand, the electrical conductivity of GO is low due to these functional groups [20].

rGO is often studied for EMI shielding applications thanks to its higher conductivity. For example, PP nanocomposites containing rGO in different proportions (0.5–5 wt%) were prepared by extrusion, and their electrical properties and EMI performance were evaluated. It has been reported that increasing the proportion of rGO increases the electrical conductivity. The nanocomposite containing 5 wt% rGO showed an EMI SE of 10.2 dB. This demonstrates that EMI can be blocked up to 90% [64]. Similar to other nanofillers, the challenge of distributing rGO within the polymer has prompted research into various kinds of polymeric structures. Oraby and co-workers [65] prepared polyurethane (PU) foams containing rGO in varying proportions. While an EMI SE of 23 dB was achieved at 5 wt% rGO content in the X-band region, it was found that rGO at this level negatively affected the mechanical properties. The optimum amount of rGO to provide acceptable mechanical properties was reported to be 3 wt%, and the EMI SE of this foam was 17 dB. Das et al. synthesised 10 wt% sulphur-doped rGO (S-RGO)-reinforced polyaniline (PANI) by in situ chemical oxidative polymerisation of aniline. Then, they used this nanopowder as a filler in paraffin. The PANI/S-RGO/Paraffin nanocomposite exhibited much higher

EMI SE (22 dB) compared with GO/Paraffin, S-RGO/Paraffin and PANI/Paraffin composites [66]. It can be concluded from this that combining nanofillers with a conductive polymer is an efficient way to improve EMI SE. Another study combining rGO and PANI was carried out by Zubair et al. rGO was coated with PANI (RGO-PANI), and this nanofiller was added (5–40 wt%) to polyvinyl chloride (PVC) to produce samples by solution casting. The authors reported that the PANI coating enhanced the interaction of rGO with PVC and its dispersion in PVC. They found that at the highest filler content (40 wt%), the EMI SE was 104 dB in the 0–20 Hz range [67]. In the study conducted by Park et al., the influence of RGO on the properties of unsaturated polyester resin (UPR) was investigated. They observed that the addition of 0.1–0.3 wt% RGO delayed the curing rate of UPR while improving its thermal stability. The nanocomposite with a percolation threshold in the range of 0.2–0.3 wt% was reported to have a conductivity of 1.9×10^{-6} S/cm at 8 GHz and an EMI SE of 9 dB [68].

GO is another nanofiller used in the preparation of polymeric shield materials. Das et al. prepared PANI-polystyrene (PS)-GO samples with a thickness of 0.85 mm by solution casting. The percolation threshold of the films was found to be 0.09 wt%, while a very efficient EMI SE performance was obtained in the X-band. A 40 dB EMI SE was measured for the PANI-PS-GO film containing 1.5 wt% GO [69]. The study revealed the synergistic effect of the combined use of conductive polymer and nanofiller. Another approach to improve the distribution of the nanofiller in the matrix, and hence the electrical and EMI shielding performance, is to graft the filler surface with suitable functional groups or polymers. Ahmad et al. functionalised GO with sulfonic acid (S-GO) to improve its dispersion in the polyether sulfone (PES) matrix and thus the properties of the nanocomposite. The interfacial polarisation and microcapacitor effects significantly increased the dielectric permittivity of PES with the addition of S-GO. The EMI SE value also gradually increased with the increasing amount of S-GO, and the nanocomposite containing 5 wt% S-GO exhibited an EMI SE of 23 dB [70].

GNPs, which are multilayer graphene structures, are also being researched for EMI applications. In particular, it has been suggested that GNPs can induce multiple reflections within the matrix due to their layered structure [71]. Conversely, GNPs tend to restack and aggregate due to their large surface area, interlayer van der Waals forces and π-π interactions. To overcome this problem, Lee et al. prepared PP nanocomposites containing 3 wt% GNPs by high-shear extrusion. The effects of screw speeds varying between 1000 and 3000 rpm were studied and compared with 200 rpm. As the shear rate increased, the size of the GNP aggregates decreased, and their interfacial polarisation improved. As a result, the increased dielectric losses improved the EMI SE. At 200 rpm, the nanocomposite was almost transparent to EMI, increasing to 5 dB at 3000 rpm [72]. Zhi et al. also evaluated the effects of graphene dispersion methods on EMI SE in graphene/bio-based epoxy nanocomposites. They used four different methods to disperse graphene: high-shear mixing, hand mixing, ultrasonication and high-shear mixing with ultrasonication. Graphene was varied as 3, 6 and 9 wt%, and 9 wt% graphene exhibited the highest EMI SE (8 dB) in all dispersion methods. The dispersion type did not provide any significant change in EMI SE [73]. The data of some nanocomposites including graphene derivatives are given in Table 9.4.

9.2.2 MXene-Reinforced Nanocomposites for EMI Shielding

MXenes are two-dimensional nanomaterials consisting of transition metal carbides, nitrides or carbonitrides with the formula $M_{n+1}X_nT_z$. In the formula, M represents transition metals, X carbon or nitrogen and T_z functional groups. MXenes are synthesised by selectively etching the A layer (IIIA or VIA elements such as Al and Si) from a MAX phase structure. Different combinations of M and X produce different MXenes [9, 17, 78]. MXenes, which exhibit hydrophilic properties due to their surface terminal groups, are water dispersible and compatible with polymer matrices. They have high thermal and electrical conductivity and, depending on the M type, can also exhibit magnetic properties [78]. MXenes are perfect candidates for EMI shielding applications due to their metallic conductivity, layered structure and considerable polarisation loss caused by excellent interlayer electrical connections. $Ti_3C_2T_z$ is the most commonly studied MXene type [9, 78].

MXenes were incorporated into various polymer matrices and the EMI shielding performance of these nanocomposites was investigated in the last few years. As discussed, polymer/MXene nanocomposites can be designed in the form of films, hydrogels, tissues, foams and blocks using various methods including solution mixing, emulsion mixing, lamination and hot pressing [79, 80]. MXenes are usually synthesised with fluoride solutions and are hydrophilic, but dried MXenes lose their hydrophilicity and become difficult to redisperse. It is known that MXene, like carbon nanostructures, tends to weakly interact with polymeric matrices and to aggregate in the structure. This problem is addressed by changing the process technique and conditions, as well as by MXene surface treatments and similar improvement strategies [81, 82]. Solution mixing is frequently chosen for its simplicity, efficiency in dispersing MXenes and ability to utilise MXene hydrophilicity for compatibility [79]. This technique is commonly used with hydrolysable or soluble polar polymers such as polyvinyl alcohol [83], polyethylene glycol [84] and PVDF. For instance, MXene nanoflakes grafted with polyaniline (PA) nanofibres were synthesised, and PVDF films with different MXene amounts were produced by solution casting. The nanocomposite films exhibited remarkable electrical conductivities around 0.2 S/cm at 6.9 vol% MXene content. While PVDF/MXene without surface modification exhibited 29 dB EMI SE in the P-band (12–18 GHz), this value increased to 38 dB with surface modification. The improvements obtained were attributed to the enhanced exfoliation of the nanoflakes by grafting MXene with PA nanofibres, the increased polarisation due to the creation of capacitor-like structures between the nanoflakes and the polymer chains, and the intrinsic effects of the PA nanofibres [85]. In another paper, Wu and co-workers fabricated PVDF/MXene nanocomposites by solution casting. They functionalised the surface of MXene with cetyltrimethylammonium bromide (MXene@CTAB) to ensure proper dispersion and interaction of MXene in the matrix. Functionalisation improved the homogeneous dispersion and compatibility of MXene in PVDF [82].

In addition to optimising the dispersion and interaction of MXene within the polymer matrix, another key challenge is to ensure the durability of the material. Depending on the application environment, EMI shielding materials may be exposed to harsh conditions such as salt, humidity and high temperatures. MXenes can oxidise to TiO_2 when exposed to high humidity and temperature, leading to a deterioration in both electrical conductivity and EMI SE of the material. To improve the service life of polymer/MXene nanocomposites as EMI shields under these conditions, it is essential to mitigate oxidation and improve environmental durability [15, 81]. Yan et al. proposed to functionalise MXene with polyhedral oligomeric silsesquioxane (POSS) to prevent oxidation of MXene and improve its interfacial interaction with the PI matrix. They prepared MXene-POSS/PI nanocomposites by the in situ polymerisation method and investigated their properties. It was reported that functionalisation with POSS improved compatibility with PI and the film with 20 wt% MXene-POSS content exhibited 12.7 dB EMI SE. In addition, the films exhibited hydrophobic properties and high flame resistance. It was reported that these multifunctional films could be used in electronic devices designed for use in extreme outdoor environments [81].

Several studies have reported on the preparation of polymer/MXene nanocomposites using different methods to achieve high EMI SE. Wan et al. reported a study about (poly(3,4-ethylenedioxythiophene)/poly(styrene sulphonate)) (PEDOT:PSS)/MXene (Ti_3AlC_2) nanocomposite films produced by the vacuum-assisted filtration technique. The produced film with 6.6 mm thickness showed an EMI SE of 40 dB [86]. In another study, Jia and co-workers developed an MXene-based shielding material to prevent secondary EMI pollution caused by reflection. They synthesised MXene-decorated polymer foam beads (MPFBs) with the assistance of polyaniline (PANI) and constructed CPC foams by assembling MPFBs into 3D accumulations with MXene networks followed by encapsulation with polydimethylsiloxane (PDMS). With this design, they achieved 23 dB total EMI SE and ultra-low reflection at 0.0225 vol% MXene content [87]. An effective method for improved EMI shielding performance was presented by Xu et al. The method involves dip coating MXenes onto a polyethyleneimine-grafted PP textile followed by vacuum-assisted compression moulding. The PP/MXene nanocomposite was reported to exhibit high electrical conductivity at 2.12 vol% MXene loading and 60 dB EMI SE in the X-band [88]. In another study using MXene as a coating, MXene-coated and cross-linked PI (C-MXene@PI) composite foams were prepared by dip coating and chemical crosslinking. With high thermal stability, the C-MXene@PI sample exhibited EMI SE in the X-band range from 22 to 62 dB [89].

The segregated structure encountered in carbon nanostructure-reinforced composites is also used in MXene-reinforced nanocomposites. Lee et al. presented a novel method for the preparation of PC/MXene nanocomposites with segregated structures. MXene was added to positively charged polycarbonate (*p*-PC) microspheres and the *p*-PC/MXene microspheres were prepared by vacuum-assisted filtration and dried. In the final step, these microspheres were hot-pressed and segregated structures were obtained. The authors emphasised that 20 dB EMI SE was achieved in the X-band at 1.15 vol% MXene loading thanks to the segregated structure, and mentioned that these nanocomposites could be used in areas such as TV broadcasting and military applications [90].

Another technique that has been investigated in recent years for the production of MXene/polymer nanocomposite EMI shielding materials is electrospinning [91–93]. Liu et al. fabricated MXene-containing poly(vinylidene fluoride-co-hexafluoropropylene) (PVDF-HFP) nanocomposites by a one-step electrospinning technique for use as wearable

TABLE 9.4 Summary of PNCs Including Graphene Derivatives

Nanocomposite	Graphene Concentration	Preparation Method	Thickness (mm)	Frequency (GHz)	EMI SE (dB)	Reference
ABS/GNP	15 wt%	Extrusion	2	8.2–12.4	30	[74]
PP/GNP	3 wt%	High-shear extrusion	2	8.2–12.4	5	[72]
PA11/PLA/GNP	10 wt%	Melt blending	2	8.2–12.4	6	[53]
WPU/GN	5 vol%	Solvent casting	–	8.2–12.4	38	[75]
Epoxy/graphene	2.5 wt%	Solution mixing	2	8.2–12.4	29	[76]
PU/rGO	5 wt%	Foaming	5	8.2–12.4	23	[65]
PVA/S-RGO	2 phr	Solvent casting	0.3	2	25	[77]
PANI/PS/GO	1.5 wt%	Solvent casting	0.85	8.2–12.4	40	[69]
PES/S-GO	5 wt%	Ex-situ polymerisation	0.2	8.2–12.4	23	[70]

Note: GN: graphene nanosheet, S-GO: sulfonic-functionalised graphene oxide, S-RGO: selectively reduced graphene oxide, WPU: waterborne polyurethane.

EMI shields. The PVDF-HFP/MXene fabric is characterised by high air permeability, hydrophobicity, corrosion resistance and flexibility. The EMI SE of the 95-μm-thick fabric was measured at approximately 160 dB/mm in the X-band, with reflection identified as the primary shielding mechanism. Given that these fabrics are intended for wearable applications, their performance when exposed to corrosive liquids was evaluated. EMI SE measurements were taken after immersion in acid, alkaline and salt solutions. The fabric demonstrated minimal loss of performance after prolonged abrasion [92].

An innovative approach that has attracted attention in recent years for the production of MXene-containing CPCs is three-dimensional (3D) printing [94, 95]. Hong et al. [95] developed polyethylene oxide (PEO)/MXene inks and designed EMI shielding materials using liquid deposition modelling (LDM) 3D printing. The EMI SE of the developed nanocomposites was investigated in the frequency range of 1.5–10 GHz. An outstanding EMI SE of ~231 dB/mm was achieved in the X-band. As can be seen, MXenes, an innovative nanofiller, are being widely investigated for use in EMI shielding applications. However, several key challenges need to be overcome to enable the large-scale production of MXene-based nanocomposites. While the toxic solvents used in MXene synthesis pose a significant risk, as with all nanofillers, the dispersion of MXenes in hydrophobic polymer matrices and oxidation problems are other issues that need to be addressed. Table 9.5 depicts a summary of Mxene based PNCs with their preparation technique and SE of the prepared samples.

9.2.3 Hybrid Nanocomposites for EMI Shielding

Many recent studies have shown that greater property enhancements can be achieved by adding a hybrid combination of carbon nanomaterials or other metallic fillers to the polymer matrix. Hybrid composites are the intended combination of various fillers at the micro- or nanoscale, enabling the creation of advanced PCs with emergent properties that

TABLE 9.5 Summary of PNCs Including MXenes

Nanocomposite	MXene Concentration	Preparation Method	Thickness (mm)	Frequency (GHz)	EMI SE (dB)	Reference
PA-MXene/PVDF	6.9 vol%	Solution mixing	–	12.4–18	38	[85]
MXene/graphene@ Fe_3O_4/PVA	25 wt%	Solution mixing	1	8.2–12.4	36	[83]
MXene/PVDF	50 wt%	Solution mixing/ compression	2	8.2–12.4	48	[96]
MXene/POSS/PI	12 wt%	In situ polymerisation	–	8.2–12.4	12	[81]
PEDOT:PSS/ MXene	90 wt%	Vacuum-assisted filtration	6.6	12.4–18.0	40	[86]
PP/MXene	2.12 vol%	Dip coating	1.93	8.2–12.4	60	[88]
MXene/PI	30 wt%	In situ polymerisation	0.11	8.2–12.4	34	[97]
MXene/BC	50 wt%	vacuum-assisted filtration	0.11	8.2–12.4	43	[98]
MXene/rCF/PLA	1 wt%	3D printing	10	8.2–12.4	30	[94]
MXene/NR	1.2 vol%	Solution mixing	0.05	8.2–12.4	34	[99]

Note: BC: bacterial cellulose, NR: natural rubber, PA-MXene: polyaniline-grafted MXene, PEDOT: poly(3,4-ethylenedioxythiophene), PI: polyimide, PP: polypropylene, PSS: poly(styrene sulfonate), rCF: recycled carbon fibre.

may not be inherent to the individual components. These composites can exhibit complex interactions between the fillers, yielding both synergistic and antagonistic effects. Maximum compatibility between the components will lead to high hybridisation and synergistic effect. The main motivations for hybrid composites are even higher property enhancement through synergistic effects, achieving multifunctionality and cost reduction [63, 100, 101].

In terms of electrical properties and EMI shielding, the main advantage of hybrid composites is that conductive composites can be prepared without compromising mechanical and thermal properties, and high conductivity is obtained at low percolation threshold. In hybrid composites prepared with multi-scale or single-scale filler, the fillers can also improve the dispersion of each other. To enhance the properties of fibre-reinforced PCs and effectively incorporate nanofillers without significantly increasing matrix viscosity or cost, it is advantageous to adopt a hybrid approach that combines microscale fibres with nanoscale fillers [63, 100, 102]. This approach optimises the performance of the composites while overcoming the challenges of viscosity and cost. For example, while the electrical conductivity of CF-reinforced composites is limited, it can be improved by using CNTs, which can build another electrical network within the matrix. The bridging effect between CF and CNTs may lead to novel electron paths [103, 104]. A similar enhancement was also reported for PEEK/CF/MXene composites [105]. In addition to multi-scale composites, the combination of nanoscale fillers is also important for EMI shielding applications. For instance, in the combined use of CNTs and GNPs, both of which are nanoscale but with different geometries, CNTs are expected to promote exfoliation of graphene layers by bridging between the layers and create a physical three-dimensional (3D) structure that facilitates the creation of a conductive pathway [106, 107].

EMI shielding materials prepared using CNT and graphene as hybrid fillers are frequently studied [106–112]. Silva et al. developed hybrid CNT/GNP-reinforced PVDF nanocomposites using melt mixing. The hybrid composites made by maintaining the total filler content constant at 3 wt% exhibited conductivities ranging from 10^{-3} S/m (PVDF/GNP) to 10^{-1} S/m (PVDF/CNT). Increasing the amount of CNT in the hybrid filler enhanced the EMI SE and assisted with the reflection mechanism. EMI SE experiments on multilayer composites revealed a performance of up to 31 dB at 15 GHz [106]. Anjos and co-workers prepared PC/ABS hybrid nanocomposites containing three different graphenes with varying lateral dimensions and thicknesses and a single MWCNT by extrusion and injection moulding. They observed that the incorporation of MWCNT improves the exfoliation and dispersion of graphene platelets in the matrix. This was attributed to the bonding of graphene and MWCNTs through π-π interactions, and to the increased viscosity with the inclusion of MWCNTs in the extrusion process. The addition of 1.5 wt% MWCNTs to 3 wt% graphene resulted in a synergistic effect and raised SE up to 20 dB in the X-band [110]. In another study, PMMA/multilayer graphene (MLG)/MWCNT and polyvinyl chloride (PVC)/MLG/MWCNT hybrid nanocomposites were fabricated by the solvent casting technique and their performances were evaluated. The PMMA nanocomposite containing 20 wt% MLG had an SE of approximately 20 dB in the X-band, while the SE of 10/10 wt% MLG/MWCNT/PMMA increased to approximately 40 dB. The synergistic effect on the shielding performance was clearly observed [113].

In another hybrid nanofiller study, the properties of CNT:CB-reinforced PP/PE nanocomposites prepared by melt blending were investigated. The results showed that CNT-containing nanocomposites exhibited higher EMI shielding performance than CNT:CB hybrid nanocomposites. Although a synergistic effect could not be achieved with hybrid doping, it was shown that the CNT:CB-doped nanocomposite is cheaper to obtain, providing a nanocomposite with an SE of 30 dB by cost calculation [114]. In a recent study, the properties of ABS nanocomposites containing CNT, CB and GNP were evaluated by Anjos et al. The nanocomposites containing 3 wt% single nanofiller, 3/3 wt% binary and 3/3/3, 2/2/2 and 1/1/1 wt% nanofiller were compounded by direct extrusion and masterbatch methods, then samples were prepared by injection moulding and material extrusion 3D printing. It was found that the processing method significantly affected the filler distribution and the nanocomposite morphology, and hence the final properties. The highest EMI SE was obtained in the 3/3/3 wt% hybrid nanocomposite in all processes. Nanocomposites prepared by direct extrusion showed higher EMI SE compared with masterbatch. Ternary nanocomposites prepared by 3D printing presented higher SE at 40 dB at 11.8 GHz compared with injection moulding [115]. Qiu et al. constructed a sandwich structure of PVDF films containing three different fillers to obtain a high-performance shielding material. The three-layer sandwich structure of GNP, nickel (Ni) and CNT-reinforced PVDF nanocomposites displayed superior EMI shielding performance due to electrical loss from conductive GNP and CNT, and magnetic loss from magnetic Ni and multiple reflections at several interfaces of the multilayer structure. The EMI SE of 0.6-mm-thick three-layer and six-layer sandwich composites at 15 GHz was measured as 41 and 46 dB, respectively. The SE value can be increased by changing the thickness and number of layers [116].

In addition to carbon nanostructures, there are also numerous investigations into the use of MXene in hybrid composites [105, 117–121]. Dehghan et al. reported in their study that the EMI SE performances of MXene/rGO/PP nanocomposites were improved by the synergistic effect of the hybrid composite [121]. Lee et al. also prepared GO-grafted MXene and reinforced PA6 with this hybrid filler. In the composites prepared with hybrid doping, synergistic effects were obtained in both mechanical properties and EMI SE compared with single composites [118].

9.3 CONCLUSION

This chapter provided an overview of EMI shielding materials based on PNCs reinforced with carbon nanostructures and MXenes. EMI can impair the performance of electronic equipment and cause potential health risks. As a result, effective shielding materials are essential to minimise the effects of this pollution. Therefore, polymer nanocomposites have been extensively studied and developed for EMI shielding. In particular, carbon nanostructures such as CB, CNFs, graphite, graphene, GO and carbon nanotubes are commonly studied nanostructures. In addition to these, MXene, an innovative alternative, has attracted attention in recent years. One of the main challenges in the use of these nanofillers is to ensure their proper dispersion within the matrix. Proper dispersion is essential to create efficient conductive networks within the matrix and to increase EMI SE. The majority of research has therefore focused on achieving high EMI shielding performance with a low percolation threshold without compromising other properties (e.g., mechanical). To this end, studies have been and are being carried out on various factors such as PNC production methods and parameters, design of different composite structures (foam, porous, segregated, laminated, 3D printed, electrospun, etc.) and surface treatment of nanofillers. In addition, the latest studies focus on hybrid and multifunctional EMI shielding materials. Polymeric electromagnetic shielding materials have the potential to be used in computers, mobile phones, satellite receivers and other electronic devices, military equipment and systems, advanced automobiles, buildings and infrastructure, energy facilities, medical devices, smart clothing and wearable devices due to their light weight, flexibility, formability and customisation properties.

REFERENCES

1. Thomas, S., Zachariah, S.M. 2022. *Nanostructured Materials for Electromagnetic Interference Shielding*. CRC Press.
2. Mamatha, G.M., Dixit, P., Krishna, R.H., Girish Kumar, S. 2024. Polymer based composites for electromagnetic interference (EMI) shielding: the role of magnetic fillers in effective attenuation of microwaves, a review. *Hybrid Advances*, 6, pp. 100200.
3. Suchea, M., Tudose, I.V., Pascariu, P., Koudoumas, E. 2020. Carbon-based nanocomposites for EMI shielding: recent advances. In: Joseph K., Wilson R., George G., editors. *Materials for Potential EMI Shielding Applications: Processing, Properties and Current Trends*, Elsevier Inc., pp. 201–211.
4. Tian, K., Hu, D., Wei, Q., Fu, Q., Deng, H. 2023. Recent progress on multifunctional electromagnetic interference shielding polymer composites. *Journal of Materials Science and Technology*, 134, pp. 106–131.

5. Xiong, X., Zhang, H., Lv, H., Yang, L., Liang, G., Zhang, J., et al. 2024. Recent progress in carbon-based materials and loss mechanisms for electromagnetic wave absorption. *Carbon*, 219, pp. 118834.
6. Ucpinar Durmaz, B., Salman, AO., Aytac, A. 2023. Electromagnetic interference shielding performances of carbon-fiber-reinforced PA11/PLA composites in the X-band frequency range. *ACS Omega*, 8, pp. 22762–22773.
7. Yan, X., Xiang, L., He, Q., Gu, J., Dang, J., Guo, J., et al. 2018. Electromagnetic interference shielding polymer nanocomposites. In: Guo, Z., Chen, Y., Lu, N.L., editors. *Multifunctional Nanocomposites for Energy and Environmental Applications*. 1st ed., Weinheim: Wiley VCH; pp. 567–601.
8. Jaroszewski, M., Thomas, S., Rane, A.V. 2018. *Advanced Materials for Electromagnetic Shielding*. 1st ed. Hoboken, NJ, USA: John Wiley & Sons, Inc.; pp. 327–365.
9. Omana, L., Chandran, A., John, R.E., Wilson, R., George, K.C., Unnikrishnan, N.V., et al. 2022. Recent advances in polymer nanocomposites for electromagnetic interference shielding: a review. *ACS Omega*, 7(30), pp. 25921–25947.
10. Zheng, S., Wang, Y., Zhu, Y., Zheng, C. 2024. Recent advances in structural design of conductive polymer composites for electromagnetic interference shielding. *Polymer Composites*, 45, pp. 43–76.
11. Jiang, D., Murugadoss, V., Wang, Y., Lin, J., Ding, T., Wang, Z., et al. 2019. Electromagnetic interference shielding polymers and nanocomposites – a review. *Polymer Reviews*, 59, pp. 280–337.
12. Mergen, Ö.B., Umut, E., Arda, E., Kara, S. 2020. A comparative study on the AC/DC conductivity, dielectric and optical properties of polystyrene/graphene nanoplatelets (PS/GNP) and multi-walled carbon nanotube (PS/MWCNT) nanocomposites. *Polymer Testing*, 90, pp. 106682.
13. Dubey, P., Gupta, M., Kundalwal, S.I. 2024. Conductive polymer nanocomposite incorporated with carbon nanotubes for effective electromagnetic interference shielding: a numerical study. *Polymer Composites*, 45(4), pp. 3576–3590.
14. Idumah, C.I. 2023. Recent advancements in electromagnetic interference shielding of polymer and MXene nanocomposites. *Polymer-Plastics Technology and Materials*, 62(1), pp. 19–53.
15. Song, P., Liu, B., Qiu, H., Shi, X., Cao, D., Gu, J. 2021. MXenes for polymer matrix electromagnetic interference shielding composites: a review. *Composites Communications*, 24, pp. 100653.
16. Nguyen, Q.D., Choi, C.G. 2024. Recent advances in multifunctional electromagnetic interference shielding materials. *Heliyon*, 10, pp. e31118.
17. Kallambadi Sadashivappa, P., Venkatachalam, R., Pothu, R., Boddula, R., Banerjee, P., Naik, R., et al. 2023. Progressive review of functional nanomaterials-based polymer nanocomposites for efficient EMI shielding. *Journal of Composites Science*, 7(2), pp. 77.
18. Tjong, SC. 2012. *Polymer Composites with Carbonaceous Nanofillers: Properties and Applications*. John Wiley & Sons.
19. Choudhury, S.N., Das, P., Bhawal, P., Pal, A., Banerji, P., Das, N.C. 2023. Double percolation behavior through the preferential distribution of conductive black in polymer blends to boost electrical properties and EMI shielding effectiveness. *Materials Today Communications*, 35, pp. 106109.
20. Poothanari, M.A., Pottathara, Y.B., Thomas, S. 2019. Carbon nanostructures for electromagnetic shielding applications. In: Thomas S., Grohen Y., Pottathara Y.B., editors. *Industrial Applications of Nanomaterials*. 1st ed., Netherlands: Elsevier; pp. 205–223.
21. Maryam, A., Rizwan, Z., Nasir, N., Ahmad, J., Imran Khan, M., Nawab, Y., et al. 2023. Optimization of percolation limit of carbon black for electromagnetic interference shielding. *Journal of Magnetism and Magnetic Materials*, 586, pp. 171164.
22. Katheria, A., Das, P., Paul, S., Nath, K., Ghosh, S.K., Das, N.C. 2023. Preferential localization of conductive filler in ethylene-co-methyl acrylate/thermoplastic polyolefin polymer blends to

reduce percolation threshold and enhanced electromagnetic radiation shielding over K band region. *Polymer Composites*, 44, pp. 1603–1616.
23. Choudhury, S.N., Nayak, J., Das, P., Pal, A., Katheria, A., Banerji, P., et al. 2024. Enhancing electrical characteristics and electromagnetic interference shielding effectiveness in thermoplastic elastomeric polymer blends by utilizing the selective distribution of conductive black. *Functional Composite Materials*, 5, pp. 4.
24. Kim, J., Kim, G., Kim, S.Y., Lee, S., Kim, Y., Lee, J., et al. 2021. Fabrication of highly flexible electromagnetic interference shielding polyimide carbon black composite using hot-pressing method. *Composites Part B: Engineering*, 221, pp. 109010.
25. Sarkar, K., Khanra, S., Sharma, S., Ghorai, S.K., Chattopadhyay, S. 2023. Self-assembled 3D-microstructured dual carbon black filled polymer nanocomposite-coated fabric for tunable electromagnetic interference shielding. *Journal of the Textile Institute*, 114(2), pp. 207–216.
26. Wu, M., Wu, F., Ren, Q., Jia, X., Luo, H., Shen, B., et al. 2022. Tunable electromagnetic interference shielding performance of polypropylene/carbon black composites via introducing microcellular structure. *Composites Communications*, 36, pp. 101363.
27. Rahaman, M. 2023. Superior mechanical, electrical, dielectric, and EMI shielding properties of ethylene propylene diene monomer (EPDM) based carbon black composites. *RSC Advances*, 13, pp. 25443–25458.
28. Nath, K., Ghosh, S.K., Katheria, A., Das, P., Das, N.C. 2022. Facile production of binary polymer/carbonic nanofiller-based biodegradable electromagnetic interference shield films with low electrical percolation threshold. *Polymer Engineering and Science*, 62, pp. 3841–3857.
29. Soares, B.G., Touchaleaume, F., Calheiros, L.F., Barra, G.M.O. 2016. Effect of double percolation on the electrical properties and electromagnetic interference shielding effectiveness of carbon-black-loaded polystyrene/ethylene vinyl acetate copolymer blends. *Journal of Applied Polymer Science*, 133(7), pp. 1–10.
30. Ravindren, R., Mondal, S., Bhawal, P., Ali, S.M.N., Das, N.C. 2019. Superior electromagnetic interference shielding effectiveness and low percolation threshold through the preferential distribution of carbon black in the highly flexible polymer blend composites. *Polymer Composites*, 40(4), pp. 1404–1418.
31. Lee, D.K., Yoo, J., Kim, H., Park, S.H. 2024. Investigation of electromagnetic interference shielding effectiveness and electrical percolation of carbon nanotube polymer composites with various aspect ratios. *Carbon Letters*, 34, pp. 133–140.
32. Shin, B., Mondal, S., Lee, M., Kim, S., Huh, Y. Il., Nah, C. 2021. Flexible thermoplastic polyurethane-carbon nanotube composites for electromagnetic interference shielding and thermal management. *Chemical Engineering Journal*, 418, pp. 129282.
33. Rani, P., Malik, R.S. 2024. Efficient electromagnetic interference shielding performance of poly (amide imide)/MWCNT-based nanocomposites. *Journal of Applied Polymer Science*, 141(22), pp. e55442.
34. Luo, J., Sun, W., Zhou, H., Zhang, Y., Wen, B., Xin, C. 2021. Bioderived and biodegradable Poly(3-hydroxybutyrate-co-3-hydroxyvalerate) nanocomposites based on carbon nanotubes: microstructure observation and EMI shielding property improvement. *ACS Sustainable Chemistry and Engineering*, 9(32), pp. 10785–10798.
35. Al-Saleh, M.H., El-Methaly, H.M. 2024. Influence of PLA/HDPE ratio and CNT content on the morphology, electrical, and EMI shielding of CNT-filled PLA/HDPE blends. *Synthetic Metals*, 304, pp. 117592.
36. Jyoti, J., Chauhan, G.S., Yang, S., Tripathi, S.K., Kim, K.H., Sandhu, M., et al. 2023. Stretchable and lightweight MWCNTs/TPU composites films with excellent electromagnetic interference shielding and dynamic mechanical properties. *Journal of Polymer Research*, 30, pp. 322.
37. Huang, B., Wang, Z., Tu, J., Liu, C., Xu, P., Ding, Y. 2023. Interfacial distribution and compatibilization of imidazolium functionalized CNTs in poly(lactic acid)/polycaprolactone composites with excellent EMI shielding and mechanical properties. *International Journal of Biological Macromolecules*, 227, pp. 1182–1190.

38. Lee, J.H., Jang, S.H., Ju, S.Y., Kim, E., Park, N.H. 2023. Electromagnetic interference shielding effects of modified carbon nanotube-nylon 12 composite films. *Polymer-Plastics Technology and Materials*, 62(12), pp. 1610–1620.
39. Tang, F., Kim, S.J., Eom, T.G., Seo, M., Song, Y.G., Park, J.H., et al. 2023. Impacts of polyketone-grafted carbon nanotube on the thermal, mechanical, and electromagnetic shielding performance of polyketone composites. *European Polymer Journal*, 201, pp. 112554.
40. Ma, R.Y., Yi, S.Q., Li, J., Zhang, J. Le., Sun, W.J., Jia, L.C., et al. 2023. Highly efficient electromagnetic interference shielding and superior mechanical performance of carbon nanotube/polydimethylsiloxane composite with interface-reinforced segregated structure. *Composites Science and Technology*, 232, pp. 109874.
41. Chen, J., Liao, X., Li, S., Wang, W., Guo, F., Li, G. 2020. A promising strategy for efficient electromagnetic interference shielding by designing a porous double-percolated structure in MWCNT/polymer-based composites. *Composites Part A: Applied Science and Manufacturing*, 138, pp. 106059.
42. Wu, J., Gao, Y. 2024. High-performance electromagnetic shielding composite materials based on carbon nanotubes and Sandwich laminate structures. *Polymer Composites*, 45(15), pp. 13639–13649.
43. Fu, H., Bai, Y., Duan, S., Zhou, H., Gong, W. 2023. Structure design of multi-layered ABS/CNTs composite foams for EMI shielding application with low reflection and high absorption characteristics. *Applied Surface Science*, 624, pp. 157168.
44. Wang, T., Yu, W.C., Zhou, C.G., Sun, W.J., Zhang, Y.P., Jia, L.C., et al. 2020. Self-healing and flexible carbon nanotube/polyurethane composite for efficient electromagnetic interference shielding. *Composites Part B: Engineering*, 193, pp. 108015.
45. Cui, C.H., Yan, D.X., Pang, H., Xu, X., Jia, L.C., Li, Z.M. 2016. Formation of a segregated electrically conductive network structure in a low-melt-viscosity polymer for highly efficient electromagnetic interference shielding. *ACS Sustainable Chemistry and Engineering*, 4(8), pp. 4137–4145.
46. Wang, Y.Y., Zhou, Z.H., Zhou, C.G., Sun, W.J., Gao, J.F., Dai, K., et al. 2020. Lightweight and robust carbon nanotube/polyimide foam for efficient and heat-resistant electromagnetic interference shielding and microwave absorption. *ACS Applied Materials and Interfaces*, 12(7), pp. 8704–8712.
47. Shao, R., Wang, G., Chai, J., Wang, G., Zhao, G. 2024. Flexible, reliable, and lightweight multiwalled carbon nanotube/polytetrafluoroethylene membranes with dual-nanofibrous structure for outstanding EMI shielding and multifunctional applications. *Small*, 20(24), pp. 2308992.
48. Maleki, A.H., Zolfaghari, A. 2023. Investigation of electrical, electromagnetic interference shielding and tensile properties of 3D-printed acrylonitrile butadiene styrene/carbon nanotube composites. *Journal of Thermoplastic Composite Materials*, 37(7), pp. 2409–2424.
49. Xu, Z., Dou, T., Wang, Y., Zuo, H., Chen, X., Zhang, M., et al. 2023. Three-dimensional-printed carbon nanotube/polylactic acid composite for efficient electromagnetic interference shielding. *Polymers*, 15(14), pp. 3080.
50. Chauhan, S., Nikhil Mohan, P., Raju, K.C.J., Ghotia, S., Dwivedi, N., Dhand, C., et al. 2023. Free-standing polymer/multiwalled carbon nanotubes composite thin films for high thermal conductivity and prominent EMI shielding. *Colloids and Surfaces A: Physicochemical and Engineering Aspects*, 673, pp. 131811.
51. Jameson, C., Devine, D.M., Keane, G., Gately, N.M. 2024. Influence of extrusion screw speed and CNT concentration on the mechanical and EMI properties of PC/ABS based nanocomposites. *Materials*, 17, pp. 2625.
52. Kaushal, A., Singh, V. 2021. Electromagnetic interference shielding response of multiwall carbon nanotube/polypropylene nanocomposites prepared via melt processing technique. *Polymer Composites*, 42(3), pp. 1148–1154.
53. Üçpınar Durmaz, B. 2023. *Karbon Katkılı İletken Polimer Kompozitlerin Üretimi, Karakterizasyonu Ve Elektromanyetik Dalga Kalkanlamada Kullanımlarının Değerlendirilmesi*. Kocaeli Üniversitesi.

54. Niu, J.H., Yang, T.B., Li, X.Y., Xu, L., Lin, H., Zhong, G.J. 2023. 3D-printed co-continuous segregated CNT/TPU composites with superb electromagnetic interference shielding and excellent mechanical properties. *Industrial and Engineering Chemistry Research*, 62(27), pp. 10498–10506.
55. Pérez-Rodríguez, S., Alegre, C., Sebastián, D., Lázaro, M.J. 2021. Emerging carbon nanostructures in electrochemical processes. In: Sadjadi, S., editor. *Emerging Carbon Materials for Catalysis*. 1st ed., Netherlands: Elsevier; pp. 353–388.
56. Mohan, L., Kumar, T.N., Karakkad, S., Krishnan, S.T. 2021. Development of cost-effective carbon nanofiber epoxy nanocomposites for lightweight wideband EMI shielding application. *IEEE Transactions on Nanotechnology*, 20, pp. 627–634.
57. Tahalyani, J., Akhtar, M.J., Kar, K.K., Chakradhary, V.K. 2022. Design and development of carbon nanofiber based lightweight flexible nanocomposite structure for EMI shielding application. *2022 IEEE Microwaves, Antennas, and Propagation Conference*, MAPCON 2022, pp. 1228–1233.
58. Chakradhary, V.K., Juneja, S., Jaleel Akhtar, M. 2020. Correlation between EMI shielding and reflection loss mechanism for carbon nanofiber/epoxy nanocomposite. *Materials Today Communications*, 25, pp. 101386.
59. Serrato, V.M., Padilla, V., Jones, D., Herrera, S., Campos, L., Serrato, I., et al. 2023. Electromagnetic interference shielding effectiveness of compression molded carbon nanofiber-reinforced polyvinylidene difluoride film. *Polymer Composites*, 44(1), pp. 592–608.
60. Chen, Z., Yin, X., Chen, H., Fu, X., Sun, Y., Chen, Q., et al. 2024. Mechanical, crystallization, rheological, and supercritical CO_2 foaming properties of polybutylene succinate nanocomposites: impact of carbon nanofiber content. *Polymers*, 16(1), pp. 28.
61. Mondal, S., Nayak, L., Rahaman, M., Aldalbahi, A., Chaki, T.K., Khastgir, D., et al. 2017. An effective strategy to enhance mechanical, electrical, and electromagnetic shielding effectiveness of chlorinated polyethylene-carbon nanofiber nanocomposites. *Composites Part B: Engineering*, 109, pp. 155–169.
62. Yuksek, M. 2020. Electromagnetic wave shielding and mechanical properties of vapor-grown carbon nanofiber/polyvinylidene fluoride composite fibers. *Journal of Engineered Fibers and Fabrics*, 15, pp. 1–12.
63. Safdari, M., Al-Haik, MS. 2018. A review on polymeric nanocomposites: effect of hybridization and synergy on electrical properties. Fauzi Ismail, A., Sean Goh, P., Editors, *Carbon-Based Polymer Nanocomposites for Environmental and Energy Applications*, Elsevier, pp. 113–146.
64. Kaushal, A., Singh, V. 2020. Effect of filler loading on the shielding of electromagnetic interference of reduced graphene oxide reinforced polypropylene nanocomposites prepared via a twin-screw extruder. *Journal of Materials Science: Materials in Electronics*, 31(14), pp. 22162–22170.
65. Oraby, H., Naeem, I., Darwish, M., Senna, M.H., Tantawy, H.R. 2022. Optimization of electromagnetic shielding and mechanical properties of reduced graphene oxide/polyurethane composite foam. *Polymer Engineering and Science*, 62(10), pp. 3075–3087.
66. Das, P., Deoghare, A.B., Ranjan Maity, S. 2022. Synergistically improved thermal stability and electromagnetic interference shielding effectiveness (EMI SE) of in-situ synthesized polyaniline/sulphur doped reduced graphene oxide (PANI/S-RGO) nanocomposites. *Ceramics International*, 48(8), pp. 11031–11042.
67. Zubair, K., Ashraf, A., Gulzar, H., Shakir, M.F., Nawab, Y., Rehan, Z.A., et al. 2021. Study of mechanical, electrical and EMI shielding properties of polymer-based nanocomposites incorporating polyaniline coated graphene nanoparticles. *Nano Express*, 2(1), pp. 010038.
68. Park, J.H., Kim, S.J., Eom, T.G., Seo, M., Song, Y.G., Tang, F., et al. 2024. Reduced graphene oxide-reinforced unsaturated polyester resin nanocomposites with improved thermal stability, mechanical modulus, and electromagnetic interference shielding performance. *Journal of Applied Polymer Science*, 141(6), pp. 1–12.

69. Das, M., Sethy, P.P., Sundaray, B. 2023. EMI shielding performance of graphene oxide reinforced polyaniline/polystyrene solution cast thin films. *Synthetic Metals*, 296, pp. 117369.
70. Ahmad, M.W., Dey, B., Sammar, A.A., Choudhury, A. 2020. Sulfonic-functionalized graphene oxide reinforced polyethersulfone nanocomposites with enhanced dielectric permittivity and EMI shielding effectiveness. *Journal of Macromolecular Science, Part A: Pure and Applied Chemistry*, 57(11), pp. 778–790.
71. Gao, Z., Dong, Q., Shang, M., Shentu, B., Wu, C. 2021. Microstructure and properties of poly(butylene terephthalate)/poly(ethylene terephthalate) composites based on carbon nanotubes/graphene nanoplatelets hybrid filler systems. *Journal of Applied Polymer Science*, 139(9), pp. 51733.
72. Lee, B., Hwang, U., Kim, J., Kim, S.H., Choi, K., Park, I.K., et al. 2023. Highly dispersed graphene nanoplatelets in polypropylene composites by employing high-shear stress for enhanced dielectric properties and frequency-selective electromagnetic interference shielding capability. *Composites Communications*, 37, pp. 101409.
73. Zhi, L.X., Majid, M.S.A., Ridzuan, M.J.M., Ismail, M.S., Azaman, M.D., Khasri, A. 2024. Effect of dispersion methods on electromagnetic interference shielding effectiveness and thermal conductivity of graphene bio-based epoxy composites. *Polymer Composites*, 45(15), pp. 13721–13736.
74. Chandra, R.B.J., Shivamurthy, B., Kumar, M.S., Prabhu, N.N., Sharma, D. 2023. Mechanical and electrical properties and electromagnetic-wave-shielding effectiveness of graphene-nanoplatelet-reinforced acrylonitrile butadiene styrene nanocomposites. *Journal of Composites Science*, 7(3), pp. 117.
75. Hsiao, S.T., Ma, C.C.M., Tien, H.W., Liao, W.H., Wang, Y.S., Li, S.M., et al. 2015. Effect of covalent modification of graphene nanosheets on the electrical property and electromagnetic interference shielding performance of a water-borne polyurethane composite. *ACS Applied Materials and Interfaces*, 7(4), pp. 2817–2826.
76. Santhosi, B.V.S.R.N., Ramji, K., Rao, N.B.R.M. 2020. Design and development of polymeric nanocomposite reinforced with graphene for effective EMI shielding in X-band. *Physica B: Condensed Matter*, 586, pp. 412144.
77. Manna, K., Srivastava, S.K., Mittal, V. 2016. Role of enhanced hydrogen bonding of selectively reduced graphite oxide in fabrication of poly(vinyl alcohol) nanocomposites in water as EMI shielding material. *Journal of Physical Chemistry C*, 120(309), pp. 17011–17023.
78. Kim, S., Byun, J. 2024. Research trends in electromagnetic shielding using MXene-based composite materials. *Journal of Korean Powder Metallurgy Institute*, 31(1), pp. 57–76.
79. Chen, X., Zhao, Y., Li, L., Wang, Y., Wang, J., Xiong, J., et al. 2021. MXene/polymer nanocomposites: preparation, properties, and applications. *Polymer Reviews*, 61(1), pp. 80–115.
80. Lang, L., Zou, Y., Wang, Y., Men, J., Cheng, J., Yan, X., et al. 2024. A review of recent advances in MXenes/polymer-based electromagnetic interference shielding materials. *Polymer Composites*, 45, pp. 1–19.
81. Yan, Y.Z., Chen, K., Moon, H.R., Park, S.S., Ha, C.S. 2024. Flexible and hydrophobic polyimide/MXene-POSS nanocomposite films for electromagnetic interference shielding. *European Polymer Journal*, 202, pp. 112655.
82. Wu, W., Zhao, W., Sun, Q., Yu, B., Yin, X., Cao, X., et al. 2021. Surface treatment of two dimensional MXene for poly(vinylidene fluoride) nanocomposites with tunable dielectric permittivity. *Composites Communications*, 23, pp. 100562.
83. Yao, Y., Jin, S., Wang, M., Gao, F., Xu, B., Lv, X., et al. 2022. MXene hybrid polyvinyl alcohol flexible composite films for electromagnetic interference shielding. *Applied Surface Science*, 578, pp. 152007.
84. Xu, D., Huang, Q., Yang, L., Chen, Y., Lu, Z., Liu, H., et al. 2023. Experimental design of composite films with thermal management and electromagnetic shielding properties based on polyethylene glycol and MXene. *Carbon*, 202, pp. 1–12.

85. Habibpour, S., Zarshenas, K., Zhang, M., Hamidinejad, M., Ma, L., Park, C.B., et al. 2022. Greatly enhanced electromagnetic interference shielding effectiveness and mechanical properties of polyaniline-grafted Ti3C2Tx MXene-PVDF composites. *ACS Applied Materials and Interfaces*, 14(18), pp. 21521–21534.
86. Wan, Y.J., Li, X.M., Zhu, P.L., Sun, R., Wong, C.P., Liao, W.H. 2020. Lightweight, flexible MXene/polymer film with simultaneously excellent mechanical property and high-performance electromagnetic interference shielding. *Composites Part A: Applied Science and Manufacturing*, 130, pp. 105764.
87. Jia, X., Shen, B., Zhang, L., Zheng, W. 2021. Construction of compressible polymer/MXene composite foams for high-performance absorption-dominated electromagnetic shielding with ultra-low reflectivity. *Carbon*, 173, pp. 932–940.
88. Xu, M.K., Liu, J., Zhang, H. B., Zhang, Y., Wu, X., Deng, Z., et al. 2021. Electrically conductive $Ti_3C_2T_x$ MXene/polypropylene nanocomposites with an ultralow percolation threshold for efficient electromagnetic interference shielding. *Industrial and Engineering Chemistry Research*, 60(11), pp. 4342–4350.
89. Zeng, Z.H., Wu, N., Wei, J.J., Yang, Y.F., Wu, T.T., Li, B., et al. 2022. Porous and ultra-flexible crosslinked MXene/polyimide composites for multifunctional electromagnetic interference shielding. *Nano-Micro Letters*, 14(59), pp. 1–16.
90. Lee, J., Hong, S., Sun, Y., Lee, S.K., Hwang, U., Nam, J.do., et al. 2023. Parasitic reaction driven facile preparation of segregated-MXene/polycarbonate nanocomposites for efficient electromagnetic interference shielding. *Surfaces and Interfaces*, 40, pp. 103101.
91. Yuan, W., Yang, J., Yin, F., Li, Y., Yuan, Y. 2020. Flexible and stretchable MXene/polyurethane fabrics with delicate wrinkle structure design for effective electromagnetic interference shielding at a dynamic stretching process. *Composites Communications*, 19, pp. 90–98.
92. Liu, L., Guo, R., Gao, J., Ding, Q., Fan, Y., Yu, J. 2022. Mechanically and environmentally robust composite nanofibers with embedded MXene for wearable shielding of electromagnetic wave. *Composites Communications*, 30, pp. 6–11.
93. Liang, W., Wu, J., Zhang, S., Zhao, P.-Y., Cong, Y., Guo, Y., et al. 2024. Porous Ti3C2Tx MXene nanosheets sandwiched between polyimide fiber mats for electromagnetic interference shielding. *Nano Research*, 17(3), pp. 2070–2078.
94. Hu, Y., Gu, J., Zhang, Y., Liu, G., Yi, X., Liu, X. 2024. FDM 3D printed MXene/recycled carbon fibre reinforced polylactic acid composites: interface optimization, toughening and enhanced electromagnetic shielding performance. *Composites Communications*, 48, pp. 101953.
95. Hong, S.Y., Sun, Y., Lee, J., Yifei, M., Wang, M., Nam, J.-D., et al. 2021. 3D printing of free-standing Ti3C2Tx/PEO architecture for electromagnetic interference shielding. *Polymer*, 236(51), pp. 124312.
96. Rajavel, K., Luo, S., Wan, Y., Yu, X., Hu, Y., Zhu, P., et al. 2020. 2D Ti3C2Tx MXene/polyvinylidene fluoride (PVDF) nanocomposites for attenuation of electromagnetic radiation with excellent heat dissipation. *Composites Part A: Applied Science and Manufacturing*, 129, pp. 105693.
97. Chu, N., Luo, C., Chen, X., Li, L., Liang, C., Chao, M., et al. 2023. Ti3C2Tx MXene/polyimide composites film with excellent mechanical properties and electromagnetic interference shielding properties. *Journal of Alloys and Compounds*, 955, pp. 170241.
98. Liu, H., Cui, Z., Luo, L., Liao, Q., Xiong, R., Xu, C., et al. 2023. Facile fabrication of flexible and ultrathin self-assembled Ti3C2Tx/bacterial cellulose composite films with multifunctional electromagnetic shielding and photothermal conversion performances. *Chemical Engineering Journal*, 454, pp. 140288.
99. Ye, Z., Zhao, D., Liu, F., Luo, J., Liu, X., Zhao, W. 2024. Flexible and highly conductive Ti3C2Tx/natural rubber composites with interconnected networks for high-performance electromagnetic interference shielding. *Composites Part A: Applied Science and Manufacturing*, 180, pp. 108067.

100. Szeluga, U., Kumanek, B., Trzebicka, B. 2015. Synergy in hybrid polymer/nanocarbon composites. A review. *Composites Part A: Applied Science and Manufacturing*, 73, pp. 204–231.
101. Ucpinar Durmaz, B., Aytac, A. 2023. The synergistic effects of micro- and nano-fillers on the properties of polyamide 11/poly (lactic acid) blend. *Journal of Thermoplastic Composite Materials*, 37(7), pp. 2329–2355.
102. Cougo, C., Pezzin, S.H., Pachekoski, W.M., Amico, S.C. 2019. Multiscale hybrid composites with carbon-based nanofillers. In: Khan A., Jawaid M., Asiri A.M., editors. *Nanocarbon and its Composites*. 1st ed., Woodhead Publishing; pp. 449–470.
103. Haghgoo, M., Ansari, R., Hassanzadeh-Aghdam, M.K. 2019. Prediction of electrical conductivity of carbon fiber-carbon nanotube-reinforced polymer hybrid composites. *Composites Part B: Engineering*, 167, pp. 728–735.
104. Patadia, M., Quinn, A., Tank, M., Jolowsky, C., Luiz, L., Psulkowski, A., et al. 2024. Enhanced multifunctionality in carbon fiber/carbon nanotube reinforced PEEK hybrid composites: superior combination of mechanical properties, electrical conductivity, and EMI shielding. *Composites Part B: Engineering*, 284, pp. 111674.
105. Yang, X., Luo, J., Ren, H., Xue, Y., Yang, C., Yuan, T., et al. 2023. Simultaneously improving the EMI shielding performances and mechanical properties of CF/PEKK composites via MXene interfacial modification. *Journal of Materials Science and Technology*, 154, pp. 202–209.
106. Silva, J.M.F., Indrusiak, T., Barra, G.M.O.., Letichevsky, S.., Silva, A.A., Soares, B.G. 2024. Hybrid carbonaceous filler as promising additives for EMI SE of PVDF-based composites: comparison between monolayered and multilayered structures. *FlatChem*, 43, 100603.
107. Saberi, M., Moradi, A., Ansari, R., Hassanzadeh-Aghdam, M.K., Jamali, J. 2024. Analyzing the electrical conductivity of hybrid carbon nanotube/graphene nanoplatelet polymer-matrix nanocomposites: an extended physics-based modeling approach. *Alexandria Engineering Journal*, 108, pp. 344–353.
108. Kang, J.S., Zhang, H.X., Yoon, K.B. 2024. Hybrid rGO-CNT/epoxy nanocomposites as electromagnetic interference shielding materials. *ACS Applied Nano Materials*, 7(9), pp. 10770–10778.
109. Ahmad, M.W., Soren, K., Dey, B., Khan, M.S., Choudhury, A. 2021. Synergistic reinforcement effect of 3D graphene@multi-walled carbon nanotube hybrid nanofiller in enhancing the electrical, EMI shielding, and mechanical properties of polyethersulfone. *International Journal of Polymer Analysis and Characterization*, 26(8), pp. 754–771.
110. dos Anjos, E.G.R., Moura, N.K., Antonelli, E., Baldan, M.R., Gomes, N.A.S., Braga, N.F., et al. 2023. Role of adding carbon nanotubes in the electric and electromagnetic shielding behaviors of three different types of graphene in hybrid nanocomposites. *Journal of Thermoplastic Composite Materials*, 36(8), pp. 3209–3235.
111. Jyoti, J., Arya, A.K. 2020. EMI shielding and dynamic mechanical analysis of graphene oxide-carbon nanotube-acrylonitrile butadiene styrene hybrid composites. *Polymer Testing*, 91, pp. 106839.
112. Xie, Z., Chen, H., Xie, M., Zhang, D., Zhao, H., Chen, W. 2024. Electrical percolation networks of MWCNT/Graphene/Polyaniline nanocomposites with enhanced electromagnetic interference shielding efficiency. *Applied Surface Science*, 655, pp. 159613.
113. Joseph, J., Munda, P.R., John, D.A., Sidpara, A.M., Paul, J. 2019. Graphene and CNT filled hybrid thermoplastic composites for enhanced EMI shielding effectiveness. *Materials Research Express*, 6(8), pp. 085617.
114. Al-Saleh, M.H. 2023. Towards a cost-effective carbon nanofillers-based composites for EMI shielding applications. *Synthetic Metals*, 293, pp. 117271.
115. Ribeiro dos Anjos, E.G., Passador, F.R., de Carvalho, A.B., Rezende, M.C., Sundararaj, U., Pessan, L.A. 2024. Advanced ternary carbon-based hybrid nanocomposites for electromagnetic functional behavior in additive manufacturing. *Applied Materials Today*, 40, pp. 102362.

116. Qi, Q., Ma, L., Zhao, B., Wang, S., Liu, X., Lei, Y., et al. 2020. An effective design strategy for the sandwich structure of PVDF/GNP-Ni-CNT composites with remarkable electromagnetic interference shielding effectiveness. *ACS Applied Materials and Interfaces*, 12(32), pp. 36568–36577.
117. Li, Z., Feng, D., Li, B., Xie, D., Mei, Y. 2023. FDM printed MXene/$MnFe_2O_4$/MWCNTs reinforced TPU composites with 3D Voronoi structure for sensor and electromagnetic shielding applications. *Composites Science and Technology*, 231, pp. 109803.
118. Lee, D.E., Lee, G.H., Son, N.R., Zhang, H.X., Yoon, K.B. 2023. Polyamide 6/MXene-grafted graphene oxide hybrid nanocomposites. *Iranian Polymer Journal (English Edition)*, 32, pp. 377–388.
119. Hu, Y., Chen, J., Yang, G., Li, Y., Dong, M., Zhang, H., et al. 2024. Highly conductive and mechanically robust MXene@CF core-shell composites for in-situ damage sensing and electromagnetic interference shielding. *Composites Science and Technology*, 246, pp. 110356.
120. Liu, H., Huang, Z., Chen, T., Su, X., Liu, Y., Fu, R. 2022. Construction of 3D MXene/silver nanowires aerogels reinforced polymer composites for extraordinary electromagnetic interference shielding and thermal conductivity. *Chemical Engineering Journal*, 427, pp. 131540.
121. Dehghan, P., Simiari, M., Gholampour, M., Aghvami-Panah, M., Amirkiai, A. 2023. Tuning the electromagnetic interference shielding performance of polypropylene cellular nanocomposites: role of hybrid nanofillers of MXene and reduced graphene oxide. *Polymer Testing*, 126, pp. 108162.

CHAPTER 10

Role of 0D, 1D and 2D Nanomaterials/Polymer Nanocomposites in Electromagnetic Interference Shielding

Raghvendra Kumar Mishra and Yedla Venkata Durga Anudeep

10.1 INTRODUCTION TO EMI SHIELDING

Electromagnetic interference (EMI) shielding is essential for safeguarding electronic devices and human health against the detrimental effects of electromagnetic radiation. With rapid technological advancements, the demand for effective shielding materials has surged, driving innovations in both conductive and insulating materials [1]. Metals and carbon-based materials are popular due to their high conductivity and superior EMI-blocking properties [2, 3]. However, the need for electrically insulating polymeric materials has gained prominence to prevent short-circuiting while maintaining high shielding effectiveness [4]. The integration of polymers with metal or carbon fillers not only enhances shielding performance but also imparts multifunctionality [3, 4]. EMI shielding operates primarily through two mechanisms: reflection and absorption of electromagnetic waves. Reflection mitigates wave penetration into sensitive regions, while absorption converts electromagnetic radiation into heat, reducing secondary pollution from reflected waves. This shielding is crucial in densely packed electronic circuits, where high-frequency signals and sources such as high-speed clock signals, processor noise and switching frequencies cause operational disturbances. The electromagnetic spectrum spans a range of wave types, defined by wavelengths, frequencies and applications. As the spectrum progresses from radio waves to gamma rays, wavelengths shorten, frequencies increase and wave energy

DOI: 10.1201/9781003512363-10

TABLE 10.1 Electromagnetic Spectrum by Wave Type, Wavelength, Frequency and Typical Uses [5]

Region	Wavelength	Frequency	Typical Uses
Radio Waves	> 1 m	< 300 MHz	Broadcasting (AM, FM radio), Communication (TV, mobile phones), Radar, Satellite
Microwaves	1 mm to 1 m	300 GHz to 300 MHz	Microwave ovens, Radar, Satellite communication, GPS, Wireless networks
Infrared (IR)	700 nm to 1 mm	300 GHz to 430 THz	Thermal imaging, Night vision, Remote controls, Heating, Astronomy
Visible Light	400 nm (violet) to 700 nm (red)	430 THz to 770 THz	Human vision, Photography, Illumination, Lasers
Ultraviolet (UV)	400 nm to 10 nm	770 THz to 30 PHz	Sterilisation, Black lights, Tanning beds, Medical treatments, Fluorescence
X-rays	0.01–10 nm	30 PHz to 30 EHz	Medical imaging (X-rays), Security screening, Material analysis
Gamma Rays	< 0.01 nm	> 30 EHz	Cancer treatment (radiotherapy), Nuclear reactions, Astrophysical phenomena

intensifies. These waves serve diverse functions, from communication (radio waves) to medical applications such as imaging and cancer treatment (X-rays and gamma rays) [2, 3], as shown in Table 10.1.

EMI adversely affects electronic systems and communication networks, necessitating effective mitigation strategies. Techniques employing phaseless electric and magnetic near-field data are instrumental in pinpointing critical EMI sources such as PCB traces, to optimise filter designs [6]. Natural sources such as lightning and solar flares significantly impact radio communications, while the corona effect in high-voltage systems generates EMI across a broad frequency range, with emissions up to 1.2 GHz. This effect can be mitigated by increasing conductor radii, smoothing surfaces, improving insulation and reducing humidity [7, 8]. Human-made sources, including commutator motors, fluorescent lights and ignition systems, also contribute to EMI, disrupting nearby electronic devices. In System-on-Chip (SoC) devices, high-speed clock signals and dense PCB layouts exacerbate EMI issues due to high-frequency switching noise, often breaching compliance margins set by regulatory bodies such as the FCC and CE [9]. Variations in processor activity further introduce voltage noise, impacting signal integrity and causing timing violations, with voltage swings averaging 4% and worst-case margins reaching 14% [10, 11]. Silicon carbide (SiC) converters, known for high efficiency and power density, face EMI challenges from fast-switching transients, especially in multi-cell configurations. Active and passive filters effectively suppress EMI across frequencies, with common-mode filters enhancing mitigation in SiC modules [12–14]. Material-based EMI shielding is essential, balancing reflection and absorption properties to minimise secondary pollution. Absorptive materials, such as concrete and conductive dielectric composites, convert electromagnetic energy into heat, achieving high shielding effectiveness [15]. Advanced materials, such as epoxy-based composites and multilayered structures, enhance performance through optimised microstructures [16, 17]. Lightweight and multifunctional materials, including silver nanowires and carbon nanofibre/polyurethane composites, offer high shielding effectiveness alongside thermal management and flexibility [18]. For instance, silver nanowire/carboxymethyl cellulose films exhibit high conductivity (3.72×10^6 S/m) and excellent shielding performance due to their layered structure [19]. Carbon nanofibre/polyurethane composites demonstrate a maximum shielding effectiveness of 71.61 dB at 2.21 mm thickness, maintaining performance under varied conditions [20]. Polypyrrole-modified cotton achieves a conductivity of 2715.8 S/m and shielding effectiveness of 28.2 dB while retaining properties after mechanical stress [21]. Aramid nanofibre/MXene composites provide dual protection against EMI and mechanical impacts, with a shielding effectiveness of 65.5 dB [22]. Nanomaterials significantly enhance shielding effectiveness through multiple absorption, reflection and scattering events at the nanoscale. Metal oxide nanoparticles in polymer composites improve mass attenuation coefficients, enhancing gamma-ray absorption [23]. Their incorporation facilitates electromagnetic wave reflection and scattering, disrupting radiation paths for improved shielding [24–26]. This chapter explores these materials, their properties, fabrication techniques and comparative performance, emphasising their transformative role in advancing EMI shielding technologies.

10.2 OVERVIEW OF NANOMATERIALS IN EMI SHIELDING

Nanomaterials have become crucial in the development of effective EMI shielding materials, addressing the escalating concerns of electromagnetic radiation pollution. Their exceptional properties, including high electrical conductivity, mechanical strength and magnetic capabilities, make them ideal for these applications, as depicted in Figure 10.1. Carbon-based structures, metals and metal oxides are the most widely utilised materials due to their superior EMI shielding performance [27, 28]. Graphene and carbon nanotubes (CNTs), in particular, are highly valued for their excellent electrical conductivity, mechanical durability and thermal stability. Incorporating these carbon nanostructures into hybrid composites with magnetite enhances EMI shielding through synergistic interactions, resulting in superior shielding effectiveness [28]. Carbon fibres (CFs) also play a critical role, contributing to lightweight and flexible composites with excellent shielding properties [29]. Metal and metal oxide nanoparticles further improve EMI shielding. For instance, Fe_3O_4 and barium hexaferrite nanoparticles (BHNs) serve as magnetic cores with conductive surfaces, enhancing interfacial polarisation and achieving shielding effectiveness exceeding 50 dB in nanocomposite films [30–32]. The integration of metal oxides with carbon materials in polymer matrices has demonstrated promising results, achieving high attenuation constants and improved conductivity [30].

Conductive polymers such as polyaniline (PANI) also offer high electrical conductivity and flexibility, which, when combined with metal nanoparticles, significantly enhance shielding efficiency (SE) across various applications [33]. Additionally, nanocellulose, a biodegradable and sustainable material, is gaining attention for EMI shielding when paired with conductive fillers [31]. Nanomaterials are categorised based on their geometric dimensions, which influence their properties and applications. Zero-dimensional (0D)

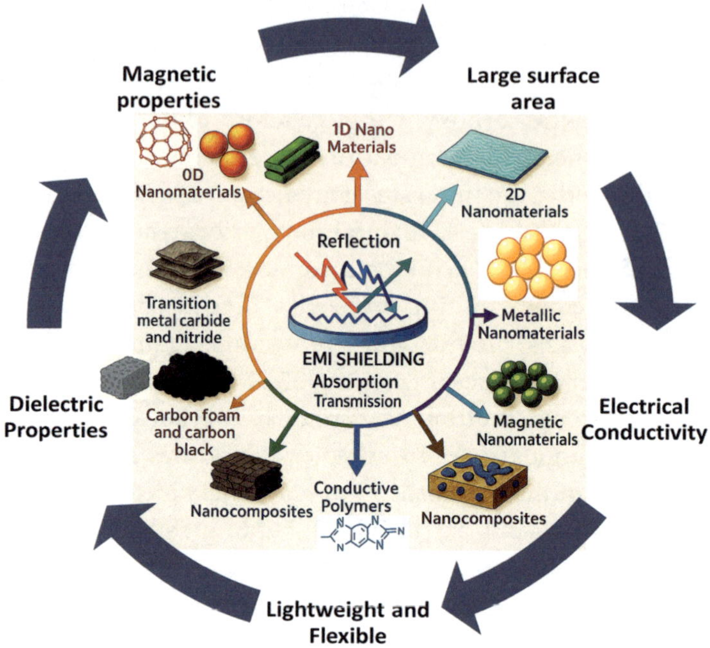

FIGURE 10.1 Material properties for effective EMI shielding: highlights the material classes, mechanisms and unique properties enabling EMI attenuation.

nanomaterials, such as quantum dots (QDs), nanoparticles and nanocrystals, exhibit quantum confinement effects that result in unique optical and electronic properties [34, 35]. One-dimensional (1D) nanomaterials, including nanowires, nanotubes and nanofibres, are characterised by high aspect ratios, making them suitable for applications in electronics, sensors and energy storage [36]. Two-dimensional (2D) nanomaterials, such as graphene, MXenes, nanosheets and transition metal dichalcogenides (TMDs), possess high surface areas and exceptional mechanical properties, making them ideal for coatings and composite materials [36, 37]. Three-dimensional (3D) nanomaterials, including nanostructured bulk materials, find applications in catalysis, biomedical devices and other fields [36]. Despite their vast potential, commercial applications of nanomaterials in EMI shielding face challenges, particularly in achieving uniform dispersion and scalability. Incompatible materials often cause nanoparticle agglomeration within polymer matrices, reducing shielding effectiveness. Techniques such as blade-coating and electrodeposition have been investigated to overcome these issues but require precise parameter control to ensure uniformity [24, 38]. Current fabrication methods, although effective, are not easily scalable without compromising material properties. Cost-effectiveness also remains a critical challenge, as affordable raw materials and production processes are essential for commercial viability [39]. While polymer nanocomposites are favoured for their affordability, they often struggle to meet high-performance standards due to dispersion inconsistencies and integration challenges [24]. Their unique properties and versatility ensure their continued importance in addressing modern challenges associated with electromagnetic radiation pollution.

10.2.1 EMI Shielding Mechanism

EMI shielding mechanisms are essential for protecting electronic devices and human health from harmful electromagnetic waves. These mechanisms – reflection, absorption and multiple scattering – utilise material properties such as electrical conductivity, magnetic permeability and dielectric loss to block or dissipate electromagnetic energy effectively. Figure 10.2 provides an overview of these mechanisms, illustrating their principles, material examples and applications, with a focus on advanced materials such as MXenes, polymers and composites for addressing modern EMI challenges [3, 40].

Reflection is a predominant mechanism occurring at the interface between the shielding material and incoming electromagnetic waves, requiring free electrons for effective wave reflection. Metals such as copper and silver, as well as carbon-based materials such as graphene and carbon nanotubes, are widely utilised. Applications include electronic enclosures, cables and antennas [41, 42]. Absorption converts electromagnetic energy into heat within the shielding material, enhanced by materials with high electrical and magnetic loss properties. Polymers containing fillers such as multi-walled carbon nanotubes (MWCNTs), MXenes and magnetic fillers such as ferrites (e.g., Fe_3O_4 and NiZn) are highly effective in automotive electronics, mobile devices and military equipment [43, 44]. Multiple scattering mechanisms arise from internal reflections and scattering within composite structures with high filler content. Materials such as MXene-polymer composites, CNT-polymer composites, foams and nanocellulose-based composites are ideal for lightweight and

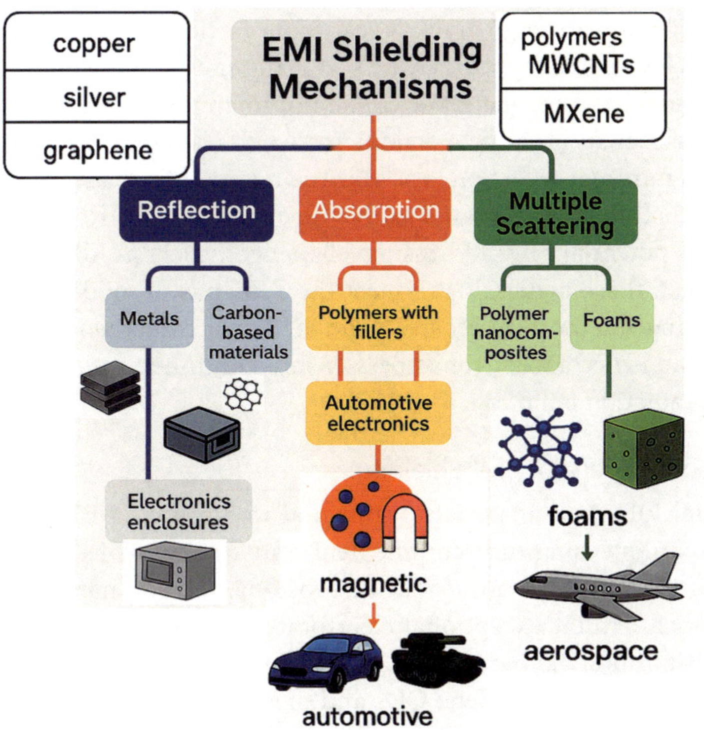

FIGURE 10.2 EMI shielding mechanisms.

flexible EMI shielding in aerospace and wearable electronics [45]. Interference suppression adjusts material conductivity or magnetic properties. Conductive inks and hybrid composites (e.g., metal-carbon or polymer-metal combinations) are suited for wearable devices and IoT (Internet of Things) systems [46]. Permeability shielding leverages materials with high magnetic permeability, such as nickel, iron oxide and permalloy, for transformers, power converters and motors [47, 48]. Hybrid materials combining reflection, absorption and scattering mechanisms enhance SE. Examples include MXene/graphene composites, layered structures and polymer-metal hybrids, critical for advanced electronics. Materials such as graphene and MXenes excel due to their layered structures, which synergise absorption and reflection [42]. Electrically conductive materials such as metallic nanoparticles and carbon-based composites prevent electromagnetic wave penetration by reflection. Additionally, porosity in materials enhances internal reflections, improving absorption efficiency. Conductive concrete, for example, exceeds 100 dB in shielding effectiveness, meeting military standards [43]. Reducing conduction losses in electronic components, such as Power Factor Correction (PFC) converters, improves EMI management. Bridgeless PFC converters demonstrate different loss profiles from conventional designs, affecting EMI emissions. This highlights the importance of optimising material conductivity [44]. Nanomaterials, including graphene and MWCNTs, significantly enhance EMI shielding due to their superior electrical conductivity and structural integrity. Effective dispersion of nanofillers in polymer matrices is critical to prevent agglomeration, ensuring optimal mechanical strength and performance. Challenges such as high production costs and fragility due to filler content remain to be addressed for practical applications [25, 49–51]. Enhancing EMI shielding through multiple scattering particles has gained research interest. Metal microparticles, such as copper, create scattering sites within composites, increasing interaction paths for electromagnetic waves and improving shielding. Multi-walled carbon nanotube composites incorporating metal microparticles demonstrate effective absorption mechanisms with minimal reflection loss [46]. The integration of advanced materials such as MXenes, conductive polymers and nanocomposites into EMI shielding applications offers significant potential. By addressing challenges such as dispersion uniformity, scalability and cost, these materials can meet the demands of modern electronics across industries, including automotive, aerospace and IoT. Their combined properties of flexibility, conductivity and mechanical robustness ensure continued innovation in addressing electromagnetic radiation pollution.

10.2.2 0D Nanomaterials for EMI Shielding

Zero-dimensional (0D) nanomaterials are pivotal in advancing EMI shielding materials due to their unique quantum confinement and structural features. These materials, restricted to less than 100 nm in all dimensions, include nanoparticles, QDs and nanospheres, which exhibit exceptional electromagnetic wave absorption and dissipation properties. Common 0D nanomaterials such as carbon quantum dots (CQDs), graphene quantum dots (GQDs), MXene QDs and metal nanoparticles (e.g., Ag, Fe_3O_4) are particularly effective in EMI shielding applications due to their high electrical conductivity, dielectric properties and magnetic loss mechanisms [52]. The discrete density of states (DOS) in 0D nanomaterials arises from their size being smaller than the Bohr exciton radius, resulting in quantised energy levels and enhanced optical and electromagnetic properties [53]. This quantum confinement enhances absorption and energy dissipation capabilities, while their high surface area and tunable morphology improve interaction with electromagnetic waves. By varying QD size, the local density of states (LDOS) can be modulated, impacting thermal radiation and EMI shielding performance [54, 55]. Defects and heterogeneous interfaces in 0D materials further enhance electromagnetic wave absorption through interfacial polarisation, multiple scattering and reflection mechanisms. Core-shell structures, such as $Ag@Fe_3O_4$, exemplify this synergy, combining reflection and absorption properties for optimised shielding. For instance, Co/CNTs@C heterostructures and $SiO_2@SiC@C$ composites demonstrate significant reflection losses of −59.1 and −59.53 dB, respectively, through defect-induced interfacial polarisation [56–58]. Cobalt-based 0D nanomaterials exhibit multi-mode magnetic resonances, improving electromagnetic wave capture and impedance matching. When embedded in carbon-based structures such as carbon nanotubes and nanofibres, these materials achieve remarkable reflection losses, as low as −73.87 dB, due to synergistic magnetic and dielectric losses [57, 59]. Tunable absorption frequencies further enhance their applicability in advanced technologies such as terahertz systems, where materials such as graphene and vanadium dioxide enable over 99.99% absorption, supporting adjustable operational frequencies [60, 61]. Lightweight 0D nanomaterials, including hollow Fe_3O_4 nanospheres, facilitate efficient absorption through interfacial polarisation and multiple reflections. Silver-adorned multi-walled carbon nanotubes (MWCNTs) achieve EMI shielding effectiveness of 81 dB, leveraging their high surface area for enhanced interaction with electromagnetic waves [62, 63]. Morphological control, such as modifying core-shell structures, allows for tailored electromagnetic properties. For example, Ag-Ni core-shell structures demonstrated a 45% enhancement in absorption loss through morphological adjustments [64]. Incorporating 0D fillers into hybrid nanocomposites with conducting polymers and nanoclays improves mechanical strength, dielectric properties and shielding performance, making them suitable for automotive electronics and advanced devices [65, 66]. Carbon black and metal nanoparticles (e.g., silver, copper, gold) contribute to conductivity and electromagnetic field enhancement. For example, carbon black establishes conductive networks by connecting carbon nanotubes, enhancing electromagnetic absorption while reducing reflection [54, 67–69]. Incorporation of silver nanoparticles in composites increases dielectric constants while minimising dielectric loss, further enhancing SE [69]. Combining 0D nanomaterials such as carbon nanoparticles with polymer matrices leads to superior mechanical and electromagnetic properties, offering robust solutions for diverse EMI shielding applications [49, 65]. The versatility and tunability of 0D nanomaterials make them integral to modern EMI shielding solutions. By addressing challenges such as filler dispersion and production costs, these materials can be optimised for various advanced technologies including aerospace, automotive electronics and IoT systems. Figure 10.3 illustrates the properties, applications and integration strategies of 0D nanomaterials in EMI shielding, emphasising their transformative potential in mitigating electromagnetic pollution [52].

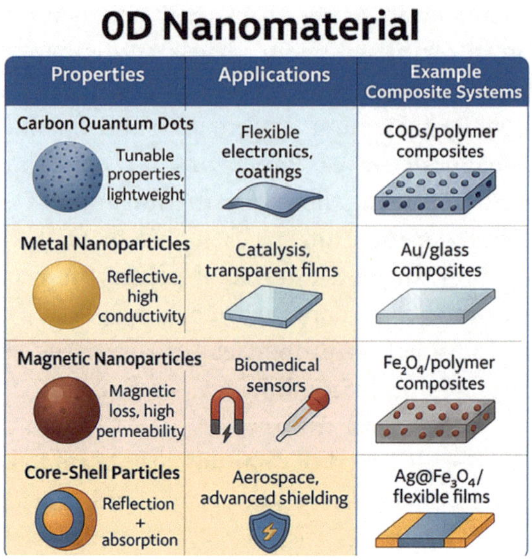

FIGURE 10.3 An overview of 0D nanomaterials for EMI shielding, emphasising their properties, applications and integration.

10.2.3 1D Nanomaterials for EMI Shielding

One-dimensional (1D) nanomaterials, characterised by confinement in two dimensions and a degree of freedom in the third, are pivotal in enhancing EMI shielding. These materials typically measure less than 100 nm along two axes, with examples including nanowires, nanorods and nanotubes. Nanowires (NWs) and nanorods, distinguished by their elongated structures and high aspect ratios, exhibit unique properties that make them ideal for EMI shielding applications. Nanorods, with higher length-to-diameter ratios, further enhance mechanical and electrical properties. Prominent 1D nanomaterials include carbon nanotubes (CNTs), metal and metal oxide nanowires and polymeric nanowires, which excel in conductivity, flexibility and lightweight characteristics, particularly for aerospace and electronic applications [52]. The high aspect ratios of 1D nanomaterials facilitate the formation of conductive networks, critical for EMI shielding. CNTs, with aspect ratios reaching up to 150, ensure rapid percolation and enhance electrical conductivity in composite materials. This property makes them exceptional conductive fillers for lightweight composites [70, 71]. Nanofibres, similar in aspect ratio, contribute to improved mechanical strength and electrical properties, supporting their integration in advanced shielding applications [72, 73]. Hybrid 1D materials, such as CNTs combined with magnetic nanoparticles (e.g., Fe_3O_4), optimise shielding through combined reflection and absorption mechanisms. For instance, Fe_3O_4-integrated multi-walled CNTs (MWCNTs) in a PVDF matrix achieved a microwave shielding effectiveness (SET) of 33.28 dB, displaying the synergistic effect of magnetic and conductive components [29, 30]. Additionally, carbon nanofibre composites have demonstrated SET values exceeding 30 dB within the X-band frequency range, underscoring their potential for robust EMI shielding [66, 74]. The synthesis and functionalisation of 1D nanomaterials enhance their performance. Techniques such as solvothermal synthesis improve

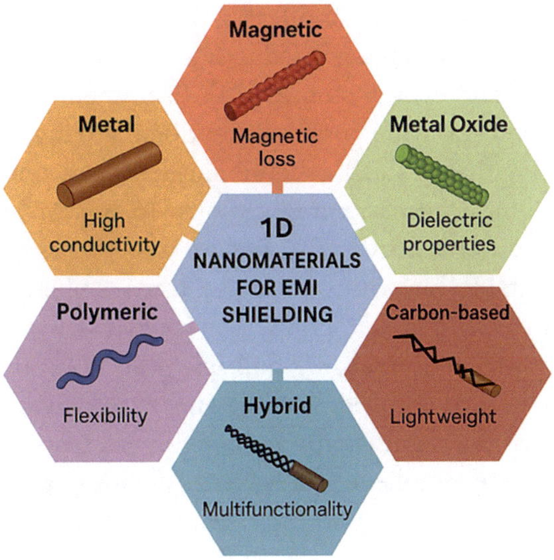

FIGURE 10.4 1D nanomaterials for EMI shielding.

dispersion within matrices, maximising their SE [29]. Conducting polymer nanocomposites, leveraging the high aspect ratios of 1D fillers, achieve effective charge transport and dielectric polarisation, enhancing electrical conductivity and EMI performance [75–77]. Magnetic nanostructures, such as Fe_3O_4 nanorings and nanotubes, exhibit unique magnetic vortex structures, improving microwave absorption through enhanced permeability and resonance. Similarly, CNT/$FeNi_3$ composites, combining CNT conductivity with magnetic properties, demonstrated significant absorption with a reflection loss of −59.80 dB at 8.5 GHz [78, 79]. Integration of 1D nanomaterials into polymer matrices produces lightweight, flexible and chemically stable composites suitable for demanding applications. CNT-based composites, for instance, achieve remarkable shielding effectiveness (up to 180 dB) while maintaining performance under mechanical stress, making them ideal for aerospace, wearable electronics and military technologies [80]. Metal nanowires, such as silver nanowires, further enhance the conductivity and EMI SE of ultra-thin films, enabling their application in flexible and transparent electronics [19, 81]. These composites demonstrate excellent repeatability under bending, ensuring reliability in dynamic environments [81]. The unique properties and high aspect ratios of 1D nanomaterials, coupled with their integration into hybrid and composite systems, enable advanced EMI shielding solutions for modern applications. From aerospace to military and wearable technologies, 1D nanomaterials ensure superior shielding performance, mechanical durability and adaptability [82, 83]. Figure 10.4 illustrates their properties, mechanisms and integration strategies, emphasising their transformative role in mitigating electromagnetic pollution.

10.2.4 2D Nanomaterials for EMI Shielding

Two-dimensional (2D) magnetic nanomaterials, defined by quantum confinement along one axis and freedom of carrier movement in the other two dimensions, are essential for EMI shielding. With sizes below 100 nm in one dimension, these materials include nanosheets, nanoplates and porous nanosheets, each offering unique advantages. Their high surface area, mechanical strength and electrical conductivity enable diverse applications, including electronics, aerospace and energy storage [84]. Prominent 2D nanomaterials such as graphene, MXenes and boron nitride possess exceptional attributes. Graphene offers high tensile strength (exceeding that of steel), exceptional electrical conductivity and flexibility, making it suitable for high-performance applications [85, 86]. MXenes, with their tunable surfaces and metallic conductivity, are well-suited for wearable electronics and energy storage devices [85]. Boron nitride, known for its high thermal stability and low dielectric loss, excels in high-temperature EMI shielding [86]. The layered structure of 2D nanomaterials introduces a "tortuous path" effect, causing multiple reflections and scattering of electromagnetic waves. This structure, coupled with the anisotropic nature of 2D materials, promotes interfacial polarisation, significantly enhancing electromagnetic wave absorption and dissipation. For instance, graphene and MXenes demonstrate superior EMI shielding capabilities through quantum effects and high electrical conductivity, achieving absorption efficiencies above 90% in the GHz range [87–89]. Nanocrystals such as manganese selenide amplify interfacial polarisation losses, further improving wave absorption. Similarly, composites incorporating 2D materials increase dielectric constants, enhancing shielding performance. TMDs such as MoS_2 and WSe_2 improve charge storage and dissipation due to their high dielectric constants, with some exceeding traditional dielectrics such as SiO_2 and Al_2O_3 [90, 91]. These properties, combined with rapid ion transport, make 2D materials ideal for EMI shielding and energy storage applications [92]. The incorporation of 2D nanomaterials into polymer matrices enhances their mechanical, thermal and electrical properties. For example, MXenes and graphene improve tensile strength, flexibility and thermal conductivity, enabling polymer composites to endure mechanical stress without compromising performance [93, 94]. Techniques such as in-situ polymerisation and solution blending ensure uniform dispersion of nanosheets, preventing agglomeration and optimising EMI shielding [51, 93]. Core-shell composites with 2D materials create multiple interfaces, enhancing dielectric loss and absorption efficiency. These materials demonstrate superior performance in mobile technologies and electric vehicles (EVs) due to their lightweight nature and high shielding effectiveness. For instance, polymer-nanoparticle composites containing 2D materials have been shown to achieve exceptional shielding effectiveness across various frequency ranges [87, 88, 91]. Despite their potential, commercial adoption of 2D nanomaterials faces challenges, including energy-intensive production processes and high costs [95, 96]. Furthermore, high filler content in composites can lead to brittleness, negatively impacting durability. Addressing these issues through cost-effective synthesis methods and optimised composite formulations will be key to unlocking the full potential of 2D nanomaterials in EMI shielding and beyond [96, 97]. Figure 10.5 illustrates the properties and applications of 2D nanomaterials, highlighting their transformative role in advanced EMI shielding solutions.

10.2.5 Comparative Analysis of 0D, 1D and 2D Nanomaterials in EMI Shielding

The evaluation of 0D, 1D and 2D nanomaterials in EMI shielding highlights the unique advantages of each dimensional category, showcasing their tailored applications based on structural and functional properties. Zero-dimensional (0D) nanomaterials, including

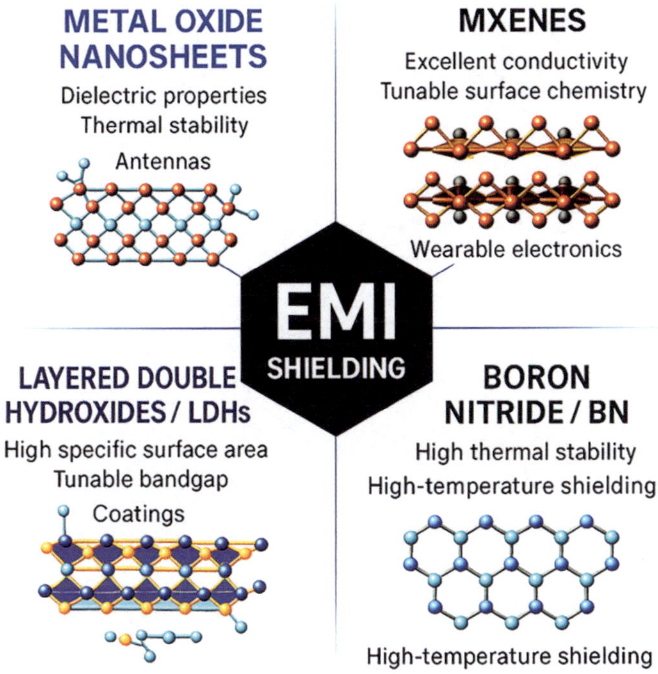

FIGURE 10.5 Properties and applications of 2D nanomaterials for EMI shielding.

nanoparticles and QDs, are characterised by high surface area and tunable dielectric properties, making them effective for enhancing composite electrical properties and EMI shielding via absorption mechanisms. These materials are typically used in combination with other nanofillers to optimise performance but are less effective as standalone shields [49]. One-dimensional (1D) nanomaterials, such as carbon nanotubes (CNTs) and nanowires, excel in flexibility, mechanical strength and electrical conductivity. These attributes make them ideal for aerospace applications and flexible electronics. Hybrid composites with CNTs enhance shielding effectiveness, achieving absorption values between 19 and 27 dB in specific frequency ranges by leveraging dielectric losses and electron-rich structures [29, 98, 99]. Two-dimensional (2D) materials, such as MXenes and graphene, surpass 0D and 1D nanomaterials in EMI shielding due to their exceptional electrical conductivity, large surface area and unique layered structures. These properties facilitate multiple internal reflections, interfacial polarisation and effective wave absorption and dissipation, resulting in shielding effectiveness exceeding 110 dB in MXene-based composites [55, 100]. Additionally, 2D materials such as boron nitride and TMDs contribute to high-performance shielding in applications requiring thermal stability and high conductivity [92]. Hybrid materials combining different dimensional nanomaterials (e.g., 0D-1D or 1D-2D) exhibit synergistic properties that enhance EMI shielding. For instance, integrating magnetite with CNTs combines the magnetic properties of magnetite with the electrical conductivity of CNTs, significantly improving shielding performance [29]. Similarly, layered structures such as reduced graphene oxide (rGO) nanoribbons combined with MXenes achieve total shielding effectiveness up to 54.1 dB by leveraging multiple

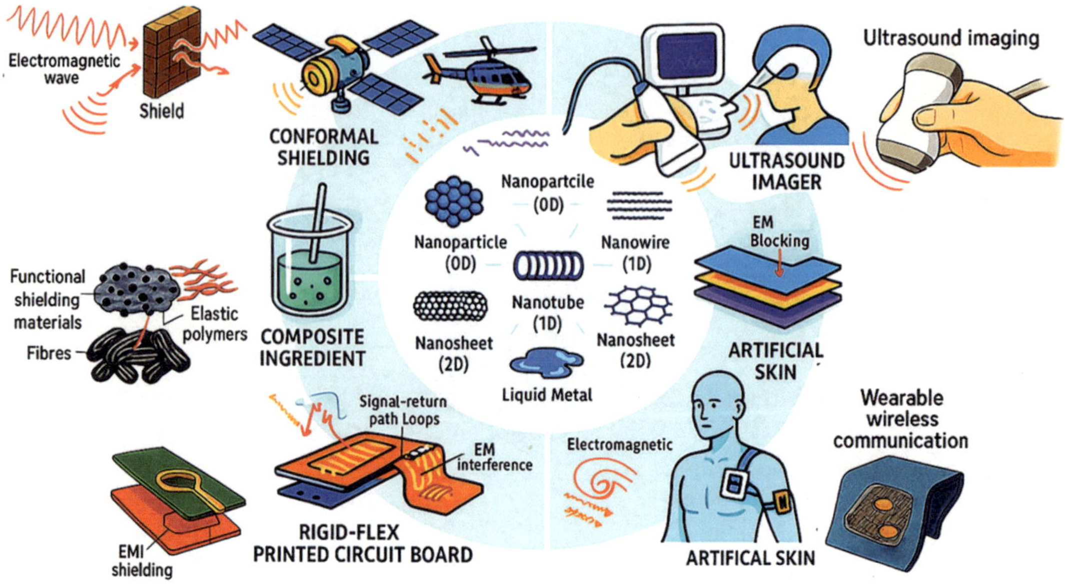

FIGURE 10.6 Comparative analysis of 0D, 1D and 2D nanomaterials in EMI shielding, Permission from [105].

scattering, interfacial polarisation and dielectric permittivity [101]. Materials such as VO_2 nanowires demonstrate switchable conductivity, achieving shielding effectiveness up to 59.4 dB in the Ku-band. Additionally, barium hexaferrites coated with polyaniline enhance interfacial polarisation and charge accumulation, further improving EMI shielding [29, 32, 102]. The superior performance of 2D nanomaterials in EMI shielding is attributed to their layered structure, high conductivity, large surface area and charge carrier mobility. Layered structure facilitates multiple reflections and scattering, enhancing absorption and dissipation of electromagnetic waves [92, 103]. High conductivity supports effective wave attenuation through both reflection and absorption mechanisms [92]. Large surface area promotes enhanced polarisation and interaction with electromagnetic waves, boosting shielding performance [55]. Charge carrier mobility improves interlayer interactions and wave absorption efficiency [26, 104]. The integration of 2D nanomaterials with conductive polymers further optimises shielding performance, enabling lightweight, flexible EMI shields suitable for commercial applications such as wearable devices and aerospace components. Techniques such as blade-coating ensure scalable production of 2D composites while maintaining superior properties [38]. Figure 10.6 illustrates the comparative advantages and applications of 0D, 1D and 2D nanomaterials, emphasising the exceptional capabilities of 2D materials in EMI shielding [102].

10.3 POLYMER MATRIX SELECTION AND PROCESSING TECHNIQUES FOR EMI SHIELDING

The selection of polymer matrices and processing techniques is crucial in optimising the mechanical properties and performance of polymer matrix composites (PMCs). The choice of matrix depends on factors such as the type of reinforcement, desired material properties and processing methods. PMCs can incorporate either thermoplastic or thermosetting matrices, each offering distinct benefits: thermoplastics provide recyclability and ease of processing, while thermosets are known for their superior thermal stability and mechanical strength [106]. Additionally, the choice between natural and synthetic fibres influences the composite's cost, mechanical properties and environmental impact. Natural fibres, in particular, promote sustainability and cost reduction, making them an attractive option for environmentally conscious applications [106, 107], as summarised in Table 10.2.

Techniques such as compression moulding and injection moulding are commonly used to fabricate PMCs, with the selection depending on part design and required properties [106, 108]. Microwave-assisted processing is gaining attention for its energy efficiency and faster curing times, particularly for thermoset composites, though challenges remain when processing thermoplastic composites with natural fibres [109]. The authors emphasise the importance of control during the curing process and discuss methods for optimising PMC manufacturing [110]. Polymers such as polyaniline, polyvinylidene fluoride (PVDF) and epoxy resins are increasingly used in EMI shielding composites due to their lightweight nature, mechanical strength and compatibility with nanomaterials. When combined with carbon-based nanomaterials, polyaniline enhances electrical conductivity and EMI shielding effectiveness [111]. PVDF composites, containing non-oxidised graphene flakes, maintain the graphene's sp^2 structure, leading to superior conductivity and SE [112]. Epoxy resins, incorporating metal nanoparticles and carbon fillers, improve EMI shielding through both absorption and reflection mechanisms [15]. Ensuring uniform dispersion of nanomaterials is crucial for optimal EMI shielding, with solution casting and melt blending being standard techniques for integrating nanomaterials into polymer matrices [24, 113]. Key fabrication methods for polymer nanocomposites include solution casting, electrospinning, in situ polymerisation and hot pressing. Solution casting is widely used for biopolymer-based nanocomposites in food packaging, enhancing barrier properties [114]. Electrospinning produces multifunctional nanofibres, such as PAN/CNT-$CoFe_2O_4$ fibres, which offer both EMI shielding and magnetic properties [115, 116]. In situ polymerisation improves interfacial bonding, enhancing the mechanical and thermal properties of composites [117]. Hot pressing applies heat and pressure to improve density and strength, making it ideal for structural applications [118]. Optimisation of polymer nanocomposites for EMI shielding is influenced by challenges like nanoparticle agglomeration, poor dispersion and the polymer-nanomaterial interface. Effective dispersion, achieved through techniques like covalent modification of nanoparticles, is critical for uniform distribution, improving conductivity and shielding performance [119]. Strategies such as creating preformed 3D architectures and multi-interface structures can enhance conductivity while maintaining lightweight properties [120]. Various strategies for EMI shielding have emerged, including the use of conductive fillers and advanced processing methods. Blends like ethylene propylene diene rubber (EPDM) and linear low-density polyethylene (LLDPE) with conductive carbon black achieve significant shielding effectiveness through a dual-step percolation process, reaching −27.80 dB. Hybrid composites combining conducting polymers, multi-walled carbon nanotubes (MWCNTs) and graphene substrates show superior EMI shielding properties, improving both mechanical and electrical performance [51]. Melt blending ensures uniform distribution of conductive fillers, lowering percolation thresholds and enhancing shielding [121].

TABLE 10.2 Polymer Matrix Selection and Processing Techniques for Nanocomposites [5]

Category	Polymer Examples	Key Properties	Processing Techniques	Applications
Thermoplastics	Polypropylene (PP), Poly(vinylidene fluoride) (PVDF), Poly(methyl methacrylate) (PMMA), Polycarbonate (PC)	Lightweight, flexible, easy to process	Melt blending, Injection moulding	EMI shielding films, lightweight panels
Thermosets	Epoxy, Polyurethane (PU), Phenolic resins	High thermal stability, strong mechanical properties	Solution casting, Compression moulding	Structural EMI shields, coatings
Elastomers	Polydimethylsiloxane (PDMS), Polyurethane elastomers	High flexibility, durability, excellent elasticity	In-situ polymerisation, Dip coating	Wearable shielding, flexible electronics
Conductive Polymers	Polyaniline (PANI), Polypyrrole (PPy), Poly(3,4-ethylenedioxythiophene) (PEDOT)	Intrinsic conductivity, tunable EMI shielding	Electrospinning, Spin coating	Sensors, conductive fabrics
Biopolymers	Chitosan, Cellulose, Starch-based polymers	Biodegradable, sustainable, environmentally friendly	Solution casting, Solvent exchange	Sustainable EMI shielding films
Hybrid Polymer Matrices	Polymer blends (e.g., PP/PC, PANI/PDMS composites)	Synergistic properties (e.g., flexibility, conductivity)	Melt blending, Layered deposition	Multifunctional EMI shielding

Applying external forces, such as magnetic or mechanical fields, can induce filler orientation, further boosting shielding performance [122]. Spectroscopic techniques and electrical conductivity measurements are essential for assessing EMI shielding effectiveness [123]. Research on conductive polymer composites, such as magnetite (particle size <30 nm) deposited on modified graphene (MMG)+PVDF via ultrasonication, demonstrates remarkable EMI shielding effectiveness of 51.74 dB at a low thickness of 3 mm, within the 12–18 GHz frequency range [124]. Multilayered EMI shielding composites with selective on-off characteristics for specific frequency ranges are created using solution printing. These composites, incorporating hexagonal boron nitride (BN) and silver-coated BN particles, enhance shielding across a broad spectrum [125].

10.4 APPLICATIONS OF NANOMATERIAL-ENHANCED EMI SHIELDING POLYMER COMPOSITES

Nanomaterial-enhanced EMI shielding polymer composites are vital for mitigating EMI from electronic devices. These composites utilise the unique properties of nanomaterials, such as graphene and polyaniline-copper nanoparticles, to significantly enhance shielding effectiveness without compromising mechanical integrity. Incorporating graphene into polymer matrices can improve EMI shielding, achieving up to 57.8 dB in the 8.2–12.4 GHz range [126]. Such composites find wide applications across various industries. In consumer electronics, they provide lightweight, flexible shielding for devices such as smartphones and laptops. Aerospace and defence industries benefit from their thermal stability and radiation resistance, crucial for satellites and military equipment. Automotive applications, including shielding for EVs and sensors, leverage high SE and reduced weight. In medical devices, biocompatible composites protect equipment such as pacemakers from EMI, while telecommunications utilise them for 5G systems. Flexible electronics, including wearable sensors, benefit from their stretchability and low weight, and energy systems use these composites for shielding batteries and renewable energy setups. In industrial settings, these materials provide EMI shielding for power inverters and control panels, demonstrating their versatility and resilience (Figure 10.7).

Polyaniline-copper nanocomposites exhibit enhanced shielding properties, especially after gamma irradiation, with varying nanoparticle concentrations improving performance [127]. Composites such as graphene nanoplatelets (GNP)/thermoplastic polyurethane (TPU), known for their high flexibility and tensile strength (6.95 MPa), are suitable for diverse applications [126]. Nanoparticles also enhance thermal conductivity, a crucial property for electronics requiring efficient heat dissipation, making these composites ideal for electronics, aerospace components and other sectors where EMI shielding and fire retardancy are essential [25]. Flexible and lightweight EMI shielding materials are increasingly important in portable electronics such as smartphones, tablets and wearables. Liquid metal and elastomeric composites, in grid-patterned thin films, provide high EMI shielding effectiveness while remaining soft and deformable [128]. These films combine liquid metal and elastomeric materials, enabling strain-tunable properties without generating secondary electromagnetic pollution. Polypyrrole-modified cotton fabrics, achieving a conductivity of 2715.8 S/m and an SE of 28.2 dB at only 0.5 mm thickness, retain lightweight and breathable

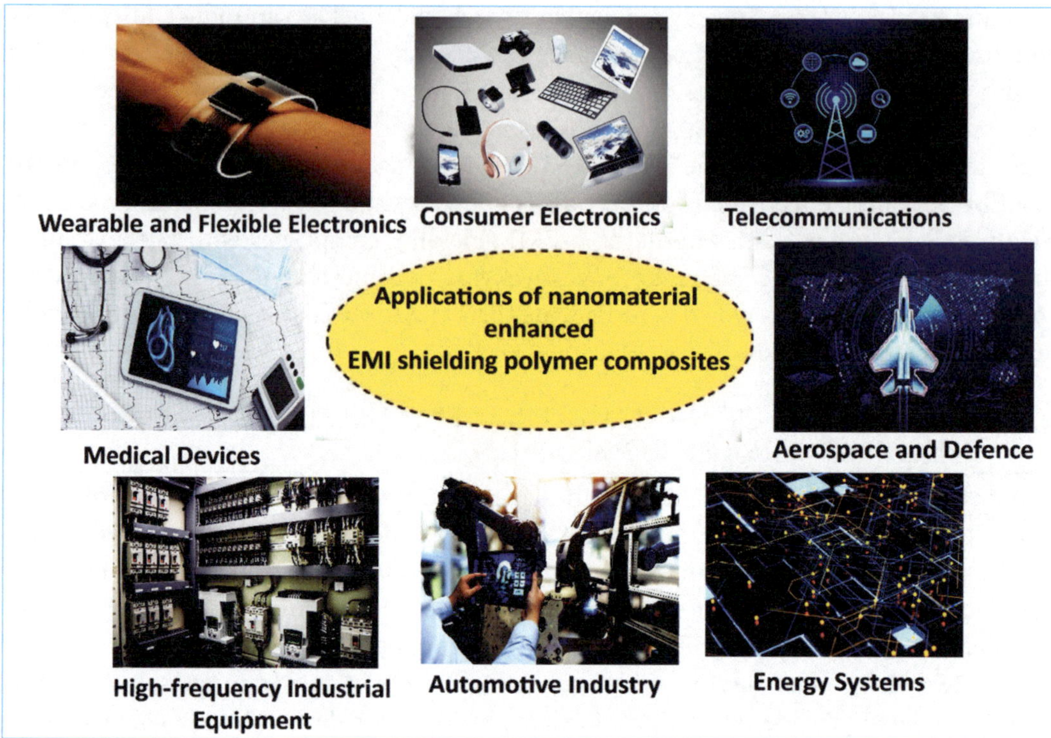

FIGURE 10.7 Applications of nanomaterial-enhanced EMI shielding polymer composites, driven by their tunable properties, lightweight and multifunctionality.

properties for wearable applications [21]. Cellulose/MXene/polyurethane composite aerogels, with an SE of 48.59 dB, offer thermal insulation and hydrophobicity, while MXene enhances electrical conductivity, making them robust for wearable devices [129]. Multifunctional composites combining aramid nanofibres and MXene exhibit an impressive SE of 65.5 dB and impact resistance, making them ideal for portable electronics [22]. In sectors such as EVs and avionics, lightweight EMI shielding materials are essential for protecting sensitive components while maintaining system efficiency. Carbon fibre-reinforced composites (CFRC) provide excellent mechanical strength and EMI shielding, achieving up to 88.27 dB effectiveness from 30 MHz to 3 GHz, crucial for EV battery casings [130]. Multifunctional materials offering EMI shielding along with thermal management and sensing capabilities address the complex demands of modern EVs [131]. In aerospace, lightweight nanocomposites enhance EMI shielding, protecting navigation and communication systems from interference [98]. Ferrite nanoparticle-infused ceramics are used in aerospace applications where both weight and performance are critical for EMI shielding [132]. In the medical field, particularly for implantable devices, EMI shielding ensures device functionality and safety in environments with increasing electromagnetic activity from wireless power transfer systems and 5G technology. Wireless devices, such as walkie-talkies, can exceed safe limits, posing risks to implantable medical devices (IMDs) [133, 134]. Biopolymer composites, such as polylactic acid combined with carbon nanotubes, offer effective EMI shielding

(~30.1 dB) while maintaining biocompatibility, ensuring the safety of IMDs in electromagnetic environments [135]. The International Commission on Non-Ionising Radiation Protection (ICNIRP) guidelines set safety limits for electromagnetic exposure, crucial when designing implantable devices [136]. Monitoring EMI characteristics ensures devices remain within safe operational limits, particularly with emerging technologies such as 5G [137]. Advanced shielding materials are essential for managing high-frequency emissions in 5G infrastructure. Carbonyl iron powder and carbon fabric composites within a polydimethylsiloxane matrix demonstrate EMI shielding effectiveness over 120 dB, with potential increases to 160 dB, making them suitable for 5G applications in the Ka-band (26–40 GHz). These materials also provide enhanced mechanical strength and thermal conductivity, which are vital for maintaining device integrity under operational stresses [138]. A metamaterial absorber designed for 5G mm-wave applications (22–43.5 GHz) achieves over 30 dB shielding effectiveness while being environmentally friendly and flexible, offering polarisation-insensitivity and optical transparency for complex electromagnetic environments [139]. Frequency Selective Surfaces (FSS) for 5G applications operating at 5 GHz provide over 40 dB shielding, with symmetric designs that effectively block EMI from both transverse electric (TE) and transverse magnetic (TM) waves. Advanced neural network-based design techniques optimise FSS configurations for enhanced shielding effectiveness under various incident angles and polarisations [140, 141].

10.5 FUTURE DIRECTIONS AND CHALLENGES

The future of 0D, 1D and 2D nanomaterials in polymer nanocomposites for EMI shielding presents both exciting prospects and challenges. As demand for effective shielding materials grows, researchers are exploring innovative nanomaterial-polymer combinations to enhance performance while navigating production complexities. Carbon-based materials, metal nanoparticles and MXenes play key roles due to their unique electrical properties [24, 49]. Effective EMI shielding relies on the reflection, absorption and multiple reflections of electromagnetic waves, necessitating the careful selection and optimisation of nanofillers [24]. Achieving uniform dispersion of nanofillers within polymer matrices is critical for maintaining mechanical strength and maximising shielding effectiveness, as poor dispersion can compromise both properties [49, 50]. However, integrating advanced nanomaterials often increases production costs, posing challenges to commercial viability, particularly when scaling up for industrial applications [50]. Hybrid composites, which combine different nanofillers, show promise in enhancing EMI shielding performance while maintaining mechanical integrity [25, 142]. Additionally, innovations such as 3D printing and microcellular foaming are being explored to optimise EMI shielding through complex structural designs [142]. The synthesis of 0D, 1D and 2D nanomaterials typically involves expensive raw materials and sophisticated fabrication techniques, driving up costs, which can affect their feasibility for large-scale production [25, 92]. While polymer nanocomposites are generally more affordable, the incorporation of high-performance nanoparticles still increases production costs [24]. Methods like sol-gel, electrospinning and chemical vapour deposition are often used for synthesising different nanomaterials, each with its own set of challenges [49, 143]. Ensuring uniform dispersion and nanoparticle stability in polymer matrices remains a significant challenge, directly impacting EMI shielding performance [49]. Environmental concerns also arise with the use of metal-based nanomaterials, particularly regarding toxicity and pollution. In contrast, biodegradable or recyclable polymer composites incorporating 0D, 1D and 2D nanomaterials offer a promising alternative, reducing environmental impacts while enhancing shielding performance [111, 127]. Metal nanoparticles may leach into ecosystems, posing risks to aquatic and terrestrial life, whereas biodegradable composites decompose naturally, reducing landfill waste while maintaining effective EMI shielding through absorption and reflection mechanisms [49]. Integrating various nanofillers, including carbon and metal oxides, optimises EMI shielding while promoting environmental sustainability [24]. Advances in doping, functionalisation and hybridisation of nanomaterials have substantially improved the performance of EMI shielding materials. For example, incorporating conductive materials such as carbon nanotubes (CNTs) into rGO enhances conductivity and prevents aggregation, improving shielding performance across a broad frequency range [144]. MXenes combined with conductive polymers such as poly(3,4-ethylenedioxythiophene):poly(styrene sulfonate) (PEDOT:PSS) exhibit exceptional shielding effectiveness, with an average SE of 66 dB at minimal thickness [38]. Furthermore, when MXenes are integrated with multi-wall carbon nanotubes (MWCNTs) in a polyimide matrix, a sandwich structure is formed, enhancing shielding performance while maintaining flexibility and thermal insulation [145]. The careful selection of nanomaterials such as MXenes and carbon-based materials is crucial in developing lightweight, cost-effective polymer nanocomposites for EMI shielding, which rely on optimised material design and processing techniques to achieve efficient performance [24].

10.6 CONCLUSIONS

The potential of 0D, 1D and 2D nanomaterials in polymer nanocomposites for EMI shielding is significant due to their unique properties. These nanomaterials enhance EMI shielding by improving absorption, reflection and multiple reflections of electromagnetic waves. 0D nanomaterials, such as carbon black and metal nanoparticles, primarily enhance shielding through absorption by improving electrical conductivity. 1D nanomaterials, including nanowires and nanotubes, facilitate charge transport, boosting the conductivity of the polymer matrix and thereby enhancing shielding effectiveness. 2D nanomaterials, such as graphene and rGO, enhance shielding by improving interfacial polarisation, which strengthens overall EMI protection. Incorporating magnetic materials such as Co_2Z hexaferrite into polymer matrices such as polyvinylidene fluoride (PVDF) further increases shielding effectiveness by improving both absorption and reflection of electromagnetic waves. The combination of various nanofillers in polymer composites can produce synergistic effects, optimising electromagnetic and mechanical properties. These advanced materials, particularly polymer-based composites, exhibit multifunctionality, making them ideal for applications in electronics, automotive and telecommunications. Improved electrical conductivity is essential for effective EMI shielding, with carbon nanotube-reinforced composites demonstrating superior conductivity and shielding performance. The use of conductive fillers such as carbon black enhances electrical percolation, as demonstrated in Printex black composites, which show significant EMI SE. Dielectric permittivity plays a critical role in absorbing electromagnetic waves, with the microstructure of these composites engineered for optimised absorption. Additionally, mechanical strength is crucial for the durability of EMI shielding materials. Hybrid composites, such as CF and carbon nanotube-reinforced polyether ether ketone (PEEK), show increased flexural strength, making them suitable for load-bearing applications. The lightweight nature of polymer composites, combined with high mechanical strength, makes them ideal for automotive and aerospace applications where weight reduction is crucial.

REFERENCES

1. Mishra, R.K., Loganathan, S., Jacob, J., Saha, P.R., & Thomas, S. Progress in polymer nanocomposites for electromagnetic shielding application. in *Modern Physical Chemistry: Engineering Models, Materials, and Methods with Applications* (ed. Reza K. Haghi, Emili Besalu, Maciej Jaroszewski, Sabu Thomas, Praveen K.M.) 1st edn, 198–237 (Apple Academic Press, New York, 2018).
2. Abbasi, H., Antunes, M. & Velasco, J. I. Recent advances in carbon-based polymer nanocomposites for electromagnetic interference shielding. *Progress in Materials Science* vol. 103, Preprint at https://doi.org/10.1016/j.pmatsci.2019.02.003 (2019).
3. Kumari, N., Meena, S., Singhal, R., Choubey, R. K. & Dwivedi, U. K. Advanced material for EMI shielding. in *Advanced Radiation Shielding Materials: Radiation and Radiological Protection* (2024). doi:10.1016/B978-0-323-95387-0.00014-5.
4. Liu, J. & Nicolosi, V. Electrically insulating electromagnetic interference shielding materials: a perspective. *Advanced Functional Materials* (2024) doi:10.1002/adfm.202407439.
5. Mishra, R. K., Thomas, M. G., Abraham, J., Joseph, K. & Thomas, S. Electromagnetic interference shielding materials for aerospace application. in *Advanced Materials for Electromagnetic Shielding* 327–365 (Wiley, 2018). doi:10.1002/9781119128625.ch15.
6. Nowak, R., Henke, A. & Frei, S. Characterization of EMI sources from reconstructed current distributions based on phase-less electric and magnetic near-field data. in *2020 International Symposium on Electromagnetic Compatibility – EMC EUROPE* 1–6 (IEEE, 2020). doi:10.1109/EMCEUROPE48519.2020.9245636.
7. Aragao, G. F., Fontgalland, G. & Dionísio de Andrade, H. Measuring potential EMI fields from electrostatic sources. in *2017 IEEE International Symposium on Antennas and Propagation & USNC/URSI National Radio Science Meeting* 2607–2608 (IEEE, 2017). doi:10.1109/APUSNCURSINRSM.2017.8073346.
8. Mardiguian, M. Generalities on radiated interference. in *Controlling Radiated Emissions by Design* 1–15 (Springer International Publishing, Cham, 2014). doi:10.1007/978-3-319-04771-3_1.
9. Huang, Q. *et al.* Accurate prediction and mitigation of EMI from high-speed noise sources using full wave solver. in *2019 IEEE International Symposium on Electromagnetic Compatibility, Signal & Power Integrity (EMC+SIPI)* 595–599 (IEEE, 2019). doi:10.1109/ISEMC.2019.8825222.
10. Thomas, R., Barber, K., Sedaghati, N., Zhou, L. & Teodorescu, R. Core tunneling: variation-aware voltage noise mitigation in GPUs. in *2016 IEEE International Symposium on High Performance Computer Architecture (HPCA)* 151–162 (IEEE, 2016). doi:10.1109/HPCA.2016.7446061.
11. Reddi, V. J. *et al.* Voltage smoothing: characterizing and mitigating voltage noise in production processors via software-guided thread scheduling. in *2010 43rd Annual IEEE/ACM International Symposium on Microarchitecture* 77–88 (IEEE, 2010). doi:10.1109/MICRO.2010.35.
12. Qing, Z. *et al.* Conducted EMI model and coupling resonant mechanism analysis of multi-cell converter system. in *2024 IEEE 10th International Power Electronics and Motion Control Conference (IPEMC 2024-ECCE Asia)* 898–901 (IEEE, 2024). doi:10.1109/IPEMC-ECCE Asia60879.2024.10567981.

13. Li, F. *et al.* Design and implementation of a new hybrid EMI filter for DC-DC converter with SiC MOSFETs. in *2023 6th Asia Conference on Energy and Electrical Engineering, ACEEE 2023* (2023). doi:10.1109/ACEEE58657.2023.10239563.
14. Zhang, Y. *et al.* Design and integration methodology of CM stripline inductors and EMI filters in SiC power modules for low conductive CM EMI. in *2024 IEEE 10th International Power Electronics and Motion Control Conference (IPEMC 2024-ECCE Asia)* 2253–2258 (IEEE, 2024). doi:10.1109/IPEMC-ECCEAsia60879.2024.10567189.
15. Albert, A. A., Parthasarathy, V. & Kumar, P. S. Review on recent progress in epoxy-based composite materials for electromagnetic interference (EMI) shielding applications. *Polymer Composites* vol. 45 Preprint at https://doi.org/10.1002/pc.27928 (2024).
16. Pavlík, M., Bereš, M. & Beňa, Ľ. The influence of various commonly used building materials on the shielding effectiveness, reflection and absorption of the electromagnetic wave. *Applied Sciences* 14, 2521 (2024).
17. Zheng, S., Wang, Y., Wang, X. & Lu, H. Research progress on high-performance electromagnetic interference shielding materials with well-organized multilayered structures. *Materials Today Physics* 40, 101330 (2024).
18. Bagotia, N. Electromagnetic (EM radiation) interference shielding material epicenter to carbon filler-based composite. in *Advanced Radiation Shielding Materials: Radiation and Radiological Protection* (2024). doi:10.1016/B978-0-323-95387-0.00005-4.
19. Dai, Y. *et al.* Ultrathin and highly flexible layered silver nanowires/carboxymethyl cellulose nanofiber nanocomposite films for electromagnetic interference shielding. *International Journal of Biological Macromolecules* 278, 134770 (2024).
20. Tahalyani, J., Akhtar, M. J. & Kar, K. K. Development of lightweight and flexible carbon nanofiber and polyurethane nanocomposite for EMI shielding applications. in *2023 IEEE Microwaves, Antennas, and Propagation Conference (MAPCON)* 1–6 (IEEE, 2023). doi:10.1109/MAPCON58678.2023.10463914.
21. Meng, L. *et al.* Lightweight, breathable and self-cleaning polypyrrole-modified multifunctional cotton fabric for flexible electromagnetic interference shielding. *International Journal of Biological Macromolecules* 274, 133347 (2024).
22. Wen, H. *et al.* Lightweight flexible composite material with superior electromagnetic interference shielding and impact resistance properties. *Composites Communications* 47, 101868 (2024).
23. Naeema, N. *et al.* Shielding performance of metal oxide nanoparticles-doped polypropylene composites against gamma rays and neutrons exposure. *Radiation Physics and Chemistry* 216, 111461 (2024).
24. Roohani, Y. & Yahyaei, H. Polymer nanocomposite films and coatings for screening electromagnetic interference pollution. in *Polymer Nanocomposite Films and Coatings* 691–727 (Elsevier, 2024). doi:10.1016/B978-0-443-19139-8.00002-4.
25. Ray, S. S., Temane, L. T. & Orasugh, J. T. *NPs for Polymer-Based EMI Shielding and Fire Retarding Nanocomposites* 119–169 (Springer, 2024). doi:10.1007/978-3-031-51924-6_6.
26. Kallambadi Sadashivappa, P. *et al.* Progressive review of functional nanomaterials-based polymer nanocomposites for efficient EMI shielding. *Journal of Composites Science* 7, 77 (2023).
27. Liu, K. *et al.* Multifunctional nanocellulose-based electromagnetic interference shielding composites: design, functionality, and future prospects. *Industrial Crops and Products* 210, 118148 (2024).
28. Jovanovic, S. *et al.* Carbon-based nanomaterials in electromagnetic interference shielding: graphene oxide, reduced graphene oxide, electrochemically exfoliated graphene, and biomass-derivated graphene. in *2024 International Conference on Manipulation, Automation and Robotics at Small Scales (MARSS)* 1–5 (IEEE, 2024). doi:10.1109/MARSS61851.2024.10612734.
29. Kaidar, B. *et al.* Magnetite-incorporated 1D carbon nanostructure hybrids for electromagnetic interference shielding. *Nanomaterials* 14, 1291 (2024).
30. Govindasamy, T., Mathew, N. K., Asapu, V. K., Subramanian, V. & Subramanian, B. Advancing X-band electromagnetic interference shielding: single-layer thin films of Fe_3O_4-MWCNT

(1:2)-PVDF meta-nanocomposite fabricated by low-cost drop-casting process. *Advanced Engineering Materials* 26 (2024).
31. Orasugh, J. T., Temane, L. T. & Ray, S. S. Nanocellulose-based conductive composites: a review of systems for electromagnetic interference shielding applications. *International Journal of Biological Macromolecules* 277, 133891 (2024).
32. Shakir, H. F. *et al.* Electromagnetic interference shielding and DFT study of barium hexaferrites nanoparticles coated polyaniline with heterointerface polarization behavior. *Ceramics International* 50, 28139–28149 (2024).
33. Chopra, A. *et al.* Leveraging $BaFe_{12}O_{19}$/PANI nanocomposite incorporated cotton fabric as an effective EMI shielding flexible material for X-band frequency. *ACS Applied Electronic Materials* 6, 2242–2257 (2024).
34. Ameta, S. C. *Nanomaterials: An Introduction. in The Science of Nanomaterials* 1–18 (Apple Academic Press, New York, 2022). doi:10.1201/9781003283126-1.
35. Almessiere, M. A., Slimani, Y., Thurkanov, A. V. & Baykal, A. Structural and morphological characterization of nanomaterials. in *Synthesis and Applications of Nanoparticles* 121–137 (Springer Nature Singapore, Singapore, 2022). doi:10.1007/978-981-16-6819-7_6.
36. Wan, F., Thakur, A. & Thakur, P. Classification of nanomaterials (carbon, metals, polymers, bioceramics). in *Integrated Nanomaterials and their Applications* 37–55 (Springer Nature Singapore, Singapore, 2023). doi:10.1007/978-981-99-6105-4_3.
37. Verma, G. Nanostructures. in *Nanostructures* 25–60 (Elsevier, 2023). doi:10.1016/B978-0-12-820048-3.00006-1.
38. Roy, S. S., Ghosh, K., Meyyappan, M. & Giri, P. K. MXene-PEDOT:PSS nanocomposite-based electromagnetic interference shields with ultrahigh absolute shielding effectiveness. *ACS Applied Nano Materials* 6, 23357–23369 (2023).
39. Harshapriya, P., Kaur, P. & Basandrai, D. Synthesis and characterization of BFO/BTO/ZnO nanocomposites for promising EMI shielding applications. *Inorganic Chemistry Communications* 156, 111186 (2023).
40. Isari, A. A., Ghaffarkhah, A., Hashemi, S. A., Wuttke, S. & Arjmand, M. Structural design for EMI shielding: from underlying mechanisms to common pitfalls. *Advanced Materials* 36 (2024).
41. Singh, A. K., Shishkin, A., Koppel, T. & Gupta, N. Porous materials for EMI shielding. in *Materials for Potential EMI Shielding Applications* 287–314 (Elsevier, 2020). doi:10.1016/B978-0-12-817590-3.00018-X.
42. Swapnalin, J. *Applications of MXenes in EMI Shielding.* 28–47 (2024). doi:10.21741/9781644902875-2.
43. Nguyen, L., Bergstrom, M. & McGaw, D. Insertion loss of conductive concrete line filters. *IEEE Letters on Electromagnetic Compatibility Practice and Applications* 2, 34–39 (2020).
44. Li, Q., Andersen, M. A. E. & Thomsen, O. C. Conduction losses and common mode EMI analysis on bridgeless power factor correction. in *2009 International Conference on Power Electronics and Drive Systems (PEDS)* 1255–1260 (IEEE, 2009). doi:10.1109/PEDS.2009.5385856.
45. Zheludev, I. S. Electrical conduction and dielectric losses. in *Physics of Crystalline Dielectrics* 455–532 (Springer US, Boston, MA, 1971). doi:10.1007/978-1-4615-8984-6_2.
46. Wang, Y., He, Q.-M., Gao, Y.-N., Yue, T.-N. & Wang, M. Achieving remarkable enhancement on electromagnetic shielding performance in multi-walled carbon nanotube/polydimethylsiloxane composites via adding a small amount of metal micro-particles as scattering points. *Composites. Part A, Applied Science and Manufacturing* 162, 107135 (2022).
47. Warden, D. & Bayazitoglu, Y. Consideration of particle wave diffraction to enhance spacecraft radiation shielding. *Journal of Quantitative Spectroscopy and Radiation Transfer* 246, 106876 (2020).
48. Mishchenko, M. I., Travis, L. D. & Lacis, A. A. *Multiple Scattering of Light by Particles: Radiative Transfer and Coherent Backscattering.* vol. 1 (2006).

49. Kausar, A., Rakha, S. A. & Ahmad, I. Electromagnetic interference shielding and ionizing radiation shielding effect of stimuli-responsive nanocomposites. in *Shape Memory Polymer-Derived Nanocomposites* 179–211 (Elsevier, 2024). doi:10.1016/B978-0-443-18504-5.00004-X.
50. Sebastian, A. & Raghavan, A. Advanced polymer composite with graphene content for EMI shielding. *International Research Journal of Advanced Engineering Hub (IRJAEH)* 2, 1334–1340 (2024).
51. Sebastian, A. & Raghavan, A. Intercalating polymer with nano stratum graphene for EMI shielding. *Suranaree Journal of Science and Technology* 30, 030161(1–9) (2024).
52. Byakodi, M. *et al.* Emerging 0D, 1D, 2D, and 3D nanostructures for efficient point-of-care biosensing. *Biosensors and Bioelectronics: X* 12, 100284 (2022).
53. Parvathi, M. M., Arivazhagan, V. & Rajesh, S. Quantum size effect on the layer by layer assembly of PbTe-InSe multilayer nanocomposite structures. *Journal of Alloys and Compounds* 646 (2015).
54. Jin, B. *et al.* Synergistic manipulation of zero-dimension and one-dimension hybrid nanofillers in multi-layer two-dimension thin films to construct light weight electromagnetic interference material. *Polymers (Basel)* 13, 3278 (2021).
55. Xiang, Z., Shi, Y., Zhu, X., Cai, L. & Lu, W. Flexible and waterproof 2D/1D/0D construction of MXene-based nanocomposites for electromagnetic wave absorption, EMI shielding, and photothermal conversion. *Nano-Micro Letters* 13, 150 (2021).
56. Zare, S. & Edalatpour, S. The quantum confinement effect on the spectrum of near-field thermal radiation by quantum dots. *Journal of Applied Physics* 130 (2021).
57. Nan, K. *et al.* Integrated design of MOFs-derived 0D/1D/2D/3D hierarchical network for high-efficiency electromagnetic wave absorption. *Carbon N Y* 224, 119049 (2024).
58. Zhang, Q. *et al.* Constructing multiple heterogeneous interfaces in one-dimensional carbon fiber materials for superior electromagnetic wave absorption. *Carbon N Y* 226, 119233 (2024).
59. Li, S., Sun, Y., Zhang, K., Jiang, X. & Yu, H. In situ fabrication of heterogeneous Co/nanoporous carbon nano-islands for excellent electromagnetic wave absorption. *Small* 20 (2024).
60. Tan, J. *et al.* Tunable perfect absorber and multifunctional logic device based on graphene metamaterial. *Results in Physics* 50, 106599 (2023).
61. Yang, G. *et al.* Tunable broadband terahertz metamaterial absorber based on vanadium dioxide. *AIP Advances* 12 (2022).
62. Tang, P. *et al.* Construction of Fe_3O_4/PANI composites with hollow 0D/1D structures for efficient microwave absorption. *Applied Surface Science* 649, 159183 (2024).
63. Rengaswamy, K., Asapu, V. K., Sundara, R. & Venkatachalam, S. Effective attenuation of electromagnetic waves by Ag adorned MWCNT-polybenzoxazine composites for EMI shielding application. *Composites Science and Technology* 223, 109411 (2022).
64. Sahoo, R., Sundara, R. & Venkatachalam, S. Morphology dependent EMI shielding performance of Ag-Ni core-shell nanowires. *Journal of Alloys and Compounds* 981, 173693 (2024).
65. Nigil, M. R. *Nanoclay-based Polymer Nanocomposites for Electromagnetic Interference Shielding* 53–78 (2022). doi:10.21741/9781644902035-3.
66. Tiwari, A. K., Kumar, A. & Said, Z. Nanomaterials for electromagnetic interference shielding application. in *Nanotechnology in the Automotive Industry* 749–772 (Elsevier, 2022). doi:10.1016/B978-0-323-90524-4.00035-9.
67. Ravi, S. N. *et al.* Carbon nanomaterials: pioneering innovations in bioimaging and biosensing technologies. *Journal of Molecular Structure* 1316, 138987 (2024).
68. Zhang, J. X. J. & Hoshino, K. Nanomaterials for molecular sensing. in *Molecular Sensors and Nanodevices* 413–487 (Elsevier, 2019). doi:10.1016/B978-0-12-814862-4.00007-7.
69. Lu, J. & Wong, C. P. Dielectric performance enhancement of nanoparticle-based materials for embedded passive applications. in *2008 58th Electronic Components and Technology Conference* 736–741 (IEEE, 2008). doi:10.1109/ECTC.2008.4550055.
70. Hsu, C.-Y. *et al.* Nanowires properties and applications: a review study. *South African Journal of Chemical Engineering* 46, 286–311 (2023).

71. Xu, Z. *et al.* Study on percolation state and electrical properties of carbon nanotube network. in *2021 4th International Conference on Circuits, Systems and Simulation (ICCSS)* 158–162 (IEEE, 2021). doi:10.1109/ICCSS51193.2021.9464190.
72. Lee, D.-K., Yoo, J., Kim, H., Kang, B.-H. & Park, S.-H. Electrical and thermal properties of carbon nanotube polymer composites with various aspect ratios. *Materials* 15, 1356 (2022).
73. Tsentalovich, D. E. *et al.* Influence of carbon nanotube characteristics on macroscopic fiber properties. *ACS Applied Materials & Interfaces* 9, 36189–36198 (2017).
74. Tahalyani, J., Akhtar, M. J., Kar, K. K. & Chakradhary, V. K. Design and Development of Carbon Nanofiber based Lightweight Flexible Nanocomposite Structure for EMI Shielding Application. in *2022 IEEE Microwaves, Antennas, and Propagation Conference (MAPCON)* 1228–1233 (IEEE, 2022). doi:10.1109/MAPCON56011.2022.10047143.
75. Petchimuthu, K., Ragupathy, B. S., Anand, J. S., Perumal, S. & Chinnapiyan, V. Recent development in one-dimensional polymer-based nanomaterials for high-performance solar cells. in *One-Dimensional Polymeric Nanocomposites* 321–336 (CRC Press, Boca Raton, 2022). doi:10.1201/9781003223764-18.
76. Gooneie, A. & Hufenus, R. Hybrid carbon nanoparticles in polymer matrix for efficient connected networks: self-assembly and continuous pathways. *Macromolecules* 51, 3547–3562 (2018).
77. Nimal, R. *et al.* One-dimensional polymeric nanocomposites for electrochemical sensors. in *One-Dimensional Polymeric Nanocomposites* 381–400 (CRC Press, Boca Raton, 2022). doi:10.1201/9781003223764-21.
78. Bai, X. *et al.* Magnetic vortex configuration in Fe_3O_4 nanorings/nanotubes for aspect ratio and magnetic orientation regulated microwave absorption. *Results in Physics* 58, 107543 (2024).
79. Cai, H. *et al.* Carbon nanotube/$FeNi_3$ nanoparticle composites for electromagnetic wave absorption. *ACS Applied Nano Materials* 7, 11302–11312 (2024).
80. Dubey, P., Gupta, M. & Kundalwal, S. I. Conductive polymer nanocomposite incorporated with carbon nanotubes for effective electromagnetic interference shielding: a numerical study. *Polymer Composites* 45, 3576–3590 (2024).
81. Ye, J. *et al.* Carbon nanotube films embedded with Cu@C nanocubes for electromagnetic interference shielding. *Journal of Polymer Science* 61, 2688–2696 (2023).
82. Manna, K., Sen Gupta, R. & Bose, S. A universal approach to 'host' carbon nanotubes on a charge triggered 'guest' interpenetrating polymer network for excellent 'green' electromagnetic interference shielding. *Nanoscale* 15, 1373–1391 (2023).
83. Raju, P., Kanaka Durga, M., Udaya Kumar, S. & Neeraja Rani, G. Microwave attenuation studies of polypyrrole-SWCNT nanocomposite films for improved EMI shielding. *ECS Journal of Solid State Science and Technology* 11, 091007 (2022).
84. Mishra, R. K. *et al.* Exploring transformative and multifunctional potential of MXenes in 2D materials for next-generation technology. *Open Ceramics* 18, 100596 (2024).
85. Mateen, A., Tag Eldin, E., Bajaber, M. A. & Sufyan Javed, M. MXene-based 2D materials for energy production, conversion and storage applications. in *Materials for Energy Production, Conversion, and Storage* 271–284 (CRC Press, Boca Raton, 2024). doi:10.1201/9781003318859-14.
86. Alqarni, A. Characterization and application of graphene and other two-dimensional materials in nanotechnology and electronics. Preprint at 10.21203/rs.3.rs-4192806/v1 (2024).
87. Hu, M. *et al.* Two-dimensional materials: emerging toolkit for construction of ultrathin high-efficiency microwave shield and absorber. *Frontiers of Physics (Beijing)* 13, 138113 (2018).
88. Zhang, J. *et al.* Design and synthesis strategies: 2D materials for electromagnetic shielding/absorbing. *Chemistry, an Asian Journal* 16, 3817–3832 (2021).
89. Zhang, X. F., Guo, J. J. & Qin, G. W. Assembled micro-nano particles with multiple interface polarizations for electromagnetic absorption at gigahertz. *Applied Physics Letters* 104 (2014).
90. Du, G., Li, C., Shan, L. & Cheng, L. High-throughput predictions of two-dimensional dielectrics with first-principles calculations. *Physical Review B* 108, 235409 (2023).

91. Yu, M. Fast-kinetics electrochemical charge storage by 2D layered materials. *ECS Meeting Abstract* MA2024-01, 979–979 (2024).
92. Saini, A., Ohlan, A., Dhawan, S. K. & Singh, K. Nanostructured two-dimensional (2D) materials as potential candidates for EMI shielding. in *Smart Materials Design for Electromagnetic Interference Shielding Applications* 465–526 (Bentham Science Publishers, 2022). doi:10.2174/9789815036428122010014.
93. Zuikafly, S. N. F., Yahaya, H., Ibrahim, M. H. & Ahmad, F. Two-dimensional nanomaterial-based polymer nanocomposites. in *Two-Dimensional Nanomaterials-Based Polymer Nanocomposites* 231–283 (Wiley, 2024). doi:10.1002/9781119905110.ch7.
94. Gao, Q., Wang, X., Schubert, D. W. & Liu, X. Review on polymer/MXene composites for electromagnetic interference shielding applications. *Advanced Nanocomposites* 1, 52–76 (2024).
95. Rasilainen, I., Lahtela, V. & Kärki, T. A review of plastic waste nanocomposites: assessment of features and applications. *Discover Nano* 19, 112 (2024).
96. You, L. *et al.* Energy storage performance of polymer-based dielectric composites with two-dimensional fillers. *Nanomaterials* 13, 2842 (2023).
97. Hooshyari, K. *et al.* Thermal, mechanical, and viscoelastic properties of two-dimensional nanomaterial-based polymer nanocomposites. in *Two-Dimensional Nanomaterials-Based Polymer Nanocomposites* 311–341 (Wiley, 2024). doi:10.1002/9781119905110.ch9.
98. Bijulin Greety D. J. & Wessley, G. J. J. A study on EMI shielding in aircraft: introduction, methods and significance of using electrospun nanocomposites. *Journal of Space Safety Engineering* 11, 150–160 (2024).
99. Mohan, L., Karakkad, S. & Krishnan, S. T. Development of lightweight carbon nanotube-based epoxy nanocomposite shield for broadband electromagnetic interference shielding: estimating shielding effectiveness and experimental validation. *Journal of Materials Science: Materials in Electronics* 32, 4437–4447 (2021).
100. Tiwari, A. *et al.* Nanomaterials for electromagnetic interference shielding applications: a review. *Nano* 17 (2022).
101. Habibpour, S. *et al.* Synergistic layered design of aerogel nanocomposite of graphene nanoribbon/MXene with tunable absorption dominated electromagnetic interference shielding. *Small* 20 (2024).
102. Liang, S. *et al.* VO_2 nanowires based composite films with broadband and high tunable electromagnetic interference shielding performance. Preprint at 10.21203/rs.3.rs-3306249/v1 (2023).
103. Dhanasekaran, A., Dhanaraj, K. & Venugopal, V. 2D transition metal dichalcogenides-based composites for electromagnetic interference shielding and microwave absorption applications. *Physica Status Solidi* 221 (2024).
104. Oliveira, F. M. & Gusmão, R. Recent advances in the electromagnetic interference shielding of 2D materials beyond graphene. *ACS Applied Electronic Materials* 2, 3048–3071 (2020).
105. Nan, Z., Wei, W., Lin, Z., Chang, J. & Hao, Y. Flexible nanocomposite conductors for electromagnetic interference shielding. *Nano-Micro Letters* 15, 172 (2023).
106. Singh, M. K., Verma, N. & Zafar, S. Conventional processing of polymer matrix composites. in *Lightweight Polymer Composite Structures* 21–66 (CRC Press, 2020). doi:10.1201/9780429244087-2.
107. Thanikodi, S. *et al.* Optimizing the selection of natural fibre reinforcement and polymer matrix for plastic composite using LS-SVM technique. *Chemosphere* 349, 140971 (2024).
108. Ngo, T.-D. Composite preparation techniques. in *Wool Fiber Reinforced Polymer Composites* 105–120 (Elsevier, 2022). doi:10.1016/B978-0-12-824056-4.00003-0.
109. Naik, T. P., Singh, I. & Sharma, A. K. Processing of polymer matrix composites using microwave energy: a review. *Composites. Part A, Applied Science and Manufacturing* 156, 106870 (2022).
110. Carlone, P., Sorrentino, L. & Kent, R. M. 7.16 Process control for polymeric composite manufacture. in *Comprehensive Composite Materials II* 337–354 (Elsevier, 2018). doi:10.1016/B978-0-12-803581-8.03926-6.

111. Mamatha, G. M., Dixit, P., Krishna, R. H. & Girish Kumar, S. Polymer based composites for electromagnetic interference (EMI) shielding: the role of magnetic fillers in effective attenuation of microwaves, a review. *Hybrid Advances* 6, 100200 (2024).
112. Lee, J. Y. *et al.* Development of non-oxidized graphene flakes (NOGF)/polyvinylidene fluoride (PVDF) composites for high-performance EMI shielding efficiency. *Molecular Crystals and Liquid Crystals* 768, 410–415 (2024).
113. Li, J. *et al.* Design, processing, and challenges of multicomponent polymer composites for improved electromagnetic interference shielding properties: a review. *Macromolecular Materials and Engineering* 309 (2024).
114. Moreira, J. B. *et al.* Preparation of the nanocomposite films for the food packaging applications. in *Biocomposites for Industrial Applications* 313–334 (Elsevier, 2024). doi:10.1016/B978-0-323-91866-4.00007-X.
115. Ni, Q. Q., Jin, X. D., Xia, H. & Liu, F. Electrospinning, processing and characterization of polymer-based nano-composite fibers. in *Advances in Filament Yarn Spinning of Textiles and Polymers* 128–148 (Elsevier, 2014). doi:10.1533/9780857099174.2.128.
116. Ranjbar, B., Foroughirad, S. & Ranjbar, Z. Polymer nanocomposite films and coatings. in *Polymer Nanocomposite Films and Coatings* 1–27 (Elsevier, 2024). doi:10.1016/B978-0-443-19139-8.00001-2.
117. Ray, S. S., Temane, L. T. & Orasugh, J. T. *Polymer Nanocomposites* 1–5 (2024). doi:10.1007/978-3-031-51924-6_1.
118. Gou, J., Zhuge, J. & Liang, F. Processing of polymer nanocomposites. in *Manufacturing Techniques for Polymer Matrix Composites (PMCs)* 95–119 (Elsevier, 2012). doi:10.1533/9780857096258.1.95.
119. Gebrekrstos, A. & Ray, S. S. Superior electrical conductivity and mechanical properties of phase-separated polymer blend composites by tuning the localization of nanoparticles for electromagnetic interference shielding applications. *Journal of Polymer Science* 61, 2567–2584 (2023).
120. Wu, X. Y. & Zhang, H. Bin. Study on structure design and electromagnetic shielding properties of polymer nanocomposites. *Acta Polymerica Sinica* 51 (2020).
121. Choudhury, S. N. *et al.* Enhancing electrical characteristics and electromagnetic interference shielding effectiveness in thermoplastic elastomeric polymer blends by utilizing the selective distribution of conductive black. *Functional Composite Materials* 5, 4 (2024).
122. Liang, C., Qiu, H., Zhang, Y., Liu, Y. & Gu, J. External field-assisted techniques for polymer matrix composites with electromagnetic interference shielding. *Science Bulletin (Beijing)* 68, 1938–1953 (2023).
123. Ray, S. S., Temane, L. T. & Orasugh, J. T. *Techniques for polymer-based EMI shielding and fire retarding characteristics measurement* 43–118 (2024). doi:10.1007/978-3-031-51924-6_5.
124. Ayub, S., Guan, B. H. & You, K. Y. Electromagnetic interference shielding mechanisms of MMG@PVDF composites for a broadband frequency range. *Materials Today Communications* 35, 106273 (2023).
125. Seol, M. *et al.* Solution printable multifunctional polymer-based composites for smart electromagnetic interference shielding with tunable frequency and on–off selectivities. *Advanced Composites and Hybrid Materials* 6, 46 (2023).
126. Kang, G.-H., Jeong, J.-R. & Kim, W.-S. Enhancing electromagnetic interference-shielding: flexible graphene nanoplate composites with high connectivity. *Applied Surface Science* 666, 160282 (2024).
127. Bekhit, M., Fathy, E. S. & Sharaf, A. Effect of gamma irradiation on properties of the synthesized PANI-Cu nanoparticles assimilated into PS polymer for electromagnetic interference shielding application. *Scientific Reports* 14, 16403 (2024).
128. Wei, Y. *et al.* Liquid metal grid patterned thin film devices toward absorption-dominant and strain-tunable electromagnetic interference shielding. *Nano-Micro Letters* 16, 248 (2024).

129. Chai, H. *et al.* Lightweight and robust cellulose/MXene/polyurethane composite aerogels as personal protective wearable devices for electromagnetic interference shielding. *International Journal of Biological Macromolecules* 271, 132435 (2024).
130. Hu, S., Wang, D., Večerník, J., Křemenáková, D. & Militký, J. Electromagnetic interference (EMI) shielding and thermal management of sandwich-structured carbon fiber-reinforced composite (CFRC) for electric vehicle battery casings. *Polymers (Basel)* 16, 2291 (2024).
131. Nguyen, Q.-D. & Choi, C.-G. Recent advances in multifunctional electromagnetic interference shielding materials. *Heliyon* 10, e31118 (2024).
132. Abramovskis, V. *et al.* High-temperature, lightweight ceramics with nano-sized ferrites for EMI shielding: synthesis, characterisation, and potential applications. *Materials* 16, 7615 (2023).
133. Hikage, T. Impact of electromagnetic interference arising from wireless power transfer upon implantable medical device. in *Wireless Power Transfer Technologies: Theory and Technologies* 295–314 (Institution of Engineering and Technology, 2024). doi:10.1049/PBPO256E_ch13.
134. Duque, J. L., Urbina, R., Pérez, M. R. & Araque, J. L. Numerical investigation of the interaction of wireless EM fields with implantable medical devices. in *2023 17th European Conference on Antennas and Propagation (EuCAP)* 1–5 (IEEE, 2023). doi:10.23919/EuCAP57121.2023.10133801.
135. Liu, T. *et al.* Green segregated honeycomb biopolymer composites for electromagnetic interference shielding biomedical devices. *Chemical Engineering Journal* 493, 152438 (2024).
136. Che, K. *et al.* Electromagnetic simulation and electromagnetic safety characteristics analysis of implantable medical devices in wireless charging processes. *Journal of Computational Methods in Sciences and Engineering* 24, 2357–2374 (2024).
137. Tsumura, F. *et al.* Interference-induced voltage estimation for EMI characteristics evaluation of implantable medical devices exposed to 5G frequency radio waves. in *2023 IEEE International Symposium on Antennas and Propagation (ISAP)* 1–2 (IEEE, 2023). doi:10.1109/ISAP57493.2023.10388721.
138. Wang, Y.-H. & Li, C.-C. Highly effective EMI shielding composites for 5G Ka-band frequencies. *Applied Materials Today* 36, 102041 (2024).
139. Li, S. *et al.* High-performance low-profile green electromagnetic interference shielding material for 5G mm-wave using characteristic mode theory. *Journal of Physics D: Applied Physics* 57, 145103 (2024).
140. Florence, E., Samuel Robinson, R., Vimal Samsingh, R. & Abirami, S. Design of a single layer single band 5G FSS as shielding materials. *IOP Conference Series: Materials Science and Engineering* 1291, 012044 (2023).
141. Shall, H., Hawess, H., Baccar, S. & Kadi, M. A neural network-empowered inverse FSS design and synthesis approach for 5G shielding applications. *IEEE Access* 12, 102288–102298 (2024).
142. Antunes, M. Recent trends in polymeric foams and porous structures for electromagnetic interference shielding applications. *Polymers (Basel)* 16, 195 (2024).
143. Rusen, E. *et al.* MXenes and polymeric colloids nanocomposites for EMI shielding. *Journal of Materials Chemistry C* 12, 11586–11593 (2024).
144. Kang, J.-S., Zhang, H.-X. & Yoon, K.-B. Hybrid rGO–CNT/epoxy nanocomposites as electromagnetic interference shielding materials. *ACS Applied Nano Materials* 7, 10770–10778 (2024).
145. Liang, W. *et al.* Construction of PI-MXene-MWCNT nanocomposite film integrating conductive gradient with sandwich structure for high-efficiency electromagnetic interference shielding in extreme environments. *Carbon (N.Y.)* 228, 119328 (2024).

CHAPTER 11

Recent Progress in Non-polymer EMI Shielding Materials

Soleyman Sahebi and Farhad Aslani

11.1 BASIC OF EMI SHIELDING FOR DIFFERENT FREQUENCIES

Electromagnetic interference (EMI) is interruption caused by external sources that impact an electrical circuit through electrostatic coupling, electromagnetic induction or conduction [1]. EMI can disrupt the performance of electronic devices, causing failures or data loss. EMI shielding aims to block or reduce this interference, ensuring the proper functioning of electronic systems. The efficiency of EMI shielding is influenced by frequency, as electromagnetic waves (EMWs) exhibit varying propagation characteristics across different frequencies. Shielding effectiveness (SE) is a measure of a material's ability to attenuate EMWs, expressed as the ratio of the incident energy to the energy transmitted through the shielding material. SE includes both absorption and reflection phenomena, with the residual energy being that which is neither absorbed nor reflected, thereby penetrating through the shield. Also, multiple scattering plays a significant role in enhancing the effectiveness of EMI shielding, particularly in materials that rely on absorption rather than mere reflection of EMWs. When EMWs encounter heterogeneous structures, such as composite materials with various fillers or complex structures, the waves are scattered multiple times within the material. This process increases the path length of the waves inside the shielding material, which in turn enhances the chances of absorption. The scattered waves interact with the material's conductive or magnetic components, dissipating their energy in the form of heat. The phenomenon of multiple scattering is especially effective at higher frequencies, as shorter wavelengths are more susceptible to scattering [2]. Mathematically, *SE* is represented as Equation (11.1):

$$SE(dB) = 20\log\left(\frac{E_{\text{incident}}}{E_{\text{transmitted}}}\right) \tag{11.1}$$

where E_{incident} is the incident electric field and $E_{\text{transmitted}}$ is the transmitted electric field.

DOI: 10.1201/9781003512363-11

The effectiveness of EMI shielding is heavily influenced by the skin depth, which determines how deeply EMWs penetrate the material; conductivity, which dictates how well the material reflects or absorbs the waves; and the thickness of the shielding, which must exceed the skin depth to ensure optimal protection across different frequencies.

Skin depth (δ) denotes the depth to which an EMW can penetrate a conductive material before its intensity is significantly diminished and is inversely proportional to the square root of the frequency (f) and the material's electrical conductivity (σ), as described by Equation (11.2):

$$\delta = \frac{1}{\sqrt{\mu \pi f \sigma}} \tag{11.2}$$

where μ is the material's permeability and f is the frequency of the EMW. As the frequency increases, the skin depth decreases, causing the wave to penetrate only the surface of the shielding material. As the frequency increases, thinner materials can be effective for shielding since the waves do not penetrate deeply.

Material conductivity plays a crucial role in determining the *SE* as highly conductive materials such as metals excel at reflecting and attenuating EMWs. Conductivity is inversely related to electrical resistivity (ρ), with the relationship described by Equation (11.3):

$$\sigma = \frac{1}{\rho} \tag{11.3}$$

where σ is the electrical conductivity (S/m) and ρ is the electrical resistivity (Ů·m). For effective EMI shielding, materials with high conductivity (low resistivity), such as metals, are preferred due to their superior ability to reflect and absorb EMWs efficiently.

Material thickness also directly influences shielding performance. At lower frequencies, thicker materials are typically required because the waves can penetrate deeper into the shield. At higher frequencies, the reduced skin depth allows thinner materials to provide effective shielding, as the waves are absorbed near the surface.

EMWs consist of magnetic field (H) and electric field (E) components. The ratio of these fields (E to H), known as wave impedance, is 377 Ω in free space [3]. EMI shielding is categorised into two regions: near field, where the distance between the radiation source and the shield is less than $k/2\pi$ (with k representing the wavelength), and far field, where the distance is greater. Near-field shielding uses electric and magnetic dipole contributions, while far-field shielding employs electromagnetic plane wave theory [4].

EMI shielding operates on the principle of reflection and absorption of EMWs. Materials used for shielding, such as metals and conductive polymers, reflect and absorb EMI, preventing it from penetrating sensitive electronic components.

The key parameters influencing EMI SE include: (1) shielding material, (2) shield thickness and (3) frequency of EMI.

Furthermore, the effectiveness of EMI shielding varies significantly across different frequencies. This means that a material or shield that provides excellent protection at one frequency may not perform as well at another frequency. The following is the basis for categorising and characterising the effect of frequency on the effectiveness of EMI shielding materials for low, medium and high-frequency ranges:

1. Low frequencies (3 kHz to 30 MHz): At low frequencies, EMI propagates through conduction and induction. The wavelength of these frequencies is long, making it easier for EMI to penetrate shielding materials. Materials with high magnetic permeability, such as mu-metal, are employed to achieve effective shielding. Mu-metal is a high-permeability alloy widely recognised for its exceptional magnetic shielding properties. It is composed of 14% iron, 5% copper, 1.5% chromium and 79.5% nickel, making it particularly effective at minimising magnetic interference by channelling magnetic fields away from sensitive components. SE in this range often relies on the thickness and continuity of the shield.
2. Medium frequencies (3 MHz to 1 GHz): At medium frequencies, EMI primarily transmits through both conduction and radiation. The wavelengths are shorter, which increases the energy and potential interference. Materials with high electrical conductivity, such as copper and aluminium, are employed to achieve effective shielding at these frequencies. These materials reflect EMWs effectively and are often applied as conductive coatings on non-metallic surfaces.
3. High frequencies (1–10 GHz): At high frequencies, EMI primarily propagates via radiation due to the shorter wavelengths and higher energy levels, which increase the ease with which EMI can penetrate shielding materials. To achieve effective shielding, advanced materials such as carbon nanotubes (CNTs), graphene composites and metalised fabrics are employed. These materials provide excellent electrical conductivity and flexibility, rendering them ideal for high-frequency applications [1].

Understanding the frequency range of EMI is crucial for selecting the appropriate shielding materials. Each frequency range interacts differently with shielding materials, influencing the choice of material based on its electrical conductivity, magnetic permeability and physical properties.

11.2 INTRODUCTION TO NON-POLYMER EMI SHIELDING MATERIALS

EMI shielding is crucial for blocking EMWs to prevent interference with electronic devices, which can be achieved through mechanisms of absorption, reflection or a combination of both. EMI SE is quantified using parameters such as SE, absorption loss (SEA) and reflection loss (SER) [5]. This chapter provides an inclusive review of non-polymer materials used for EMI shielding, addressing both functional and multifunctional structural materials.

In EMI shielding, materials can broadly be classified into organic and inorganic categories, each offering distinct advantages. Organic materials, including carbon-based substances such as graphene, graphite and CNTs, are highly regarded for their excellent electrical conductivity and outstanding absorption properties. These characteristics make

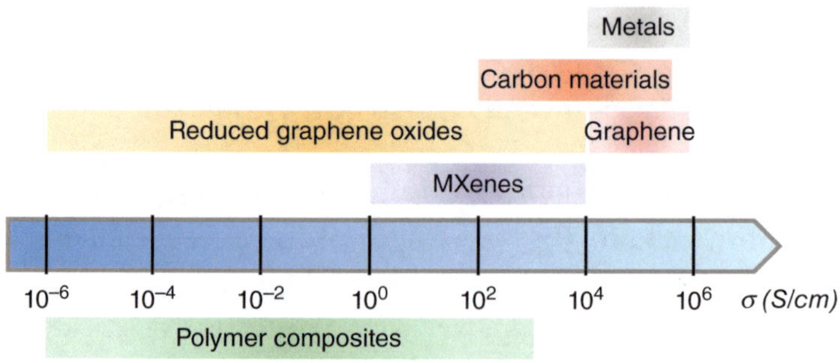

FIGURE 11.1 Conductivity ranges of various raw materials for EMI shielding. Reproduced [8] in accordance with the Creative Commons Attribution License.

them particularly suitable for lightweight and flexible shielding solutions. These materials are derived from carbon structures and are noted for their versatile mechanical properties and superior absorption of EMWs. In contrast, inorganic materials, such as metals (e.g., copper, aluminium and nickel) and ceramics (e.g., ferrites and silicon carbide), are valued for their high conductivity, durability and stability across a wide range of frequencies. Metals offer excellent reflection and mechanical strength, while ceramics are increasingly used for their thermal stability and capacity to absorb EMWs at specific frequencies. The choice between organic and inorganic materials depends on the desired properties, such as flexibility, weight and the frequency range of interference [2, 6].

Metals, including copper, aluminium and nickel, have traditionally been used for EMI shielding due to their excellent electrical conductivity and robust mechanical properties. Carbon-based materials, such as graphite, graphene and CNTs, are recognised for their exceptional electrical conductivity and superior absorption efficiency. Meanwhile, ceramics such as ferrites and silicon carbide are gaining prominence in EMI shielding applications because of their stability and their capability to absorb EMWs across a range of frequencies [7]. Also, cement-based materials become effective EMI shields when combined with conductive components such as CFs or steel fibres (SF) [5]. Hybrid and composite materials, which combine metals, carbons and polymers, are emphasised for their ability to offer tailored properties that enhance EMI shielding performance. The structure of these composites, including fibre arrangement and the use of nanofillers, is critical to their effectiveness [2].

Figure 11.1 shows the conductivity ranges of various EMI shielding materials. The graph indicates that metals and carbon materials, including graphene, exhibit high electrical conductivity (σ) in the range of 10^4 to 10^6 S/cm. Reduced graphene oxides (GO) and MXenes occupy the intermediate range, while polymer composites and ceramics have the lowest conductivity.

11.3 METALLIC-BASED EMI SHIELDING MATERIALS

Despite significant research into alternative materials for EMI shielding, metal sheets remain the primary choice due to their high SE. Although metals exhibit exceptional performance in EMI shielding, ongoing studies aim to develop alternatives that overcome their limitations. Nevertheless, many of these emerging materials have yet to replicate the combined benefits of metals, particularly their superior electrical conductivity and efficiency in attenuating EMI [9]. However, despite their effectiveness, metallic materials such as copper, aluminium and silver come with several limitations. Weight is a significant drawback, especially in applications where lightweight materials are required, such as in aerospace and portable devices. Moreover, metals are prone to corrosion, particularly when exposed to harsh environments, which can degrade their long-term performance. This is evident with materials such as copper, which, while highly conductive, is vulnerable to oxidation. Cost is another limitation, particularly for high-performance metals such as silver, which, despite its superior conductivity, is expensive and often impractical for large-scale applications. These challenges have prompted research into alternatives such as metal-polymer composites that aim to balance shielding performance with reduced weight, corrosion resistance and cost-effectiveness [10].

Metals come in various forms, including traditional metal sheets such as copper and aluminium, which offer excellent SE but are often heavy and prone to corrosion. Metal foils and thin metal coatings on substrates such as polymers and textiles provide a lighter and more flexible alternative while maintaining high SE [11]. Additionally, metal nanoparticles, such as those of silver, iron or copper, are incorporated into composites to enhance their EMI shielding properties, offering a balance of conductivity and reduced weight [12, 13].

Metal-polymer composites combine the benefits of metals and polymers, creating materials that are both conductive and flexible. Furthermore, metal-coated fabrics and textiles are increasingly used for wearable electronics and other applications where flexibility and lightweight are crucial. Advanced forms such as layered metal structures and metal foams increase multiple reflections and scattering of EMWs, significantly boosting their SE [14]. These diverse forms of metal-based materials are continually being optimised to improve their shielding performance, mechanical properties and long-term stability, making them well suited for various practical applications. However, modifying metallic materials often reduces their electrical conductivity, negatively impacting SE. Therefore, modifications must maintain high SE while addressing the drawbacks of metal sheets [15].

EMI shielding is fundamentally achieved through three main mechanisms: reflection, absorption and multiple reflections. Reflection occurs when conductive materials with free charge carriers reflect EMWs, while absorption involves the transformation of EMW energy into heat through electric and magnetic dipoles. Multiple reflections are particularly significant in porous materials, which enhance EMI shielding by causing EMWs to reflect multiple times within the material's structure [13]. Previous research has identified three factors contributing to EMW leakage through metallic shielding enclosures:

1. Conductive penetrations such as pipes and wires, which allow EMWs to bypass the shielding layer.
2. Apertures, such as small gaps between shielding panels.
3. The specific kind of material employed for shielding.

To reduce the impact of conductive penetrations, supplementary shielding measures, such as gaskets, should be implemented. The study by Hoeft et al. on nickel-on-copper plating and aluminium, examining frequencies from 1 kHz to 20 GHz, revealed that at lower frequencies, the presence and dimensions of apertures are more significant than the choice of shielding material. Gaps between metal panels can function as apertures through which EMWs escape. Careful assembly and the application of gaskets can help to mitigate this issue [16].

Also, to understand the factors influencing SE, it is essential to correlate this performance with a material's electrical, magnetic, dimensional and microstructural attributes. For example, for nickel, a ferromagnetic and conductive material, this correlation involves examining its SE, magnetic properties, electrical resistivity and microstructure. Table 11.1 presents these data for filamentous nickel with diameters of 0.4, 2 and 20 μm [17].

Based on Table 11.1, the EMI SE of filamentous nickel increases as the diameter increases (0.4, 2 and 20 μm), with the smallest diameter exhibiting the highest SE, coercive force and hysteresis energy. Greater EMI SE is associated with smaller diameters and larger precise boundary areas. The primary effect is the skin effect rather than grain size, resistivity or the number of grains across the diameter [5].

Traditional metals such as copper and aluminium are commonly used for EMI shielding due to their high electrical conductivity. However, these metals are heavy, prone to corrosion and difficult to fabricate into the desired shapes. Consequently, research has focused on developing novel materials and composites that overcome these limitations while maintaining high SE. For most metallic sheets and other forms, including foam and nanoparticles, the SE is predominantly due to absorption. For instance, at 1 GHz, the SE due to absorption (SEA) is 43 dB, while the SE due to reflection (SER) is only 5 dB [5]. This absorption dominance persists despite the aluminium foil's thinness and high conductivity. This phenomenon is partly attributed to the remarkably high relative electric permittivity, measured at 54,800 at 2 kHz. This elevated permittivity arises from the interaction between a minor proportion of free electrons and the surrounding atoms, which promotes polarisation. This polarisation significantly enhances the material's capacity for absorption, contributing to its effectiveness in EMI shielding.

TABLE 11.1 Interaction Between EMI Shielding Effectiveness and Magnetic, Electrical and Microstructural Properties

Features	**Nickel Nanofibre**	**Nickel Fibre A**	**Nickel Fibre B**
Diameter (μm)	0.4	2	20
EMI SE (1–2 GHz), 19 vol.% filler	92	72	5
Nickel magnetic hysteresis loss (J/m^3)	7370	4930	1020
Magnetic coercive field (kA/m)	16.9	4.7	0.5
Resistivity (DC), 19 vol.% filler in composite (Ω.cm)	1.5×10^{-3}	2.9×10^{-4}	5.3×10^{-2}
Grain size (μm)	0.016	0.018	2
Grain count across diameter	9	110	10

Source: Modified from [17, 18].

Among various metals, silver, copper and nickel will be briefly discussed in the context of EMI shielding. These metals can be utilised in various forms, such as nanoparticles, nanowires, metal films or composites, to effectively enhance shielding performance. Silver is highly effective for EMI shielding because of its superior conductivity and cost efficiency. Silver matrix composites are typically classified into two categories: silver nanoparticles and silver nanowires. Silver nanowires have demonstrated superior EMI SE at lower content levels compared to silver nanoparticles. To enhance the weak interaction between silver nanowires, innovative structures such as silver nanotubes with a core-shell design and graphene/silver nanowire/graphene sandwich structures have been developed. These configurations offer high bending resistance, light transmittance and improved EMI SE. For example, a silver nanowire/polymer composite film with a thickness of just 20 μm achieved an EMI SE of 33 dB. The creation of 3D conductive networks using silver nanowires, particularly when combined with aerogels or gels, has also proven effective in boosting EMI shielding while maintaining mechanical flexibility [19]. Silver performs well in harsh conditions due to its oxidation resistance, especially when compared to other metals. While silver's cost is higher than alternatives such as copper or aluminium, its benefits, including easier activation conditions and durability, make it a strong candidate for industrial and commercial applications [20].

Copper is a cost-effective metal with excellent conductivity, making it a strong candidate for high-performance EMI shielding. However, copper is more susceptible to oxidation than silver and nickel, which can reduce its conductivity. Most metals are utilised for EMI shielding by being incorporated into composite and polymer matrices, where they are bonded to enhance conductivity and SE. In recent studies, two methods are highlighted for connecting copper layers to polymer substrates:

1. Modify the polymer surface using thiosulphate ($S_2O_3^{2-}$) to promote copper deposition, with the help of activators such as gold or palladium [21].
2. Graft amino groups onto the polymer, commonly using silane or chitosan, to improve copper plating. Activators such as gold, silver or palladium are then employed to catalyse the process, with potassium chloride identified as a potential new activator [22, 23].

Selecting the correct reducing agent and precursor is essential when converting divalent copper cations into copper nanoparticles. For example, using copper sulphate with a weak reducing agent may result in the formation of Cu_xS rather than copper nanoparticles. Cu_xS, a conductive material, can enhance the water and thermal stability of composites. Li et al. successfully developed copper sulphide (Cu_xS)/PAN nanofibre mats with an impressive thickness of only 423 μm. These mats demonstrated an EMI SE of 31 dB, along with an exceptional EMI SE per unit density of 16,655.92 dB·cm^2/g, which notably outperformed other materials of similar specifications [24].

Polyphenols and flavonoids are commonly used as inducers in the formation of copper nanowires, which are typically synthesised through a chemical liquid-phase method. This process produces reddish-brown flocs that are filtered and dried to yield copper nanowires.

It is essential to maintain a neutral pH during the reaction to prevent oxidation. This approach can also be utilised to create mixed-metal nanowires or nanorods, such as a Cu-Ni nanowire network [25]. During reduction, copper can form low-valence compounds that enhance shielding when combined with metallic copper. Although different shapes of conductive fillers can be created using acid etching and surfactants, the process is complex and not ideal for mass production. Copper is a versatile metal matrix material for EMI shielding, but its controllability needs further enhancement in the future [19].

Nickel, a ferromagnetic metal known for its high hardness, ductility, oxidation resistance and conductivity, is widely utilised in EMI shielding due to its affordability. However, the carcinogenic nature of nickel compounds, classified as Group 1 carcinogens by the IARC, limits their applicability in medical applications and wearable devices. The nickel coating process closely resembles that of silver and copper. However, when hypophosphorous acid is used as a reducing agent, phosphorus can be incorporated into the nickel coating, which subsequently lowers its conductivity. Therefore, it is essential to control phosphorus deposition, often achieved by adjusting the pH of the reaction solution. Furthermore, researchers have developed nickel-iron-phosphorus alloy composites to enhance EMI shielding performance [26]. A recent development involves a polyvinylidene fluoride (PVDF)/Ni composite, characterised by a uniformly closed microporous structure, achieved through foaming techniques. This design lowers material density while increasing EM wave loss, resulting in improved conductivity with reduced volume. Additionally, an ultra-thin metal film composed of Ni and Ag, with a thickness of just 2 μm, has been created, offering high versatility and excellent EMI SE [27]. Nickel is a magnetic material with good electrical conductivity, making it highly effective for EMI shielding. Its electromagnetic synergy significantly enhances EMI SE, requiring only a small amount to achieve optimal results. Although nickel's reduction process is simpler compared to copper, its high density and carcinogenic nature limit its applications. As a result, nickel is often used in small quantities and combined with other metals and conductive materials. Table 11.2 shows an overview of various metallic shielding materials and composites, organised by their SE.

Magnetic metals such as nickel, iron, cobalt, steel and specialised alloys like mu-metal are crucial for EMI shielding. Nickel and iron are highly effective for shielding applications owing to their strong magnetic properties and excellent electrical conductivity. Similarly,

TABLE 11.2 Summary of Some Metallic Shielding Materials/Composites, Arranged by Shielding Effectiveness

Material – Matrix	Thickness	Frequency	Shielding Effect	Reference
AgNWs/cellulose film	44.5 (μm)	8.2–12.4 GHz	101 dB	[28]
Ag/Cu/Ni	–	8.2–12.4 GHz	85–93 dB	[15]
Nylon/nickel	100 (μm)	8.2–12.4 GHz	77 dB	[29]
Zn-Al arc thermal metal spray	100–700 μm	1 kHz to 10 GHz	50–120 dB	[30]
Ni foam	2 mm	8.2–12.4 GHz	45–70 dB	[31]
Cu coating	3 mm	8.2–12.4 GHz	61–65 dB	[32]
Ag coating	200 nm	2.45–18 GHz	40–60 dB	[33]

cobalt and steel are commonly utilised in such applications due to their reliable performance in EMI attenuation. Mu-metal, a nickel-iron alloy characterised by its high magnetic permeability, is commonly employed in sensitive environments for shielding against low-frequency EMI. For example, CNT-coated FeNi particles prove to be more effective as fillers in polymer-matrix composites than when either material is used independently. Likewise, combining mu-metal and graphite flakes within a polymer-matrix composite offers superior shielding performance compared to using each component on its own. The presence of a magnetic component enhances SE, provided the composite's electrical resistivity remains below 10 Ω cm, facilitating efficient eddy current paths [34, 35].

While it is commonly believed that high conductivity primarily enables EMI shielding through reflection, it can also lead to significant absorption. This is evident in aluminium, the predominant metal used for EMI shielding [36]. Iron-based materials such as ferrites, ferric oxide (Fe_2O_3) and magnetite (Fe_3O_4) are effective for EMI shielding applications. Performance results indicate that spinel ferrites like $NiFe_2O_4$ demonstrate a total SE ranging from 47 to 68 dB, with particle sizes between 10 and 30 nm. Additionally, varying the thickness from 0.1 to 0.7 mm enhances the SE from 24 to 68 dB. Ferric oxide (Fe_2O_3) exhibits excellent microwave absorption properties, particularly in popcorn-like and coin-like morphologies of α-Fe_2O_3. Magnetite (Fe_3O_4) displays diverse morphologies, including nanorods and dendritic forms, with SE results of $RL_{-min}= -25$ dB at 2.5 GHz with a matching thickness of 3 mm [9]. Iron-based materials, with their unique magnetic and dielectric properties, provide effective EMI shielding across various applications. However, challenges such as high density, processing difficulties and narrow absorption bandwidths remain.

Hollow and porous metal-based materials such as aluminium foams are noted for their lightweight properties and high SE due to increased multiple reflections within their porous structures. For example, aluminium foams with porosities up to 93.08% demonstrate SE ranging from 25 to 75 dB across different frequencies, making them suitable for various shielding applications [37]. Metal-based composites are also explored for their potential in EMI shielding. These composites often incorporate metal oxides, carbides and nanoparticles to enhance both SE and mechanical properties. One notable example is the PVA (polyvinyl alcohol)/Fe_2O_3 composite, which achieves an SE of 45.2 dB with a 200 nm Ag coating. Such composites offer a balance of high SE and robust mechanical properties, making them viable for commercial applications [33].

The development of transparent EMI shielding materials also represents a significant advancement. Indium tin oxide (ITO), metal nanowires and hybrid composites with CNTs and graphene are designed for applications requiring visibility, such as microwave oven windows, while still providing effective shielding against lower-frequency EMWs. Multilayer composites like indium zinc oxide (IZO)/Ag demonstrate SE above 45 dB and optical transparency over 70%, highlighting their suitability for these specialised applications [31]. For transparent EMI shielding, the challenge lies in scalability and balancing transparency, electrical conductivity, mechanical strength and environmental durability.

Metal coatings, flexible metal/polymer composites and thin metal coatings have been developed to meet the demands of modern electronics and wearable devices. Techniques like arc thermal metal spraying (ATMS) and electroless plating allow for the creation of thin, effective shielding layers on existing surfaces, combining the benefits of high SE and flexibility. Lee et al. used ATMS to apply copper (Cu) and zinc-aluminium (Zn-Al) coatings of varying thicknesses (100, 300, 500 and 700 μm) on sandblasted glass surfaces for EMI shielding. The performance was assessed across a frequency range of 14 kHz to 18 GHz. Remarkably, ATMS coatings, even at a thickness of 100 μm, achieved SE comparable to that of 3-mm-thick steel and copper plates. Among them, Zn-Al coatings performed better due to the presence of fewer microcracks [30]. These developments in EMI shielding materials are crucial for protecting electronic devices and ensuring human health by mitigating the adverse effects of electromagnetic radiation.

In conclusion, metal-based materials are highly efficient for EMI shielding, primarily due to their superior electrical conductivity; however, they are hindered by disadvantages such as bulkiness, susceptibility to corrosion and challenges in fabrication. Despite extensive research into alternatives, metal sheets are still widely used because of their superior SE. Coatings or composite materials that combine metals and polymers have shown potential for improving overall performance, offering enhanced durability, corrosion resistance and lighter weight. These materials not only improve SE but also provide increased mechanical flexibility. This makes them more versatile and suitable for a broader range of applications compared to conventional metal-based solutions. Thin conductive metal coatings on polymer substrates improve SE and maintain flexibility, whereas adding polymer particles or fibres to metal matrices only slightly enhances SE, raising questions about the practicality and efficacy of such composites. Future research should optimise the balance between SE and mechanical strength, address long-term performance and explore novel fabrication techniques to develop superior EMI shielding materials that could replace traditional metal plates, offering high effectiveness and durability.

11.4 2D NANOMATERIALS MATERIALS FOR EMI SHIELDING

Two-dimensional (2D) materials have drawn considerable attention for their potential in EMI shielding, thanks to their distinctive properties such as high surface area, adjustable electronic properties and exceptional mechanical strength. The groundbreaking research on graphene has sparked interest in other 2D materials including MXenes, hexagonal boron nitride (h-BN), black phosphorus, transition metal dichalcogenides (TMDs) and layered double hydroxides (LDHs) [38, 39].

These 2D materials are characterised by their excellent electrical and thermal conductivity, along with superior mechanical properties, making them highly suitable for EMI shielding applications. The high aspect ratio and extensive surface area enable efficient interactions with EMWs, thereby enhancing SE. Furthermore, these materials are lightweight and flexible, overcoming the challenges associated with traditional metallic shields, which tend to be heavy and inflexible. The number of studies focusing on 2D-material-based shielding materials has risen significantly, from 69 to 514 over the past six years, underscoring their unique properties and growing significance in research [6].

Graphene, the most extensively researched 2D material, is renowned for its outstanding electrical conductivity and mechanical strength. A comprehensive discussion on 2D carbon-based materials, including graphene, will be provided under the subheading

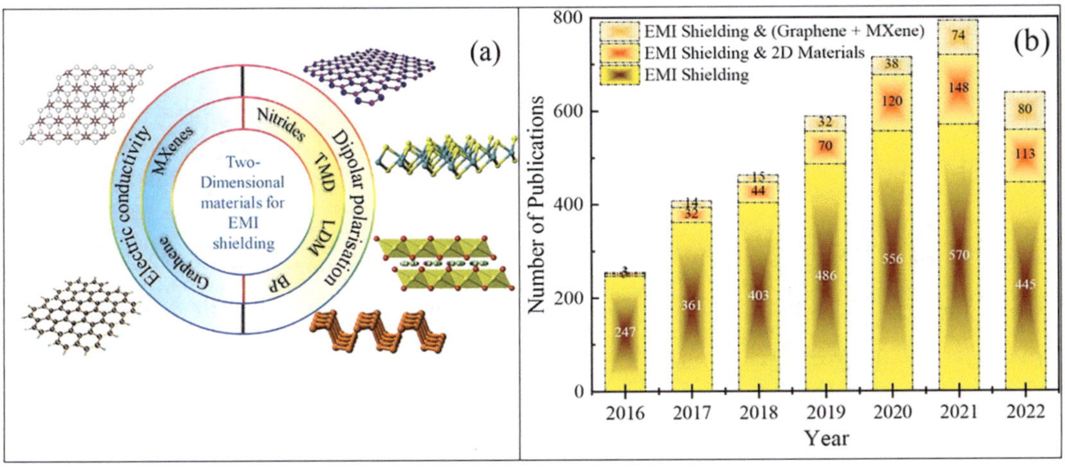

FIGURE 11.2 (a) illustrates two-dimensional materials employed for EMI shielding applications and (b) the significant contributions of 2D materials in this field. Reproduced [6] in accordance with Creative Commons Attribution License.

"2D Carbon-Based Materials" within the section on carbon-based materials for EMI shielding. Nevertheless, in this section focus will be on other 2D materials, specifically divided into MXenes and semiconductors [40]. Figure 11.2(a) shows various two-dimensional materials employed for EMI shielding applications, showcasing their diverse structural forms and properties, and (b) the significant contributions of these 2D materials in advancing the field of EMI shielding [6].

11.4.1 MXenes

2D MXenes are composed of transition-metal carbides, nitrides and/or carbonitrides, described by the general formula M*n*+1X*n*T*x*. In this formula, M denotes an early transition metal, such as titanium (Ti), zirconium (Zr), vanadium (V), niobium (Nb), tantalum (Ta) or molybdenum (Mo); X represents carbon and/or nitrogen; and Tx corresponds to the surface functional groups present on the material [38]. MXenes are classified based on the transition-metal element present (e.g., Ti, V, Nb, Ta). Figure 11.3 shows the periodic table highlighting the compositions of MXenes [41].

Since the initial report highlighting the EMI shielding potential of 2D $Ti_3C_2T_x$ MXene in 2016, extensive research has been conducted, resulting in over 100 publications within three years [40]. The review paper by Iqbal et al. highlights the superior properties of MXenes, which make them exceptionally well suited for EMI shielding applications [38]. These properties include their high electrical conductivity, which is crucial for reflection loss, and their large surface area, which enhances absorption loss. The surface chemistry of MXenes enables the creation of diverse structural designs, including compact laminates, porous foams, layer-by-layer assemblies and aerogels, all of which further enhance their SE [38].

A key advantage of MXenes is their ability to form composites and hybrids with other materials, enhancing their intrinsic properties. For example, MXene composites with

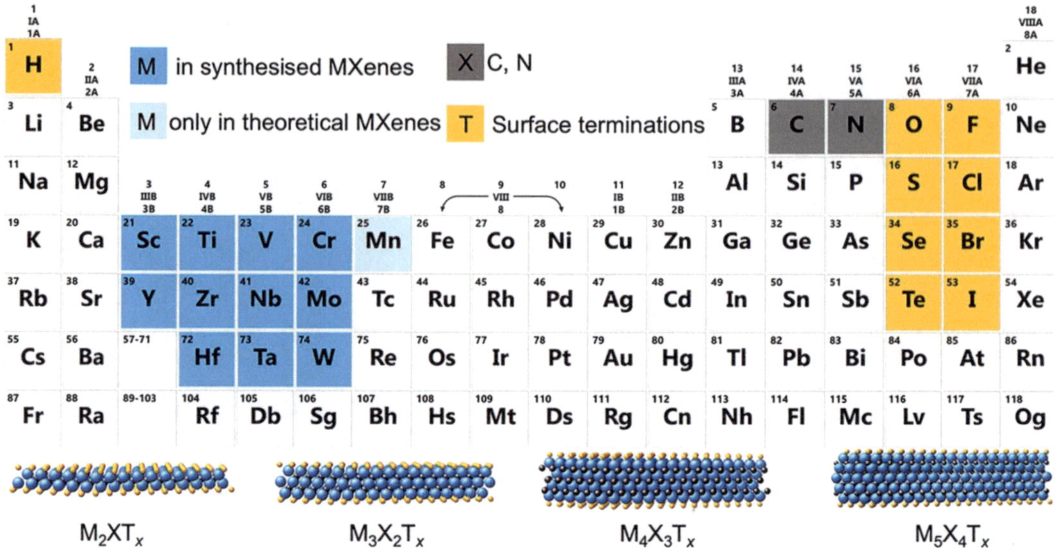

FIGURE 11.3 Periodic table highlighting the compositions of MXenes, with elements colour-coded. Schematics of four common MXene structures are shown at the bottom. Provided by Prof. Babak Anasori, IUPUI. Reproduced from [41] in accordance with Creative Commons Attribution License.

polymers, carbon nanofibres (CNFs), CNTs and other conductive materials exhibit improved mechanical strength, flexibility and environmental stability while maintaining high EMI SE. There is also development of MXene-based flexible and wearable textiles, which are particularly important for applications in smart electronics and personal protection against EMI. An example of a highly effective MXene material is $Ti_3C_2T_x$, which has demonstrated an EMI SE ranging from 48 to 92 dB in the X-band frequency range (8.2–12.4 GHz) with thicknesses varying from 1.5 to 45 μm [40]. This remarkable performance is attributed to its excellent electrical conductivity (≈5000 S $^{cm-1}$) and laminate architecture, which improves the reflection and absorption of EMWs.

While 2D materials like MXenes have shown exceptional promise in EMI shielding, their large-scale production remains a critical factor for practical applications. Research on scalable synthesis methods, such as chemical vapour deposition (CVD) and etching techniques, has made significant progress, with recent developments in batch production of MXenes showing potential for industrial-scale manufacturing [38]. However, challenges still exist in ensuring uniformity and control over surface terminations during large-scale fabrication. In terms of stability, MXenes have demonstrated environmental sensitivity, particularly to oxidation, which can be mitigated by surface functionalisation or encapsulation techniques. Compatibility with different substrates, including polymers and textiles, has been successfully explored through the development of MXene-polymer composites and flexible, wearable textiles, offering a wide range of applications in electronics and personal protection [40].

Figure 11.3 highlighting the elements involved in the composition of MXenes. Transition metals used in synthesised MXenes are shown in blue (Zirconium (Zr), Titanium (Ti),

Hafnium (Hf), Niobium (Nb), Vanadium (V), Tantalum (Ta), Molybdenum (Mo), Chromium (Cr) and Tungsten (W)), while those theorised but not yet synthesised are in light blue (Scandium (Sc), Yttrium (Y)). Carbon and nitrogen, which form the non-metallic components, are marked in grey, and surface-termination elements like oxygen, fluorine and chlorine are shown in yellow. The figure also includes schematics of typical MXene structures, indicating how these elements combine to form different MXene compositions with specific surface terminations that influence their properties [41].

MXenes demonstrate exceptional EMI shielding performance owing to their superior metallic conductivity, surpassing other conducting materials and composites at comparable thicknesses. While synthesis methods have been developed for over 30 different MXenes, only a few, including Ti_3C_2, Ti_2C, Mo_2TiC_2, $Mo_2Ti_2C_3$ and Ti_3CN, have been extensively studied. Investigating additional MXenes is crucial for gaining insight into how their elemental composition, layered structure and transition-metal arrangement impact their EMI shielding properties [38].

The potential of MXenes as the next generation of EMI shielding materials lies in their excellent properties and versatility in forming effective composites and hybrids. Ongoing research and development in this field are expected to lead to new material solutions for various advanced applications requiring efficient EMI shielding. For instance, transparent EMI shielding materials allow for visual monitoring of protected systems, maintaining aesthetic and functional properties in applications such as smart windows, touch screens, wearable devices and medical implants [42]. However, MXenes can be prone to oxidation, which may affect their long-term stability, though surface functionalisation and protective coatings have been shown to mitigate these issues in various environmental conditions [41].

11.4.2 Semiconductors EMI Shielding Materials

Semiconductors are 2D materials with electrical conductivity values that fall between those of conductors (like metals) and insulators (like ceramics). This intermediate conductivity arises because semiconductors have a band gap that is small enough to allow electrons to transfer from the valence band to the conduction band when energy is provided (e.g., thermal energy or photon absorption). The key property of semiconductors is their band gap (typically ranging from about 0.1 to 4 eV). Semiconductors can be elemental, such as silicon (Si) and germanium (Ge), or compound, such as gallium arsenide (GaAs) and silicon carbide (SiC). Table 11.3 and Figure 11.4 present a list of semiconducting materials that

TABLE 11.3 List of Semiconducting Materials Identified as Potential Fillers for Use in EMI Shielding Applications

Band Gap (eV)	**Materials**	**Category**
0.95–2	Cu_2S, CuO, FeS_2, MoS_2, MnO_2, CdSe, RuS_2, OsS_2	Low band gap
1.3–2.5	Fe_3O_4, CuS, CoO, Fe_2O_3, CdS, WS_2	Medium band gap
2–3.2	Cu_2O, V_2O_5, WO_3	Medium band gap
3–4	$BaTiO_3$, MoO_3, TiO_2, $SrTiO_3$, SnO_2, ZnO, ZnS	High band gap

Source: Modified from [43] in accordance with the Creative Commons Attribution License.

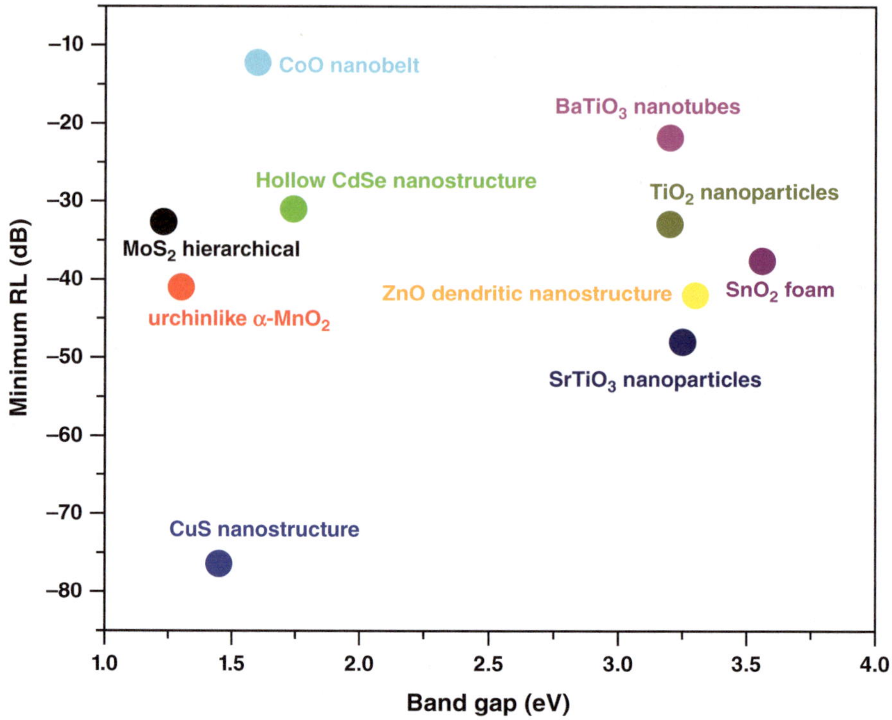

FIGURE 11.4 List of semiconductors showing their minimum reflection loss (RL) in dB versus their band gap in eV as fillers for EMI shielding applications. Reproduced [43] in accordance with Creative Commons Attribution License.

could serve as potential fillers for EMI shielding applications. Figure 11.4 also shows the list of semiconductors used as fillers, displaying their minimum reflection loss (RL) in dB versus their band gap in eV.

For example, MoS_2-based composites exhibit a good balance of conductivity and flexibility, making them effective at mid-frequency ranges. However, they require higher filler loading (25 wt.%) for optimal performance. **Reduced GO** on MoS_2/PVDF shows a minimum RL of –26.11 dB at 11.36 GHz with a thickness of 2.5 mm for 25 wt.% MoS_2 [44]. These findings suggest that hierarchical MoS_2 nanospheres are suitable candidates for high-performance absorbing materials.

Doping semiconductors with conductive elements or combining them with other materials, such as graphene or metallic nanoparticles, can significantly improve their shielding efficiency by enhancing conductivity and absorption capabilities. Hybridisation of semiconductors with polymers and carbon-based materials has also been explored, resulting in flexible, lightweight composites with improved electromagnetic absorption and mechanical strength [43, 45]. A study indicates that CuS-based composites are the optimal choice for EMI shielding owing to their superior SE, wide frequency applicability and robust dielectric and interfacial polarisation properties. They also require precise synthesis and uniform dispersion in the matrix. The synthesis and characterisation of CuS

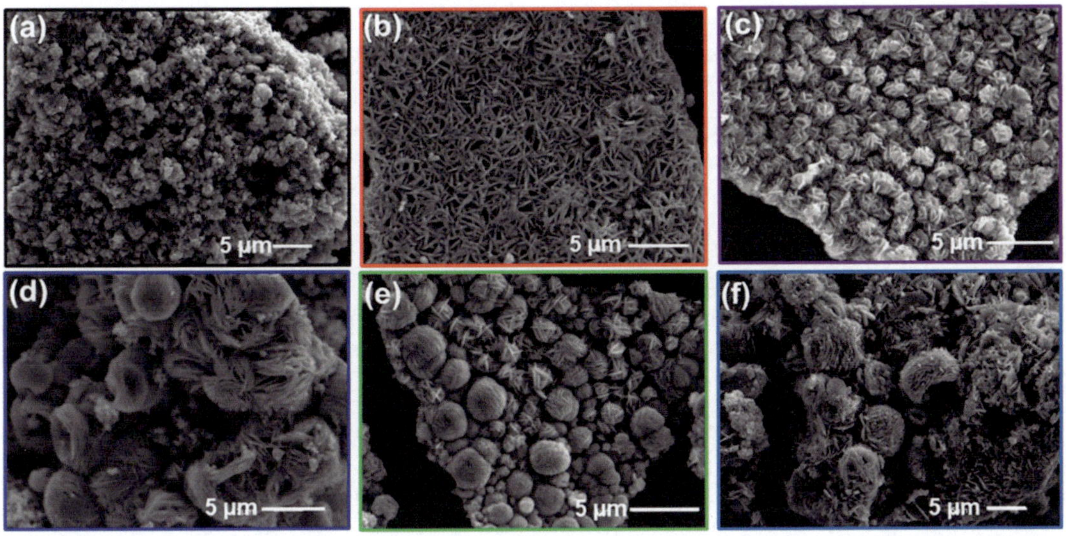

FIGURE 11.5 (a–f) SEM images of synthesised CuS particles at various reaction temperatures: (a) 120 °C, (b) 140 °C, (c) 160 °C, (d) 170 °C, (e) 180 °C and (f) 200 °C, all under a consistent reaction duration of one hour. Reproduced [46] in accordance with Creative Commons Attribution License.

nanostructures demonstrated that varying the precursor ratios significantly impacted their EMI SE. For instance, with a shield thickness of 3.5 mm and 20 wt.% CuS loading, the CuS–paraffin composite achieved a minimum RL value of –76.4 dB at 12.64 GHz. The outstanding microwave absorption performance is attributed to the dielectric losses and interfacial polarisation of CuS, resulting from its heterogeneous structure [46]. Figure 11.5(a–f) shows SEM images of CuS particles synthesised at different temperatures under a constant reaction time of one hour.

Further, Hu et al. investigated PANI/flower-like CuS composites for EMI shielding. Compared to pure CuS and PANI, the total SE for PANI/flower-like CuS composites increased. When incorporated into a paraffin matrix, the PANI–CuS composites (with 50 wt.% CuS) achieved an optimal SET of –45.2 dB at 2.78 GHz for a 3 mm thickness. Notably, the PANI/CuS composite (20 wt.%) incorporated into the paraffin matrix displayed enhanced SE, with values below –18 dB across the frequency range of 300 kHz to 3 GHz. The enhanced EMI SE is primarily attributed to the interaction of mobile charge carriers in CuS with electromagnetic fields. Additionally, the compatible bonding and robust π–π interactions between PANI and CuS significantly contribute by facilitating efficient charge transfer at their interface, further improving overall shielding performance [14].

Semiconductors offer significant advantages for EMI shielding, including tunable electrical conductivity, high dielectric properties and the ability to form lightweight, flexible and nanostructured composites that enhance microwave absorption through increased surface area and multiple reflections. They are particularly suitable for applications requiring a balance of SE and mechanical properties, such as portable and wearable electronics. However, their disadvantages include complex and costly production processes, potential performance limitations due to lower intrinsic conductivity compared to metals and issues with agglomeration and restacking, especially in two-dimensional forms. Additionally, their effectiveness can be influenced by environmental factors like humidity and temperature, which may restrict their use in certain conditions [47].

Semiconductors for EMI shielding also have specific limits and contributions when compared to traditional materials such as metals and carbon-based materials. Unlike metals, which primarily rely on reflection due to their high conductivity, semiconductors contribute to EMI shielding through a combination of absorption and reflection, thanks to their dielectric properties and tunable conductivity. This makes them particularly useful in applications where both shielding and absorption are required, such as in sensitive electronic devices. However, semiconductors generally offer lower SE compared to metals. Enhancing the shielding efficiency of semiconductors can be achieved through doping with conductive elements or by hybridising them with materials such as graphene or metallic nanoparticles, which improve both their conductivity and absorption properties. Compared to carbon-based materials like graphene and CNTs, semiconductors provide a broader range of tunable properties and can be more easily integrated into diverse polymer matrices. Their unique applications include use in flexible and transparent shielding materials, wearable electronics and smart windows, where traditional materials might fail to provide the required combination of transparency, flexibility and SE [45].

11.5 CARBON-BASED EMI SHIELDING MATERIALS

Carbon-based materials have emerged as highly effective EMI shielding agents due to their unique electrical, thermal and mechanical properties. Carbon-based materials such as graphite nano-platelets (GNPs), CNTs, graphene derivatives (e.g., GO, reduced graphene oxide (rGO), graphene nano-platelets (GNPs)) and carbon nanowires have shown excellent EMI SE. Among the various allotropes of carbon, GO, CNTs, graphite and carbon black (CB) are the most studied and utilised for EMI shielding applications. Carbon allotropes like graphene and CNTs are especially favoured due to their high electrical conductivity and their capability to form conductive networks within polymer matrices, which enhances both the reflection and absorption of EMWs, significantly improving the overall shielding performance [48]. Their exceptional electrical conductivity, coupled with their lightweight nature and flexibility, makes them highly suitable for a diverse array of applications, ranging from aerospace technologies to wearable electronics, where performance and adaptability are critical.

Combining carbon materials with polymers or metals can further enhance their EMI shielding performance. For instance, composites with polyaniline (PANI), polyvinylidene fluoride (PVDF) and polydimethylsiloxane (PDMS) have been studied extensively. Innovative composites such as 3D graphene-CNT hybrids and MXene-rGO hybrids have demonstrated exceptional EMI shielding capabilities [39]. These materials benefit from their unique porous structures and high electrical conductivities. Combining carbon materials with metals or metal oxides and MXenes further enhances EMI shielding performance due to the synergistic effects of the conductive and magnetic properties. These hybrid composites often show improved mechanical properties and thermal stability, making them suitable for a wide range of applications.

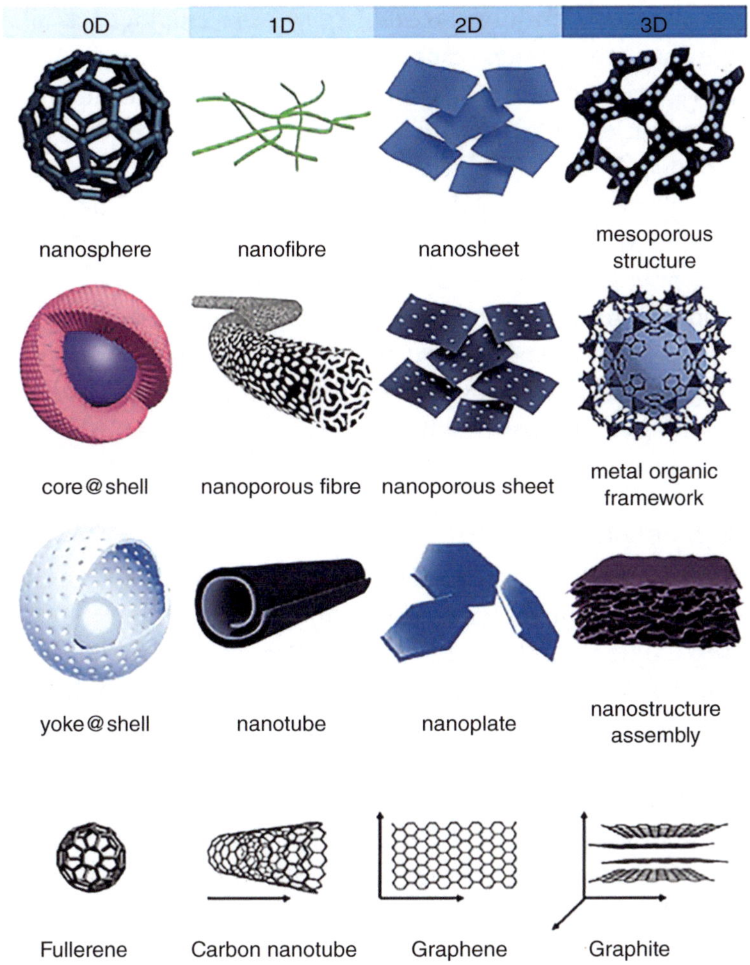

FIGURE 11.6 Categorisation of 0D, 1D, 2D and 3D nanostructures. Reproduced from [49] in accordance with Creative Commons Attribution License.

In this section, the classification of carbon-based materials into zero-dimensional (0D), one-dimensional (1D), two-dimensional (2D) and three-dimensional (3D) categories provides a structured approach to understanding their unique properties and applications. This classification is based on the distinct morphologies and dimensionalities of the materials, which influence their electrical, mechanical and electromagnetic properties, thereby affecting their shielding performance. Figure 11.6 shows the categorisation of 0D, 1D, 2D and 3D nanostructures, with examples at the bottom.

11.5.1 0D Carbon Structure

Zero-dimensional (0D) carbon structures are nanomaterials confined to the nanoscale in all three dimensions, essentially existing as point-like particles. Examples include carbon dots (CDs), fullerenes (C60) and graphene quantum dots (GQDs), which exhibit unique properties due to their high surface area and quantum mechanical effects.

In one study, polyaniline (PAni) nanocomposites doped with fullerene (C60) were synthesised at 0°C using a template-free method. Aniline served as the monomer, while fullerene (C60) and titanium dioxide (TiO_2) were employed as fillers. Fullerene (C60)-doped PAni nanocomposites, especially those treated with CSA, showed significantly improved electrical conductivity and morphology, which likely enhanced their EMI shielding capabilities [50]. In another study, nanocomposite films reinforced with GQDs were developed. The incorporation of GQDs significantly enhanced the thermal stability, dielectric properties and EMI SE of the nanocomposites. The maximum EMI SE achieved was 9.8 dB in the X-band and 8.9 dB in the Ku-band, indicating the potential of these materials for practical EMI shielding applications [51].

Zero-dimensional (0D) carbon structures show potential in EMI shielding due to their high surface area and quantum mechanical properties. However, their current effectiveness may not be sufficient for industrial applications that require higher shielding performance. Additionally, the complex and costly synthesis processes may limit their scalability, suggesting that further research is needed to enhance their practicality for widespread use.

11.5.2 1D Carbon Structure

One-dimensional (1D) carbon nanostructures, such as CNTs, CFs, carbon nanofibres and carbon nanowires, are characterised by their tubular configurations. Among these, CFs are extensively employed in EMI shielding applications due to their remarkable combination of properties, including superior tensile strength, excellent heat resistance and lightweight nature, which make them highly effective and versatile in such applications [52]. For instance, an epoxy composite incorporating Ni-coated CF and CNT foam with 2.35 vol% hybrid fillers demonstrated an EMI SE of 40.8 dB, effectively doubling the SE of the original composite [53].

CNTs have exceptionally high aspect ratios (length-to-diameter ratio), around 10^3. This large aspect ratio allows CNTs to behave as superior reinforcing materials in polymer matrices compared to traditional CF. However, during processing, CNTs often break into shorter segments, decreasing their aspect ratio and thus reducing their effectiveness in reinforcing the composite. The dispersion method, especially those involving sonication or shear mixing, often contributes to this breakage [54]. A study compared three types of MWNTs with aspect ratios of 60, 173 and 400. The composite with the highest aspect ratio (400) exhibited the best performance in terms of electrical percolation and EMI SE. Also, as the aspect ratio increases, the percolation threshold (the minimum MWNT content required to form a conductive network) decreases, indicating that fewer MWNTs are needed to achieve conductivity in composites with high aspect ratios [55]. EMI shielding in CNT composites occurs through a combination of reflection and absorption of EMWs. The higher aspect ratio of MWNTs results in greater absorption, attributed to their increased electrical conductivity and capacitance, which together enhance the overall EMI SE [56].

Proper dispersion of CNTs is also crucial to achieving high-performance composites. Insufficient dispersion results in poor network formation and significantly diminishes the mechanical and electrical benefits of CNTs. Alignment of CNTs, particularly for creating anisotropic properties in composites, remains difficult. Misalignment or waviness reduces the reinforcement capability of CNTs by a factor of 50–200, according to studies [54]. The interaction between CNTs and the surrounding polymer matrix, often referred to as interfacial bonding, is another critical aspect of composite performance. Weak bonding limits stress transfer between the CNTs and the matrix. Functionalisation techniques, such as oxidation or surface treatments, are used to enhance interfacial bonding. However, these methods can disrupt the CNT structure and reduce their mechanical strength. Higher aspect ratio CNTs tend to agglomerate, making it difficult to maintain uniform dispersion, which is crucial for network formation. Achieving this balance between high aspect ratio and good dispersion remains a challenge in composite production. While high aspect ratio CNTs provide significant advantages for EMI shielding and mechanical reinforcement, challenges related to dispersion, alignment and processing must be addressed to fully realise their potential [54, 57].

A study by Yingying Yu et al. suggests a hierarchical porous structure within CNT/carbon composites, where both nanoscale pores within the CNT film and microscale pores present in the amorphous carbon layer collectively contribute to the enhancement of EMI SE. The network of CNTs forms a dense, interconnected structure that facilitates multiple internal reflections of EMWs. The composite's network formation enables multiple internal reflections of EM waves within the porous structure, significantly increasing SEA while reducing SER. The results for the interconnected CNT network indicate an impressive EMI SE of 61.4 dB, much higher than the 36.5 dB for pristine CNT films [58]. The high cost of high-purity, well-aligned CNTs with the necessary aspect ratio makes large-scale production for EMI shielding materials expensive. The fabrication of CNT composites involves complex processing techniques to ensure proper alignment and network formation, further adding to the challenges.

Research suggests that integrating CFs with metals or metal oxides improves the absorption of electromagnetic (EM) waves [59]. Another study explored CF-reinforced plastics and metal-based composites, focusing on their potential for EMI shielding applications. The research involved weaving stainless steel, copper or nickel wire meshes (wraps) into CF-reinforced plastics (wefts) to improve their EMI SE [60]. The investigation revealed that the EMI SE, along with reflection and absorption losses, varied based on the type of metal wire mesh used. The samples exhibited varying attenuation behaviours depending on the side from which the input was applied. Despite this variation, all samples achieved an average SE exceeding 94 dB, with the lowest recorded value being 82 dB. Notably, the inclusion of nickel and copper significantly enhanced the shielding performance, demonstrating superior effectiveness in attenuating EMI.

Another highly conductive 1D carbon structure is CNTs, particularly multi-walled carbon nanotubes (MWCNTs).

A recent study developed a thin, lightweight orthogonal pattern of spinnable MWCNTs to evaluate its effectiveness in EMI shielding. The orthogonal pattern, achieved through precise alignment of the MWCNTs, demonstrated efficient attenuation of EMI within the X-band frequency range. The material achieved an EMI SE of 19.2 dB, with a thickness-averaged specific EMI SE of 73,633 dB·cm^2/g at a thickness of 4.48 mm. This work underscores the significant potential of MWCNTs in the design of high-performance, lightweight materials for EMI shielding applications [61].

Various inorganic compounds, particularly oxides, have been combined with CNTs to enhance electromagnetic (EM) shielding applications, leveraging their magnetic properties to improve EM wave absorption [62]. Yan et al. designed hierarchical hollow frameworks for electromagnetic (EM) wave absorption by integrating interconnected nitrogen-doped CNT networks and CoS_2 particles. These structures were synthesised using a metal-organic framework (ZIF-67)-derived method. The process involved two stages of pyrolysis in reducing and inert gases, which allowed the CNTs to be intricately woven into the framework, while CoS_2 nanoparticles were attached to the surface. By optimising the filler loadings within paraffin-filler mixtures, the EM SE of CoS_2/nitrogen-doped CNT composites was significantly enhanced. Composites containing a 50% loading of CoS_2/nitrogen-doped CNTs exhibited outstanding EM wave absorption capabilities within the X-band frequency range [63].

1D carbon structures also show significant promise in EMI shielding due to their high tensile strength, lightweight nature and heat resistance. However, despite their potential, the effectiveness of these materials in EMI shielding can be inconsistent and highly dependent on factors like composite structure, filler type and material alignment. Additionally, the complex synthesis and precise alignment required for optimal performance, particularly in hierarchical composites like nitrogen-doped CNTs with CoS_2 particles, can make these materials challenging to produce at scale.

11.5.3 2D Carbon Allotropes

2D carbon nanosheets such as graphene, GO and reduced graphene oxide (rGO) are two-dimensional (2D) layered materials with a honeycomb structure formed through covalent bonding between sp^2-hybridised carbon atoms. Graphene has an exceptional strength-to-weight ratio and excellent electron mobility, making it a potential alternative to semiconductor materials. GO and rGO derivatives have a similar layered structure to graphene but differ in oxygen functionalities and electrical properties. rGO, with minimal oxygen functionalities, shows superior electrical properties [64]. Table 11.4 summarises various doped graphene-based materials used for EMI shielding, detailing their microstructures, conductivity, EMI SE and corresponding frequency ranges.

Doping graphene with various heteroatoms introduces defects that alter its electrical and mechanical properties, improving reflection and absorption capabilities. Heteroatom doping replaces certain carbon atoms in the graphene lattice with atoms from another element. This adjustment modifies the charge distribution and electronic properties of the carbon network, leading to a change in its work function and enhancing its adaptability for specific applications. The four common dopants for graphene are boron, nitrogen, phosphorus and sulphur, each altering graphene's properties to enhance its functionality for specific applications [64]. Doping graphene introduces defects in its lattice structure, creating charged centres that induce dipolar polarisation, thereby improving complex permittivity and coercivity. When combined with magnetic particles, doped graphene enhances internal polarisation relaxation and conduction losses, converting electromagnetic energy into thermal energy. The combination of dielectric loss from magnetic components and

TABLE 11.4 Top-Performing Carbon-Based Composites for EMI Shielding

Materials	Density (g/cm³)	Thickness (mm)	EMI Shielding (dB)	Reference
Graphene/CNT/Fe_2O_3/PEDOT	–	0.6	133.2	[65]
Graphene-aerogel coated with PEDOT	0.0762	1.5	91.9	[66]
MCMBs/MWCNTs/Fe_3O_4	0.5	0.5	80	[57]
CNT/WPU (waterborne polyurethane)	1.2	0.8	80	[67]
CNT/CF/pyrolytic carbon	0.93	2	72.8	[68]
Ag@C core-shell hybrid sponges	0.00382	3	70.1	[69]
CNT macrofilms	0.1	0.004	67	[70]
CNT/GTR	–	2.6	66.9	[71]
CNT/PS	–	2	63.6	[72]
CNT/Fe_3O_4/PF	0.0441	2	62	[73]
Graphene-coated PU sponge	0.03	60	57.7	[74]
CNT/PVDF	0.79	2	56.72	[75]
CNT/graphene/PDMS	–	2.75	55	[76]
Iodine-doped Graphene paper	–	0.01	52.2	[77]
Graphene/γ-Fe_2O_3/PANI	–	2	51	[78]
Aerogel-like carbon	0.121	10	51	[79]
Short-CF/EVA/NBR	–	10	50	[80]
Graphite/HANF	1.79	0.06	47.4	[81]
CF/SiC	1.39	3	44	[82]
Macroscopic carbon grids	0.28	0.3	44.5	[83]
CB/PE composite	–	2.8	43	[84]
CNT/PC	1.1	2	43	[85]
Graphene/PS	–	2.5	45.1	[86]
Graphene/WPU	–	1	34	[87]

Source: Reproduced from [88] in accordance with the Creative Commons Attribution License.

conduction loss from doped graphene significantly enhances the material's effectiveness in attenuating EMWs and achieving optimal impedance matching [89].

Researchers have expanded the electromagnetic absorption (EMA) bandwidth of doped graphene through various approaches. Utilising lightweight graphene foams with a 3D porous structure enhances dielectric losses, while combining graphene with magnetic metal oxides like Fe_3O_4 improves both impedance matching and absorption [90]. N-doping is especially effective, as it significantly increases the EMA bandwidth by improving electronic properties and introducing specific nitrogen configurations, such as pyrrolic-N, pyridinic-N and graphitic-N [91].

Nitrogen-doped graphene (NG) is highly effective in enhancing graphene's magnetic properties, particularly with pyrrolic-N, which significantly boosts magnetisation and microwave absorption. This makes NG a promising material for developing lightweight, microwave-absorbing applications due to its improved impedance matching, polarisation relaxation and defect creation. Utilising urea as the doping precursor, researchers discovered that NG derived from flower-like GO demonstrated outstanding performance, with an RL of –21.7 dB at 16.8 GHz, a 2.3 GHz bandwidth and a 2 mm thickness. They also observed that a higher carbon-to-oxygen (C/O) ratio in the doped material enhanced absorption properties. The study emphasised the significant impact of GO precursor morphology on the effectiveness of NG-based microwave absorbers [92].

Enhancing the EMW absorption capabilities of graphene can be achieved through heteroatom doping, particularly when nitrogen is introduced. This is due to nitrogen's similar atomic size to carbon, higher electronegativity and the presence of lone electron pairs. Graphene materials are ideal candidates for low-density microwave absorbers due to their tunable electrical properties and effective polarisation relaxation, enhanced further by doping. Doped graphene addresses the common limitations of conventional EMI shielding materials, including inadequate microwave absorption, restricted bandwidth and the need for high filler content. It provides advantages like enhanced dielectric loss, better dipole polarisation, optimised impedance matching and the ability to achieve effective shielding with lower filler content [64].

The integration of GO and rGO into composite materials, such as polymers and ferrite nanoparticles, significantly enhances impedance matching and dielectric loss, thus improving EMI shielding at relatively low filler contents. For instance, the combination of rGO with ZnO nanoparticles produces a lightweight nanocomposite with superior dielectric loss and enhanced impedance matching, achieving an impressive SE of 38 dB compared to the individual components. The synergistic interaction between rGO's conductive network and ZnO's polarisation centres contributes to superior shielding performance, even at low filler contents. This integration enables the creation of efficient, lightweight and flexible EMI shielding materials ideal for modern technological applications [93].

11.5.4 3D Carbon Structures

3D carbon structures like foams, aerogels and sponges offer low density, high flexibility and strong mechanical properties. They regulate permittivity and induce multiple scatterings or reflections. 3D structures provide high conductivity and numerous interfaces for enhanced shielding performance. Examples include graphene aerogels and CNT-modified carbon aerogels [48].

The **porous nature** of 3D carbon structures allows for **multiple scattering** and reflections of EMWs, which increases the absorption and reduction of these waves, thereby improving shielding performance. Moreover, the **interconnected network** within these structures provides extensive conductive pathways, crucial for achieving high electrical conductivity, which is essential for effective EMI shielding [94]. The hierarchical porous structures in graphene composites are highly effective in enhancing EMI shielding performance. Porosity allows for multiple internal reflections and scattering of EMWs. For example, porous graphene films show improved dielectric loss and multiple reflections, resulting in an EMI SE of up to 43.8 dB. The enhanced performance is attributed to the fine-tuned balance between electrical conductivity and structural porosity, providing excellent shielding at low material thicknesses and densities [95].

Also, the interconnected nature of porous carbon structures, such as in CNT/graphene composites, is crucial for maintaining mechanical integrity while improving conductivity. These networks form a continuous conductive path that enhances electron mobility, crucial for EMI shielding, while also providing strength and flexibility. Research on CNT/graphene composites has shown that incorporating CNTs into porous graphene structures not only improves EMI shielding from 50 to 60 dB but also boosts mechanical resilience, allowing these materials to maintain their structural integrity under mechanical stress [58]. The porosity and interconnected networks of these materials offer high flexibility, low density and strong mechanical resilience. This combination enables the materials to maintain structural integrity while providing significant EMI shielding. For instance, **graphene-CNT hybrid foams** and **carbon aerogels** have demonstrated excellent EMI SE while retaining lightweight and flexible mechanical properties. These features are highly desirable for applications in industries such as aerospace, where both mechanical durability and efficient EMI shielding are critical [94].

3D network-structured graphene foams (GFs) are increasingly recognised as highly effective alternatives to conventional materials for EMI shielding. This growing recognition highlights the potential of GFs in advancing the development of next-generation EMI shielding solutions. These foams combine graphene and CNTs, creating a three-dimensional conductive network that provides excellent EMI shielding. For example, Wang et al. reported on advanced 3D nano-micro architectures tailored for smart electromagnetic devices, which emphasised the strategic assembly processes, effective electromagnetic (EM) responses and functional relationships [96]. These structures use advanced assembly strategies, demonstrate EM responses and achieve optimised processing-structure-function relationships, attaining a maximum EMI SE of 83 dB with a film thickness of 300 μm.

Additionally, Xu et al. developed 3D nanoporous graphene materials that demonstrated a maximum EMI SE of 55 dB at 18 GHz. This high performance is due to the material's porous structure, which enhances both reflection and absorption of EM waves [97].

High-performance EMI shielding has also been developed by assembling carbon nanostructures through carbon-carbon (C-C) bonding, effectively connecting 1D CNTs with 2D graphene. Zhou et al. demonstrated the synergistic properties of macro-assembled graphene-CNT composite films. The study showed that adding 20 wt.% CNTs to graphene films substantially improved their properties, enhancing thermal conductivity from 510 W/m·K to 1154 W/m·K and increasing electrical conductivity from 1.78×10^5 S/m to 2.74×10^5 S/m. Moreover, the EMI SE of the graphene films improved notably, rising from approximately 50 dB to around 60 dB with the addition of CNTs. This research underscores the potential of combining different carbon nanostructures to achieve superior EMI shielding performance [98]. Figure 11.7(a) illustrates the mechanism of the bridging effect of CNTs to graphene sheets and (b) the diagram of the EMI shielding mechanism of FCC-GF.

Yu et al. (2020) also investigated composites created by depositing a carbon matrix co-decorated with Fe nanoparticles (NPs) and CNTs onto the surface of glass fibres (GLF). This 3D composite exhibited a remarkable EMI SE of 48 dB within the frequency range (8.2–12.4 GHz). Additionally, the composite achieved an impressive specific EMI SE of 347.8 dB·cm^3/g, highlighting its potential as a high-performance material for EMI

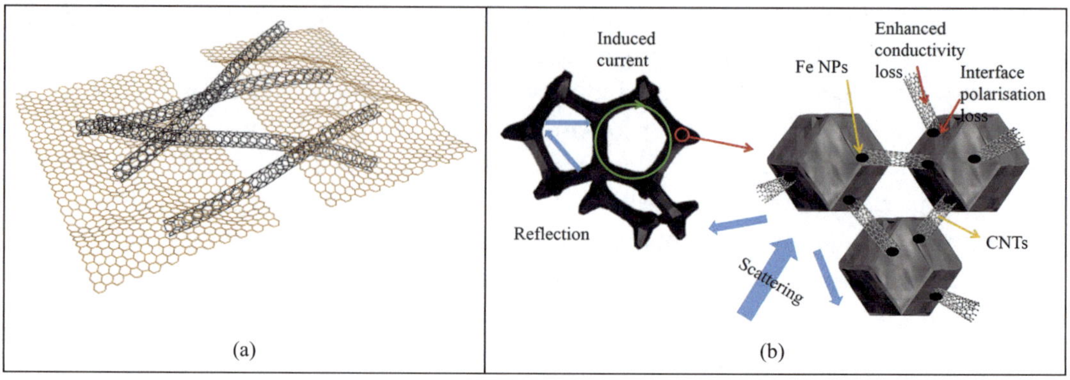

FIGURE 11.7 (a) The mechanism of the bridging effect of CNTs on graphene sheets and (b) schematic diagram of EMI shielding mechanism of FCC-GF. Reproduced [98, 99] in accordance with Creative Commons Attribution License.

mitigation [99]. Nickel foams were treated and used as templates for the growth of graphene foam through a CVD process. The study introduced a novel lightweight and highly efficient EMI shielding material, consisting of a carbon matrix co-decorated with Fe nanoparticles and CNTs, integrated into a 3D graphene foam structure. The resulting composite exhibited high EMI SE and low density, making it an ideal candidate for practical applications in EMI shielding.

Graphene aerogels stand out due to their ultra-lightweight and highly porous nature, which makes them highly effective for EMI shielding. The aerogel structure supports efficient absorption and multiple reflections of EMWs within its porous network. Research has demonstrated that graphene aerogels can be combined with other conductive materials or polymers to further enhance their SE. These aerogels maintain high thermal stability and mechanical strength, making them suitable for harsh environments. Specific examples include graphene aerogels that achieve an EMI SE of up to 47.8 dB at a thickness of 5 mm, displaying their superior performance compared to traditional materials. Zhou et al. proposed an innovative carbon aerogel synthesised from biomass for EMI shielding material. The aerogels, which are cellulose-derived carbon, formed a porous, conductive network structure that provided an effective propagation path for EMWs. In the X-band frequency range, this aerogel, made entirely from biomass, demonstrated an outstanding EMI SE of 93.1 dB, despite having an exceptionally low density of just 0.079 g/cm^3. This study highlights the potential of biomass-derived carbon aerogels as highly effective, lightweight materials for EMI shielding applications, offering a sustainable alternative with impressive performance [100]. The combination of high EMI SE, lightweight nature and superior mechanical properties makes these materials suitable for a broad range of uses, from consumer electronics to aerospace applications.

While 3D carbon structures like foams and aerogels show significant promise for EMI shielding due to their low density, high flexibility and strong mechanical properties, there are challenges that need to be addressed for their broader application. One critical issue is the complexity and cost associated with production processes, such as CVD for graphene foams or the precise assembly required for CNT-graphene composites. Their long-term durability, especially under harsh conditions, also requires further investigation. The integration of these materials into large-scale industrial applications may face scalability challenges, as maintaining uniformity and performance across larger surfaces could be difficult.

11.6 CERAMICS

Traditional metal, carbon and polymer composites offer good EMI shielding but are weaker in mechanical strength and thermal stability. Effective EMI control depends on the required shielding level, frequency and cost. Ceramics provide high strength, wear resistance and thermal stability but typically lack microwave shielding capabilities. Combining ceramics with conductive fillers enhances EMI shielding in harsh environments. Recent studies have explored carbon nanomaterials and ceramic composites like graphene-ceramics, SiCN-based composites and ferrites for improved EMI shielding [101]. Ceramics and ceramic composites are widely studied for their ability to absorb EMWs because of their outstanding thermal stability, mechanical properties and resistance to corrosion and oxidation.

Some studies categorise MXenes as ceramics due to their shared properties of high thermal stability and mechanical strength. However, in this work, MXenes are classified under 2D nanomaterials. Despite these commonalities, MXenes are 2D materials with a layered structure and superior conductivity, consisting of transition-metal carbides, nitrides or carbonitrides. In contrast, traditional ceramics are inorganic, electrically insulating, non-metallic materials like alumina and zirconia. Thus, the unique characteristics of MXenes differentiate them from traditional ceramics, justifying their placement in the 2D nanomaterials category.

The custom development of innovative EMI shielding materials through 3D printing has only emerged in recent years, yet advancements in research have demonstrated significant potential across various applications and market opportunities. 3D printing EMI shielding solutions can surpass the limitations of traditional methods, providing enhanced applicability and versatility. Integrating advanced 3D printing techniques with novel materials is expected to drive breakthroughs in developing next-generation EMI shielding solutions for future engineering applications [8].

With the introduction of 3D printing, some studies have utilised this technology to fabricate ceramics for EMI shielding. 3D printing ceramics commonly include materials such as aluminium oxide (alumina), zirconium oxide (zirconia), aluminium nitride, silicon carbide and silicon nitride [102]. Within this category of ceramics, there are two primary groups: oxides and non-oxides, along with various types of composites. The oxides group contains metal oxides such as zirconia, alumina and silica. In contrast, the non-oxide ceramics group comprises materials like silicon boride, silicon carbide and silicon nitride, among others. These technical ceramics are highly valued for industrial applications due to their superior electrical, mechanical, thermal and chemical properties compared to most metals and polymers [7].

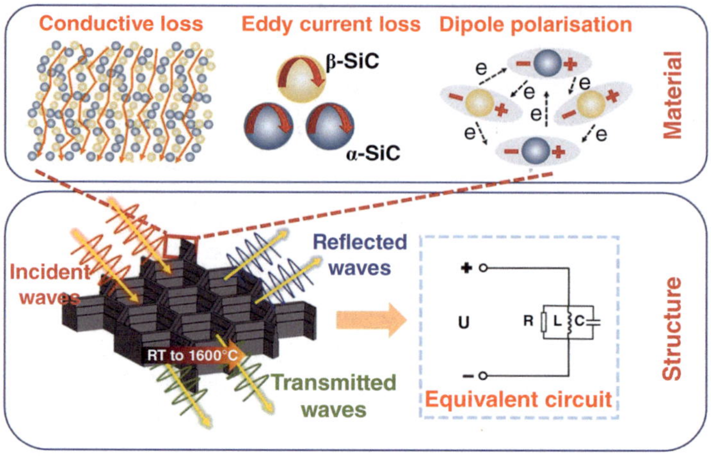

FIGURE 11.8 Diagram showcasing the wide-range microwave absorption capabilities of the SiC honeycomb meta-structure. Reproduced [103] in accordance with Creative Commons Attribution License.

Wang et al. also developed a large-scale SiC honeycomb meta-structure through material-extrusion 3D printing, aimed at achieving ultra-broadband wave absorption. The SiC slurry was prepared and extruded using an additive manufacturing system based on material extrusion. The printed structures were then subjected to precursor infiltration and pyrolysis (PIP) to create the final SiC honeycomb meta-structure. The SiC honeycomb meta-structure demonstrated a broadband EM performance with a –10 dB effective bandwidth of 34.64 GHz across a frequency range at a 60° incident angle. This performance remained stable under both in situ 1000 °C and ex situ 1600 °C erosion in air. The SiC honeycomb meta-structure exhibited excellent absorption characteristics, making it a promising candidate for advanced engineering applications requiring effective EM wave management at elevated temperatures [103]. Figure 11.8 shows the illustration for the wide-range microwave absorption capabilities of the SiC honeycomb ceramic meta-structure.

Li et al. categorised EMW absorption materials into dielectric-loss and magnetic-loss types, based on their respective loss mechanisms. Metals and polymers as single-component materials are generally not favoured due to their inherent limitations, especially at high temperatures. Many ceramic composites exhibit effective EMW absorption above 300 °C. Despite significant progress, challenges persist in creating adequate conducting networks in ceramic matrices, optimising the relationship between EMW absorption and physical properties, and developing flexible, lightweight EMW-absorbent materials [7].

Ceramics offer excellent mechanical strength and thermal stability, but their lack of intrinsic microwave shielding capabilities limits their use in EMI applications. Combining ceramics with conductive fillers, such as carbon nanomaterials, has improved their performance, but challenges remain in creating efficient conducting networks within ceramic matrices. Recent developments like SiC honeycomb meta-structures show promise, yet balancing absorption efficiency with flexibility and weight is still a hurdle. Future research should focus on optimising filler integration and utilising 3D printing to enhance ceramics' potential in EMI shielding.

11.7 METAMATERIALS

Several types of electromagnetic shielding have been developed, each tailored for specific needs such as electric or magnetic fields. Some shielding solutions are effective for both field types. Near-field shielding primarily uses metallic and ferrite shields. While metallic shields are effective for magnetic near fields, they suffer from significant losses, are heavy and can be costly. Additionally, their broadband characteristics cause them to block a wide range of the EM spectrum, resulting in a lack of frequency selectivity. Unlike traditional metallic shields, metamaterials deliver effective near-field magnetic shielding while providing the added benefit of frequency selectivity. This capability allows for adjustable performance, which is not achievable with metallic or ferrite shields [104]. Effectively shielding high-intensity, low-impedance and low-frequency EM waves remains a persistent challenge, prompting continuous exploration of materials that can provide reliable protection against such waves. Metamaterials have shown significant promise in this regard, proving to be effective not only in shielding microwaves, which were the initial focus of research, but also in providing protection against very low-frequency EM waves. This ongoing exploration highlights the potential of metamaterials as versatile solutions for a broader range of EM wave frequencies [105].

The concept of metamaterials can be traced back to the late 19th and early 20th centuries, with notable contributions from researchers such as Jagadish Chandra Bose and Karl Ferdinand Lindman. In 1967, Victor Veselago laid the theoretical foundation by describing materials with a negative index of refraction. Veselago subsequently introduced the concept of a negative refractive index in his seminal work [106]. The concept was later exemplified using metallic wires and split-ring resonators, which effectively illustrated negative permittivity and permeability [107]. Later, Smith and his team experimentally validated these structures in the microwave spectrum by combining both configurations to create left-handed materials (LHMs) [108]. These LHMs possess unique properties derived from their engineered "meta-atomic structure" rather than the constituent materials themselves. Since the introduction of LHMs, numerous applications of metamaterials have been explored, including invisibility cloaks [109], antennas and EMW absorbers [109]. In 2008, Landy et al. proposed and realised the first microwave absorbers using metamaterials [110].

There are several effective methods for using metamaterials for shielding purposes. Researchers have explored various strategies, but four main approaches are discussed as follows [104]:

- The first approach to shielding involves using transformation optics to bend electromagnetic (EM) waves around an object, allowing them to rejoin on the opposite side. This method ensures that the EM wave remains unchanged, with no reflection, absorption or distortion, making it appear as if it has passed through the object without interference. Initially discussed for cloaking, this approach effectively shields the object from EM waves. It has demonstrated near-perfect cloaking for microwave frequencies and has been extended to other frequency spectrums [111].

- The second approach to shielding involves using an anti-polarised material to cover the target material. This coating cancels the polarisation of incoming EM waves, preventing reflection and restoring the polarisation on the other side. This makes the wave appear unchanged, as if it passed through the material without interference. Designing materials with anti-polarisation capabilities is essential for this method.
- The third method of shielding entails developing robust reflectors capable of reflecting all incoming EM waves, thereby preventing their absorption or transmission. These "perfect reflectors" block EM waves from passing through the target material, providing effective shielding. The target material can either function as a perfect reflector itself or be covered with specially designed reflectors. Research has shown that metamaterials can achieve near-perfect reflection, highlighting their potential for this shielding approach.
- The fourth method of shielding a material from EM waves involves absorbing the incoming waves entirely. This principle is currently being implemented in modern stealth technology for defence applications. By completely absorbing the incoming waves, the object remains undetected as there are no reflections for detectors to analyse. This method requires perfect absorbers, and various technologies and materials have been reported to achieve near-perfect absorption. Metamaterials have demonstrated the capability to serve as perfect or near-perfect absorbers, making them effective for this type of EM wave shielding [104].

These materials possess ordered macroscopic structures arranged in three-dimensional (3D), two-dimensional (2D) or one-dimensional (1D) configurations, which can be periodic or non-periodic. The unique properties of metamaterials arise from both their structure and composition, allowing them to respond to electromagnetic (EM) waves in ways that conventional materials cannot. They are characterised by their magnetic permeability (μ) and electrical permittivity (ε), and their refractive index is derived from the product of these two parameters. When both permittivity and permeability are negative, metamaterials exhibit a negative refractive index, leading to unusual phenomena such as backward wave propagation. Metamaterials offer an advanced solution for EMI shielding due to their ability to selectively interact with specific frequency ranges, providing high levels of absorption and reflection.

Metamaterials are highly effective in EMI shielding due to their ability to manipulate EMWs in unconventional ways. They can achieve high levels of absorption and reflection, making them suitable for a variety of applications, including stealth technology and electromagnetic compatibility. They offer significant advantages for electromagnetic shielding across a broad range of frequencies. Their ability to provide tailored electromagnetic responses through engineered structures makes them ideal for advanced shielding applications. Advancements in the design, fabrication and application of metamaterials are expected to further improve their efficiency and broaden their utilisation across various technological sectors. Ramesh Kumar Mishra and Suwarna Datar successfully developed a high-bandwidth metamaterial microwave absorber (MMA) tailored for X-band applications. This innovation is aimed at enhancing both EMI shielding and stealth capabilities, offering significant advancements in these critical areas. They identified the primary challenge in designing metamaterial structures as increasing bandwidth (BW) while ensuring the material remains lightweight and highly absorbent for effective EMI shielding and stealth. Figure 11.9 shows the structural configuration of a microwave absorber incorporating resistive double square loop arrays [112].

High-performance EMI shielding windows currently rely on transparent conductive materials, including microscale metallic grids, networks, nanowires and transparent conductive oxides, to effectively reflect microwave and radio waves. Although these materials

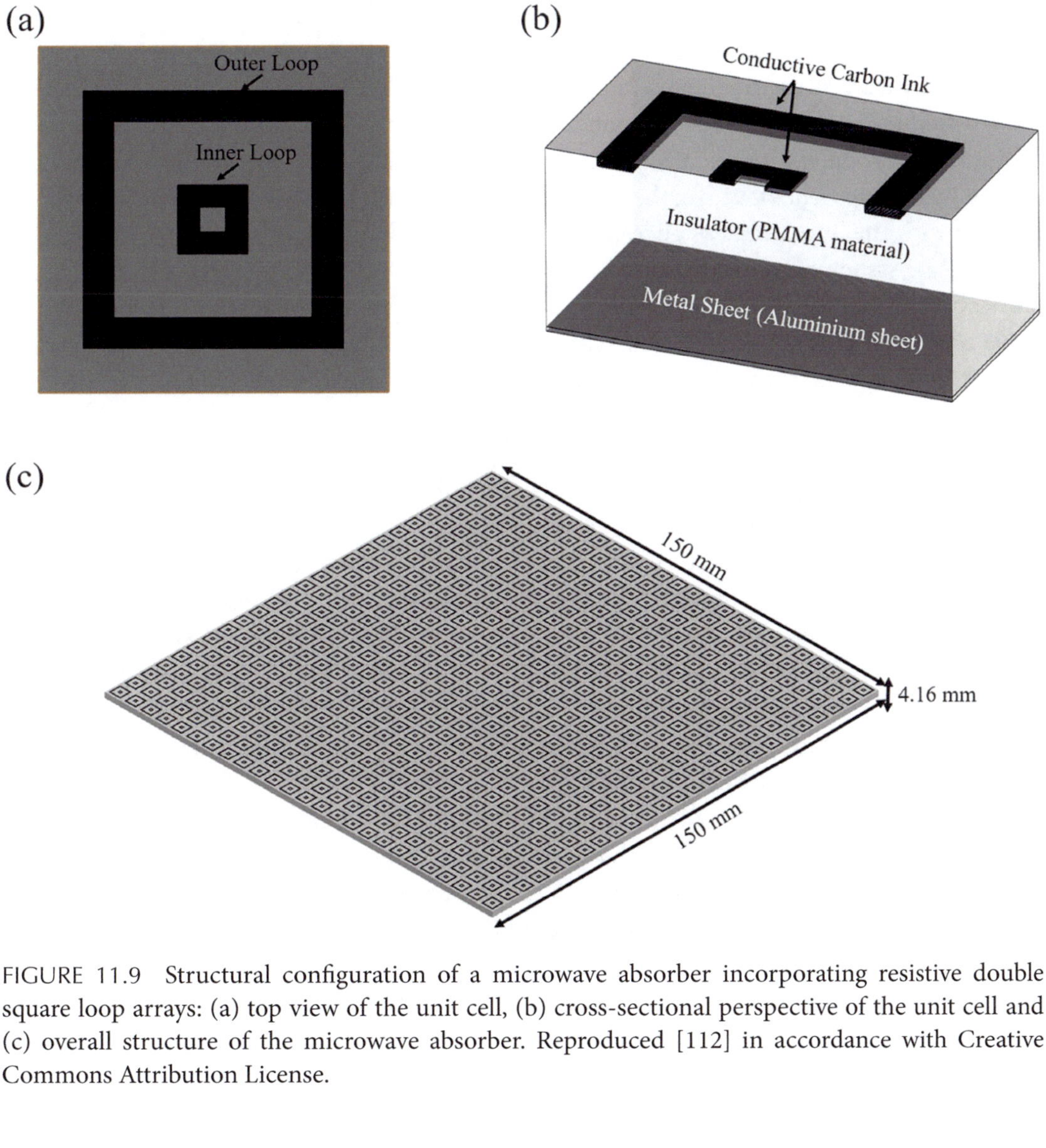

FIGURE 11.9 Structural configuration of a microwave absorber incorporating resistive double square loop arrays: (a) top view of the unit cell, (b) cross-sectional perspective of the unit cell and (c) overall structure of the microwave absorber. Reproduced [112] in accordance with Creative Commons Attribution License.

provide strong shielding performance and high optical transparency, they cannot eliminate the strong background scattering signals caused by secondary EM radiation pollution in space. Y. Zhang et al. introduced a high-performance optical window for broadband EMI shielding, designed using a metamaterial absorber. Their work demonstrated an innovative, highly transparent EMI shielding window with outstanding broadband absorption properties [113].

The window incorporates double split circular ring (SCR) elements, resulting in two moderate resonant peaks. It employs ITO to generate high ohmic loss, thereby broadening the absorption bandwidth. The window achieves over 90% absorptivity across the 7.8–18.0 GHz frequency range and demonstrates a measured SE exceeding 18.25 dB within the 7.0–18.0 GHz range. Additionally, it maintains an average optical transmittance of around 73.10% in the visible–near-infrared (Vis–NIR) region (400–1500 nm) [113]. These properties make the window ideal for use in EMI shielding optical windows and electronic displays across various fields, including aeronautics, medical, civilian and research facilities.

The scalability of metamaterials presents a significant challenge, particularly as the frequency increases. As the frequency increases, the dimensions of the required metamaterials must be reduced, leading to a corresponding decrease in the bulk conductivity of the metal utilised. Additionally, the metal's plasma frequency, which occurs in the UV region, can induce insulating behaviour, further complicating the design. These reduced dimensions, coupled with the limitations of current nanofabrication facilities, make it difficult to create the intricate structures needed, such as split-ring resonator (SRR) designs. As a result, the fabrication and integration of these devices for high-frequency applications become formidable obstacles. This challenge underscores the need for advancements in nanofabrication techniques to enable the practical use of metamaterials in high-frequency domains [104].

11.8 CEMENTITIOUS MATERIALS FOR EMI SHIELDING

Cement-based materials are promising for EMI shielding due to their versatility and structural properties. Several studies have suggested and tested cementitious composites for their SE. Büyüköztürk et al. used a free-space measurement method to calculate the complex permittivity of materials like Teflon, Lexan, Bakelite and Portland cement concrete, verifying results with open-field measurements. Their study found SE values of 2.28 dB, 3.74 dB, 7.25 dB and 5.77 dB, respectively, within the 8–18 GHz range [114]. Cement-based materials naturally possess inherent conductivity due to ion transfer within their porous microstructure. Enhancing this conductivity with various fillers makes these materials highly effective for EMI shielding [115].

The SE of concrete alone is insufficient for effective EMI shielding, prompting the development of cementitious materials with added fillers. By incorporating conductive fillers such as CNTs, graphite or metallic particles into the cement matrix, these composites can effectively attenuate EMWs. The inclusion of these fillers enhances the electrical conductivity of the cement, enabling it to absorb and reflect electromagnetic radiation, thereby providing robust EMI shielding. This makes cement-based composites particularly useful in building walls and other infrastructure components, where they can serve dual functions of structural integrity and protection against electromagnetic pollution [116].

When assessing the effectiveness of EMI shielding in cementitious composites, it is essential to consider both long-term durability and environmental factors, as these significantly impact the material's shielding performance over time and under varying conditions. One of the critical durability concerns is the composite's resistance to chemical deterioration, such as sulphate and acid attacks. Exposure to aggressive environments, particularly water-containing chemicals such as sulphates and chlorides, can degrade the mechanical properties and reduce the SE of these materials. For example, studies have shown that incorporating CNTs into engineered cementitious composites improves their resistance to sulphate and acid attacks while maintaining their EMI shielding properties. The inclusion of CNTs not only strengthens the composite but also reduces expansion and weight loss due to chemical attack, enhancing long-term durability in harsh environments [117, 118].

Environmental factors such as moisture, temperature fluctuations and UV exposure also play a role in the degradation of EMI shielding performance. For example, the interaction between humidity and temperature can cause thermal expansion and contraction, leading to microcracks, which may disrupt the conductive network necessary for EMI shielding. High temperatures can also reduce the effectiveness of certain fillers like CFs or metallic particles, especially in composites exposed to fluctuating thermal cycles. Moreover, the resistance of these composites to corrosion, particularly for metallic fillers, is crucial for maintaining long-term shielding performance. Combining carbon-based materials with magnetic fillers or biochar has been proposed to enhance both EMI shielding and environmental resistance [119].

Research shows that adding conductive fillers like CFs, graphite and CNTs can enhance the SE of cement-based materials. These composites have been found to improve the electrical conductivity and attenuation of EMWs, making them effective for EMI shielding applications. However, the cost of high-conductivity fillers and the need for precise dispersion methods remain challenges. Further research is needed to develop cost-effective and efficient cementitious composites for EMI shielding. Filler materials for cement-based EMI shielding can be categorised into three main groups:

1. **Metallic and Magnetic fillers:**

   Steel and nickel fibres and metallic nanoparticles, such as copper and aluminium, can enhance electrical conductivity and provide effective EMI shielding in cementitious composites. Magnetic materials can be metals like ferromagnetic steel and nickel, or ceramics, such as ferrimagnetic iron oxide. Ceramics are particularly appealing because they are non-conductive and do not present galvanic corrosion issues. Ceramics will be discussed in a separate section.

   Magnetic fillers like ferrite particles and magnetic nanoparticles (e.g., Fe_3O_4) can be added to improve conductive properties. Magnetic materials are highly effective for microwave absorption because the AC magnetic field interacts with magnetic dipoles in the material, causing them to flip and generate dipole friction. This friction mechanism helps absorb the radiation. Therefore, adding magnetic admixtures to cement-based materials can significantly enhance their ability to absorb EMWs [116].

He et al. enhanced the SE of cementitious composites by using nano-Fe_3O_4 fluids, which distribute better than traditional Fe_3O_4 powder. Testing under the 8–18 GHz range revealed that the composite with 5 wt.% Fe_3O_4 exhibited the highest SE, primarily due to EMW absorption [120]. The liquid form of Fe_3O_4 outperformed the powder form in SE, attributed to better dispersion. Moreover, the nano-Fe_3O_4 fluid enhanced cement hydration, resulting in improved early compressive strength. Although magnetic particles improve the SE of cementitious composites, their high density can increase the composite's weight. Thus, it is essential to carefully control the particle volume in the mix to maintain a desirable overall density.

Guowei Ma et al. develop a cementitious composite incorporating waste copper slag with EMW absorption properties. The mixture with 25% copper slag exhibited the best balance of printability and structural stability. The optimal composite (25% copper slag and 6% copper powder) showed a peak RL value of 10.2 dB and a 3.48 GHz absorption bandwidth. Increased copper powder content beyond 6% led to reduced absorption due to the percolation threshold effect. The findings highlight the potential for using solid waste in creating functional materials for EMI shielding and construction automation [121].

Magnetic admixtures are generally more expensive than conductive ones; for example, nickel-coated CF costs more than uncoated CF. Cost-effectiveness is crucial for cement-based materials. Magnetic admixtures like steel also tend to have higher densities compared to conductive ones like carbon. However, low density is typically not a requirement for civil infrastructure. Combining conductive and magnetic admixtures offers a balance of cost and performance, making it an attractive option [116].

2. **Carbon-Based Materials:**

GO, CNTs, CB and CFs are highly effective in enhancing SE in cement matrices. CNTs are valued for their high electrical conductivity and mechanical strength, making them excellent for improving SE when properly dispersed. GO offers excellent SE due to its high conductivity and surface area, which enhance the composite's electrical properties. CFs create conductive networks within the cementitious material, significantly boosting its EMI shielding properties [116].

Fu and Chung conducted early experiments using carbon filaments (0.1 μm diameter, >100 μm length) to enhance EMW reflection in conductive concrete for automatic highways. They tested mixes with 0.5, 1.0 and 1.5 wt.% carbon filaments, surface-treated with ozone gas for better bonding. SE tests at 1 GHz showed that increasing carbon filament content improved SE, with the 1.5 wt.% mix achieving the highest SE, and the primary shielding mechanism was EMW reflectivity [122].

A significant number of experiments on EMI shielding cementitious composites have emphasised the use of high-conductivity fillers, such as graphite, to enhance their SE. Graphite is preferred for its excellent electrical conductivity. Guan et al. analysed several mixes with graphite and found that SE and conductivity increase with higher graphite content. A composite containing 30 wt.% graphite and measuring 3 mm in thickness achieved an SE ranging from 10 to 40 dB within the frequency range of 200–1600 MHz. Another study showed that graphite coatings, made by suspending graphite particles in water or alcohol and applying them to composites, improved electrical conductivity and SE. A 4.4 mm coating provides an SE of 22.3 dB at 1.0 GHz and 25.6 dB at 1.5 GHz. Using styrene-butadiene latex and silica fume helps disperse graphite, further enhancing SE [123].

In addition, incorporating CB particles into cementitious composites enhances their SE. A 10-mm-thick composite with 3.0 vol.% CB achieved an SE of 6–8 dB in the 2–8 GHz range. Adding a secondary absorbent increased SE to 15 dB by decreasing conductivity, which improved impedance matching and increased EMW absorption [70]. Dai et al. developed an EMI shielding cementitious composite using high-structure CB fillers to achieve superior SE. High-structure CB, characterised by its extensive branching, high electrical conductivity and large specific surface area, was preferred over the less commonly utilised low-structure CB. Various mixes with different CB content were fabricated to compare shielding and mechanical properties against a control mix with 0 wt.% CB. Tests conducted over 8.0–18.0 GHz and 18.0–26.5 GHz showed that conductivity increased with higher CB content, while compressive strength decreased significantly, particularly beyond 3 wt.% CB. The reduced compressive strength was attributed to the higher water absorption by CB due to its large surface area [124].

In cementitious composites, GO is frequently utilised as a conductive filler due to its high electrical conductivity, large surface area and capability to dampen EMWs via internal reflections. Zhao et al. employed GO powder, derived through a modified Hummer's method, and dispersed it in cement mixes using ultrasonication. They tested various composites with different GO contents for SE in the 8.2–12.4 GHz range. Their findings indicated that both compressive and flexural strengths increased with ageing, with SE peaking at 0.08 wt.% GO content before declining. This suggests that GO enhances both mechanical properties and SE, although the reasons for SE fluctuation were not provided [125].

The composite with GO achieved an SE of 11–15 dB within the 8.2–12.4 GHz range. Both materials show increased SE with rising frequency. Despite the GO composite containing only 0.08 wt.% particles, it offers higher SE. Another composite, featuring 2 wt.% SF and 10 wt.% GO, exhibits significantly greater SE, roughly six times higher, owing to the synergistic interaction between SF and GO, which enhances the conductive network. This composite maintains an SE of 33–48 dB over the 0.8–8.4 GHz range. These findings indicate that adding only carbon particles is insufficient for achieving adequate SE, highlighting the need for secondary fillers [126].

CFs are favoured for their lower manufacturing cost compared to CNTs. Zhang and Sun created cementitious composites with varying CF amounts and compared them to composites with SF [127]. Results showed that SE increased with higher CF and SF content. However, while more SF increased EMW reflectivity, more CF initially increased reflectivity before decreasing due to impedance mismatch. The maximum CF content was limited to 1% due to cost.

MWCNTs comprise several folded graphene layers and are frequently used due to their high electrical conductivity. Although MWCNTs are highly conductive, composites containing only these fillers do not meet practical SE standards. To enhance SE, researchers have investigated the addition of secondary fillers to MWCNT-enhanced cementitious composites. Nam and Lee varied the silica fume (SF) and fly ash (FA) content while maintaining 0.6 wt.% MWCNT in their experiment. Previous research indicated that SF is effective in dispersing MWCNTs, which could potentially improve the composite's SE. The composite with 0.6 wt.% MWCNTs, SF and FA had the lowest SE, ranging from 1 to 55 dB [128].

3. **Hybrid Composites:**
Combining CFs with metallic or magnetic particles can create a synergistic effect, enhancing both the mechanical and shielding properties of the composite. Mazzoli et al. aimed to enhance the SE of cementitious composites by combining GO with metal fibres. They achieved this by mixing GO microparticles with brass-coated SF to develop a highly conductive network. Their experiment tested different mixes: one without conductive fillers, one with only GO, one with only SF and one with both GO and SF, assessing their mechanical and EMI shielding properties [126]. EMI shielding tests within the 0.8–7.8 GHz range showed that SE increased with higher GO content, but the combination of GO and SF had a more profound effect. The composite with 10 wt.% GO and 2 wt.% SF achieved stable SE between 40 and 50 dB, due to the enhanced conductive network and synergistic effect of the fillers, leading to higher internal refraction and attenuation of EMWs.
For EMWs to penetrate the material, the composite must have a certain level of porosity. Lv et al. sought to maximise EMW absorption in cementitious composites by increasing their porosity using hollow glass microspheres (HGM). However, increased porosity can compromise the composite's strength, necessitating an optimal balance. To achieve this, the researchers combined graphene nano-platelets (GN) with HGM. They created and tested various composite mixes for SE and mechanical properties, finding that the GN and HGM combination achieved the desired balance of strength and EMW absorption. Based on the results, there is an optimal level of graphene nano-platelets (GN) for maximum EMW absorption; exceeding this level negatively impacts SE. The study also indicates that EMW absorption cannot be solely controlled by varying the specimen thickness, as it depends on multiple factors, including EMW frequency. The authors concluded that for composites with 40 vol.% HGM and 0.2 wt.% GN, the optimal thickness for EMW absorption within the 2–18 GHz frequency range is between 20 and 30 mm [129].

To enhance SE, researchers have incorporated CF with secondary fillers like ferro ferric oxide (Fe_3O_4), a naturally occurring magnetite known for its soft magnetic properties and high surface energy. Liu et al. studied the effects of combining CF and Fe_3O_4 nanoparticles on SE. They fabricated composites using Fe_3O_4 nanoparticles prepared via a solvothermal method and manually cut CFs (3–5 mm). Various mixes with 1, 3 and 5 wt.% Fe_3O_4 were created, maintaining a 0.4% CF content. SE tests on 7-mm-thick specimens revealed that Fe_3O_4 inclusion enhanced filler distribution and the overall SE of the composites [130]. The SE results show that increasing Fe_3O_4 content enhances the composite's SE, with the 5% Fe_3O_4 mix reaching a maximum SE of 29.8 dB, primarily due to absorption. The enhanced SE is attributed to the synergistic effect of the fillers, which causes multiple reflections and scattering of EMWs.

A composite containing 1% CF exhibited an SE of 30–45 dB, which increased with frequency. Conversely, a composite with 0.4% CF and ferrite achieved about half the SE of the 1% CF composite, underscoring the importance of substantial CF content for high SE. However, higher CF percentages lead to increased costs. The composite with SF showed an SE of 55–70 dB, almost double that of the CF composite, partly due to its higher filler content (3% vs. 1% CF). These results indicate that future studies could investigate higher CF percentages combined with other fillers to enhance SE [130, 131].

Summary, the analyses show that adding conductive fillers enhances the SE of cementitious composites, but their high cost limits usage, leading to SE levels below minimum standards. Composites with well-formed conductive networks can effectively reflect EMWs and improve SE. Fibrous fillers contribute to creating robust conductive networks, while particle fillers both absorb and reflect EMWs. Combining fibres and fillers can significantly improve SE, provided the fillers are well dispersed within the composite. Proper dispersion requires effective dispersing agents and careful experimentation to find the best medium. The properties of composites with the best shielding performance are summarised in Table 11.5.

Although conventional concrete used in the industry exhibits poor shielding properties, high-strength concrete combined with steel reinforcement has been shown to provide significantly better EMI SE. To enhance SE, high-conductivity fillers such as nano/microfibres and different carbon allotropes can be incorporated.

GO is recognised for its outstanding shielding capabilities among carbon-based composites. Its superior performance is due to the high electrical conductivity of GO combined with the large surface area of its fine particles, which enhance electrical conductivity and promote multiple reflections of EMWs within the composite material. Magnetic nano-Fe_3O_4 fluid-based

TABLE 11.5 Summary of Optimal Shielding Properties in Cementitious Composites [115]

Primary Filler	Frequency Range	SE	Secondary Fillers	Thickness
Nano-Fe_3O_4 magnetic fluid	8–18 GHz	−5 to −35 dB (Reflection loss)	–	20 mm
SF	10 kHz to 1 MHz	40–120 dB	Carbon powder, taconite and wire mesh structures	3 ft
MWCNT	1.7–2.6 GHz	60–80 dB	(Three-layered structure with concrete middle layer)	50 mm
GO particles	0.8–7.8 GHz	25–50 dB	Straight, brass-coated SF	2.5 m
CF	8–18 GHz	20–40 dB	–	30 mm

composites also show good SE due to better particle dispersion, although they still fall short compared to GO-based composites. In fibre-based cementitious composites, SE increases with overall conductivity. Composites with high-conductivity fibres such as CFs and MWCNTs demonstrate good SE but are costly. So far, no cost-benefit analysis has been conducted on these composites. Moreover, a composite consisting of SF, carbon powder and taconite demonstrated high SE but only exceeded standards when cast around a steel mesh. Most novel materials demand a high volume of costly fillers to meet SE standards. Further research is needed to develop cost-effective EMI shielding composites with adequate SE.

Combining carbon and steel admixtures is not advisable due to the risk of galvanic corrosion in steel. However, mixing carbon and iron oxide does not pose this problem. For instance, using GO and iron oxide together is effective. Despite being less conductive, GO's high permittivity aids in radiation absorption. Another example is combining CF with Fe_3O_4 (magnetite) particles. Conductive magnetic admixtures are beneficial as they enable eddy current flow in response to magnetic fields, enhancing radiation absorption and reflection due to increased conductivity [116].

11.9 RECOMMENDATIONS AND FUTURE RESEARCH

Based on the advantages and limitations outlined in Table 11.6, future research should focus on addressing the challenges associated with each class of EMI shielding materials. For metals, research could explore innovative corrosion-resistant coatings to improve durability while maintaining high conductivity. Conducting polymers, although cost-effective and easily synthesised, require advancements in flexibility and dielectric properties to enhance their application range. For carbon materials, efforts should be directed towards improving mechanical properties and scaling up production techniques. In the case of

TABLE 11.6 Benefits and Challenges of Commonly Studied Categories of EMI Shielding Materials

Material	Advantages	Limitations
Metals	High electrical and thermal conductivity, Ductile	High density, Susceptible to corrosion without protection
Conducting polymers	Easy synthesis process, Low cost, High conductivity	Poor flexibility, Low dielectric properties
Carbon materials	Good flexibility, Lightweight, High electrical conductivity	Lack of bulk production techniques, Low mechanical properties
MXenes	Good mechanical flexibility, High metallic conductivity	Shielding mechanism not fully understood, Needs cost-effective bulk production
Cementitious materials	High durability, Low cost, Good thermal stability	Low electrical conductivity, Brittle nature
Semiconductors	High electrical conductivity, Tunable electronic properties	Expensive, Limited thermal stability
Metamaterials	Tailored electromagnetic properties, High tunability	Complex fabrication, High cost, Limited scalability
Ceramic	High thermal stability, Good corrosion resistance	Brittle, Low electrical conductivity

MXenes, a deeper understanding of their shielding mechanisms and cost-effective bulk production methods are essential. Cementitious materials would benefit from research aimed at enhancing electrical conductivity without compromising their structural integrity. Meanwhile, semiconductor research should focus on reducing costs and increasing thermal stability, while metamaterials require advancements in fabrication techniques and scalability. Finally, for ceramics, improving their electrical conductivity while maintaining thermal stability and corrosion resistance could significantly broaden their applicability in EMI shielding.

Given the advancements in non-polymer EMI shielding materials, several areas of future research and recommendations can be proposed to further enhance the performance and applicability of these materials. One promising possibility is the integration of 3D and 4D printing technologies. The introduction of 3D printing has revolutionised the fabrication of complex and customised structures. Utilising 3D printing to fabricate ceramics for EMI shielding has shown promising results. Future research should focus on optimising 3D printing techniques for non-polymer materials to achieve precise control over the material composition and microstructure. This optimisation could enhance the SE and mechanical properties of the composites. Additionally, 4D printing adds the dimension of time to 3D printed objects, allowing them to change shape or properties in response to external stimuli. Exploring 4D printing for EMI shielding materials could lead to the development of adaptive and responsive shielding solutions. These materials could dynamically adjust their properties to provide optimal shielding across a range of frequencies and environmental conditions.

Furthermore, the development of multifunctional composites is crucial. Combining different types of fillers, such as carbon-based materials, metals and ceramics, can create hybrid composites with enhanced SE and mechanical properties. Research should focus on understanding the synergistic effects of these fillers and optimising their combinations for maximum performance. There is also a growing demand for flexible and lightweight EMI shielding materials, especially for wearable electronics and portable devices. Future studies should aim to develop composites that maintain high SE while being flexible and lightweight. Future research should prioritise cost-effective large-scale production, material stability and multifunctional properties (such as combining EMI shielding with mechanical strength or thermal insulation). Expanding into novel hybrid materials and environmentally friendly alternatives could align with sustainable engineering goals.

REFERENCES

1. S. Geetha, K. K. Satheesh Kumar, C. R. K. Rao, M. Vijayan, and D. C. Trivedi, 2009. "EMI shielding: methods and materials – a review," *Journal of Applied Polymer Science*, 112(4), pp. 2073–2086.
2. X. Y. Wang, S. Y. Liao, Y. J. Wan, P. L. Zhu, Y. G. Hu, T. Zhao, R. Sun, and C. P. Wong, 2022. "Electromagnetic interference shielding materials: recent progress, structure design, and future perspective," *Journal of Materials Chemistry C*, *10*(1), pp. 44–72.
3. H. S. Nalwa, 1997. *Handbook of Organic Conductive Molecules and Polymers*, Conductive Polymers: Transport, Photophysics and Applications, vol. 4: Wiley Publishers.

4. Y. Wang and X. Jing, 2005. "Intrinsically conducting polymers for electromagnetic interference shielding," *Polymers for Advanced Technologies*, vol. 16, no. 4, pp. 344–351.
5. D. D. L. Chung, 2000. "Materials for electromagnetic interference shielding," *Journal of Materials Engineering and Performance*, *9*, pp. 350–354.
6. S. Kumari, J. Dalal, V. Kumar, A. Kumar, and A. Ohlan, 2023. "Emerging two-dimensional materials for electromagnetic interference shielding application," *International Journal of Molecular Sciences*, *24* (15), p. 12267.
7. W. Li, Z. Yu, Q. Wen, Y. Feng, B. Fan, R. Zhang, and R. Riedel, 2023. "Ceramic-based electromagnetic wave absorbing materials and concepts towards lightweight, flexibility and thermal resistance," *International Materials Reviews*, *68*(5), pp. 487–520.
8. J. Liu, M. Y. Yu, Z. Z. Yu, and V. Nicolosi, 2023. "Design and advanced manufacturing of electromagnetic interference shielding materials," *Materials Today*, *66*, pp. 245–272.
9. V. Shukla, "Review of electromagnetic interference shielding materials fabricated by iron ingredients," *Nanoscale Advances*, vol. 1, no. 5, pp. 1640–1671, 2019.
10. Y. Yao *et al.*, "Polymer-based lightweight materials for electromagnetic interference shielding: a review," *Journal of Materials Science*, vol. 56, pp. 6549–6580, 2021.
11. J. C. Shu, W. Q. Cao, and M. S. Cao, "Diverse metal–organic framework architectures for electromagnetic absorbers and shielding," *Advanced Functional Materials*, vol. 31, no. 23, p. 2100470, 2021.
12. H. Duan, P. He, H. Zhu, Y. Yang, G. Zhao, and Y. Liu, "Constructing 3D carbon-metal hybrid conductive network in polymer for ultra-efficient electromagnetic interference shielding," *Composites Part B: Engineering*, vol. 212, p. 108690, 2021.
13. D. Wanasinghe and F. Aslani, "A review on recent advancement of electromagnetic interference shielding novel metallic materials and processes," *Composites Part B: Engineering*, vol. 176, p. 107207, 2019.
14. X. S. Hu and Y. Shen, "Fabrication of novel polyaniline/flowerlike copper monosulfide composites with enhanced electromagnetic interference shielding effectiveness," *Journal of Applied Polymer Science*, vol. 134, no. 34, p. 45232, 2017.
15. H. Ji, R. Zhao, N. Zhang, C. Jin, X. Lu, and C. Wang, "Lightweight and flexible electrospun polymer nanofiber/metal nanoparticle hybrid membrane for high-performance electromagnetic interference shielding," *NPG Asia Materials*, vol. 10, no. 8, pp. 749–760, 2018.
16. L. O. Hoeft, "Using unconventional materials for electromagnetic shielding," in *2013 IEEE International Symposium on Electromagnetic Compatibility*, 2013: IEEE, pp. 118–123.
17. X. Shui and D. D. L. Chung, "Nickel filament polymer-matrix composites with low surface impedance and high electromagnetic interference shielding effectiveness," *Journal of Electronic Materials*, vol. 26, pp. 928–934, 1997.
18. X. Shui and D. D. L. Chung, "Magnetic properties of nickel filament polymer-matrix composites," *Journal of Electronic Materials*, vol. 25, pp. 930–934, 1996.
19. J. Chang, H. Zhai, Z. Hu, and J. Li, "Ultra-thin metal composites for electromagnetic interference shielding," *Composites Part B: Engineering*, vol. 246, p. 110269, 2022.
20. X. Li, Y. Qu, X. Wang, H. Bian, W. Wu, and H. Dai, "Flexible graphene/silver nanoparticles/aluminum film paper for high-performance electromagnetic interference shielding," *Materials & Design*, vol. 213, p. 110296, 2022.
21. Y. Wang *et al.*, "Facile preparation of a high-quality copper layer on epoxy resin via electroless plating for applications in electromagnetic interference shielding," *Journal of Materials Chemistry C*, vol. 5, no. 48, pp. 12769–12776, 2017.
22. Y. Lu and L. Xue, "Electromagnetic interference shielding, mechanical properties and water absorption of copper/bamboo fabric (Cu/BF) composites," *Composites Science and Technology*, vol. 72, no. 7, pp. 828–834, 2012.
23. L. Zhu, S. Bi, H. Zhao, L. Hou, Y. Xu, and Y. Lu, "Cu–Ni–Gd coating with improved corrosion resistance on linen fabric by electroless plating for electromagnetic interference shielding," *Journal of Materials Science: Materials in Electronics*, vol. 29, pp. 16348–16358, 2018.

24. H. Li *et al.*, "CuxS/PAN 3D nanofiber mats as ultra-lightweight and flexible electromagnetic interference shielding materials," *Macromolecular Materials and Engineering*, vol. 304, no. 12, p. 1900482, 2019.
25. R. Li, S. Wang, P. Bai, B. Fan, B. Zhao, and R. Zhang, "Enhancement of electromagnetic interference shielding from the synergism between Cu@Ni nanorods and carbon materials in flexible composite films," *Materials Advances*, vol. 2, no. 2, pp. 718–727, 2021.
26. M.-Q. Wang, J. Yan, S.-G. Du, J.-S. Zhao, B. Wang, and J.-Q. Ding, "Microstructure and electromagnetic interference shielding effectiveness of electroless Ni-P alloy coating on PVC plastic," *Fibers and Polymers*, vol. 15, no. 6, pp. 1175–1181, 2014.
27. X. Liu *et al.*, "Highly flexible electromagnetic interference shielding films based on ultrathin Ni/Ag composites on paper substrates," *Journal of Materials Science*, vol. 56, pp. 5570–5580, 2021.
28. C. Liang, K. Ruan, Y. Zhang, and J. Gu, "Multifunctional flexible electromagnetic interference shielding silver nanowires/cellulose films with excellent thermal management and joule heating performances," *ACS Applied Materials & Interfaces*, vol. 12, no. 15, pp. 18023–18031, 2020.
29. Y. Xu *et al.*, "Flexible and highly conductive sandwich nylon/nickel film for ultra-efficient electromagnetic interference shielding," *Applied Surface Science*, vol. 455, pp. 856–863, 2018.
30. H.-S. Lee, H.-B. Choe, I.-Y. Baek, J. K. Singh, and M. A. Ismail, "Study on the shielding effectiveness of an arc thermal metal spraying method against an electromagnetic pulse," *Materials*, vol. 10, no. 10, p. 1155, 2017.
31. W. M. Kim *et al.*, "The electromagnetic interference shielding effect of indium–zinc oxide/silver alloy multilayered thin films," *Thin Solid Films*, vol. 473, no. 2, pp. 315–320, 2005.
32. P. Lv *et al.*, "Metal-based bacterial cellulose of sandwich nanomaterials for anti-oxidation electromagnetic interference shielding," *Materials & Design*, vol. 112, pp. 374–382, 2016.
33. H.-R. Kim, B.-S. Kim, and I.-S. Kim, "Fabrication and EMI shielding effectiveness of Ag-decorated highly porous poly(vinyl alcohol)/Fe_2O_3 nanofibrous composites," *Materials Chemistry and Physics*, vol. 135, no. 2–3, pp. 1024–1029, 2012.
34. J. Wu and D. D. L. Chung, "Combined use of magnetic and electrically conductive fillers in a polymer matrix for electromagnetic interference shielding," *Journal of Electronic Materials*, vol. 37, pp. 1088–1094, 2008.
35. A. V. Menon, G. Madras, and S. Bose, "Magnetic alloy-MWNT heterostructure as efficient electromagnetic wave suppressors in soft nanocomposites," *ChemistrySelect*, vol. 2, no. 26, pp. 7831–7844, 2017.
36. H. Guan and D. D. L. Chung, "Radio-wave electrical conductivity and absorption-dominant interaction with radio wave of exfoliated-graphite-based flexible graphite, with relevance to electromagnetic shielding and antennas," *Carbon*, vol. 157, pp. 549–562, 2020.
37. Z. Xu and H. Hao, "Electromagnetic interference shielding effectiveness of aluminum foams with different porosity," *Journal of Alloys and Compounds*, vol. 617, pp. 207–213, 2014.
38. A. Iqbal, P. Sambyal, and C. M. Koo, "2D MXenes for electromagnetic shielding: a review," *Advanced Functional Materials*, vol. 30, no. 47, p. 2000883, 2020.
39. P. Song *et al.*, "Honeycomb structural rGO-MXene/epoxy nanocomposites for superior electromagnetic interference shielding performance," *Sustainable Materials and Technologies*, vol. 24, p. e00153, 2020.
40. F. Shahzad *et al.*, "Electromagnetic interference shielding with 2D transition metal carbides (MXenes)," *Science*, vol. 353, no. 6304, pp. 1137–1140, 2016.
41. Y. Gogotsi and B. Anasori, *The Rise of MXenes*, vol. 13: ACS Publications, 2019, pp. 8491–8494.
42. Y. Deng *et al.*, "Transparent electromagnetic interference shielding materials using MXene," *Carbon Energy*, p. e593, 2024.
43. K. Sushmita, G. Madras, and S. Bose, "Polymer nanocomposites containing semiconductors as advanced materials for EMI shielding," *ACS Omega*, vol. 5, no. 10, pp. 4705–4718, 2020.

44. X.-J. Zhang *et al.*, "Self-supported construction of three-dimensional MoS_2 hierarchical nanospheres with tunable high-performance microwave absorption in broadband," *The Journal of Physical Chemistry C*, vol. 120, no. 38, pp. 22019–22027, 2016.
45. T. Yamazaki, Y. Takano, and S. Honma, "Technology to form EMI-shielding film on semiconductor packages using sputter deposition process," *Semiconductor Packaging and Assembly Technologies Supporting Evolution of Various Devices*, vol. 71, no. 6, p. 19, 2016.
46. S. He *et al.*, "Controllable fabrication of CuS hierarchical nanostructures and their optical, photocatalytic, and wave absorption properties," *ChemPlusChem*, vol. 78, no. 3, pp. 250–258, 2013.
47. P. Kumar, "Ultrathin 2D nanomaterials for electromagnetic interference shielding," *Advanced Materials Interfaces*, vol. 6, no. 24, p. 1901454, 2019.
48. R. Kumar *et al.*, "Recent progress on carbon-based composite materials for microwave electromagnetic interference shielding," *Carbon*, vol. 177, pp. 304–331, 2021.
49. M. Byakodi *et al.*, "Emerging 0D, 1D, 2D, and 3D nanostructures for efficient point-of-care biosensing," *Biosensors and Bioelectronics: X*, vol. 12, p. 100284, 2022.
50. N. Mokhtar, *Synthesis and Characterization of Fullerene Doped Polyaniline/Titanium Dioxide Nanocomposites for Electromagnetic Interference Shielding*, University of Malaya (Malaysia), 2018.
51. P. Rani, B. Ahamed, and K. Deshmukh, "Electromagnetic interference shielding properties of graphene quantum-dots reinforced poly(vinyl alcohol)/polypyrrole blend nanocomposites," *Journal of Applied Polymer Science*, vol. 137, no. 45, p. 49392, 2020.
52. S. Tian *et al.*, "Electromagnetic interference shielding effectiveness of a flexible carbon fiber felt containing in situ grown hafnium carbide nanowires and nanobelts," *Ceramics International*, vol. 45, no. 15, pp. 19513–19516, 2019.
53. H. Duan *et al.*, "Effect of carbon nanofiller dimension on synergistic EMI shielding network of epoxy/metal conductive foams," *Composites Part A: Applied Science and Manufacturing*, vol. 118, pp. 41–48, 2019.
54. R. Andrews and M. C. Weisenberger, "Carbon nanotube polymer composites," *Current Opinion in Solid State and Materials Science*, vol. 8, no. 1, pp. 31–37, 2004.
55. D.-K. Lee, J. Yoo, H. Kim, and S.-H. Park, "Investigation of electromagnetic interference shielding effectiveness and electrical percolation of carbon nanotube polymer composites with various aspect ratios," *Carbon Letters*, vol. 34, no. 1, pp. 133–140, 2024.
56. Y. Jiang *et al.*, "3D printing of ultralight MWCNT@OCNF porous scaffolds for high-efficiency electromagnetic interference shielding," *Carbohydrate Polymers*, vol. 314, p. 120945, 2023.
57. A. Chaudhary, R. Kumar, S. Teotia, S. K. Dhawan, S. R. Dhakate, and S. Kumari, "Integration of MCMBs/MWCNTs with Fe_3O_4 in a flexible and light weight composite paper for promising EMI shielding applications," *Journal of Materials Chemistry C*, vol. 5, no. 2, pp. 322–332, 2017.
58. Y. Yu *et al.*, "Enhanced electromagnetic interference shielding properties of CNT/carbon composites by designing a hierarchical porous structure," *Nanomaterials*, vol. 14, no. 13, p. 1099, 2024.
59. Q. Wang, S. Xiao, S. Q. Shi, S. Xu, and L. Cai, "Self-bonded natural fiber product with high hydrophobic and EMI shielding performance via magnetron sputtering Cu film," *Applied Surface Science*, vol. 475, pp. 947–952, 2019.
60. A. Tugirumubano, S. J. Vijay, S. H. Go, L. K. Kwac, and H. G. Kim, "Characterization of electromagnetic interference shielding composed of carbon fibers reinforced plastics and metal wire mesh based composites," *Journal of Materials Research and Technology*, vol. 8, no. 1, pp. 167–172, 2019.
61. D. W. Lee *et al.*, "Orthogonal pattern of spinnable multiwall carbon nanotubes for electromagnetic interference shielding effectiveness," *Carbon*, vol. 152, pp. 33–39, 2019.
62. Z. Liu, H. Xing, Y. Liu, H. Wang, H. Jia, and X. Ji, "Hydrothermally synthesized Zn ferrite/multi-walled carbon nanotubes composite with enhanced electromagnetic-wave absorption performance," *Journal of Alloys and Compounds*, vol. 731, pp. 745–752, 2018.

63. J. Yan, Y. Huang, X. Han, X. Gao, and P. Liu, "Metal organic framework (ZIF-67)-derived hollow CoS_2/N-doped carbon nanotube composites for extraordinary electromagnetic wave absorption," *Composites Part B: Engineering*, vol. 163, pp. 67–76, 2019.
64. R. Kumar *et al.*, "Heteroatom doping of 2D graphene materials for electromagnetic interference shielding: a review of recent progress," *Critical Reviews in Solid State and Materials Sciences*, vol. 47, no. 4, pp. 570–619, 2022.
65. S.-H. Lee, D. Kang, and I.-K. Oh, "Multilayered graphene-carbon nanotube-iron oxide three-dimensional heterostructure for flexible electromagnetic interference shielding film," *Carbon*, vol. 111, pp. 248–257, 2017.
66. Y. Wu *et al.*, "Ultralight graphene foam/conductive polymer composites for exceptional electromagnetic interference shielding," *ACS Applied Materials & Interfaces*, vol. 9, no. 10, pp. 9059–9069, 2017.
67. Z. Zeng *et al.*, "Thin and flexible multi-walled carbon nanotube/waterborne polyurethane composites with high-performance electromagnetic interference shielding," *Carbon*, vol. 96, pp. 768–777, 2016.
68. X. Liu *et al.*, "Fabrication and electromagnetic interference shielding effectiveness of carbon nanotube reinforced carbon fiber/pyrolytic carbon composites," *Carbon*, vol. 68, pp. 501–510, 2014.
69. Y. J. Wan, P. L. Zhu, S. H. Yu, R. Sun, C. P. Wong, and W. H. Liao, "Anticorrosive, ultralight, and flexible carbon-wrapped metallic nanowire hybrid sponges for highly efficient electromagnetic interference shielding," *Small*, vol. 14, no. 27, p. 1800534, 2018.
70. Z. P. Wu *et al.*, "Electromagnetic interference shielding of carbon nanotube macrofilms," *Scripta Materialia*, vol. 64, no. 9, pp. 809–812, 2011.
71. L.-C. Jia, Y.-K. Li, and D.-X. Yan, "Flexible and efficient electromagnetic interference shielding materials from ground tire rubber," *Carbon*, vol. 121, pp. 267–273, 2017.
72. M. Arjmand, T. Apperley, M. Okoniewski, and U. Sundararaj, "Comparative study of electromagnetic interference shielding properties of injection molded versus compression molded multi-walled carbon nanotube/polystyrene composites," *Carbon*, vol. 50, no. 14, pp. 5126–5134, 2012.
73. Q. Li *et al.*, "Open-cell phenolic carbon foam and electromagnetic interference shielding properties," *Carbon*, vol. 104, pp. 90–105, 2016.
74. B. Shen, Y. Li, W. Zhai, and W. Zheng, "Compressible graphene-coated polymer foams with ultralow density for adjustable electromagnetic interference (EMI) shielding," *ACS Applied Materials & Interfaces*, vol. 8, no. 12, pp. 8050–8057, 2016.
75. H. Wang *et al.*, "3D network porous polymeric composites with outstanding electromagnetic interference shielding," *Composites Science and Technology*, vol. 125, pp. 22–29, 2016.
76. L. Kong *et al.*, "Electromagnetic wave absorption properties of graphene modified with carbon nanotube/poly(dimethylsiloxane) composites," *Carbon*, vol. 73, pp. 185–193, 2014.
77. Y.-J. Wan, P.-L. Zhu, S.-H. Yu, R. Sun, C.-P. Wong, and W.-H. Liao, "Graphene paper for exceptional EMI shielding performance using large-sized graphene oxide sheets and doping strategy," *Carbon*, vol. 122, pp. 74–81, 2017.
78. A. P. Singh *et al.*, "Encapsulation of γ-Fe_2O_3 decorated reduced graphene oxide in polyaniline core–shell tubes as an exceptional tracker for electromagnetic environmental pollution," *Journal of Materials Chemistry A*, vol. 2, no. 10, pp. 3581–3593, 2014.
79. L. Yuan-Qing, S. Y. Abdul, P. Kyriaki, and L. Kin, "Lightweight and highly conductive aerogel-like carbon from sugarcane with superior mechanical and EMI shielding properties," *ACS Sustainable Chemistry & Engineering*, vol. 3, 1419–1427, 2015.
80. M. Rahaman, T. K. Chaki, and D. Khastgir, "High-performance EMI shielding materials based on short carbon fiber-filled ethylene vinyl acetate copolymer, acrylonitrile butadiene copolymer, and their blends," *Polymer Composites*, vol. 32, no. 11, pp. 1790–1805, 2011.
81. Y. Liu *et al.*, "PTEN deficiency confers colorectal cancer cell resistance to dual inhibitors of FLT3 and aurora kinase A," *Cancer Letters*, vol. 436, pp. 28–37, 2018. Liu, Y. et al.,. Hydrated

aramid nanofiber network enhanced flexible expanded graphite films towards high EMI shielding and thermal properties. *Composites Science and Technology, 168*, pp. 28–37, 2018.
82. L. Chen *et al.*, "Mechanical and electromagnetic shielding properties of carbon fiber reinforced silicon carbide matrix composites," *Carbon*, vol. 95, pp. 10–19, 2015.
83. Z. Chen *et al.*, "Semi-transparent biomass-derived macroscopic carbon grids for efficient and tunable electromagnetic shielding," *Carbon*, vol. 139, pp. 271–278, 2018.
84. M. H. Al-Saleh and U. Sundararaj, "X-band EMI shielding mechanisms and shielding effectiveness of high structure carbon black/polypropylene composites," *Journal of Physics D: Applied Physics*, vol. 46, no. 3, p. 035304, 2012.
85. S. Pande, A. Chaudhary, D. Patel, B. P. Singh, and R. B. Mathur, "Mechanical and electrical properties of multiwall carbon nanotube/polycarbonate composites for electrostatic discharge and electromagnetic interference shielding applications," *RSC Advances*, vol. 4, no. 27, pp. 13839–13849, 2014.
86. D. X. Yan *et al.*, "Structured reduced graphene oxide/polymer composites for ultra-efficient electromagnetic interference shielding," *Advanced Functional Materials*, vol. 25, no. 4, pp. 559–566, 2015.
87. S.-T. Hsiao *et al.*, "Lightweight and flexible reduced graphene oxide/water-borne polyurethane composites with high electrical conductivity and excellent electromagnetic interference shielding performance," *ACS Applied Materials & Interfaces*, vol. 6, no. 13, pp. 10667–10678, 2014.
88. S. Gupta and N.-H. Tai, "Carbon materials and their composites for electromagnetic interference shielding effectiveness in X-band," *Carbon*, vol. 152, pp. 159–187, 2019.
89. R. Shu *et al.*, "Facile synthesis of nitrogen-doped reduced graphene oxide/nickel-zinc ferrite composites as high-performance microwave absorbers in the X-band," *Chemical Engineering Journal*, vol. 384, p. 123266, 2020.
90. P. Liu *et al.*, "Broadening electromagnetic absorption bandwidth: design from microscopic dielectric-magnetic coupled absorbers to macroscopic patterns," *Physica Status Solidi*, vol. 214, no. 12, p. 1700589, 2017.
91. Z. Li *et al.*, "Solvothermal synthesis of nitrogen-doped graphene decorated by superparamagnetic Fe_3O_4 nanoparticles and their applications as enhanced synergistic microwave absorbers," *Carbon*, vol. 115, pp. 493–502, 2017.
92. L. Quan, F. X. Qin, D. Estevez, W. Lu, H. Wang, and H. X. Peng, "The role of graphene oxide precursor morphology in magnetic and microwave absorption properties of nitrogen-doped graphene," *Journal of Physics D: Applied Physics*, vol. 52, no. 30, p. 305001, 2019.
93. A. K. Singh *et al.*, "Lightweight reduced graphene oxide-ZnO nanocomposite for enhanced dielectric loss and excellent electromagnetic interference shielding," *Composites Part B: Engineering*, vol. 172, pp. 234–242, 2019.
94. D. A. David *et al.*, "Porous carbon materials and their composites for electromagnetic interference (EMI) shielding: the state-of-the-art of technologies," in *Handbook of Porous Carbon Materials*: Springer, 2023, pp. 669–702.
95. D. Lai, X. Chen, G. Wang, X. Xu, and Y. Wang, "Highly conductive porous graphene film with excellent folding resilience for exceptional electromagnetic interference shielding," *Journal of Materials Chemistry C*, vol. 8, no. 26, pp. 8904–8916, 2020.
96. H. Kashani, M. Giroux, I. Johnson, J. Han, C. Wang, and M. Chen, "Unprecedented electromagnetic interference shielding from three-dimensional bi-continuous nanoporous graphene," *Matter*, vol. 1, no. 4, pp. 1077–1087, 2019.
97. Z. Xu *et al.*, "The preparation of carbonized silk cocoon-Co-graphene composite and its enhanced electromagnetic interference shielding performance," *Composites Part A: Applied Science and Manufacturing*, vol. 119, pp. 111–118, 2019.
98. E. Zhou *et al.*, "Synergistic effect of graphene and carbon nanotube for high-performance electromagnetic interference shielding films," *Carbon*, vol. 133, pp. 316–322, 2018.

99. C. Yu *et al.*, "Fe nanoparticles and CNTs co-decorated porous carbon/graphene foam composite for excellent electromagnetic interference shielding performance," *Journal of Alloys and Compounds*, vol. 820, p. 153108, 2020.
100. Z.-H. Zhou *et al.*, "Structuring dense three-dimensional sheet-like skeleton networks in biomass-derived carbon aerogels for efficient electromagnetic interference shielding," *Carbon*, vol. 152, pp. 316–324, 2019.
101. P. Kumari, P. Tripathi, S. P. Singh, and D. Kumar, "Electromagnetic shielding using ceramic materials," in *Materials for Potential EMI Shielding Applications*: Elsevier, 2020, pp. 315–331.
102. W. Yang *et al.*, "3D printing of PDC-SiOC@SiC twins with high permittivity and electromagnetic interference shielding effectiveness," *Journal of the European Ceramic Society*, vol. 41, no. 11, pp. 5437–5444, 2021.
103. W. Wang, Z. Li, X. Gao, Y. Huang, and R. He, "Material extrusion 3D printing of large-scale SiC honeycomb metastructure for ultra-broadband and high temperature electromagnetic wave absorption," *Additive Manufacturing*, vol. 85, p. 104158, 2024.
104. Y. S. Choudhary and N. Gomathi, "Metamaterials as shielding materials," *Advanced Materials for Electromagnetic Shielding: Fundamentals, Properties, and Applications*, pp. 367–391, 2018.
105. S. A. Schelkunoff, *Electromagnetic waves*: D. Van Nostrand Company, Inc, 1943.
106. V. G. Veselago, "Electrodynamics of media with simultaneously negative electric permittivity and magnetic permeability," in *Advances in Electromagnetics of Complex Media and Metamaterials*: Springer, 2002, pp. 83–97.
107. J. B. Pendry, A. J. Holden, D. J. Robbins, and W. J. Stewart, "Magnetism from conductors and enhanced nonlinear phenomena," *IEEE Transactions on Microwave Theory and Techniques*, vol. 47, no. 11, pp. 2075–2084, 1999.
108. T. J. Cui, D. R. Smith, and R. Liu, *Metamaterials*. Springer, 2010.
109. R. Liu, C. Ji, J. J. Mock, J. Y. Chin, T. J. Cui, and D. R. Smith, "Broadband ground-plane cloak," *Science*, vol. 323, no. 5912, pp. 366–369, 2009.
110. N. I. Landy, S. Sajuyigbe, J. J. Mock, D. R. Smith, and W. J. Padilla, "Perfect metamaterial absorber," *Physical Review Letters*, vol. 100, no. 20, p. 207402, 2008.
111. W. Cai, U. K. Chettiar, A. V. Kildishev, and V. M. Shalaev, "Optical cloaking with metamaterials," *Nature Photonics*, vol. 1, no. 4, pp. 224–227, 2007.
112. J.-Y. Jeong *et al.*, "Fabrication and characterization of resistive double square loop arrays for ultra-wide bandwidth microwave absorption," *Scientific Reports*, vol. 11, no. 1, p. 12767, 2021.
113. Y. Zhang, H. Dong, N. Mou, L. Chen, R. Li, and L. Zhang, "High-performance broadband electromagnetic interference shielding optical window based on a metamaterial absorber," *Optics Express*, vol. 28, no. 18, pp. 26836–26849, 2020.
114. O. Büyüköztürk, T.-Y. Yu, and J. A. Ortega, "A methodology for determining complex permittivity of construction materials based on transmission-only coherent, wide-bandwidth free-space measurements," *Cement and Concrete Composites*, vol. 28, no. 4, pp. 349–359, 2006.
115. D. Wanasinghe, F. Aslani, G. Ma, and D. Habibi, "Advancements in electromagnetic interference shielding cementitious composites," *Construction and Building Materials*, vol. 231, p. 117116, 2020.
116. D. D. L. Chung, "A review of microwave absorption and reflection by cement-based materials, with emphasis on electromagnetic interference shielding and admixture effects," *Advanced Functional Materials*, p. 2408220, 2024.
117. Z. Zhang, Y. Ji, and W. Ji, "Durability performance investigation for engineering fiber cementitious composites (ECC)," *Polymers*, vol. 15, no. 4, p. 931, 2023.
118. N. Bheel, B. S. Mohammed, M. S. Liew, and N. A. W. A. Zawawi, "Durability behaviours of engineered cementitious composites blended with carbon nanotubes against sulphate and acid attacks by applying RSM modelling and optimization," *Buildings*, vol. 13, no. 8, p. 2032, 2023.
119. W. Chen *et al.*, "Simultaneously enhanced mechanical and electromagnetic interference shielding properties of steel slag-recycled carbon fiber cementitious composites via wet-grinding process," *Materials and Structures*, vol. 56, no. 10, p. 182, 2023.

120. Y. He, L. Lu, K. Sun, F. Wang, and S. Hu, "Electromagnetic wave absorbing cement-based composite using Nano-Fe_3O_4 magnetic fluid as absorber," *Cement and Concrete Composites*, vol. 92, pp. 1–6, 2018.
121. G. Ma, J. Sun, L. Wang, F. Aslani, and M. Liu, "Electromagnetic and microwave absorbing properties of cementitious composite for 3D printing containing waste copper solids," *Cement and Concrete Composites*, vol. 94, pp. 215–225, 2018.
122. X. Fu and D. D. L. Chung, "Radio-wave-reflecting concrete for lateral guidance in automatic highways," *Cement and Concrete Research*, vol. 28, no. 6, pp. 795–801, 1998.
123. H. Guan, S. Liu, Y. Duan, and J. Cheng, "Cement based electromagnetic shielding and absorbing building materials," *Cement and Concrete Composites*, vol. 28, no. 5, pp. 468–474, 2006.
124. Y. Dai, M. Sun, C. Liu, and Z. Li, "Electromagnetic wave absorbing characteristics of carbon black cement-based composites," *Cement and Concrete Composites*, vol. 32, no. 7, pp. 508–513, 2010.
125. D. Zhao, J. Chen, Q. Gao, and H. Y. Ge, "Graphene oxide/cement composites for electromagnetic interference shielding," in *Materials Science Forum*, 2015, vol. 809: Trans Tech Publications, pp. 485–489.
126. A. Mazzoli *et al.*, "Effect of graphene oxide and metallic fibers on the electromagnetic shielding effect of engineered cementitious composites," *Journal of Building Engineering*, vol. 18, pp. 33–39, 2018.
127. X. Zhang and W. Sun, "Electromagnetic shielding and absorption properties of fiber reinforced cementitious composites," *Journal of Wuhan University of Technology-Materials Science Edition*, vol. 27, no. 1, pp. 172–176, 2012.
128. I. W. Nam and H.-K. Lee, "Synergistic effect of MWNT/fly ash incorporation on the EMI shielding/absorbing characteristics of cementitious materials," *Construction and Building Materials*, vol. 115, pp. 651–661, 2016.
129. X. Lv, Y. Duan, and G. Chen, "Electromagnetic wave absorption properties of cement-based composites filled with graphene nano-platelets and hollow glass microspheres," *Construction and Building Materials*, vol. 162, pp. 280–285, 2018.
130. Z. Liu, H. Ge, J. Wu, and J. Chen, "Enhanced electromagnetic interference shielding of carbon fiber/cement composites by adding ferroferric oxide nanoparticles," *Construction and Building Materials*, vol. 151, pp. 575–581, 2017.
131. Z.-J. Wang, K.-Z. Li, and C. Wang, "Freezing–thawing effects on electromagnetic wave reflectivity of carbon fiber cement based composites," *Construction and Building Materials*, vol. 64, pp. 288–292, 2014.

CHAPTER 12

Future Materials and Technology for EMI Shielding Applications

Reenu Elizabeth John, Rejo Mathew Joseph, N.V. Unnikrishnan, and Anoop Chandran

12.1 INTRODUCTION

Rapid progress in the field of digital technology has brought a great leap in the miniaturisation of electronic devices. The extensive implementation of advanced technologies has resulted in electromagnetic interference (EMI), a new kind of environmental contamination. EMI describes the deterioration in the performance of electronic devices by internal or external electromagnetic radiation [1]. These disruptions adversely affect the operation of sophisticated electronic devices, thereby posing an enormous challenge to a variety of industries, including consumer electronics, healthcare, aircraft, telecommunications and defence [2].

Interference radiations generally originate from human-made sources such as power lines, radio transmitters or electronic circuits, or from natural ones like lightning or solar activity. However, extensive dependence on high-frequency electronics has brought a steep escalation in the threat posed by EMI. Additionally, this issue is further worsened by the growth of wireless communication and the Internet of Things (IoT). Hence, sophisticated EMI mitigation strategies are the necessity of the hour. The capacity of electronic equipment to function in its intended electromagnetic environment without causing unacceptable electromagnetic disturbances to other devices is known as electromagnetic compatibility [3]. Therefore, adequate EMI shielding is not just a technological necessity but also a safety and regulatory measure.

In today's era, the rapid digitalisation and widespread demand for densely integrated devices lead to detrimental effects due to perturbations induced by EMI. Proper EMI shielding has become a mandatory requirement to shield sophisticated electronic equipment and ensure the hassle-free operation of devices. Additionally, these radiations not only impair electronic devices but also adversely affect human health and biological systems, leading to

DOI: 10.1201/9781003512363-12

cancers such as brain tumours, leukaemia and breast cancer [4]. The proper choice of material for EMI shielding stands as the foremost priority of the day. Usually, materials with high conductivity and magnetic properties are chosen for blocking EMI radiations [5].

The conventional strategy for EMI shielding predominantly relies on metallic materials, owing to their superior electromagnetic attenuation capabilities. Metals like aluminium, copper and nickel, due to their high conductivity and dielectric constant, contribute to a high EMI shielding efficiency [6]. Unfortunately, even though metals exhibit great shielding performance, drawbacks such as corrosion, high surface reflection and poor processability have limited their wider deployment in next-generation portable, lightweight and wearable electronic devices.

Recent research is increasingly focused on the development of lightweight shielding materials that seamlessly integrate additional desirable properties such as transparency, flexibility and mechanical durability into a single multifunctional material [7]. Even though transparent metal films and metal oxide layers are good candidates for shielding microwaves, achieving portable flexibility in miniaturised electronic devices remains a great challenge. The fast-paced developments of conducting polymers with various conductive fillers like metal powders, carbon-based fillers such as graphene, reduced graphene oxide, carbon nanotubes and inorganic nanoparticles have brought about rapid progress within the class of EMI shielding materials. Despite their potential, this class of materials also faces challenges in delivering high shielding performance at minimal thickness, along with preserving desirable mechanical and optical properties.

Despite the large number of review articles that deal with the development of EMI shielding materials, most publications focus on existing technologies, greatly limiting discussions on the future outlook of this field. Although such reviews usually cover material properties, synthesis technologies and the current performance of these products, they rarely tackle broader technological details or foresee combinations of next-generation strategies. This chapter presents a detailed review of the most recent developments in advanced EMI shielding materials and showcases cutting-edge advancements that have substantially improved performance and realised multifunctionality. Beyond this, it provides a clear path towards next-generation shielding solutions based on breakthroughs in novel materials and materials fabrication techniques, along with eco-friendly, sustainable strategies. It also highlights the transformative capability of ML and AI to refine material design and tune the properties of future shielding materials with precision. Together, these insights lay a strategic roadmap for developing adaptable, efficient and environmentally responsible EMI shielding technologies specifically designed to meet the demanding requirements of future electronic and communication systems.

12.2 CURRENT EMI SHIELDING MATERIALS

12.2.1 Metallic Materials

Metals and metallic coatings are crucial materials for EMI shielding, capable of protecting sensitive electronic devices from disruptive EM (electromagnetic) waves, thereby ensuring their optimal performance and reliability. The excellent conductivity of metals makes them suitable for effectively reflecting and absorbing EM waves over a broad range of frequencies. Furthermore, their excellent mechanical strength and thermal stability make them suitable for use in harsh environmental conditions. Even though metals exhibit certain limitations such as high weight, corrosion vulnerability, manufacturing complexity and high cost, they still have the upper hand in EMI shielding applications owing to their excellent shielding performance.

Compared to metals, alloys play a critical role in EMI shielding due to their superior corrosion resistance, electrical conductivity and mechanical strength. Jang and his co-workers elaborately studied the electrical conductivity and SE of copper (Cu), copper-zinc (Cu-Zn) and copper-Nickel (Cu-Ni) coatings of varying thicknesses [8]. They observed that increasing the thickness of the coating plays a crucial role in shielding EM waves. Among the coatings, Cu-Zn exhibited the highest shielding efficiency of 80 dB at 1 GHz, compared to Cu and Cu-Ni, which showed SE values of 12 dB and 68 dB, respectively. Results indicated that a 100 μm Cu-Zn coating meets EMI shielding requirements, while Cu and Cu-Ni need greater thickness for optimal performance.

The ability to tailor alloy compositions allows for the optimisation of shielding performance based on specific application needs. Li and his team prepared an Mg-8Li-3Y-2Zn alloy through vacuum melting and studied its tensile conductivity and EMI shielding properties [9]. They observed that α-Mg in the alloys increased the tensile strength to 263.7 MPa and elongation reached 45.3%. The addition of Y and Zn improved the SE of the alloy, maintaining values between 76.9 and 106.7 dB in the frequency range of 30–4500 MHz. Zhang et al. synthesised a novel alloy, Mg98.5Zn0.5Y, using friction-stir processing (FSP) [10]. They studied the effect of FSP on the EMI shielding properties of the sample. The studies revealed that the long-period stacking ordered (LPSO) phase is the predominant structure that undergoes significant grain refinement. This grain refinement and fragmentation of the LPSO phase resulted in modification of the mechanical properties of the alloy. The ultimate tensile strength was found to be 358.2 MPa with an elongation of 15.1%. This reflected a 344% increase in strength and a 733% improvement in ductility compared to the as-cast material. The alloys were found to have a conductivity of 11.6 Ms m^{-1}. Simultaneously, the EMI SE was maintained between 70 and 110.4 dB over a broad frequency range (30–4500 MHz).

In another work, Hui et al. used a low-speed accumulative roll-bonding (ARB) process to synthesise Cu-based Cu/Ni/Cu and Ni-based Ni/Cu/Ni composites with different numbers of layers [11]. Enhancement in ARB pass resulted in the breaking and distribution of reinforcing layers inside the matrix. This led to a decrease in grain size, while enhancing the hardness of the Ni and Cu layers. Hardness improved by 105 HV in Ni and 24 HV in Cu from pass 1 to pass 7, although with ARB pass 7 it was observed that the hardness values were fluctuating. EMI shielding increased with increased Ni/Cu interfacial barriers, up to 55 dB in Ni-based composites and 59 dB in Cu-based composites. Increased layer stacking produced better distribution of reinforcement and enhanced shielding. However, it was observed that optimum shielding efficiency was achieved in Ni/Cu/Ni/Cu/Ni/Cu/Ni and Cu/Ni/Cu/Ni/Cu/Ni/Cu composites, and Ni/Cu/Ni while Cu/Ni/Cu composites exhibited the highest hardness growth.

Building on the principles of integrating conductive fillers into structured matrices for enhanced performance, Sharma and his co-workers developed silver-plated graphene-modified carbon foams with EMI shielding effectiveness (SE) of 73.5 dB in the X-band, an electrical conductivity of 180 S cm^{-1} and a thermal conductivity of 6.0 W/m·K, along with enhanced compressive strength of 6.5 MPa [12]. These remarkable properties are attributed to the synergistic effect of graphene nanoplatelets and uniform silver electroplating, which together form a conductive, thermally efficient and mechanically robust porous structure, enabling their use in sophisticated applications.

Despite their shielding efficiency, metal composites face a major drawback in practical applications: they are prone to corrosion, which reduces the longevity of the shield and increases maintenance costs [13]. A critical research focus involves developing lightweight metal composites capable of maintaining high SE while reducing weight, as this would represent a significant milestone for industrial adoption. Furthermore, the creation of corrosion-resistant coatings to extend the lifespan of metal shields is essential for their efficient deployment in high-end applications such as aerospace, telecommunications and military infrastructure.

12.2.2 Conductive Polymers-Based Composite Materials

A good EMI shielding material should be characterised by low density, flexibility, thin thickness and good conductivity, and respond to a wide range of frequencies. With the ongoing trend towards the miniaturisation of electronic devices, there is an increasing demand for lightweight and flexible shielding solutions. As a result, polymeric composites incorporated with conductive or magnetic fillers have been widely utilised as efficient EMI shielding materials. Polymers can be commonly categorised into two types: insulating polymers and intrinsically conducting polymers (ICPs). Insulating polymers such as polystyrene (PS), poly(vinylidene fluoride) (PVDF), polypropylene (PP), poly(methyl methacrylate) (PMMA), poly(vinyl alcohol) (PVA), polyethylene (PE), poly(vinylpyrrolidone) (PVP) and epoxy have low conductivity, which can be enhanced by adding conducting fillers. The other group, including polyaniline (PANI), polypyrrole (PPy), poly(3,4-ethylene dioxythiophene):polystyrenesulfonate (PEDOT:PSS) and polythiophene (PTP), are generally ICPs, and they play a major role in EMI shielding studies [14]. The major advantage of incorporating polymers as EMI shielding candidates is that, contrary to traditional metal-based shielding materials, they offer benefits such as corrosion resistance, flexibility, ease of processing and tuneable conductivity. A few of the recent works on polymer composites are reviewed here.

For instance, Raju and his team developed a polymer nanocomposite by incorporating single-walled carbon nanotubes (SWCNTs) into a polypyrrole (PPy) matrix [15]. The addition of SWCNTs enhanced dielectric losses and improved impedance matching, resulting in better EM wave attenuation. As a result, the composite achieved an SE of 32.1 dB in the X-band frequency range (8.2–12.4 GHz). Additionally, a minimum reflection loss of −45.54 dB and a broad absorption bandwidth of 3.1 GHz were recorded at 5 wt.% SWCNT loading, making the material suitable for lightweight, broadband EMI shielding applications.

Mohan et al. explored the shielding properties of PANI-graphene composites and investigated the role of graphene in establishing effective EMI shielding [16]. The composite film, with a thickness of 10 μm, exhibited an SE of 57 dB in the S-band (2–4 GHz) and X-band (8–12 GHz) frequency ranges. An optimal graphene loading of 2% was found to significantly enhance the absorption-dominated shielding mechanism, thereby contributing to the overall performance of the material. Chen et al. advanced the concept of polyaniline-based composites by integrating carbon nanofibres into a 3D aerogel architecture [17]. This structure achieved an outstanding EMI SE of 85.4 dB and a specific EMI SE of 791.2 $dB{\cdot}cm^3{\cdot}g^{-1}$ in the X-band. Furthermore, the aerogels demonstrated superior flame-retardant behaviour and thermal protection, with a peak heat-release rate of just 7.8 $W{\cdot}g^{-1}$ and total heat release of 0.58 $kJ{\cdot}g^{-1}$, making them promising candidates for next-generation multifunctional shielding applications.

Bora et al. explored a different avenue by leveraging MXene's conductive properties within a crosslinked PEDOT:PSS matrix for EMI shielding. An SE of 41 dB at just 6 μm thickness and an absolute SE of 89,924 $dB{\cdot}cm^2{\cdot}g^{-1}$ was obtained [18]. This is almost nine times that of pristine PEDOT:PSS and over three times that of MXene alone. Enhanced electrical conductivity and increased absorption sites from the interconnected network contribute to superior absorption-dominated shielding, positioning the material as a strong candidate for real-time EMI shielding in fields such as telecommunications, healthcare, defence and aerospace. In another work, Li and his team employed defect engineering to enhance the EMI shielding performance of materials [19]. They intercalated amine molecules, specifically *p*-phenylenediamine (PPD), with varying sizes and chemical affinities into MXene aerogels. The introduction of PPD expanded the interlayer spacing (d-spacing) from 0.960 to 1.642 nm and increased Ti–Ti layer defects, thereby enhancing internal polarisation loss. As a result, the PPD@MXene aerogel demonstrated absorption-dominated shielding behaviour with a high SE of 50.4 dB. This work provides valuable insights into the design of interlayer-engineered MXene materials for efficient EMI attenuation.

To tackle the growing demand for high-performance and eco-friendly EMI shielding materials, Ghosh and his team fabricated a phosphorus-doped multilayer $Ti_3C_2T_x$ MXene/PEDOT:PSS (P-$Ti_3C_2T_x$@PP) hybrid aerogel [20]. The aerogel structure, characterised by interconnected conductive networks and heteroatom-induced dipole polarisation, enabled it to achieve a remarkable SE of approximately 52 dB at 18 GHz, with an average of 47.6 dB across the Ku-band. It was observed that the introduction of phosphorus enhanced polarisation losses, improved impedance matching with free space and reduced relative permittivity. In another work, Naqvi and his team explored the role of surface functionalisation in enhancing the compatibility of $Ti_3C_2T_x$ MXenes with hydrophobic epoxy matrices [21]. For this purpose, the surface of MXenes was functionalised with alkylated 3,4-dihydroxy-L-phenylalanine ligands. This modification led to a dramatic improvement in composite performance, achieving an impressive electrical conductivity of 8200 $S{\cdot}m^{-1}$ and an exceptional SE of 100 dB at 110 GHz and 61 dB at 8.2 GHz. Additionally, the thermal conductivity of the nanocomposite increased to 1.37 W/mK, and tensile toughness improved by 300%, reaching 271 $KJ{\cdot}m^{-3}$. This research demonstrated the critical importance of targeted surface functionalisation in optimising interfacial interactions, thereby paving the path for improved network formation and unlocking the multifunctional characteristics of 2D-material-based composites.

Although conducting polymers with fillers are widely used for electromagnetic shielding applications, they exhibit significantly lower SE compared to metals. Additionally, these polymers face environmental stability issues, which compromise the long-term durability of the shielding material. However, the development of hybrid composites through advanced material engineering, such as integrating multiscale conductive fillers (e.g., carbon nanotubes and metal oxides) or employing surface-functionalised reinforcements, can substantially enhance SE while addressing stability challenges.

12.2.3 Hybrid Materials

Hybrid fillers are a combination of carbon-based fillers with other fillers such as metal oxides, ferrites, metallic particles etc. Hybrid fillers provide excellent mechanical and physical properties, and they are also proficient in tuning the permittivity, permeability and electrical and thermal conductivities of composites. A suitable combination of the magnetic characteristics of metal-based fillers and the dielectric properties of carbon-based fillers provides better attenuation of EM waves. Through a solution processing method, a hybrid composite of PVDF with 3 wt.% of layered graphene (LG) and varying concentrations of nickel spinel ferrites (NSF) was studied [22]. The PVDF/LG composite showed improved SE of 25–45 dB for the frequency range 1–12 GHz, compared to ~0 dB for neat PVDF. Incorporation of 15 wt.% NSF further increased shielding to 30–53 dB due to effective FLG–NSF network formation. However, at 30 wt.% NSF, agglomeration led to reduced performance (12–43 dB). Expanding on the use of carbon-based and magnetic fillers in polymer matrices, Kazmi and his team adopted oxidative polymerisation of aniline in the presence of binary core materials (NFO@CNTs and NFO@RGO) to synthesise ternary composites (PANI/NFO@CNTs and PANI/NFO@RGO), nickel ferrite ($NiFe_2O_4$) being labelled as NFO [23]. The composite PANI/NFO@CNTs exhibited an SE of 22.36 dB at 2 mm thickness, while PANI/NFO@RGO displayed a higher SE value of 31 dB at 2 mm and 43 dB at 4 mm thickness. The enhanced shielding properties can be attributed to a combination of good impedance matching, effective dielectric and magnetic loss mechanisms and improved electrical conductivity, all of which facilitate multiple reflections and scattering of incident EM waves. The outcome highlights the potential of these lightweight, flexible and low-filler-content composites for advanced EMI shielding applications.

Although polymer/rGO composites are widely used as shielding materials, the aggregation of rGO layers hampers their conductivity and EMI shielding performance. To address this, a composite film of polyvinyl alcohol/reduced graphene oxide/iron hydroxide (PVA/rGO/α-FeOOH) was developed [24]. The dielectric PVA, conductive rGO and magnetic α-FeOOH components work synergistically to shield against electromagnetic fields. The films exhibit an electrical conductivity of 6.92 S m^{-1} at 4 wt.% rGO/FeOOH. Additionally, the films achieved an SE of 28.27 dB, and their thermal conductivity increased to 2.23 W/m·K, enhancing long-term stability. The composite also displayed improved mechanical properties, with a Young's modulus of 3.03 GPa and a tensile strength of

43.25 MPa. This work highlights the potential of this material as an effective EMI shielding solution with additional benefits such as thermal stability and structural integrity.

Recently, Katheria et al. fabricated a new class of multifunctional polymer composite incorporating MXene and Fe_3O_4-*g*-C_3N_4 into an ethyl methacrylate (EMA) matrix, fabricated via solution mixing [25]. The resulting composite demonstrated a hierarchical structure that synergistically combined the high electrical conductivity of MXene with the magnetic and structural properties of Fe_3O_4-*g*-C_3N_4. Characterisation revealed a notable EMI SE of –36 dB, enhanced thermal conductivity of 0.98 W/m·K and strong mechanical integrity. These outcomes underscore the composite's suitability for EMI shielding and thermal management, offering promising applications in electronics, aerospace and telecommunications. Further, Yin and his team focused on improving the SE by synthesising a multifunctional CNF/MXene/$ZnFe_2O_4$@PANI composite film using a vacuum-assisted filtration method [26]. The Janus architecture of the film enabled seamless integration of conductive and magnetic layers, achieving efficient EMI shielding and thermal management. The prepared samples exhibited a high SE of 45.75 dB with rapid Joule heating, reaching 235.8 °C within 50 seconds at an applied voltage of 9 V. This is attributed to the formation of a well-connected MXene-based conductive network. The samples also showed improved mechanical properties, with a tensile strength of 27.24 MPa and a fracture strain of 6.21%, making them suitable for durable, high-performance EMI shielding applications with integrated thermal response capabilities.

A major drawback in fabricating hybrid EMI shielding materials lies in the complexity of processing and manufacturing, which often leads to inconsistencies in composite quality due to factors like filler-dispersion heterogeneity and curing-parameter variations. Therefore, research to simplify fabrication techniques such as solvent-free methods or 3D printing and to establish standardised testing protocols is critical to ensure reliability. Additionally, integrating AI-driven process optimisation and scalable nanomanufacturing techniques could bridge the gap between laboratory innovations and industrial adoption.

12.3 EMERGING MATERIALS FOR EMI SHIELDING

12.3.1 Carbon-Based Nanomaterials

Carbon-based fillers such as carbon nanotubes (CNTs), graphene, graphene oxide (GO), carbon fibres (CF) and carbon black (CB) are widely used in combination with polymers due to their unique ability to provide excellent electrical conductivity, EMI shielding and structural enhancement. For instance, Liu and his co-workers developed a well-dispersed polyurethane and single-walled carbon nanotube (PU/SWCNT) EMI shielding composite material by a solvent-casting method [27]. The fabricated material exhibited an SE of 17 dB in the X-band (8.2–12.4 GHz). Nonetheless, it was found that the reflecting characteristic was dominated by the material's absorption capacity at high SWCNT loading. In another work, Raju et al. synthesised polyaniline (PANI)/SWCNT nanocomposites through ultrasonically induced in situ polymerisation, achieving a high EMI SE of 32.80 dB at 11 GHz with 99.999% absorption [28]. Nguyen and his co-workers investigated the SE of SWCNT and its composites with poly(3,4-ethylene dioxythiophene): poly(styrene sulfonate)

(SWCNT/PEDOT:PSS) in the X-band frequency range, prepared through vacuum-filtering and dip-coating methods [29]. The material exhibited an SE of 32.1–55.53 dB with thicknesses ranging from 0.24 to 2.12 μm. Additionally, the composites demonstrated outstanding stability under challenging conditions. The conducting materials' synergistic spatial arrangement is responsible for their better performance, which holds promise for flexible, lightweight EMI shielding in next-generation electronics.

In 2020, Mei and his co-workers, in an effort to develop reliable and high-performance EMI shielding materials, incorporated CNT films into an epoxy matrix [30]. The multilayer structure of CNT films positively impacted the SE of the composite, successfully achieving an SE of 52.31 dB. Additionally, the prepared composite exhibited improved mechanical strength and enhanced impact-energy absorption, thereby modifying delamination resistance, making it suitable for practical applications. For the purpose of fabricating EMI shielding materials in areas of wearable, defence and aerospace applications, Feng et al. incorporated graphene along with CNT and polyurethane [31]. This resultant GC/PU film exhibits high tensile strength of 42.9 MPa, ultra-high conductivity of 2087.2 S m^{-1} and EMI shielding efficiency of 37.7 dB at 130 μm and 72.1 dB at 520 μm in the X-band frequency range. Additionally, it demonstrates efficient electric heating (127.5 °C at 5.0 V within 20 seconds) and flame-retardant characteristics, making it suitable for anti-icing and de-icing applications.

With an effort to design sophisticated materials with superior material-distribution uniformity, enhanced electrical conductivity and robust EM wave attenuation capabilities for versatile applications, Mingjie and his team introduced a "popcorn making" mimicking method for fabricating rGO covered in laser induced graphene (LIG) embedded with nickel ferrite (NiFe2O4) nanoparticles (8.2–12.4 GHz, X-band) [32]. The prepared rGO/LIG@$NiFe_2O_4$ composite exhibited an SE of 36 dB at 70 μm on a single side and 51 dB at 166 μm on double-sided films, with enhanced absorption ratios of 75% and 73%, respectively.

In another work, Xie and his team fabricated highly conductive and mechanically stable yarns by modifying polyamide yarns with polyethyleneimine to boost their interaction with GO sheets [33]. Followed by reduction with hydroiodic acid, the resulting yarns exhibited improved GO loading and coating stability. These conductive yarns maintained high conductivity of 4.7×10^3 S m^{-1} even after 10,000 bending cycles and were capable of enduring various textile processes such as embroidering, sewing and weaving. The woven textiles demonstrated outstanding EMI SE of 66.7 dB for a thickness of 0.615 mm. Moreover, the EMI shielding remained stable after 5000 bending cycles, which emphasises their long-term durability. An additional feature of this work is that it offers a viable strategy for large-scale production of multifunctional yarns for EMI shielding, thermal regulation and de-icing applications.

Solution-casting method was employed to fabricate titanium dioxide (TiO_2) on poly(vinyl alcohol) (PVA) films reinforced with rGO [34]. The composite films exhibited an SE of 55.8 dB in the X-band frequency range (8–12 GHz). In a recent work, Mohamed and his team performed an in-depth study on the mechanical, thermal and EMI shielding properties of polyvinyl alcohol/reduced graphene oxide/iron hydroxide

(PVA/rGO/α-FeOOH) nanocomposite films [24]. The uniform distribution of FeOOH facilitates tailor-tuning of electrical conductivity, enabling a high electrical conductivity of 6.92 S m^{-1} (˃10^6 times that of PVA) with an ultralow percolation threshold of 1.31 wt.%, one of the smallest reported percolation thresholds in this regard. The films exhibited an SE of 28.27 dB, with the dominant shielding mechanism being absorption. Additionally, their thermal conductivity reached 2.23 W mK^{-1}, nine times higher than pure PVA, ensuring long-term durability by reducing degradation risks.

CB is a carbon-based filler with a crystalline structure and short-range graphitic order. It is composed of carbon atoms fused into aggregates. CBs are used along with polymers and elastomers to enhance electrical, mechanical and optical properties. Kumari and her co-workers utilised fluoroelastomers with conductive CB through an electrospinning method [35]. The resultant composite demonstrated an SE of 72 dB in the X-band frequency range. Moreover, the material is suitable for defence and stealth technology applications due to its reduced radar-cross-section.

Tian et al. provided a universal approach for efficient and large-scale preparation of lightweight and multifunctional polymer composite foams [36]. They fabricated polypropylene/CF/CB composite foam bonded with continuous CF-reinforced polypropylene prepregs via high-pressure foam injection moulding. The addition of CB enhanced the thermal stability and flame retardance of the PP composites, making them suitable for applications requiring fire resistance. The synergistic effect of CF and CB not only improved the mechanical properties but also contributed to the overall durability of the foams [37]. The synthesised foams exhibited an EMI-shielding performance of 65.4 dB at 10 wt.% CF and 15 wt.% CB, signifying a scalable approach for lightweight, multifunctional polymer foams.

The present fifth generation of wireless communication has advanced rapidly, bringing with it the promise of lower latency and faster transmission in the IoT era. To meet these requirements, Wang et al. prepared a composite material designed to improve SE in the Ka-band (26–40 GHz) frequency range [38]. They integrated carbonyl iron powder (CIP) and CF in a polydimethylsiloxane (PDMS) matrix. The outstanding SE of ≥120 dB in the Ka-band can be attributed to the collective EM-wave reflections within the CF structure and the enhanced absorption properties of CIP. The shielding could be increased up to 160 dB by varying the CF content. Additionally, the composite material is promising for 5G Ka-band shielding applications, as it also possesses thermal stability greater than 300 °C and a thermal conductivity greater than 5 W m^{-1} K^{-1}.

Recently, in another innovative work, Wang and his team used a 3D-printing technique to design lightweight and structurally controllable reduced graphene oxide (rGO) embedded in iron-based metal-organic frameworks [39]. The frameworks exhibited high EMI SE, with 44.95 dB in the K-band (18–27 GHz) and 53.90 dB in the Ka-band (27–40 GHz).

The widespread use of carbon composites in polymers as shielding materials faces certain challenges such as non-uniform dispersion of fillers in the polymer matrix, production costs and potential health risks during inhalation at the fabrication site. Hence, the adoption of safer-handling practices and regulations, along with extensive research on optimising the dispersion technique, is mandatory for fabricating EMI shielding materials for high-end applications.

12.3.2 MXenes

2D MXenes are transition metal carbides, nitrides or carbonitrides having the general formula $M_{n+1}X_nT_x$. Here, M represents a transition metal such as Ti, Zr or V; X is carbon or nitrogen; and T represents the functional group on the surface of MXene. MXenes are suitable candidates for suppressing EMI due to their enhanced electrical conductivity, large surface area, lightweight nature and processability. Moreover, the versatile surface chemistry of these materials enables them to be shaped into different morphologies, including dense laminates, layer-by-layer architectures and porous ones. Owing to their structural versatility, MXenes have been the centre of attention as the most suitable lightweight materials for EMI shielding, with an upsurge of research interest. $Ti_3C_2T_x$ is a form of MXene, namely a titanium carbide with surface functional groups (T_x) such as —OH, —O or —F. It is one of the most extensively researched MXenes.

Li et al. synthesised multilayered sandwich-like morphology of Ti_2CT_x with high surface functional groups (e.g., —OH, —O, —F), capable of improving electromagnetic (EM) attenuation and absorption efficiency [40]. It has a high electrical conductivity of 0.3 S cm^{-1}, responsible for its superior EMI SE. The material has an average EMI SE of 61.1 dB in the X-band (8.2–12.4 GHz), with a maximum SE of ~70 dB at only 0.8 mm thickness. Ti_2CT_x shows better shielding performance compared to Ti_2AlC MAX phase (46.2 dB). In contrast to graphene, which involves a complicated reduction process, Ti_2CT_x prepared through a straightforward chemical etching process with HCl + LiF, retaining its surface functional groups and enhancing processability. This renders it a more convenient substitute for graphene-based EMI shielding materials.

Liang and co-workers fabricated a high-performance polyimide-MXene-multi-wall carbon nanotube (PI-MXene-MWCNT) nanocomposite film through a controlled electrospinning-spraying technique [41]. The outer layers consist of PI electrospun fibre films, while the central layer incorporates MXene and MWCNT to improve EMI shielding efficiency without increasing filler content. The film exhibited an excellent EMI SE of 66.8 dB and absolute shielding efficiency of 13,153.8 dB cm^2 g^{-1}. Furthermore, the prepared samples sustained steady shielding capability under extreme conditions and exhibited excellent flexibility with 35.4% elongation at break, making them resilient and durable candidates for EMI shielding in harsh industrial environments.

In another work, Wang et al. annealed $Ti_3C_2T_x$/epoxy nanocomposites prepared through ionic-intercalation sonication-assisted exfoliation followed by thermal reduction at 200 °C [42]. Structural examination proved the formation of ultra-thin and highly crystalline MXene nanosheets with enhanced electrical conductivity (105 S m^{-1} at 15 wt.%) and EMI shielding performance (41 dB in X-band, a 37% increase compared to non-annealed MXene composites). Moreover, the shielding was dominated by absorption with strong conductive networks and multiple reflections. Mechanical properties achieved maximum values at 5 wt.% loading, with Young's modulus being 4.32 GPa and hardness 0.29 GPa. In comparison with graphene- and CNT-based EMI composites, $Ti_3C_2T_x$ MXene provides better shielding with reduced filler loading and easier processability.

In another work, Li et al. used interlayer- and defect-engineering to enhance electrical conductivity and EMI shielding of $Ti_3C_2T_x$ MXene [19]. They synthesised MXene aerogels with tunable d-spacing and amine molecules as intercalants and cross-linkers. The intercalation of *p*-phenylenediamine (PPD) increased the d-spacing from 0.960 to 1.642 nm, enhancing defect density and polarisation loss. The PPD@MXene aerogel achieved an absorptivity of 0.92 and an SE of 50.4 dB, providing insights into interlayer-engineered MXene shielding materials.

Bai and his team studied the exceptional EMI shielding performance and piezoelectric sensing capabilities of a dual-network structure of 2D transition metal carbides and nitrides (MXene) and bacterial cellulose-thermoplastic polyurethane (BC-TPU) foams [43]. Anisotropic MXene/BC aerogels were capable of achieving a high conductivity of 1912 S m^{-1}, an ultra-high EMI of 86 dB and an absolute SE of 63,608 dB cm^2 g^{-1}. The structure gained flexibility by incorporating an interpenetrating neuro-like TPU network through a controlled coagulation method. The final product, labelled as MXene/BC-TPU foams, exhibited remarkable elasticity and fatigue resistance over a broad strain range of −80 to 80%. Additionally, the foams maintained an outstanding SE of 76 dB, with 86.8% retention after 5000 compression–release cycles and 69.7% retention after 100 stretch–release cycles, making them highly suitable for EMI shielding and motion monitoring in wearable electronics.

Tao and his co-workers constructed a film using poly(vinyl alcohol-co-ethylene) (PVA-co-PE) nanofibres as a substrate and twill-structured nylon 6 fabric as a filter. The PVA-co-PE/silver nanowire (AgNW) and MXene, labelled as PVA/Ag/M, was synthesised via vacuum filtration, templating and hot pressing [44]. At 2.5 wt.% of AgNW/MXene, the film exhibited an SE of 101.6 dB in the frequency range 8–20 GHz. Strong shielding performance, which enables 95% wave absorption and simultaneously increases wave collisions and internal reflections, is ascribed to the synergistic impact of the twill gradient structure and its high conductivity.

The structural versatility of MXenes enables their widespread application in EMI shielding materials. By leveraging internal scattering and optimising impedance matching, MXenes can be engineered to transition from reflection-dominant to absorption-dominant shielding regimes. However, realising these advanced shielding properties requires meticulous and precise architectural design, particularly when employing multilayered configurations. Moreover, the large-scale production of defect-free MXene films and the fabrication of complex structural configurations present significant technological challenges. Achieving cost-effective development of MXene-based materials with enhanced durability under harsh environmental conditions, while maintaining superior electromagnetic SE, necessitates thorough and sustained research efforts.

12.3.3 Biomass-Derived Materials

In response to challenges such as high density, limited flexibility and environmental non-sustainability associated with traditional EMI shielding materials, bio-based alternatives have emerged as a promising solution. These materials are not only biodegradable but also inherently flexible and compatible with a wide range of processing techniques. Moreover, their performance can be significantly enhanced through functionalisation with conductive fillers such as graphene, carbon nanotubes or metal nanoparticles. Additionally, bio-based materials often allow for lightweight and cost-effective fabrication, aligning well with current trends in sustainable technology. Furthermore, the incorporation of such materials opens new avenues for the development of next-generation, environmentally responsible and high-performance electronic devices. Shielding materials obtained from renewable resources like wood, bamboo, lignin and cellulose have attained wide adoption as next-generation shielding materials.

Wood-based materials can be effectively used for EMI shielding by manipulating their natural porous structure and their ability to incorporate conductive fillers. A vacuum impregnation process adopted by Jiang and his team to construct a sandwich structure of MXene/wood (MWC) exhibited a tensile strength of 68.1 MPa, high electrical conductivity of 1858 Sm^{-1} and excellent EMI SE of 32.7 dB at a thickness of 0.38 mm [45]. The high electrical conductivity of MXene and the natural microchannels in wood allow multiple internal reflections of EM waves, enhancing absorption and dissipation. Moreover, these samples exhibited fire-retardant properties, which are attributed to the synergistic effect of MXene's two versatile features: flame retardancy and densification treatment, collectively improving its self-extinguishing properties. High-performance EMI shielding of 40 dB, robust flexural strength of 70.1 MPa and self-extinguishing flame-retardant properties were achieved by a wood-bamboo hybrid material modified with copper nanoparticles [46]. The multilayer structure leverages bamboo's natural toughness and copper's conductivity, ensuring efficient EM-wave absorption, while the hydrophobic copper coating ensures durability.

In another work, bamboo was combined with 10 wt.% of CNT and recycled high-density polyethylene (HDPE) through hot pressing at a temperature of 150 °C and pressure of 10 MPa for 30 minutes to obtain flexible and lightweight shielding materials [47]. It is noteworthy that the prepared samples exhibited very high electrical conductivity of the order of 10^4 S cm^{-1} and achieved an EMI SE of 49.6 dB in the 8.2–12.4 GHz frequency range. The homogeneous dispersion of CNTs formed a conductive network which in turn enhanced the SE of the samples. Besides shielding capabilities, the composite exhibited strong mechanical properties, with a tensile strength of 30.5 MPa, along with thermal stability and corrosion resistance, making it suitable for corrosive environments. Considering the relative cost of CNT, Miao and his team reduced the CNT content to 2% and integrated them into bamboo charcoal, cellulose nanofibrils and polydimethylsiloxane to form a composite aerogel by freeze drying [48]. Owing to the good porosity and conductivity of the membrane, a large part of the EM wave was absorbed, thereby reducing secondary reflection pollution. The resulting biomass membrane exhibits excellent EMI shielding performance of 39.5 dB, good flame-retardant ability and hydrophobicity of 114.8°.

Aiming to fully harness the capabilities of cellulose-based structures, a freeze-thaw method was adopted to fabricate MXene cellulose nanofibre composite films [49]. The strength and flexibility of the prepared samples were demonstrated by a tensile strength of 8.6 MPa, toughness of 20.3 MJ m^{-3} and a Young's modulus of 2.8 GPa. Additionally, the films possess high electrical conductivity of 165.9 S cm^{-1} and an EMI SE of 54.7 dB, blocking 99.74% of EM waves, making them highly effective for shielding applications. Finite element method (FEM) simulations show that the alignment and orientation of the cellulose layers can be strategically tailored for next-generation shielding materials in flexible electronic devices.

Adopting a different technique, Zhou et al. fabricated an inorganic-filler-free, fully organic, highly conducting cellulose-nanofibre composite paper containing poly(3,4-ethylenedioxythiophene):poly(styrene sulfonate) (PEDOT:PSS) doped with lithium bis(trifluoromethanesulfonyl) imide (Li-TFSI) [50]. The resultant LPPC composite papers maintained an effective EMI SE of 68.27 dB at a thickness of only 47 μm. It was observed that the inclusion of Li-TFSI enhanced the crystallinity and oxidation level of PEDOT chains, improving conductivity to 29.67 $S{\cdot}cm^{-1}$. Furthermore, the prepared samples displayed environmental stability with 97% retention of EMI shielding and a mechanical strength of 44.42 MPa, which makes them suitable for applications in wearable electronics. The study reveals that the filler-free organic composite is a biocompatible solution for next-generation EMI shielding materials, characterised by exceptional conductivity, strength and flexibility.

Lignin is another prominent and widely used renewable and biodegradable biomass material for EMI shielding. Its porous structure, abundant functional groups such as hydroxyl and carboxyl, and aromatic benzene ring enable suitable bonding with fillers, enhancing the EMI shielding properties [51]. Moreover, the porous architecture of lignin facilitates impedance matching, which allows EM waves to enter the material and undergo multiple internal reflections and absorption, effectively diminishing their intensity.

Hu and his team developed a lignin-based polyurethane composite (FeCL-PU) by incorporating KH550-modified Fe_3O_4 nanoparticles, functionalised carbon nanotubes and lignin into a polyurethane matrix [52]. It was reported that the synergistic effects between the Fe_3O_4, CNT and lignin enhanced the EMI shielding efficiency and tensile strength. The improvement was attributed to the formation of a three-dimensional conductive network, interfacial polarisation and magnetic losses contributed by Fe_3O_4 and CNTs, which prolonged EM-wave propagation paths.

Liu et al. integrated MXene with lignin nanoparticles to create fire-retardant EMI shielding films. The composite demonstrated flame resistance and an EMI SE of 40 dB, making it suitable for aerospace and military applications where thermal stability is critical [53].

To further unlock the potential of lignin, Yan et al. adopted vacuum-assisted adsorption, freeze-drying and thermal annealing to synthesise PEG encapsulated in nickel-foam/lignin/rGO (PLGN) dual-network scaffolds. The prepared samples exhibited an SE of up to 69.9 dB at 8.2–12.4 GHz [54]. It was observed that the synergy of GO and nickel foam played a major role in weakening EM waves, which were dissipated through continuous reflection and absorption.

To bring out the inherent advantages of lignin in advanced shielding applications, Yu and his team utilised lignin as a sustainable welding agent to simultaneously enhance thermal management and EMI shielding capabilities [55]. The lignin-modified GO films undergo multi-step high-temperature annealing, where lignin's unique properties enable three key advancements: first, its hydrogen and covalent bonding with GO creates compact, ordered precursor films; second, during annealing, lignin acts as a molecular welder, repairing graphene sheet defects and bridging adjacent sheets to form larger continuous networks;

and third, this structural optimisation yields exceptional performance metrics – a 30.3% improvement in thermal conductivity (1462 W m^{-1} K^{-1}), 50% higher electrical conductivity of 1.49 × 10^6 S m^{-1} and an outstanding EMI SE of 70.3 dB. This lignin-enabled approach successfully addresses two critical challenges in modern electronics – heat dissipation and electromagnetic pollution – through a single, scalable material solution with immediate applications in high-power devices, flexible electronics and aerospace systems, demonstrating how sustainable biomass can drive next-generation technological advancements.

Considering the need for eco-friendly alternatives to conventional synthetic or metallic shields, bio-based materials derived from renewable resources have emerged as promising candidates for green technology-based EMI shielding. These materials significantly reduce concerns related to environmental impact and waste management. However, the large-scale production of high-performance bio-composites remains a major challenge. Furthermore, greater emphasis must be placed on enhancing the stability of these materials under harsh environmental conditions, and innovative material designs should be adopted to achieve multifunctional performance.

12.3.4 Liquid Metals

Amid the growing concerns of the enormous detrimental impacts of EM waves, novel materials with much advanced properties have become a research imperative. Specifically, the key parameters "flexibility and stretchability" find their way more important in engineering advanced EMI shielding materials. The advent of LM, a material with the conductivity of metallic compounds and the fluidity and compliance of liquid fluids, in the field of EMI shielding materials has the potential to replace the traditionally used flexible materials. Moreover, conventional flexible EMI shielding materials exhibit a break in conductivity paths when stretched, leading to a decrease in the shielding efficiency of the material. Hence, the incorporation of LM amalgam is a promising candidate for the fabrication of stretchable EMI shielding materials.

Gallium-based or gallium-indium eutectic metals make up the majority of room temperature LM. Furthermore, to enhance the tensile property, electrical conductivity, shielding property, self-healing property and thermal stability of the LM, incorporating LM and polymer nanofillers as composites is utilised. Another major advantage of using LM as a coating for stretchable electromagnetic shielding is that they can be conveniently and directly used on devices without the need for patterning or integration. Zhang et al. fabricated a substrate coated with an amalgam of $GaIn_{24.5}$, Ni particles and gallium oxide, which exhibited excellent EMI SE of over 75 dB at frequencies of 100 kHz–18 GHz owing to its excellent conductivity of 2.4 × 10^6 S m^{-1} [56]. This is much higher than the conductivities of samples including carbon- and metal-based materials of the order of 10^5 S m^{-1}. Furthermore, this material can satisfy military shielding requirements (>60 dB) even at strain up to 75%. Owing to the good adhesion and super flexibility of the material and the continuous conductive path even under extreme stretching conditions, they are a promising material for stretchable shielding applications.

In another work, Chen and his co-workers fabricated Al_2O_3@LM/borox-crosslinked PVA (Al_2O_3@LM/BCP) aerogel synthesised through directional freeze drying [57]. The core shell structure of Al_2O_3@LM lowers the surface tension of LM and improves its compatibility with the PVA matrix. Owing to the superior conductivity of LM and the freeze-dried porous structure of the aerogel, the composite exhibited remarkable conductivity of 312.5 S m^{-1}, an SE of 80.9 dB and an exceptionally good SE_A /SE_B ratio of 83.9%. The porous structure acts as resonators, providing suitable impedance matching that permits more EM waves to penetrate the aerogel, enhancing the absorption–reflection–reabsorption process and leading to increased wave absorption.

In an effort to fabricate flexible and durable EMI shielding materials, Sun et al. fabricated a heterostructure nanocomposite of reduced graphene oxide (rGO) with LM, termed S-rGO/LM [58]. The material successfully achieved an SE of 80 dB at a thickness of 33 μm and, on increasing the thickness to 67 μm, attained an SE of 100 dB in the X-band frequency range (8–12.4 GHz). Furthermore, this material maintained an SE above 70 dB on exposure to extreme temperatures, chemical environments and mechanical strain, positioning it as a promising material for advanced wearable applications in sophisticated areas.

Even though nanometre-thin MXene shows excellent EMI SE, it lacks the stretchability requirements needed for sophisticated applications [59]. To address this issue, Sambyal et al. [60] employed a vacuum-assisted filtration method to develop flexible and lightweight composite films of MXene and Galinstan LM nanoparticles. The incorporation of LM between the MXene layers drastically enhances the electrical conductivity (4.9×10^5 S m^{-1}) and endows an excellent SE of 66.28 dB for a minimal thickness of 15 μm in the X-band frequency range (8.2–12.4 GHz).

In another work, Jiang and his co-workers fabricated MXene/LM composites using a solvent-assisted dispersion method [61]. The coating of thickness 20 μm exhibited an extraordinary SE of 105 dB, which is 1.6 times greater than the coating prepared through conventional dry-powder mixing. The additional feature of the prepared composite is that the enhanced EMI SE remains stable even when the sample is exposed to 15 days of continuous heat treatment at 70 °C, prolonged mechanical deformation and flame exposure. This proves the durability and reliability of the sample for practical applications.

The exceptional electrical conductivity and flexibility of LM offer them a unique position in the field of EMI shielding applications. However, their practical use is constrained by challenges related to handling, safety, cost and long-term performance. Another critical issue associated with LM is their potential toxicity and hazardous impact on the environment, which necessitates careful management and containment. Since LM are often denser than polymeric composites, strenuous research efforts are required to develop lightweight LM-based composites or hybrid materials that maintain high SE while reducing overall weight. Future advances in this area will depend heavily on innovative structural designs, improved encapsulation techniques to prevent leakage and oxidation, and the development of multifunctional, environmentally benign composites that combine excellent EMI shielding with enhanced mechanical and chemical stability. Table 12.1 provides a comparative overview of various materials and their key properties which are relevant to EMI shielding applications, aiding the evaluation and selection of appropriate shielding materials.

TABLE 12.1 Comparative Summary of EMI Shielding Materials and their Key Properties

Material/Coating	Processing Method	Conductivity	Mechanical/thermal Properties	EMI shielding Effectiveness (SE)	Frequency Range	Remarks	Reference
Metallic Materials							
Cu, Cu-Zn, Cu-Ni coatings	Arc thermal spray	–	–	Cu: 12 dB Cu-Zn: 80 dB Cu-Ni: 68 dB (at 100 μm thickness)	1 GHz	Cu-Zn showed highest SE with thinner coating	[8]
Mg-8Li-3Y-2Zn alloy	Vacuum melting + Heat treatment	–	UTS: 263.7 MPa Elongation: 45.3%	SE: 76.9–106.7 dB	30–4500 MHz	Y & Zn improved EMI SE significantly	[9]
Mg98.5Zn0.5Y	Friction stir processing (FSP)	11.6 $MS \cdot m^{-1}$	UTS: 358.2 MPa Elongation: 15.1%	SE: 70–110.4 dB	30–4500 MHz	344% increase in strength, 733% in ductility	[10]
Cu/Ni/Cu and Ni/Cu/Ni multilayers	Accumulative Roll Bonding (ARB)	–	Ni: +105 HV Cu: +24 HV (Pass 1 to 7)	Ni-based: up to 55 dB Cu-based: up to 59 dB	–	Best SE with 7-layer stacking structures	[11]
Ag-plated graphene-modified carbon foam	Silver electroplating	180 $S\ cm^{-1}$	Compressive strength: 6.5 MPa Thermal cond.: 6.0 W/m·K	SE: 73.5 dB	X-band	Synergistic effect of Ag plating and graphene	[12]
Conductive Polymers Based Composite Materials							
PPy-SWCNT nanocomposite	Nanocomposite fabrication	–	–	SE: 32.1 dB	8.2–12.4 GHz (X-band)	Enhanced dielectric losses and impedance matching for better wave attenuation	[15]
PANI-Graphene composites	–	–	Thickness: 10 μm	SE: 57 dB	2–4 GHz (S-band) and 8–12 GHz (X-band)	2% graphene loading enhanced absorption-dominated shielding	[16]
Carbon nanofibre/Polyaniline aerogel	–	–	–	SE: 85.4 dB Specific SE: 791.2 $dB \cdot cm^3 \cdot g^{-1}$	X-band	Superior flame-retardant and thermal protection	[17]

MXene/PEDOT:PSS composite	Crosslinked polymer film	–	–	SE: 41 dB (6 μm thickness) Specific SE: 89,924 dB·cm^2·g^{-1}	–	Absorption-dominated shielding with superior conductivity	[18]
PPD@MXene aerogel	Defect engineering	–	–	SE: 50.4 dB	–	Enhanced interlayer spacing and defects for better EMI shielding	[19]
P-$Ti_3C_2T_x$@PP hybrid aerogel	–	–	–	SE: ~52 dB at 18 GHz Avg. 47.6 dB in Ku-band	18 GHz (Peak), Ku-band	Phosphorus doping improved polarisation and impedance matching	[20]
Surface-functionalised $Ti_3C_2T_x$ MXenes in epoxy	Surface functionalisation	8200 S·m^{-1}	Tensile toughness: 271 KJ·m^{-3}	SE: 100 dB at 110 GHz 61 dB at 8.2 GHz	8.2 GHz (Low-end), 110 GHz (High-end)	Dramatic improvement in conductivity, EMI SE and mechanical properties	[21]
			Hybrid Materials				
PVDF/LG composite with NSF addition	Solution processing method	–	–	25–45 dB (PVDF/LG), 30–53 dB (PVDF/LG+NSF)	1–12 GHz	Improved shielding with NSF addition; agglomeration at 30 wt.% NSF reduced SE	[22]
PANI/NFO@CNTs, PANI/NFO@RGO composites	Oxidative polymerisation of aniline with core materials	22.36 dB at 2 mm (PANI/NFO@CNTs), 43 dB at 4 mm (PANI/NFO@RGO)	–	22.36 dB (PANI/NFO@CNTs), 43 dB (PANI/NFO@RGO)	–	Enhanced SE due to good impedance matching and loss mechanisms	[23]
PVA/rGO/α-FeOOH composite	Solution processing	6.92 S m^{-1} (at 4 wt.% rGO/FeOOH)	Young's modulus: 3.03 GPa, tensile strength: 43.25 MPa, thermal conductivity: 2.23 W/m·K	28.27 dB	–	Shows high tensile strength, enhanced conductivity and thermal stability	[24]

(*Continued*)

TABLE 12.1 Comparative Summary of EMI Shielding Materials and their Key Properties (Continued)

Material/Coating	Processing Method	Conductivity	Mechanical/thermal Properties	EMI shielding Effectiveness (SE)	Frequency Range	Remarks	Reference
MXene and Fe_3O_4-g-C_3N_4 composite	Solution mixing	–	Enhanced thermal conductivity: 0.98 W/m·K	−36 dB	–	Strong EMI shielding and thermal conductivity with good mechanical integrity	[25]
CNF/MXene/ $ZnFe_2O_4$@PANI composite	Vacuum-assisted filtration	–	Tensile strength: 27.24 MPa, fracture strain: 6.21%, rapid Joule heating	45.75 dB	–	High EMI shielding, rapid Joule heating with good mechanical performance and integrated thermal response	[26]
			Carbon-Based Nanomaterials				
PU/SWCNT	Solvent casting	–	–	17 dB	X-band (8.2–12.4 GHz)	Reflecting characteristic dominated by absorption at high SWCNT loading	[27]
PANI/SWCNT	Ultrasonically induced in situ polymerisation	–	–	32.80 dB	11 GHz	99.999% absorption	[28]
SWCNT/ PEDOT:PSS	Vacuum filtering, dip coating	–	Outstanding stability	32.1–55.53 dB	X-band	High stability and flexibility	[29]
CNT films in epoxy matrix	–	–	Improved mechanical strength and impact energy absorption	52.31 dB	–	Enhanced delamination resistance	[30]
GC/PU	–	2087.2 S m^{-1}	High tensile strength (42.9 MPa), flame-retardant	37.7 dB (130 μm), 72.1 dB (520 μm)	X-band	Suitable for anti-icing and de-icing applications	[31]
rGO/LIG@$NiFe_2O_4$	"Popcorn making" method	–	–	36 dB (70 μm), 51 dB (166 μm)	X-band (8.2–12.4 GHz)	High absorption ratios	[32]

Polyamide yarns with GO	Modified with polyethyleneimine, reduction with hydroiodic acid	4.7×10^3 S m^{-1}	Bending resistance (10,000 cycles)	66.7 dB	0.615 mm thickness	Suitable for large-scale production	[33]
TiO_2/PVA films reinforced with rGO	Solution casting	–	–	55.8 dB	X-band (8–12 GHz)	–	[34]
PVA/rGO/FeOOH	–	6.92 S m^{-1}	High thermal conductivity (2.23 W mK^{-1})	28.27 dB	–	–	[24]
			MXenes				
Ti_2CT_x	Chemical etching (HCl + LiF)	0.3 S cm^{-1}	–	61.1 dB (avg.), ~70 dB (max)	X-band (8.2–12.4 GHz)	High SE with minimal thickness; superior to Ti_2AlC MAX phase	[40]
PI-MXene-MWCNT	Electrospinning-spraying	–	35.4% elongation at break	66.8 dB, 13,153.8 dB·cm^2·g^{-1}	–	Excellent flexibility and shielding in extreme conditions	[41]
$Ti_3C_2T_x$/Epoxy	Ionic intercalation, sonication, thermal reduction	105 S m^{-1} (at 15 wt.% MXene)	Young's modulus: 4.32 GPa; Hardness: 0.29 GPa	41 dB (X-band)	–	Enhanced conductivity and shielding with reduced filler loading	[42]
PPD-intercalated $Ti_3C_2T_x$ aerogel	Intercalation with p-phenylenediamine	–	–	50.4 dB, Absorptivity: 0.92	–	Tunable d-spacing and defect density for improved performance	[19]
MXene/BC-TPU foam	Coagulation method	1912 S m^{-1}	Elasticity: −80 to 80%; Fatigue resistance	86 dB (initial), 76 dB (after 5000 cycles)	–	Dual-network structure for stretchable and compressible shielding	[43]
PVA-co-PE/AgNW/MXene	Vacuum filtration, templating, hot pressing	–	–	101.6 dB	8–20 GHz	Low filler content with high SE and strong absorption	[44]

(Continued)

TABLE 12.1 Comparative Summary of EMI Shielding Materials and their Key Properties (Continued)

Material/Coating	Processing Method	Conductivity	Mechanical/thermal Properties	EMI shielding Effectiveness (SE)	Frequency Range	Remarks	Reference
			Biomass-Derived Materials				
MXene/wood (MWC)	Vacuum impregnation (sandwich structure)	1858 $S \cdot m^{-1}$	Tensile strength: 68.1 MPa; Fire retardancy	32.7 dB	–	High EMI SE, flame retardancy, robust flexural strength	[45]
Wood-bamboo hybrid material modified with copper nanoparticles	–	–	Flexural strength: 70.1 MPa; Self-extinguishing flame retardancy	40 dB	–	Multilayer structure, hydrophobic coating for durability	[46]
Bamboo/CNT/ HDPE composite	Hot pressing (150 °C, 10 MPa, 30 min)	10^4 $S \cdot cm^{-1}$	Tensile strength: 30.5 MPa; Thermal stability, corrosion resistance	49.6 dB	8.2–12.4 GHz	Lightweight, anti-corrosive, high shielding efficiency	[47]
Bamboo charcoal/ cellulose nanofibrils/PDMS composite aerogel	Freeze drying	–	High porosity, hydrophobic (114.8° contact angle)	39.5 dB	–	Flame-retardant, good EMI shielding performance	[48]
MXene/cellulose nanofibre composite films	Freeze-thaw intercalation	165.9 $S \cdot cm^{-1}$	Tensile strength: 8.6 MPa; Young's modulus: 2.8 GPa; Toughness: 20.3 MJ m^{-3}	54.7 dB	–	High flexibility, excellent shielding, suitable for flexible electronics	[49]
Cellulose nanofibre composite paper (PEDOT:PSS doped with Li-TFSI)	–	29.67 $S \cdot cm^{-1}$	Tensile strength: 44.42 MPa; Environmental stability (97% retention of EMI SE)	68.27 dB	–	Biocompatible, suitable for wearable electronics	[50]
Lignin-based polyurethane composite (FeCL-PU)	–	–	Synergistic effects of CNT, Fe_3O_4 and lignin for EMI shielding	–	–	Improved EMI shielding efficiency, tensile strength	[52]

MXene/lignin nanoparticles composite	–	–	–	40 dB	–	Fire-retardant, suitable for aerospace and military applications	[53]
Nickel foam/lignin/rGO dual-network scaffold	Vacuum-assisted adsorption, freeze-drying, thermal annealing	–	–	69.9 dB	8.2–12.4 GHz	Dual-network scaffold, excellent EMI shielding and energy harvesting	[54]
Lignin-modified GO films	High-temperature annealing	1.49×10^6 S·m^{-1}	Thermal conductivity: 1462 W·m^{-1}·K^{-1}; 30.3% improvement in thermal conductivity	70.3 dB	–	Addresses heat dissipation and EMI shielding in high-power devices	[55]
			Liquid Metals				
GaIn24.5-Ni particles-Gallium oxide coating	Fabrication of substrate coating with GaIn24.5 amalgam	2.4×10^6 S m^{-1}	Excellent flexibility and adhesion, continuous conductive path	>75 dB	100 kHz to 18 GHz	High conductivity and flexibility, meets military shielding requirements	[56]
Al_2O_3@LM/ - borax crosslinked PVA aerogel composite	Directional freeze drying	312.5 S m^{-1}	Superior mechanical properties, freeze-dried porous structure	80.9 dB	–	Enhanced absorption-reflection-reabsorption process for improved EMI shielding	[57]
Reduced graphene oxide (rGO) with liquid metal (S-rGO/LM)	Heterostructure nanocomposite fabrication	–	Maintains stability under extreme conditions (temperature, strain)	80 dB (33 μm), 100 dB (67 μm)	80–12.4 GHz	Promising for wearable applications due to stability and high performance	[58]
MXene with Galinstan liquid metal nanoparticles	Vacuum-assisted filtration	4.9×10^5 S m^{-1}	Flexible, lightweight	66.28 dB	8.2–12.4 GHz	High EMI shielding with minimal thickness, improved electrical conductivity	[60]
MXene/liquid metal composite	Solvent-assisted dispersion	–	Malleable, printable, bondable, enhanced wettability	105 dB	–	Exceptional durability, stable EMI shielding even under heat and deformation	[61]

12.4 FUTURE OF EMI SHIELDING: CHALLENGES AND FUTURE DIRECTIONS

EMI presents a significant challenge for future generations as digital communication technologies become increasingly integrated into daily life. Emerging threats posed by advanced electronic systems highlight the limitations of current EMI shielding technologies. To protect electronic devices and biological systems from these evolving risks, rapid advancements in novel EMI shielding materials and technologies are inevitable. This section will provide a brief overview of the latest technological advancements made towards this, followed by a detailed discussion on the challenges faced by existing shielding technologies as electronics and communication systems continue to evolve rapidly. Potential solutions to these issues and future directions for overcoming emerging challenges are discussed. After this, the sustainability considerations for future EMI shielding technologies, followed by the possibilities of using artificial intelligence (AI) and machine learning (ML) in future materials with optimised properties development, are explored.

12.4.1 Technological Innovations

Recent developments in shielding materials focus not only on enhancing SE but also on scalability, multifunctionality and the preference for absorption-dominated mechanisms over traditional reflection-based technologies. Future nanomaterials must exhibit efficiency across multiple frequency bands to support the diverse operational requirements of next-generation electronic systems. This section explores key technological innovations in three major areas: (1) Nanotechnology and Nanocoatings, (2) 3D Printing and Additive Manufacturing and (3) Smart Adaptive Materials for EMI Shielding. Their advantages, limitations and prospects are discussed.

12.4.1.1 Nanotechnology and Nanocoatings

The development of nanocoatings has significantly advanced EMI shielding research. Various nanocoatings with optimised properties have been reported [62, 63]. Among them, graphene-based coatings have garnered significant attention due to their superior EMI shielding performance, attributed to high electrical conductivity and large surface area. Their application in skin-mounted biosensors and smart contact lenses showcases mechanical flexibility, optical transparency and exceptional shielding potential [64, 65]. However, real-world applications of graphene-based coatings face significant hurdles, particularly in achieving large-scale uniform deposition and mitigating environmental degradation caused by prolonged exposure to humidity and heat. Beyond graphene coatings, carbon nanotube (CNT) composites, Fe_3O_4 nanoparticles and MXene-based coatings have also demonstrated promising EMI shielding potential. MXenes, a class of 2D transition metal carbides and nitrides, exhibit remarkable electrical conductivity and enhanced surface area, facilitating absorption-dominated EMI shielding with minimal reflectivity. When integrated into polymer matrices, foams or aerogels, MXenes offer a lightweight alternative to graphene coatings [66, 67]. However, MXenes are prone to oxidation and have batch-to-batch inconsistencies in surface terminations which affect their commercialisation and reliability when used in sensitive EMI shielding devices. To address these issues, techniques to reduce oxidative instability (surface passivation, protective coating etc.) must be used, along with standardised production protocols to ensure consistent material quality. Similarly, CNT-reinforced composite films have excellent electrical conductivity and mechanical strength which make them ideal for EMI shielding applications. Reports indicate the development of ultra-thin, flexible CNT-based EMI shields utilising techniques such as spray coating and layer-by-layer deposition assembly, particularly in wearable shielding textiles and biomedical monitoring devices [68, 69]. Nevertheless, high manufacturing costs and health concerns related to airborne CNTs suggest the requirement for further research to address these issues. The incorporation of magnetic Fe_3O_4 nanomaterials into EMI shielding coatings has also gained traction, with research highlighting their tuneable properties and potential use in space-grade shielding materials [70]. However, their thermal instability and agglomeration during processing would affect reproducibility. To solve these issues, further research focused on optimised processing conditions and the development of highly advanced nanostructure architectures is essential.

Looking ahead, research on EMI shielding materials should focus on multifunctional, durable and environmentally sustainable materials with high shielding efficiency under diverse environmental conditions. Technological innovations in broadband-tuneable nanostructures could provide protection over multiple frequency bands. One of the major obstacles in EMI shielding material optimisation is the reliance on trial-and-error methodologies that are widely adopted by many researchers currently. The integration of ML and AI into materials development could significantly accelerate the optimisation process. AI-driven approaches can enhance existing nanomaterials and coatings while facilitating the discovery of next-generation materials with superior shielding capabilities. Emerging materials such as black phosphorus and borophene present novel possibilities for EMI shielding. Black phosphorus exhibits a tuneable direct bandgap and anisotropic dielectric properties, making it suitable for frequency-selective EMI shielding purposes in flexible electronics and optoelectronic applications [71]. Borophene, on the other hand, possesses exceptional metallic conductivity, pronounced plasmonic behaviour and highly directional charge transport, which makes it a potential candidate for future ultra-thin, lightweight, high-performance EMI shielding [72]. However, despite their immense promise, the practical development of an effectively functioning shield from them could be challenging, as black phosphorus is highly air-sensitive while borophene has scalability issues and complex synthesis processes. Addressing these challenges is crucial for commercial viability. In future, efforts to improve the performance of magnetic absorbers for EMI shielding through structural-ordering engineering, such as tailoring magnetic anisotropy and domain configurations, would become more popular. This could lead to the development of the ability to overcome intrinsic constraints like Snoek's limit, leading to broadening effective frequency bands [73]. Thus, next-generation nanomaterials and coating-based EMI shielding solutions must integrate advanced materials, optimised synthesis techniques and computational modelling tools to overcome current challenges and maximise performance, scalability and durability.

12.4.1.2 3D Printing and Additive Manufacturing in EMI Shielding

Additive manufacturing is a revolutionary method of developing materials layer by layer based on a digital design. Compared to traditional techniques for material development, additive manufacturing, particularly 3D printing, enables rapid prototyping, creation of complex geometries with internal structures and minimised material waste [74]. This approach is perfectly suited for developing materials with tailored properties like gradient composition. The continuous advancements in additive manufacturing offer exciting opportunities for highly specialised EMI shielding architectures with enhanced functionalities. This section explores the advantages, limitations and future research directions of 3D printing and additive manufacturing in the development of next-generation EMI shielding materials.

3D printing allows precise control over geometry, porosity and stacking sequences, facilitating the integration of both absorption and reflection mechanisms within a single shielding platform. Recent reports highlight the use of additive manufacturing, especially 3D printing, in the development of high-performance EMI shielding materials, particularly polymer nanocomposites [75, 76]. Several promising approaches have emerged recently. Porous aerogels, MXenes and polymer matrices have been successfully 3D-printed into lightweight structures with intricate geometries that provide effective EMI protection [77, 78]. Combining 3D printing with spray coating has enabled the fabrication of multi-layered EMI shielding materials, enhancing overall effectiveness [79]. Researchers are developing 3D-printed conductive lattice frameworks with high surface area, excellent mechanical properties and optimised internal reflection, particularly for aerospace and automotive applications [80, 81]. Given the growing demand for EMI shielding in the automotive sector, 3D-printed transparent EMI shields for automotive displays and sensors have gained traction. A novel patent integrates EMI shielding with thermal dissipation, demonstrating promising applications in automotive radar [82]. AI-driven material design for 3D-printed structures has optimised materials properties, structural configurations and shielding efficiency [83].

Despite its advantages, 3D printing faces several challenges that hinder its widespread implementation in EMI shielding applications. The first is material stability, meaning 3D printing techniques are currently limited to a specific range of materials, restricting the diversity and efficiency of EMI shielding performance. Next are structural issues like void formation, layer delamination and defects associated with current 3D-printed materials, which can compromise SE. Manufacturing limitation associated with 3D printing is yet another challenge. The current capabilities of 3D printing limit the fabrication of multi-material structures, reducing flexibility in tailoring shielding properties. To overcome these challenges, researchers must develop advanced multi-material printing techniques and standardised processing protocols to ensure defect-free structures. Post-processing techniques like infiltration and surface coatings must be employed to enhance the performance of 3D-printed EMI shielding materials. Looking forward, AI-assisted optimisation of printing parameters and adaptive shielding designs will enable the fabrication of customised, frequency-tuneable and scalable EMI shielding solutions tailored to next-generation electronic and communication systems. The field of additive manufacturing will be transformed once 4D printing, which is still in its early stages, becomes more accessible and cost-effective [84]. 4D printing will enable the development of smart adaptive shielding materials that dynamically adjust their properties based on temperature, external fields and mechanical stress. The integration of 4D printing with AI-driven design strategies has the potential to revolutionise EMI shielding, paving the way for stimuli-responsive shielding materials that evolve in real time to accommodate diverse operational conditions in futuristic electronic systems.

12.4.1.3 Smart and Adaptive Materials

Smart materials, also known as intelligent or responsive materials, have the ability to dynamically respond to environmental stimuli such as temperature, humidity, mechanical stress or electromagnetic fields. These materials have the exciting potential for developing adaptive EMI shielding systems that can tune their properties in real time. Recent studies have shown the potential of smart EMI shields using phase-change materials that modulate EM wave absorption, piezoelectric polymers that alter shielding behaviour in response to mechanical deformation, and succulent-inspired smart switchable MXene coating film with on and off states [85–87]. Despite these developments, the field remains in its early stages and requires further exploration to meet the demands of emerging applications.

Smart EMI shielding is vital in advanced areas such as wearable electronics, aerospace systems and military technology, where devices are exposed to multiple frequency ranges and field intensities. Based on this, currently, efforts are under way for the development of smart coatings capable of real-time impedance matching and thereby selectively enhancing absorption or reflection, enabling dynamic and frequency-specific EMI protection. Future research could also explore the possible integration of self-healing capabilities into smart shielding materials. Some notable efforts have already been reported [88, 89]. Such systems, once perfected, would not only enhance performance across multiple frequency bands but also ensure durability and extended operational life, which are critical requirements in modern electronic systems. Additionally, the development of orientation-dependent shielding materials may help in realising technologies for directional signal blocking, reconfigurable electronic enclosures, smart antennas and even advanced stealth. Smart active metamaterials are going to play a vital role in this context, since they offer subwavelength structuring, tuneable properties and multifunctional behaviour, all of which are vital for highly adaptive EMI shielding. Another significant leap forward would be the integration of smart EMI shielding materials with IoT devices, which will necessitate the fabrication of environment-aware electromagnetic management. By linking such responsive materials to sensors and AI-based analytics, real-time EMI monitoring and automatic shielding adjustment could be implemented across a wide range of systems, from smart homes and vehicles to industrial facilities.

Despite their promising capabilities, smart material-based EMI shielding technologies face several critical challenges. Fabrication complexity and cost are the most significant among them. The integration of various functional materials and the need for precise structuring make fabrication processes both intricate and expensive. Furthermore, the use of advanced materials like metamaterials for smart EMI shields remains financially unviable for large-scale consumer or industrial use. These challenges can only be addressed if either existing fabrication technologies are modified or novel technologies are developed to realise low-cost and energy-efficient fabrication with high precision and yield. Another issue is the slow response time and limited durability of many smart materials that are currently in use. Stimuli-responsive materials, such as phase-change polymers, suffer from slow actuation speeds and limited repeatability, which affect their real-time performance [90, 91]. The most pressing issue in the future would be the development of systems and technologies with minimum energy usage to address the growing global energy demand. Considering the fact that future technology will demand not only high adaptability but also autonomous operation, miniaturisation and energy efficiency, research focused on the development of smart, low-power-actuated EMI shielding systems is expected to surge in the immediate future.

12.4.2 Environmental and Sustainability Considerations

Significant sustainability and environmental issues are associated with traditional EMI shielding materials and their manufacturing methods. Many traditional shielding solutions depend mainly on non-renewable resources such as metals (copper, aluminium etc.) and petroleum-based polymers, whose extraction and refinement require a large amount of energy and are likely to produce harmful by-products and large greenhouse gas emissions [92, 93]. Manufacturing techniques like high-temperature treatments, chemical vapour deposition and solvent-based syntheses, which are popular in industry, commonly use hazardous chemicals and release volatile organic compounds or produce persistent waste streams that are harmful to the environment and human health. As many composite materials used for EMI shielding are made of complex, non-biodegradable materials that are challenging to separate and recycle, their end-of-life disposal is also problematic. This often results in excessive landfill accumulation and the depletion of valuable resources. These factors highlight the urgent requirement to develop more sustainable production methods and improve circularity throughout the material lifecycle.

Using engineered micro/nanostructures and dynamic functionalities, emerging advanced EMI shielding solutions consisting of smart materials and metamaterials could be tailored to demonstrate superior EMI shielding properties. However, the use of these novel materials could create environmental problems. For example, these advanced materials occasionally contain rare or scarce elements, such as lanthanides, which are linked to supply risk and environmentally harmful mining practices [94, 95]. Furthermore, the synthesis processes of these advanced materials, which frequently entail intricate layering or nanoscale patterning, may employ hazardous materials or generate waste nanoparticles, which could cause bioaccumulation and thereby pose ecological risks. In addition, it is often difficult to verify the long-term durability and recyclability of such complex materials, which makes material recovery more difficult and thus increases the possibility of electronic waste accumulation. This highlights how important it is to conduct comprehensive life cycle assessments and take sustainability into account at an early stage of development.

To address the challenges mentioned, it is important to adopt a comprehensive green materials development framework. Making use of recyclable and bio-based polymers such as nanocellulose, lignin composites or polylactic acid can increase end-of-life biodegradability and reduce dependency on fossil fuels, thereby supporting circular economic models. Fabrication techniques that give importance to low-energy, solvent-free or water-based methods help to curtail hazardous emissions during the manufacturing process. Finding environmental impact hotspots and successfully balancing performance, cost and sustainability goals could be made easier by incorporating life cycle assessments from the very beginning of material design. Focusing on creating scalable, low-waste fabrication processes and substituting plentiful, environmentally friendly raw materials for scarce, critical raw materials are crucial for smart and metamaterials. Furthermore, creating modular composites that are disassemblable makes it easier to recover and reuse them at the end of their useful lives. This would help to reduce e-waste. Moreover, the promotion of collaboration among scientific researchers, industry leaders, policymakers and environmental groups is crucial in developing sustainability standards, certification programmes and policy incentives that encourage the adoption of eco-conscious materials. In the future, the combination of AI and ML is going to hasten the development of novel sustainable materials and streamline production procedures to have the least possible negative effects on the environment. Also, the development of self-healing or self-repairing EMI shielding materials could be made possible by advances in synthetic biology and biomimicry, which would significantly increase product lifespans and thereby reduce waste. Additionally, localised, on-demand production can be facilitated by utilising additive manufacturing techniques in conjunction with recycled and bio-derived feedstocks, reducing material waste and transportation emissions. Material lifecycle management could be completely transformed by the creation of closed-loop material ecosystems that integrate intelligent recycling technologies and real-time environmental monitoring. For next-generation EMI shielding technologies to have a truly sustainable future, such integrated strategies are imperative for maximising the advantages of advanced materials like smart and metamaterials without aggravating social or environmental concerns.

12.4.3 Integration with AI and Machine Learning

Traditional methods in materials science, such as first-principles calculations (density functional theory, molecular dynamics) and continuum mechanics simulations, encounter significant computational and theoretical difficulties when tackling the multiple challenges that arise in designing new materials, especially those intended for high-efficiency EMI shielding [96]. The exploration space, containing innumerable elemental compositions and microstructural configurations, extends to immensely large scales (about 10^{100} possibilities), making systematic experimental trials and detailed computational screening impractical. In addition, high-throughput, reliable experimental datasets characterising key functional properties, which include complex permittivity, magnetic permeability, skin depth and SE across frequency ranges, are limited due to technical restrictions in precision and reproducibility. Use of data-centric artificial intelligence approaches and ML algorithms, including supervised learning, reinforcement learning and generative frameworks, holds considerable promise in addressing the aforementioned challenges. By employing statistical methods, Bayesian models and surrogate-based strategies, these approaches can detect hidden relationships within complex, high-dimensional datasets, which would help to link structure, processing parameters and performance of materials such as polymers, smart materials and nanocoatings [97, 98]. This could accelerate inverse design and optimisation through dimensionality reduction and efficient exploration of parameter spaces.

Recent reports have illustrated the substantial role played by AI and ML tools in solving inverse design tasks typical of metamaterial engineering, where the focus is mainly on tuning unit cell geometries, constituent materials and hierarchical design in order to achieve specific properties such as negative refraction, anisotropic scattering and frequency-selective absorption [99–101]. For instance, Gaussian process regression along with Bayesian optimisation has been successful in optimising acoustic band gaps and absorption properties in metamaterials [102, 103]. This highlights the potential for applying analogous probabilistic and active learning techniques to develop metamaterials suitable for EMI shielding. Moreover, the integration of ML with additive manufacturing methods (3D printing), such as fused filament deposition and selective laser melting, realises real-time tuning of process parameters, defect identification through multi-sensor data fusion and digital twin-enabled lifecycle control [96]. Advancements of this type improve control over material deposition and microstructure formation, thereby aiding the creation of architected composites and multifunctional metamaterial lattices precisely engineered for broadband EMI shielding with low insertion loss in GHz frequency regimes.

In the future, the combination of AI-driven materials informatics, multiscale computational techniques and emerging quantum computing approaches is set to transform EMI shielding materials development. The current constraints posed by limited, unstandardised and fragmented high-throughput experimental and simulation data can be addressed by federated learning systems and augmented datasets generated with physics-informed neural networks (PINNs) [103]. These neural networks successfully combine data-driven discovery with first-principles knowledge by embedding physical laws derived from Maxwell's equations, constitutive material relations and symmetry properties. Such a hybrid modelling approach would enhance predictive performance, reduce the necessary training data volume and improve robustness when generalising beyond the training domain. In addition, combining quantum ML with classical continuum electrodynamics simulations may reveal new quantum phenomena and emergent behaviours in low-dimensional materials and topological systems that are pertinent to EMI shielding. When combined, this ecosystem of AI-enhanced materials will allow for the targeted design and adaptive manipulation of complex microstructures, interface characteristics and metamaterial resonator arrays to maximise SE, weight, durability and multifunctionality. These developments are crucial for next-generation aerospace, telecom and consumer electronics applications exposed to increasingly complex electromagnetic environments. Ultimately, this will pave the way for intelligent, self-optimising EMI shielding materials that harness the synergy of AI and advanced manufacturing technologies to meet evolving high-performance standards.

12.5 CONCLUSIONS

Future EMI shielding materials and technologies will mainly be developed based on three fundamental pillars – revolutionary technological innovations, the transformative power of AI and ML, and a strong commitment to sustainability and environmentally responsible practices. This chapter makes it clear that achieving advanced EMI shielding performance cannot be considered solely a materials challenge, but a multidisciplinary endeavour that integrates cutting-edge fabrication techniques, computational design and eco-conscious methodologies. Recent technological advances, such as the development of novel nanomaterials and nanocoatings, the increased use of 3D printing and additive manufacturing methods and research on smart adaptive materials including metamaterials, have triggered the emergence of multifunctional, lightweight and scalable EMI shielding solutions. Innovations such as these provide precise control over the microstructure of materials and their interface characteristics, which is helpful in maximising SE over diverse frequency bands while ensuring durability and reduced weight. However, research to minimise the complexity and cost associated with fabricating these advanced materials is still ongoing. Besides, the development and use of novel materials tailored for EMI shielding applications are also key to advancing EMI shielding solutions.

These developments are further strengthened by the growing influence of AI and ML within the materials design ecosystem. The integration of PINNs and federated learning frameworks allows hybrid modelling that merges first-principles physics with extensive experimental data. This enhances predictive accuracy and generalisability while at the same time curtailing data requirements. AI-driven design methodologies expedite the identification of novel material compositions, optimise structural configurations and enable dynamic EMI shielding via adaptive, intelligent materials interfaced with IoT networks. Such computational revolutions not only accelerate advanced materials development but also support autonomous operation, miniaturisation and energy efficiency, which are critical factors for future electronic systems. Sustainable practices must be prioritised in the development of future EMI shielding materials. Fabrication processes that ensure low energy consumption, solvent-free or water-based methods and the use of abundant, environmentally benign raw materials will help to alleviate ecological impacts. Incorporating a lifecycle assessment right from the design phase enables identification and minimisation of environmental hotspots. Additionally, the development of modular and easily disassembled composite structures enhances both recoverability and recyclability, thus addressing the challenge of electronic waste. With the help of smart recycling technologies and real-time monitoring, closed-loop material systems are set to reshape the way we approach circular material management.

As illustrated in Figure 12.1, the synergistic integration of technological innovation, AI and ML-enabled material optimisation and sustainable practices is the only way in which the full potential of next-generation EMI shielding can be realised. Together, these pillars lay out a holistic path that not only focuses on performance but also considers how easily materials can be produced, how they respond to changing needs and how they align with environmental goals. This integrated approach will be significant in meeting the future demands of aerospace, telecommunications, consumer electronics and other sectors that are about to be exposed to increasingly complex electromagnetic environments in the coming decades. Ultimately, these efforts will lay the groundwork for intelligent, self-optimising EMI shielding materials and systems that are crucial to fulfilling evolving functional requirements while adhering to global sustainability goals.

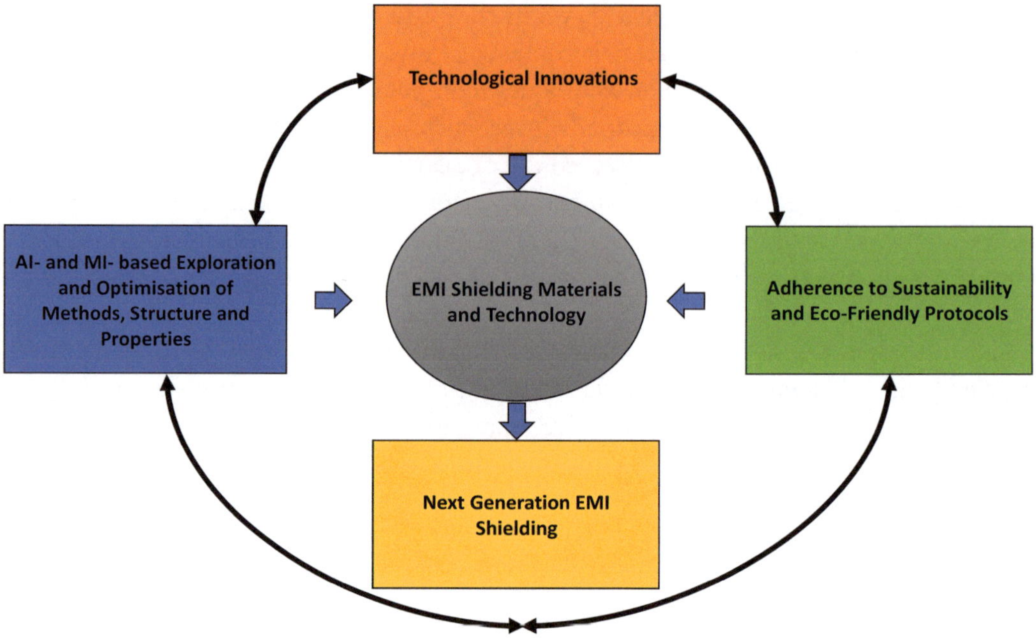

FIGURE 12.1 Core drivers of next-generation EMI shielding.

REFERENCES

1. Bhawal, P., Ganguly, S., Das, T.K., Mondal, S., Choudhury, S. and Das, N.C., 2018. Superior electromagnetic interference shielding effectiveness and electro-mechanical properties of EMA-IRGO nanocomposites through the in-situ reduction of GO from melt blended EMA-GO composites. *Composites Part B: Engineering*, 134, pp. 46–60.
2. Sankaran, S., Deshmukh, K., Ahamed, M.B. and Pasha, S.K., 2018. Recent advances in electromagnetic interference shielding properties of metal and carbon filler reinforced flexible polymer composites: a review. *Composites Part A: Applied Science and Manufacturing*, 114, pp. 49–71.
3. Paul, C.R., Scully, R.C. and Steffka, M.A., 2022. *Introduction to Electromagnetic Compatibility*. John Wiley & Sons.
4. Wang, X.Y., Liao, S.Y., Wan, Y.J., Zhu, P.L., Hu, Y.G., Zhao, T., Sun, R. and Wong, C.P., 2022. Electromagnetic interference shielding materials: recent progress, structure design, and future perspective. *Journal of Materials Chemistry C*, 10(1), pp. 44–72.
5. Shukla, V., 2019. Review of electromagnetic interference shielding materials fabricated by iron ingredients. *Nanoscale Advances*, 1(5), pp. 1640–1671.
6. Orasugh, J.T. and Ray, S.S., 2023. Functional and structural facts of effective electromagnetic interference shielding materials: a review. *ACS Omega*, 8(9), pp. 8134–8158.
7. Phiri, R., Rangappa, S.M., Siengchin, S., Oladijo, O.P. and Ozbakkaloglu, T., 2024. Advances in lightweight composite structures and manufacturing technologies: a comprehensive review. *Heliyon*, 10 (21), p. e39661.
8. Jang, J.M., Lee, H.S. and Singh, J.K., 2020. Electromagnetic shielding performance of different metallic coatings deposited by arc thermal spray process. *Materials*, 13 (24), p. 5776.
9. Li, J., Zhong, L., Wang, J., Feng, Z., Qu, Y. and Xu, R., 2024. Synergistic improvement of mechanical and electromagnetic shielding properties of a Mg-Li-Y-Zn alloy following heat treatment. *Journal of Magnesium and Alloys*, 13 (3), pp. 1243–1257.

10. Zhang, G., Dong, J., Li, Z., Feng, Z., Tan, J., Chen, X., Yi, J. and Pan, F., 2025. Improvement in electromagnetic shielding effectiveness and mechanical properties of ultrafine Mg98.5Zn0.5Y alloy via friction stir processing. *Journal of Magnesium and Alloys*, 13 (9), pp. 4395–4411.
11. Ma, H., Wang, L., Tang, J., Chen, Y. and Sun, S., 2025. Research on microstructural evolution, hardness, and electromagnetic shielding capabilities of Ni/Cu multilayered composites. *Materials Characterization*, 220, p. 114660.
12. Sharma, A., Kumar, R., Dwivedi, N., Gupta, G., Srivastava, A.K. and Mondal, D.P., 2025. Electroplating of silver on graphene-modified carbon foam for high thermal conductivity and outstanding electromagnetic interference shielding performance. *Journal of Alloys and Compounds*, 1010, p. 177080.
13. Xu, R., Sheng, S., Hu, C., Xie, H., Huang, B., Lou, P., Fei, H.F. and Zhang, Z., 2025. Lightweight flexible and efficient electromagnetic shielding composites based on silver-coated glass microspheres. *ACS Omega*, 10(11), pp. 11516–11524.
14. Omana, L., Chandran, A., John, R.E., Wilson, R., George, K.C., Unnikrishnan, N.V., Varghese, S.S., George, G., Simon, S.M. and Paul, I., 2022. Recent advances in polymer nanocomposites for electromagnetic interference shielding: a review. *ACS Omega*, 7(30), pp. 25921–25947.
15. Raju, P., Durga, M.K., Kumar, S.U. and Rani, G.N., 2022. Microwave attenuation studies of polypyrrole-swcnt nanocomposite films for improved EMI shielding. *ECS Journal of Solid State Science and Technology*, 11(9), p. 091007.
16. Mohan, R.R., Prasannakumar, A.T., John, J. and Varma, S.J., 2023. Electromagnetic shielding performance of polyaniline-nanorod/graphene-nanoflake hybrid films in the S-and X-bands. *New Journal of Chemistry*, 47(5), pp. 2565–2574.
17. Chen, M., Zhu, J., Zhang, K., Zhou, H., Gao, Y., Fan, J., Chen, R. and Wang, H.L., 2025. Carbon nanofiber/polyaniline composite aerogel with excellent electromagnetic interference shielding, low thermal conductivity, and extremely low heat release. *Nano-Micro Letters*, 17(1), p. 80.
18. Bora, P.J., Anil, A.G., Ramamurthy, P.C. and Tan, D.Q., 2020. MXene interlayered crosslinked conducting polymer film for highly specific absorption and electromagnetic interference shielding. *Materials Advances*, 1(2), pp. 177–183.
19. Li, R., Huangfu, Y., Liu, L., Hu, J., Zeng, D., Wang, Y., Fan, D., Zhang, R. and Zhao, B., 2025. Intercalation-induced interlayer and defect engineering in $Ti_3C_2T_x$ MXene for ultralow-reflection electromagnetic interference shielding. *ACS Nano*, 19 (2), pp. 2777–2787.
20. Ghosh, K., Roy, S.S. and Giri, P.K., 2025. Phosphorus-doped Ti3C2Tx MXene/PEDOT:PSS aerogel-based electromagnetic interference shields with unique combination of high green index and shielding effectiveness. *Materials Today Advances*, 26, p. 100576.
21. Naqvi, S.M., Hassan, T., Iqbal, A., Jung, S., Jeong, S., Zaman, S., Zafar, U., Hussain, N., Cho, S. and Koo, C.M., 2025. Surface functionalization of $Ti_3C_2T_x$ MXenes in epoxy nanocomposites: enhancing conductivity, EMI shielding, thermal conductivity, and mechanical strength. *ACS Applied Materials & Interfaces*, 17(13), pp. 20149–20161.
22. Ahmed, I., Jan, R., Khan, A.N., Gul, I.H., Khan, R., Javed, S., Akram, M.A., Shafqat, A., Cheema, H.M. and Ahmad, I., 2020. Graphene-ferrites interaction for enhanced EMI shielding effectiveness of hybrid polymer composites. *Materials Research Express*, 7(1), p. 016304.
23. Kazmi, S.J., Rehman, S.U., Nadeem, M., Rehman, U.U., Hussain, S. and Manzoor, S., 2024. Effect of carbon allotropes and thickness variation on the EMI shielding properties of PANI/NFO@CNTs and PANI/NFO@RGO ternary composite systems. *Physical Chemistry Chemical Physics*, 26(13), pp. 10168–10182.
24. Mohamed, Z., Al-Asbahi, B.A., Al-Hada, N.M. and El-Shamy, A.G., 2025. Three in one: a high tensile/nano mechanical nature, thermal conductivity, and green electromagnetic interference (EMI) shielding properties of polyvinyl alcohol/reduced-graphene oxide/iron hydroxide (PVA/rGO/α-FeOOH) nanocomposite films. *Surfaces and Interfaces*, 57, p. 105761.
25. Katheria, A., Das, P., Nayak, J., Roy, B., Pal, A., Biswas, S. and Das, N.C., 2025. MXene and Fe_3O_4 decorated g-C_3N_4 incorporated high flexible hybrid polymer composite for enhanced electrical conductivity, EMI shielding and thermal conductivity. *Next Materials*, 6, p. 100292.

26. Yin, L., Zhang, J., Luo, J., Shi, Y., Yu, B., Zhang, S. and Zhou, K., 2025. Janus-inspired alternating architecture CNF/MXene/ZnFe2O4@PANI composite films with outstanding electromagnetic interference shielding and Joule heating. *Journal of Materials Science and Technology*, 223, pp. 275–286.
27. Liu, Z., Bai, G., Huang, Y., Ma, Y., Du, F., Li, F., Guo, T. and Chen, Y., 2007. Reflection and absorption contributions to the electromagnetic interference shielding of single-walled carbon nanotube/polyurethane composites. *Carbon*, 45(4), pp. 821–827.
28. Raju, P., Rani, G.N., Kumar, S.U., Joseph, A. and Raju, K.J., 2022. Ultrasonically induced in situ polymerization of PANI-SWCNT nanocomposites for electromagnetic shielding applications. *Journal of Materials Science: Materials in Electronics*, 33(8), pp. 5138–5148.
29. Nguyen, T.T.L., Cho, S.J., Ko, J., Nguyen, D.C.T., Kim, M.W., Kim, N.D., Lee, D.S. and Joo, Y., 2024. Exceptional electromagnetic interference shielding using single-walled carbon nanotube/conductive polymer composites films with ultrathin, lightweight properties. *Carbon*, 230, p. 119567.
30. Mei, H., Lu, M., Zhou, S. and Cheng, L., 2021. Enhanced impact resistance and electromagnetic interference shielding of carbon nanotubes films composites. *Journal of Applied Polymer Science*, 138(11), p. 50033.
31. Feng, R., Zhu, W., Yang, W., Li, S., Zhang, C., Li, Z., Du, S. and Li, Y., 2025. Scalable production of flexible and multifunctional graphene-based polymer composite film for high-performance electromagnetic interference shielding. *Carbon*, 233, p. 119875.
32. Liu, M., Wang, Z., Song, Z., Wang, F., Zhao, G., Zhu, H., Jia, Z., Guo, Z., Kang, F. and Yang, C., 2024. A popcorn-inspired strategy for compounding graphene@$NiFe_2O_4$ flexible films for strong electromagnetic interference shielding and absorption. *Nature Communications*, 15(1), p. 5486.
33. Xie, L., Min, P., Ye, L., He, P., Yin, G., Jin, M., Zhang, H.B. and Yu, Z.Z., 2024. Interfacial enhancement enables highly conductive reduced graphene oxide-based yarns for efficient electromagnetic interference shielding and thermal regulation. *Carbon*, 230, p. 119655.
34. Araújo, M.K., Carvalho, A.S., Santos, A.R., Padrón-Hernández, E. and Falcão, E.H., 2025. Influence of graphene oxide and reduced graphene oxide on TiO_2-reinforced flexible poly(vinyl alcohol) films for electromagnetic interference shielding. *Journal of Alloys and Compounds*, 1010, p. 177671.
35. Kumari, P.Y., Sawant, K.K., Subash, A. and Kandasubramanian, B., 2025. Synergistic effects of carbon black and electrospun FKM fibers for superior EMI shielding. *Materials Letters*, 381, p. 137751.
36. Tian, D., Hou, J., Liang, J. and Chen, J., 2025. Fabrication of polypropylene/carbon fiber/carbon black composite foam bonded with continuous carbon fiber reinforced polypropylene prepregs via high-pressure foam injection molding. *Composites Part B: Engineering*, 291, p. 112006.
37. Prashanth, G.K., Gadewar, M., Lalithamba, H.S., Rao, S., Rashmi, K.V., Yatish, K.V., Swamy, M.M., Bhagya, N.P. and Ghosh, M.K., 2024. Synthesis, and applications of carbon-integrated polymer composites and foams: a concise review. *Inorganic Chemistry Communications*, p. 112614.
38. Wang, Y.H. and Li, C.C., 2024. Highly effective EMI shielding composites for 5G Ka-band frequencies. *Applied Materials Today*, 36, p. 102041.
39. Wang, Y., Luo, J., Hao, G., Di, J., Liu, G., Wang, S. and Jiang, W., 2025. 3D printing of graphene frameworks decorated with magnetic components for enhanced electromagnetic interference shielding. *Composites Part A: Applied Science and Manufacturing*, 188, p. 108588.
40. Li, X., Yin, X., Liang, S., Li, M., Cheng, L. and Zhang, L., 2019. 2D carbide MXene $Ti_2C\ T_x$ as a novel high-performance electromagnetic interference shielding material. *Carbon*, 146, pp. 210–217.
41. Liang, W., Wu, J., Zhang, S., Zhao, P.Y., Zuo, X. and Wang, G.S., 2024. Construction of PI-MXene-MWCNT nanocomposite film integrating conductive gradient with sandwich structure for high-efficiency electromagnetic interference shielding in extreme environments. *Carbon*, 228, p. 119328.

42. Wang, L., Chen, L., Song, P., Liang, C., Lu, Y., Qiu, H., Zhang, Y., Kong, J. and Gu, J., 2019. Fabrication on the annealed $Ti_3C_2T_x$ MXene/epoxy nanocomposites for electromagnetic interference shielding application. *Composites Part B: Engineering*, 171, pp. 111–118.
43. Bai, C., Jia, S., Chen, W., Li, L., Zhang, Y., Min, P., Yu, Z.Z. and Zhang, H.B., 2025. Dual-network MXene/polyurethane composite foams for both stretchable and compressible electromagnetic interference shielding and strain sensors. *ACS Applied Materials & Interfaces*, 17(7), pp. 11108–11116.
44. Tao, D., Yang, C., Chen, C., Yan, K., You, H., Wang, W. and Wang, D., 2025. Highly flexible and ultralight PVA-co-PE-AgNW/MXene composite film with low filling for multistage electromagnetic interference shielding. *Small*, 21(7), p. 2411752.
45. Jiang, Y., Ru, X., Che, W., Jiang, Z., Chen, H., Hou, J. and Yu, Y., 2022. Flexible, mechanically robust and self-extinguishing MXene/wood composite for efficient electromagnetic interference shielding. *Composites Part B: Engineering*, 229, p. 109460.
46. Zhou, Y., Liu, S., Zhu, M., He, Y., Zhang, J., Yu, Y., Yu, W. and Huang, Y., 2024. Multifunctional wood-bamboo hybrid materials for effective electromagnetic interference shielding. *Construction and Building Materials*, 440, p. 137450.
47. Yang, X., Qin, Y., Peng, L., Pan, M. and Xu, H., 2023. Lightweight and anti-corrosive carbon nanotubes (CNTs)/bamboo fiber/HDPE composite for efficient electromagnetic interference shielding. *Colloids and Surfaces A: Physicochemical and Engineering Aspects*, 672, p. 131746.
48. Miao, Y., Lin, J., Wang, E., Liang, Y., Li, W., Dai, C., Huang, J. and Zhang, W., 2023. Electrically conductive bamboo charcoal@cellulose nanofibrils based composite membranes designed for electromagnetic interference shielding and flame retardant. *Industrial Crops and Products*, 206, p. 117713.
49. Liu, D., Li, L., Gong, Q., Du, G., Xue, Y., Li, G. and Li, S., 2025. Ultrathin Ti3C2Tx MXene/cellulose nanofiber composite film for enhanced mechanics & EMI shielding via freeze-thaw intercalation. *Applied Surface Science*, 684, p. 161877.
50. Zhou, M., Zhang, S. and Zhang, X., 2025. Filler-free cellulose nanofiber composite papers with excellent mechanical properties for efficient electromagnetic interference shielding. *International Journal of Biological Macromolecules*, p. 140562.
51. Fazeli, M., Mukherjee, S., Baniasadi, H., Abidnejad, R., Mujtaba, M., Lipponen, J., Seppälä, J. and Rojas, O.J., 2024. Lignin beyond the status quo: recent and emerging composite applications. *Green Chemistry*, 26(2), pp. 593–630.
52. Hu, W., Zhang, J., Liu, B., Zhang, C., Zhao, Q., Sun, Z., Cao, H. and Zhu, G., 2021. Synergism between lignin, functionalized carbon nanotubes and Fe_3O_4 nanoparticles for electromagnetic shielding effectiveness of tough lignin-based polyurethane. *Composites Communications*, 24, p. 100616.
53. Liu, Y., Zhao, X., Liu, Z., Sun, B., Liu, X., Zhao, R., Liu, B., Sun, Z., Men, Y., Hu, W. and Shao, Z.B., 2024. Functionalized lignin nanoparticles assembled with MXene reinforced polypropylene with favorable UV-aging resistance, electromagnetic shielding effects and superior fire-safety. *International Journal of Biological Macromolecules*, 265, p. 130957.
54. Yan, R., Huang, Z., Chen, Y., Zhang, L. and Sheng, X., 2024. Phase change composite based on lignin carbon aerogel/nickel foam dual-network for multisource energy harvesting and superb EMI shielding. *International Journal of Biological Macromolecules*, 277, p. 134233.
55. Yu, J., Lin, W., Cheng, H., Zhang, Q. and Chen, F., 2025. High-performance graphene films for thermal management and electromagnetic shielding prepared through high-temperature annealing. *Colloids and Surfaces A: Physicochemical and Engineering Aspects*, p. 136692.
56. Zhang, M., Zhang, P., Wang, Q., Li, L., Dong, S., Liu, J. and Rao, W., 2019. Stretchable liquid metal electromagnetic interference shielding coating materials with superior effectiveness. *Journal of Materials Chemistry C*, 7(33), pp. 10331–10337.
57. Chen, Y., Luo, Z., Shi, X., Jiang, Y., Zhang, Y. and Xu, Z., 2025. Liquid metal-encapsulated spherical Al_2O_3/PVA aerogel composites for high-efficiency electromagnetic interference shielding and enhanced mechanical properties. *Materials Science and Engineering B*, 314, p. 118062.

58. Sun, Y., Han, X., Guo, P., Chai, Z., Yue, J., Su, Y., Tan, S., Sun, X., Jiang, L. and Heng, L., 2023. Slippery graphene-bridging liquid metal layered heterostructure nanocomposite for stable high-performance electromagnetic interference shielding. *ACS Nano*, 17(13), pp. 12616–12628.
59. Zhu, J., Li, J., Tong, Y., Hu, T., Chen, Z., Xiao, Y., Zhang, S., Yang, H., Gao, M., Pan, T. and Cheng, H., 2024. Recent progress in multifunctional, reconfigurable, integrated liquid metal-based stretchable sensors and standalone systems. *Progress in Materials Science*, 142, p. 101228.
60. Sambyal, P., Iqbal, A., Hong, J., Kim, M.K., Kim, I.D. and Koo, C.M., 2023. Conductive MXene composites with liquid metal for high-performance electromagnetic interference shielding. *Materials Chemistry and Physics*, 295, p. 127184.
61. Jiang, H., Yuan, B., Guo, H., Pan, F., Meng, F., Wu, Y., Wang, X., Ruan, L., Zheng, S., Yang, Y. and Xiu, Z., 2024. Malleable, printable, bondable, and highly conductive MXene/liquid metal plasticine with improved wettability. *Nature Communications*, 15(1), p. 6138.
62. Abdolmajidi, M., Cheraghi, A., Joshaghani, A.H. and Mazaheri, H., 2022. Design, processing and evaluation of CuO, MgO and TiO_2 nano-coatings absorbing electromagnetic waves based on epoxy paint. *Nano-Structures & Nano-Objects, 32*, p. 100911.
63. Rahman, M.M., Khan, K.H., Parvez, M.M.H., Irizarry, N. and Uddin, M.N., 2025. Polymer nanocomposites with optimized nanoparticle dispersion and enhanced functionalities for industrial applications. *Processes*, 13(4), p. 994.
64. Xu, K., Cai, Z., Luo, H., Lu, Y., Ding, C., Yang, G., Wang, L., Kuang, C., Liu, J. and Yang, H., 2024. Toward integrated multifunctional laser-induced graphene-based skin-like flexible sensor systems. *ACS Nano*, 18(39), pp. 26435–26476.
65. Liu, X., Ye, Y., Ge, Y., Qu, J., Liedberg, B., Zhang, Q. and Wang, Y., 2024. Smart contact lenses for healthcare monitoring and therapy. *ACS Nano*, 18(9), pp. 6817–6844.
66. Zhou, J., Sui, Y., Wu, N., Han, M., Liu, J., Liu, W., Zeng, Z. and Liu, J., 2024. Recent advances in MXene-based aerogels for electromagnetic wave absorption. *Small*, 20 (49), p. 2405968.
67. Sun, Z., Yu, J., Liu, C., Jiang, X. and Wang, X., 2024. Foam materials for applications of electromagnetic shielding and microwave absorption. *Materials Research Bulletin*, p. 113001.
68. Safdar, F., Ashraf, M., Javid, A. and Iqbal, K., 2022. Polymeric textile-based electromagnetic interference shielding materials, their synthesis, mechanism and applications – a review. *Journal of Industrial Textiles*, 51(5_suppl), pp. 7293S–7358S.
69. Erdem, Ö., Derin, E., Zeibi Shirejini, S., Sagdic, K., Yılmaz, E.G., Yildiz, S., Akceoglu, G.A. and İnci, F., 2022. Carbon-based nanomaterials and sensing tools for wearable health monitoring devices. *Advanced Materials Technologies*, 7(3), p. 2100572.
70. Liu, Y., Wang, L., Liu, Y., Zhang, F. and Leng, J., 2024. Recent progress in shape memory polymer composites: driving modes, forming technologies, and applications. *Composites Communications*, p. 102062.
71. Mu, H., Yu, W., Yuan, J., Lin, S. and Zhang, G., 2022. Interface and surface engineering of black phosphorus: a review for optoelectronic and photonic applications. *Materials Futures*, 1(1), p. 012301.
72. Joshi, D.J., Malek, N.I. and Kailasa, S.K., 2022. Borophene as a rising star in materials chemistry: synthesis, properties and applications in analytical science and energy devices. *New Journal of Chemistry*, 46(10), pp. 4514–4533.
73. Wang, S., Xu, J., Li, W., Sun, S., Gao, S. and Hou, Y., 2022. Magnetic nanostructures: rational design and fabrication strategies toward diverse applications. *Chemical Reviews*, 122(6), pp. 5411–5475.
74. Tabassum, T. and Mir, A.A., 2023. A review of 3D printing technology – the future of sustainable construction. *Materials Today: Proceedings*, 93, pp. 408–414.
75. Qureshi, N., Dhand, V., Subhani, S., Kumar, R.S., Raghavan, N., Kim, S. and Doh, J., 2024. Exploring conductive filler-embedded polymer nanocomposite for electrical percolation via electromagnetic shielding-based additive manufacturing. *Advanced Materials Technologies*, 9(17), p. 2400250.

76. Verma, S., Dhangar, M., Mili, M., Bajpai, H., Dwivedi, U., Kumari, N., Khan, M.A., Bhargaw, H.N., Hashmi, S.A.R. and Srivastava, A.K., 2022. Review on engineering designing of electromagnetic interference shielding materials using additive manufacturing. *Polymer Composites*, 43(7), pp. 4081–4099.
77. Yasir, M. and Kim, S.R., 2025. 3D-structured polymer composites using advanced templating strategies for EMI shielding. *Polymer Composites*, pp. 1–46.
78. Matias, M.L., Pereira, C., Almeida, H.V., Jana, S., Panigrahi, S., Menda, U.D., Nunes, D., Fortunato, E., Martins, R. and Nandy, S., 2024. 3D printed MXene architectures for a plethora of smart applications. *Materials Today Advances*, 23, p. 100512.
79. Abbas, R., Ur Rehman, U., Sultan, N., Ghazanfar, U., Ali, T. and Nadeem, M., 2024. 3D printed lightweight honeycomb vent structures with subsequent coating of silver nanowires for efficient electromagnetic interference (EMI) shielding. *Heliyon*, 10(9), p. e30429.
80. Khan, N. and Riccio, A., 2024. A systematic review of design for additive manufacturing of aerospace lattice structures: current trends and future directions. *Progress in Aerospace Sciences*, 149, p. 101021.
81. Banupriya, R., Jeevan, T.P., Divya, H.V., Yashas Gowda, T.G. and Manjunath, G.A., 2025. 3D-printed graphene-reinforced composites: opportunities and challenges. *Polymer Composites*, 46(2), pp. 1250–1266.
82. Brandenburg, S.D., Zimmerman, D.W. and Gupta, N., Aptiv Technologies AG, 2024. Electromagnetic interference shielding and thermal management systems and methods for automotive radar applications. U.S. Patent 12, 137,515.
83. Živković, M., Žujović, M. and Milošević, J., 2023. Architectural 3D-printed structures created using artificial intelligence: a review of techniques and applications. *Applied Sciences*, 13(19), p. 10671.
84. Bodaghi, M., Wang, L., Zhang, F., Liu, Y., Leng, J., Xing, R., Dickey, M.D., Vanaei, S., Elahinia, M., Hoa, S.V. and Zhang, D., 2024. 4D printing roadmap. *Smart Materials and Structures*, 33(11), p. 113501.
85. He, Y.J., Shao, Y.W., Xiao, Y.Y., Yang, J.H., Qi, X.D. and Wang, Y., 2022. Multifunctional phase change composites based on elastic MXene/silver nanowire sponges for excellent thermal/solar/electric energy storage, shape memory, and adjustable electromagnetic interference shielding functions. *ACS Applied Materials & Interfaces*, 14(4), pp. 6057–6070.
86. Singh, P.P., Nair, R.S. and Khatua, B.B., 2024. Boosting green EMI shielding and piezoelectric energy generation by defect-driven microstructure manipulation. *Journal of Materials Chemistry C*, 12(31), pp. 12075–12089.
87. Li, C., Li, D., Zhang, M., You, B., Wu, Z., Tao, Y., Sun, Y., Wu, L. and Mo, X., 2024. Succulent-inspired implicit structural change for smart "ON/OFF" switchable and flexible EMI shielding coating. *ACS Applied Materials & Interfaces*, 16(10), pp. 12939–12950.
88. Cheng, W., Zhang, Y., Tao, Y., Lu, J., Liu, J., Wang, B., Song, L., Jie, G. and Hu, Y., 2021. Durable electromagnetic interference (EMI) shielding ramie fabric with excellent flame retardancy and self-healing performance. *Journal of Colloid and Interface Science*, 602, pp. 810–821.
89. Zhao, B., Bai, Z., Lv, H., Yan, Z., Du, Y., Guo, X., Zhang, J., Wu, L., Deng, J., Zhang, D.W. and Che, R., 2023. Self-healing liquid metal magnetic hydrogels for smart feedback sensors and high-performance electromagnetic shielding. *Nano-Micro Letters*, 15(1), p. 79.
90. Du, F., Wang, S., Chen, Z. and Li, Q., 2024. Stimuli responsive actuators: recent advances. *Journal of Materials Chemistry C*, 12, pp. 8217–8242.
91. Liu, F. and Urban, M.W., 2010. Recent advances and challenges in designing stimuli-responsive polymers. *Progress in Polymer Science*, 35(1–2), pp. 3–23.
92. Lizundia, E., Luzi, F. and Puglia, D., 2022. Organic waste valorisation towards circular and sustainable biocomposites. *Green Chemistry*, 24(14), pp. 5429–5459.
93. Hassan, M.W., Elshazly, T.M. and Ponnamma, D., 2024. Cellulose-inspired approaches to sustainable EMI shielding materials: a comprehensive review. *International Journal of Biological Macromolecules*, p. 132920.

94. Bullock, A., 2024. Hybridized luminescent lanthanide super crystals used as spectral convertors to increase photoconductivity in opto-electronic devices. *Journal of Nanotechnology and Nanomaterials*, 5(1), pp. 26–30.
95. Muchlis, A.M.G., Yang, C., Tsai, Y.T., Ummartyotin, S. and Lin, C.C., 2023. Multiresponsive self-healing lanthanide fluorescent hydrogel for smart textiles. *ACS Applied Materials & Interfaces*, 15(39), pp. 46085–46097.
96. López, C., 2023. Artificial intelligence and advanced materials. *Advanced Materials*, 35(23), p. 2208683.
97. Reyes-Luna, J.F., Chang, S., Tuck, C. and Ashcroft, I., 2023. A surrogate modelling strategy to improve the surface morphology quality of inkjet printing applications. *Journal of Manufacturing Processes*, 89, pp. 458–471.
98. Samadian, D., Muhit, I.B. and Dawood, N., 2024. Application of data-driven surrogate models in structural engineering: a literature review. *Archives of Computational Methods in Engineering*, pp. 1–50.
99. Cerniauskas, G., Sadia, H. and Alam, P., 2024. Machine intelligence in metamaterials design: a review. *Oxford Open Materials Science*, 4(1), p. itae001.
100. Zheng, X., Zhang, X., Chen, T.T. and Watanabe, I., 2023. Deep learning in mechanical metamaterials: from prediction and generation to inverse design. *Advanced Materials*, 35(45), p. 2302530.
101. Chen, M.K., Liu, X., Sun, Y. and Tsai, D.P., 2022. Artificial intelligence in meta-optics. *Chemical Reviews*, 122(19), pp. 15356–15413.
102. Wray, P.R., Paul, E.G. and Atwater, H.A., 2024. Optical filters made from random metasurfaces using Bayesian optimization. *Nanophotonics*, 13(2), pp. 183–193.
103. Bollineni, R.K., Ahmed, M.S., Shahab, S. and Mirzaeifar, R., 2024. Nacre-like block lattice metamaterials with targeted phononic band gap and mechanical properties. *Journal of the Mechanical Behavior of Biomedical Materials*, 154, p. 106511.

Index

Page numbers in *italic* indicate figures
Page numbers in **bold** indicate tables.

F

G

J

K

L

M

N

For Product Safety Concerns and Information please contact our EU
representative GPSR@taylorandfrancis.com Taylor & Francis Verlag GmbH,
Kaufingerstraße 24, 80331 München, Germany

Batch number: 10615548

Printed by Printforce, the Netherlands